Chromatin Protocols

METHODS IN MOLECULAR BIOLOGY™

John M. Walker, SERIES EDITOR

137. **Developmental Biology Protocols, Volume III**, edited by *Rocky S. Tuan and Cecilia W. Lo, 2000*
136. **Adrenergic Receptor Protocols**, edited by *Curtis A. Machida, 2000*
135. **Glycoproteins Methods and Protocols:** *The Mucins*, edited by *Anthony P. Corfield, 2000*
134. **T Cell Protocols:** *Development and Activation*, edited by *Kelly P. Kearse, 2000*
133. **Gene Targeting Protocols**, edited by *Eric B. Kmiec, 1999*
132. **Bioinformatics Methods and Protocols**, edited by *Stephen Misener and Stephen A. Krawetz, 1999*
131. **Flavoprotein Protocols**, edited by *S. K. Chapman and G. A. Reid, 1999*
130. **Transcription Factor Protocols**, edited by *Martin J. Tymms, 1999*
129. **Integrin Protocols**, edited by *Anthony Howlett, 1999*
128. **NMDA Protocols**, edited by *Min Li, 1999*
127. **Molecular Methods in Developmental Biology:** *Xenopus and Zebrafish*, edited by *Matt Guille, 1999*
126. **Developmental Biology Protocols, Volume II**, edited by *Rocky S. Tuan and Cecilia W. Lo, 2000*
125. **Developmental Biology Protocols, Volume I**, edited by *Rocky S. Tuan and Cecilia W. Lo, 2000*
124. **Protein Kinase Protocols**, edited by *Alastair Reith, 1999*
123. ***In Situ* Hybridization Protocols (2nd ed.)**, edited by *Ian A. Darby, 1999*
122. **Confocal Microscopy Methods and Protocols**, edited by *Stephen W. Paddock, 1999*
121. **Natural Killer Cell Protocols:** *Cellular and Molecular Methods*, edited by *Kerry S. Campbell and Marco Colonna, 1999*
120. **Eicosanoid Protocols**, edited by *Elias A. Lianos, 1999*
119. **Chromatin Protocols**, edited by *Peter B. Becker, 1999*
118. **RNA–Protein Interaction Protocols**, edited by *Susan R. Haynes, 1999*
117. **Electron Microscopy Methods and Protocols**, edited by *M. A. Nasser Hajibagheri, 1999*
116. **Protein Lipidation Protocols**, edited by *Michael H. Gelb, 1999*
115. **Immunocytochemical Methods and Protocols (2nd ed.)**, edited by *Lorette C. Javois, 1999*
114. **Calcium Signaling Protocols**, edited by *David G. Lambert, 1999*
113. **DNA Repair Protocols:** *Eukaryotic Systems*, edited by *Daryl S. Henderson, 1999*
112. **2-D Proteome Analysis Protocols**, edited by *Andrew J. Link 1999*
111. **Plant Cell Culture Protocols**, edited by *Robert D. Hall, 1999*
110. **Lipoprotein Protocols**, edited by *Jose M. Ordovas, 1998*
109. **Lipase and Phospholipase Protocols**, edited by *Mark H. Doolittle and Karen Reue, 1999*
108. **Free Radical and Antioxidant Protocols**, edited by *Donald Armstrong, 1998*
107. **Cytochrome P450 Protocols**, edited by *Ian R. Phillips and Elizabeth A. Shephard, 1998*
106. **Receptor Binding Techniques**, edited by *Mary Keen, 1999*
105. **Phospholipid Signaling Protocols**, edited by *Ian M. Bird, 1998*
104. **Mycoplasma Protocols**, edited by *Roger J. Miles and Robin A. J. Nicholas, 1998*
103. **Pichia Protocols**, edited by *David R. Higgins and James M. Cregg, 1998*
102. **Bioluminescence Methods and Protocols**, edited by *Robert A. LaRossa, 1998*
101. **Mycobacteria Protocols**, edited by *Tanya Parish and Neil G. Stoker, 1998*
100. **Nitric Oxide Protocols**, edited by *Michael A. Titheradge, 1998*
99. **Human Cytokines and Cytokine Receptors**, edited by *Reno Debets and Huub Savelkoul, 1999*
98. **Forensic DNA Profiling Protocols**, edited by *Patrick J. Lincoln and James M. Thomson, 1998*
97. **Molecular Embryology:** *Methods and Protocols*, edited by *Paul T. Sharpe and Ivor Mason, 1999*
96. **Adhesion Protein Protocols**, edited by *Elisabetta Dejana and Monica Corada, 1999*
95. **DNA Topoisomerases Protocols:** *II. Enzymology and Drugs*, edited by *Mary-Ann Bjornsti and Neil Osheroff, 1999*
94. **DNA Topoisomerases Protocols:** *I. DNA Topology and Enzymes*, edited by *Mary-Ann Bjornsti and Neil Osheroff, 1999*
93. **Protein Phosphatase Protocols**, edited by *John W. Ludlow, 1998*
92. **PCR in Bioanalysis**, edited by *Stephen J. Meltzer, 1998*
91. **Flow Cytometry Protocols**, edited by *Mark J. Jaroszeski, Richard Heller, and Richard Gilbert, 1998*
90. **Drug–DNA Interaction Protocols**, edited by *Keith R. Fox, 1998*
89. **Retinoid Protocols**, edited by *Christopher Redfern, 1998*
88. **Protein Targeting Protocols**, edited by *Roger A. Clegg, 1998*
87. **Combinatorial Peptide Library Protocols**, edited by *Shmuel Cabilly, 1998*
86. **RNA Isolation and Characterization Protocols**, edited by *Ralph Rapley and David L. Manning, 1998*
85. **Differential Display Methods and Protocols**, edited by *Peng Liang and Arthur B. Pardee, 1997*
84. **Transmembrane Signaling Protocols**, edited by *Dafna Bar-Sagi, 1998*
83. **Receptor Signal Transduction Protocols**, edited by *R. A. John Challiss, 1997*
82. **Arabidopsis Protocols**, edited by *José M Martinez-Zapater and Julio Salinas, 1998*
81. **Plant Virology Protocols:** *From Virus Isolation to Transgenic Resistance*, edited by *Gary D. Foster and Sally Taylor, 1998*
80. **Immunochemical Protocols (2nd. ed.)**, edited by *John Pound, 1998*
79. **Polyamine Protocols**, edited by *David M. L. Morgan, 1998*
78. **Antibacterial Peptide Protocols**, edited by *William M. Shafer, 1997*
77. **Protein Synthesis:** *Methods and Protocols*, edited by *Robin Martin, 1998*

METHODS IN MOLECULAR BIOLOGY™

Chromatin Protocols

Edited by

Peter B. Becker

Adolf-Butenandt-Institut-Molekularbiologie,
Ludwig-Maximilians-Universität,
München, Germany

Humana Press ✹ Totowa, New Jersey

© 1999 Humana Press Inc.
999 Riverview Drive, Suite 208
Totowa, New Jersey 07512

All rights reserved. No part of this book may be reproduced, stored in a retrieval system, or transmitted in any form or by any means, electronic, mechanical, photocopying, microfilming, recording, or otherwise without written permission from the Publisher. Methods in Molecular Biology™ is a trademark of The Humana Press Inc.

All authored papers, comments, opinions, conclusions, or recommendations are those of the author(s), and do not necessarily reflect the views of the publisher.

This publication is printed on acid-free paper. ∞
ANSI Z39.48-1984 (American Standards Institute) Permanence of Paper for Printed Library Materials.

Cover illustration: Fig. 9 from Chapter 10, "Analysis of Chromatin by Scanning Force Microscopy," by Sanford H. Leuba and Carlos Bustamante.

Cover design by Patricia F. Cleary.

For additional copies, pricing for bulk purchases, and/or information about other Humana titles, contact Humana at the above address or at any of the following numbers: Tel.: 973-256-1699; Fax: 973-256-8341; E-mail: humana@humanapr.com; Website: http://humanapress.com

Photocopy Authorization Policy:
Authorization to photocopy items for internal or personal use, or the internal or personal use of specific clients, is granted by Humana Press Inc., provided that the base fee of US $10.00 per copy, plus US $00.25 per page, is paid directly to the Copyright Clearance Center at 222 Rosewood Drive, Danvers, MA 01923. For those organizations that have been granted a photocopy license from the CCC, a separate system of payment has been arranged and is acceptable to Humana Press Inc. The fee code for users of the Transactional Reporting Service is: [0-89603-665-0 (hardcover)/99 $10.00 + $00.25].

Printed in the United States of America. 10 9 8 7 6 5 4 3 2 1

Library of Congress Cataloging in Publication Data

Main entry under title:

Methods in molecular biology™.

Chromatin Protocols / edited by Peter B. Becker.
    p. cm. — (Methods in molecular biology™ ; v. 119)
  ISBN 0-89603-665-0 (alk. paper)
  1. Chromatin—Laboratory manuals. I. Becker, Peter B.
 II. Series: Methods in Molecular Biology (Totowa, NJ) ; 119.
  QH599.C455   1999
  572.8'7—dc21   99-10955
                                                                                      CIP

Preface

More than 40 years after the discovery of the nucleosome as the fundamental unit of chromatin, the multifaceted problem of how variations in chromatin structure affect the activity of the eukaryotic genome has not been solved. However, during the past few years research on chromatin structure and function has gained considerable momentum, and impressive progress has been made at the level of concept development as well as filling in crucial detail. The structure of the nucleosome has been visualized at unprecedented resolution. Powerful multisubunit enzymes have been identified that alter histone/DNA interactions in ways that expose regulatory sequences to factors initiating and regulating such nuclear processes as transcription. Though the importance of posttranslational modifications of histones, notably their acetylation, has long been known, the finding that a number of *bona fide* regulators increase transcription by acetylating nucleosomes has lent new support to the old idea that the process of gene regulation is intimately related to the nature of the chromatin environment. A wealth of nonhistone proteins contribute to a continuum of structures with distinct biochemical properties and varying degrees of DNA condensation. Perhaps the most important conclusion from a large number of studies is a fresh appreciation of the dynamic nature of chromatin structure, the built-in flexibility providing the basis for regulation. Needless to say, the lessons we learn from analyzing gene function in the context of chromatin have implications for the applied biosciences concerned with the sustained and regulated expression of transgenes.

The progress in our understanding of chromatin has resulted from methodological development at two fronts: (1) the establishment and optimization of efficient in vitro systems for the reconstitution of nucleosomes, nucleosomal arrays, complex chromatin, and entire nuclei, along with imaginative assays open avenues for the biochemical analysis of chromatin structure and function; and (2) the establishment of methods, such as genomic footprinting and various crosslinking techniques, that enable the characterization of specific sites in chromatin of unperturbed nuclei with respect to protein/DNA interactions, protein/protein interactions, and protein modifications.

Chromatin Protocols presents a collection of powerful methods that have already proven to be valuable, and are likely to be still more widely used in future chromatin studies. The protocols have been carefully recorded by those researchers most intimately involved in their development. I wish to thank all the authors for sharing their "tricks of the trade" with the research community and hope that our volume of *Chromatin Protocols* will help many to join in the exciting endeavor of unraveling the secrets of how chromatin structure regulates the functioning of the eukaryotic genome.

Peter B. Becker

Contents

Preface v
Contributors xi

1 Expression and Purification of Recombinant Histones and Nucleosome Reconstitution
   Karolin Luger, Thomas J. Rechsteiner, and Timothy J. Richmond 1

2 Preparation and Analysis of Positioned Nucleosomes
   Vasily M. Studitsky 17

3 Site-Directed Chemical Probing of Histone–DNA Interactions
   David R. Chafin, Kyu-Min Lee, and Jeffrey J. Hayes 27

4 Base-Pair Resolution Mapping of Nucleosomes In Vitro
   Andrew Flaus and Timothy J. Richmond 45

5 Equilibrium and Dynamic Nucleosome Stability
   Jonathan Widom 61

6 Nucleosome Structure and Dynamics: The DNA Minicircle Approach
   Ariel Prunell, Mohamed Alilat, and Filomena De Lucia 79

7 Analysis of Linker Histone Binding to Mono- and Dinucleosomes
   Simon Chandler and Alan P. Wolffe 103

8 Quantitative Analysis of Chromatin Higher-Order Organization Using Agarose Gel Electrophoresis
   Jeffrey C. Hansen, Terace M. Fletcher, and J. Isabelle Kreider 113

9 Analytical Ultracentrifugation of Chromatin
   Jeffrey C. Hansen and Cynthia L. Turgeon 127

10 Analysis of Chromatin by Scanning Force Microscopy
   Sanford H. Leuba and Carlos Bustamante 143

11 In Vivo Mapping of Nucleosomes Using Psoralen–DNA Crosslinking and Primer Extension
   Ralf Erik Wellinger, Renzo Lucchini, Reinhard Dammann, and José M. Sogo 161

12 Preparation of Chromatin Assembly Extracts
from *Xenopus* Oocytes
David John Tremethick 175

13 Preparation of Chromatin Assembly Extracts
from Preblastoderm *Drosophila* Embryos
Edgar Bonte and Peter B. Becker 187

14 A Solid-Phase Approach for the Analysis
of Reconstituted Chromatin
Raphael Sandaltzopoulos and Peter B. Becker 195

15 Reconstitution and Analysis of Hyperacetylated Chromatin
Wladyslaw A. Krajewski and Peter B. Becker 207

16 Assembly of Mitotic Chromosomes in *Xenopus* Egg Extract
**Anne-Elisabeth de la Barre, Michel Robert-Nicoud,
and Stefan Dimitrov** 219

17 Nucleotide Excision Repair Coupled to Chromatin Assembly
**Pierre-Henri Gaillard, Danièle Roche,
and Geneviève Almouzni** 231

18 Photolyase: *A Molecular Tool to Characterize Chromatin
Structure in Yeast*
**Magdalena Livingstone-Zatchej, Bernhard Suter,
and Fritz Thoma** 245

19 Transcriptional and Structural Analyses of Isolated
SV40 Chromatin
Ulla Hansen 261

20 In Vitro Replication of Chromatin Templates
Claudia Gruss 291

21 Analysis of HMG-14/-17-Containing Chromatin
Yuri V. Postnikov and Michael Bustin 303

22 Identification and Analysis of Native Nucleosomal Histone
Acetyltransferase Complexes
Patrick A. Grant, Shelley L. Berger, and Jerry L. Workman 311

23 Analysis of Nucleosome Disruption by ATP-Driven
Chromatin Remodeling Complexes
**Tom Owen-Hughes, Rhea T. Utley, David J. Steger,
Joshua M. West, Sam John, Jacques Côté,
Kristina M. Havas, and Jerry L. Workman** 319

24 Nucleosome Remodeling Factor NURF and In Vitro Transcription
of Chromatin
Gaku Mizuguchi and Carl Wu 333

Contents

25 An SDS-PAGE-Based Enzyme Activity Assay for the Detection and Identification of Histone Acetyltransferases
    James E. Brownell, Craig A. Mizzen, and C. David Allis 343

26 Analysis of DNaseI Hypersensitive Sites in Chromatin by Cleavage in Permeabilized Cells
    Rein Aasland and A. Francis Stewart 355

27 Mapping of Nucleosome Positions in Yeast
    Magdalena Livingstone-Zatchej and Fritz Thoma 363

28 Analysis of DNA Topology in Yeast Chromatin
    Randall H. Morse 379

29 DNA Methyltransferases as Probes for Chromatin Structure in Yeast
    Michael P. Kladde, Mai Xu, and Robert T. Simpson 395

30 Restriction Nucleases as Probes for Chromatin Structure
    Philip D. Gregory, Slobodan Barbaric, and Wolfram Hörz 417

31 Genomic Footprinting Using Nucleases
    Luca Cappabianca, Hélène Thomassin, Raymond Pictet, and Thierry Grange 427

32 *In Situ* Analysis of Chromatin Proteins During Development and Cell Differentiation Using Flow Cytometry
    Didier Grunwald, Claude Gorka, Sandrine Curtet, and Saadi Khochbin 443

33 Mapping DNA Target Sites of Chromatin Proteins In Vivo by Formaldehyde Crosslinking
    Helen Strutt and Renato Paro 455

34 Mapping DNA Interaction Sites of Chromosomal Proteins: Crosslinking Studies in Yeast
    Andreas Hecht, Sabine Strahl-Bolsinger, and Michael Grunstein 469

35 UV Laser Footprinting and Protein–DNA Crosslinking: Application to Chromatin
    Dimitri Angelov, Saadi Khochbin, and Stefan Dimitrov 481

36 An In Vivo UV Crosslinking Assay That Detects DNA Binding by Sequence-Specific Transcription Factors
    Alan Carr and Mark D. Biggin 497

Index 509

Contributors

REIN AASLAND • *Department of Molecular Biology, University of Bergen, Bergen, Norway*
MOHAMED ALILAT • *Institut Jacques Monod, Centre National de la Recherche Scientifique et Universite Denis Diderot, Paris, France*
C. DAVID ALLIS • *Department of Biology, University of Rochester, Rochester, NY*
GENEVIÈVE ALMOUZNI • *Research Section UMR, Institut Curie, Paris, France*
DIMITRI ANGELOV • *Institute of Solid State Physics, Bulgarian Academy of Sciences, Sofia, Bulgaria*
SLOBODAN BARBARIC • *Laboratory of Biochemistry, Faculty of Food Technology and Biotechnology, University of Zagreb, Zagreb, Croatia*
PETER B. BECKER • *Adolf-Butenandt-Institut-Molekularbiologie, Ludwig-Maximillians-Universität, München, Germany*
SHELLEY L. BERGER • *The Wistar Institute, Philadelphia, PA*
MARK D. BIGGIN • *Department of Molecular Biophysics and Biochemistry, Yale University, New Haven, CT*
EDGAR BONTE • *Gene Expression Programme, European Molecular Biology Laboratory, Heidelberg, Germany*
JAMES E. BROWNELL • *Department of Biology, University of Rochester, Rochester, NY*
CARLOS BUSTAMANTE • *Howard Hughes Medical Institute, Institute of Molecular Biology, University of Oregon, Eugene, OR*
MICHAEL BUSTIN • *Laboratory of Molecular Carcinogenesis, National Cancer Institute, National Institutes of Health, Bethesda MD*
LUCA CAPPABIANCA • *Institut Jacques Monod du CNRS, Universite Paris 7, Paris, France*
ALAN CARR • *Department of Molecular Biophysics and Biochemistry, Yale University, New Haven, CT*
DAVID R. CHAFIN • *Department of Biochemistry and Biophysics, University of Rochester Medical Center, Rochester, NY*

SIMON CHANDLER • *National Institute of Child Health and Human Development, National Institutes of Health, Bethesda, MD*

JACQUES CÔTÉ • *Laval University Cancer Research Center, Hotel-Dieu de Quebec, Quebec City, Canada*

SANDRINE CURTET • *Laboratoire de Biologie Moleculaire du Cycle Cellulaire, Institut Albert Bonniot, La Tronche, France*

ANNE-ELISABETH DE LA BARRE • *Laboratoire de Biologie Moleculaire du Cycle Cellulaire, Institut Albert Bonniot, La Tronche, France*

FILOMENA DE LUCIA • *Institut Jacques Monod, Centre National de la Recherche Scientifique et Universite Denis Diderot Paris 7, Paris, France*

REINHARD DAMMANN • *Institut für Zellbiologie, Eidgenossische Technische Hochschule, Hönggerberg, Zürich, Switzerland*

STEFAN DIMITROV • *Laboratoire de Biologie Moleculaire du Cycle Cellulaire, Institut Albert Bonniot, La Tronche, France*

ANDREW FLAUS • *Institut für Molekularbiologie und Biophysik, Eidgenossische Technische Hochschule, Hönggerberg, Zürich, Switzerland*

TERACE M. FLETCHER • *Department of Biochemistry, Milton Hershey Medical Center, The Pennsylvania State University, Hershey, PA*

PIERRE-HENRI GAILLARD • *Imperial Cancer Research Fund, South Mimms, UK*

CLAUDE GORKA • *Laboratoire de Biologie Moleculaire du Cycle Cellulaire, Institut Albert Bonniot, La Tronche, France*

THIERRY GRANGE • *Institut Jacques Monod du CNRS, Universite Paris 7, Paris, France*

PATRICK A. GRANT • *Department of Biochemistry and Molecular Biology, The Pennsylvania State University, University Park, PA*

PHILIP D. GREGORY • *Institut für Physiologische Chemie, Universitat München, München, Germany*

MICHAEL GRUNSTEIN • *Department of Biological Chemistry, UCLA School of Medicine and the Molecular Biology Institute, University of California, Los Angeles, CA*

DIDIER GRUNWALD • *Laboratoire de Biologie Moleculaire du Cycle Cellulaire, Institut Albert Bonniot, La Tronche, France*

CLAUDIA GRUSS • *Division of Biology, University of Konstanz, Konstanz, Germany*

JEFFREY C. HANSEN • *Department of Biochemistry, The University of Texas Health Science Center, San Antonio, TX*

Contributors

ULLA HANSEN • *Division of Molecular Genetics, Dana Faber Cancer Institute, Boston, MA*

KRISTINA M. HAVAS • *Department of Biochemistry and Molecular Biology, The Pennsylvania State University, University Park, PA*

JEFFREY J. HAYES • *Department of Biochemistry, University of Rochester Medical Center, Rochester, NY*

ANDREAS HECHT • *Max-Planck Institut für Immunbiologie, Freiburg, Germany*

WOLFRAM HÖRZ • *Institut für Physiologische Chemie, Universitat München, München, Germany*

SAM JOHN • *Department of Biochemistry and Molecular Biology, The Pennsylvania State University University Park, PA*

SAADI KHOCHBIN • *Laboratoire de Biologie Moleculaire du Cycle Cellulaire, Institut Albert Bonniot, La Tronche, France*

MICHAEL P. KLADDE • *Department of Biochemistry and Molecular Biology, The Pennsylvania State University, University Park, PA*

WLADYSLAW A. KRAJEWSKI • *Laboratory of Biochemistry, Institute of Developmental Biology, Russian Academy of Sciences, Moscow, Russia*

J. ISABELLE KREIDER • *Department of Biochemistry, The University of Texas Health Science Center, San Antonio, TX*

KYU-MIN LEE • *Department of Biochemistry and Biophysics, University of Rochester Medical Center, Rochester, NY*

SANFORD H. LEUBA • *Department of Physics and Astronomy, Arizona State University, Tempe, AZ*

MAGDALENA LIVINGSTONE-ZATCHEJ • *Institut für Zellbiologie, Eidgenossische Technische Hochschule, Hönggerberg, Zürich, Switzerland*

RENZO LUCCHINI • *Institut für Zellbiologie, Eidgenossische Technische Hochschule, Hönggerberg, Zürich, Switzerland*

KAROLIN LUGER • *Institut für Molekularbiologie und Biophysik, Eidgenossische Technische Hochschule, Hönggerberg, Zürich, Switzerland*

GAKU MIZUGUCHI • *Laboratory of Molecular and Cell Biology, National Cancer Institute, National Institutes of Health, Bethesda, MD*

CRAIG A. MIZZEN • *Department of Analytical Chemistry, Genentech Inc., San Francisco, CA*

RANDALL H. MORSE • *Wadsworth Center, New York State Department of Health, Albany, NY*

Tom Owen-Hughes • *Department of Biochemistry and Molecular Biology, The Pennsylvania State University, University Park, PA*
Renato Paro • *Zentrum für Molekulare Biologie, University of Heidelberg, Heidelberg, Germany*
Raymond Pictet • *Institut Jacques Monod du CNRS, Universite Paris 7, Paris, France*
Yuri V. Postnikov • *Laboratory of Molecular Carcinogenesis, National Cancer Institute, National Institutes of Health, Bethesda MD*
Ariel Prunell • *Institut Jacques Monod, Centre National de la Recherche Scientifique et Universite Denis Diderot, Paris, France*
Thomas J. Rechsteiner • *Institut für Molekularbiologie und Biophysik, Eidgenossische Technische Hochschule, Hönggerberg, Zürich, Switzerland*
Timothy J. Richmond • *Institut für Molekularbiologie und Biophysik, Eidgenossische Technische Hochschule, Hönggerberg, Zürich, Switzerland*
Michel Robert-Nicoud • *Dyogen, INSERM U309, Institut Albert Bonniot, La Tronche, France*
Danièle Roche • *Research Section UMR, Institut Curie, Paris, France*
Raphael Sandaltzopoulos • *Laboratory of Molecular and Cell Biology, National Cancer Institute, National Institutes of Health, Bethesda, MD*
Robert T. Simpson • *Department of Biochemistry and Molecular Biology, The Pennsylvania State University, University Park, PA*
José M. Sogo • *Institut für Zellbiologie, Eidgenossische Technische Hochschule, Hönggerberg, Zürich, Switzerland*
David J. Steger • *Department of Biochemistry and Molecular Biology, The Pennsylvania State University, University Park, PA*
A. Francis Stewart • *Gene Expression Programme, European Molecular Biology Laboratory, Heidelberg, Germany*
Sabine Strahl-Bolsinger • *Lehrstohl für Zellbiologie und Pflanzenphysiologie, Universitat Regensburg, Regensburg, Germany*
Helen Strutt • *Programme in Developmental Genetics, University of Sheffield, Sheffield, UK*
Vasily M. Studitsky • *Department of Biochemistry, Wayne State University Medical School, Detroit, MI*
Bernhard Suter • *Institut für Zellbiologie, Eidgenossische Technische Hochschule, Hönggerberg, Zürich, Switzerland*

Contributors

FRITZ THOMA • *Institut für Zellbiologie, Eidgenossische Technische Hochschule, Hönggerberg, Zürich, Switzerland*

HÉLÈNE THOMASSIN • *Institut Jacques Monod du CNRS, Universite Paris 7, Paris, France*

DAVID JOHN TREMETHICK • *John Curtain School of Medical Research, Australian National University, Canberra, Australia*

CYNTHIA L. TURGEON • *Department of Biochemistry, The University of Texas Health Science Center, San Antonio, TX*

RHEA T. UTLEY • *Department of Biochemistry and Molecular Biology, The Pennsylvania State University, University Park, PA*

RALF ERIK WELLINGER • *Institut für Zellbiologie, Eidgenossische Technische Hochschule, Hönggerberg, Zürich, Switzerland*

JOSHUA M. WEST • *Department of Biochemistry and Molecular Biology, The Pennsylvania State University, University Park, PA*

JONATHAN WIDOM • *Biochemistry Department, Northwestern University, Evanston, IL*

ALAN P. WOLFFE • *National Institute of Child Health and Human Development, National Institutes of Health, Bethesda, MD*

JERRY L. WORKMAN • *Department of Biochemistry and Molecular Biology, The Pennsylvania State University, University Park, PA*

CARL WU • *Laboratory of Molecular and Cell Biology, National Cancer Institute, National Institutes of Health, Bethesda, MD*

MAI XU • *Department of Biochemistry and Molecular Biology, The Pennsylvania State University, University Park, PA*

1

Expression and Purification of Recombinant Histones and Nucleosome Reconstitution

Karolin Luger, Thomas J. Rechsteiner, and Timothy J. Richmond

1. Introduction

In vitro studies on nucleosome core particles (NCPs) and nucleosomes have generally been limited to the use of histone proteins isolated from chromatin. Numerous reliable and well-established methods have been described of obtaining single histone proteins in significant quantity (e.g., **refs.** *1* and *2*, and references therein). Briefly, the histone complexes (histone octamer, or histone tetramer and histone dimer) are isolated from "long chromatin," which is extracted from nuclei. The histone complexes can be further fractionated into individual histone proteins. This approach suffers from several disadvantages. First, the procedure is time-consuming and depends on the availability of fresh tissue or blood from the organism of choice. Second, histone proteins isolated from natural sources are often degraded by contaminating proteases *(3)*. Third, histone isotypes and posttranslational modifications of histone proteins give rise to heterogeneity. The extent of heterogeneity and modification strongly depend on the type and developmental state of the tissue from which chromatin is isolated and can vary significantly between different batches. Fourth, and most important, only naturally occurring histone proteins can be obtained by this method.

The availability of large amounts of naturally occurring mutants, or of new site-directed mutants of the highly conserved histone proteins, will be extremely valuable in our attempts to reconcile the observed functions and biophysical properties of the NCP with the recently determined atomic structure *(4)*. The ability to express all four histone proteins in bacteria has allowed us to develop a method for the mapping of nucleosome position to base pair resolution *(5)* and has been instrumental in the structure determination of the NCP at high resolution *(4)*. In comparison to yeast expression systems, yields are high, protease activity is low, and purification does not rely on the presence of histidine-tags or other fusions *(6,7)*.

This section describes the overexpression of histones H2A, H2B, H3, and H4, both as full length proteins and corresponding trypsin-resistant "globular domains" (as defined in *[8]*). A simple and efficient purification protocol yields large amounts of homogenous protein in denatured form. The methods for refolding and purification of histone octamer and for assembly and purification of nucleosome core particles using 146 bp of DNA are described, together with a protocol for a high-resolution gel shift assay to monitor the purity and homogeneity of the final core particle preparation.

2. Materials

2.1. Histone Expression

1. pET-histone expression plasmids *(2)* and transformation-competent cells of the expression strain BL21(DE3) pLysS *(9)*.
2. 2X TY-AC media: 16% (w/v) bacto-tryptone, 10% (w/v) yeast extract, and 5% (w/v) NaCl, supplemented with 100 µg/L ampicillin and 25 µg/L chloramphenicol.
3. AC agar plates: 10% (w/v) bacto-tryptone, 5% (w/v) yeast extract, 8% (w/v) NaCl, and 1.5 % (w/v) Agar, supplemented with 100 µg/L ampicillin and 25 µg/L chloramphenicol.
4. IPTG: $0.4M$ Isopropyl-ß-D-thiogalactopyranoside in water; pass through 0.2-µm sterile filter, store frozen in aliquots.
5. Wash buffer: 50 mM Tris-HCl, pH 7.5, 100 mM NaCl, 1 mM Na-EDTA, 1 mM benzamidine. Shortly before use, add 5 mM 2-mercaptoethanol.

2.2. Histone Purification

1. Wash buffer: as in **Subheading 2.1.**, item **5**.
2. TW buffer: wash buffer with 1% (v/v) Triton X-100.
3. Unfolding buffer: $7M$ guanidinium HCl, 20 mM Tris-HCl, pH 7.5, 10 mM DTT. Pass through 0.4-µm filters before use.
4. Amberlite MB3 or similar ion exchange resin for batch deionization of urea stock solutions.
5. SAU-1000: $7M$ urea (deionized), 20 mM sodium acetate, pH 5.2, 1 M NaCl, 5 mM 2-mercaptoethanol, 1 mM Na-EDTA. Pass through 0.4-µm filters before use.
6. SAU-200: $7M$ urea (deionized), 20 mM sodium acetate, pH 5.2, $0.2M$ NaCl, 5 mM 2-mercaptoethanol, 1 mM Na-EDTA. Pass through 0.4-µm filters before use.
7. SAU-600: $7M$ urea (deionized), 20 mM sodium acetate, pH 5.2, $0.6M$ NaCl, 5 mM 2-mercaptoethanol, 1 mM Na-EDTA. Pass through 0.4-µm filters before use.
8. Gel filtration column XK-50 (Pharmacia, Uppsala, Sweden), packed with Sephacryl S-200 high-resolution gel filtration resin (Pharmacia). Gel bed: 5-cm diameter, 75-cm height.

9. An HPLC system equipped with a TSK SP-5PW HPLC column, 2.15 cm × 15.0 cm (Toyo Soda Manufacturing Company, Tokyo, Japan).
10. Dialysis tubing, molecular weight cutoff 6–8 kDa, widths 5 cm and 2.5 cm. Prepare according to the supplier and rinse thoroughly with distilled water before use.
11. Standard SDS-PAGE equipment; 18% SDS gels for analysis of protein fractions *(10)*.

2.3. Histone Octamer Reconstitution

1. Unfolding buffer: as in **Subheading 2.2.3.**
2. Refolding buffer: $2M$ NaCl, 10 mM Tris-HCl, pH 7.5, 1 mM Na-EDTA, 5 mM 2-mercaptoethanol.
3. Gel filtration column HiLoad 16/60 Superdex 200 prep grade (Pharmacia), equipped with UV-detector and fraction collector; at 4°C (*see* **Note 7**).
4. Standard SDS-PAGE equipment (*see* **Subheading 2.2., step 11**).
5. Concentration device suitable for 1–25 mL volumes (e.g., Sartorius ultrathimble, Sartorius AG, Göttingen, Germany; or Centricon 10, Amicon AG, Beverley, MA).

2.4. Nucleosome Core Particle Reconstitution

1. Purified histone octamer at a concentration of at least 0.75 mg/mL, in refolding buffer.
2. DNA of length greater than 138 bp, with a known concentration (at least 3 mg/mL).
3. A peristaltic pump with a double pump head, capable of maintaining a flow rate of approx 2–6 mL/min (e.g., Gilson Minipuls 3 peristaltic pump, equipped with tubing with 2.5 mm inner diameter; Gilson Medical Electronics SA, Villier-leBel, France); or two peristaltic pumps.
4. A reconstitution flask with connected tubing, as shown in **Fig. 1**.
5. Buffers for reconstitution:
   RB-high: $2M$ KCl, 10 mM Tris-HCl, pH 7.5, 1 mM EDTA, 1 mM DTT
   RB-low: $0.25M$ KCl, 10 mM Tris-HCl, pH 7.5, 1 mM EDTA, 1 mM DTT.

2.5. Nucleosome Core Particle Purification

2.5.1. Purification by HPLC-Ion Exchange Chromatography

1. TES-250: $0.25M$ KCl, 10 mM Tris-HCl, pH 7.5, 0.5 mM EDTA.
2. TES-600: $0.6M$ KCl, 10 mM Tris-HCl, pH 7.5, 0.5 mM EDTA.
3. A HPLC apparatus equipped with a TSK DEAE-5PW HPLC column, 2.15 × 15.0 cm, or a TSK DEAE-5PW HPLC column, 7.5 × 75 mm (Toyo Soda Manufacturing); preferably at 4°C.

2.5.2. Purification by Preparative Gel Electrophoresis; High-Resolution Gel Shift Assay

1. Model 491 Prep Cell (Bio-Rad Laboratories, Richmond, CA) with a standard power supply, connected to a UV detector and a fraction collector, and equipped with a peristaltic pump.
2. Gel running buffer: 0.20X TBE (1X TBE: 89 mM Tris-HCl, 89 mM boric acid, and 2.5 mM EDTA).

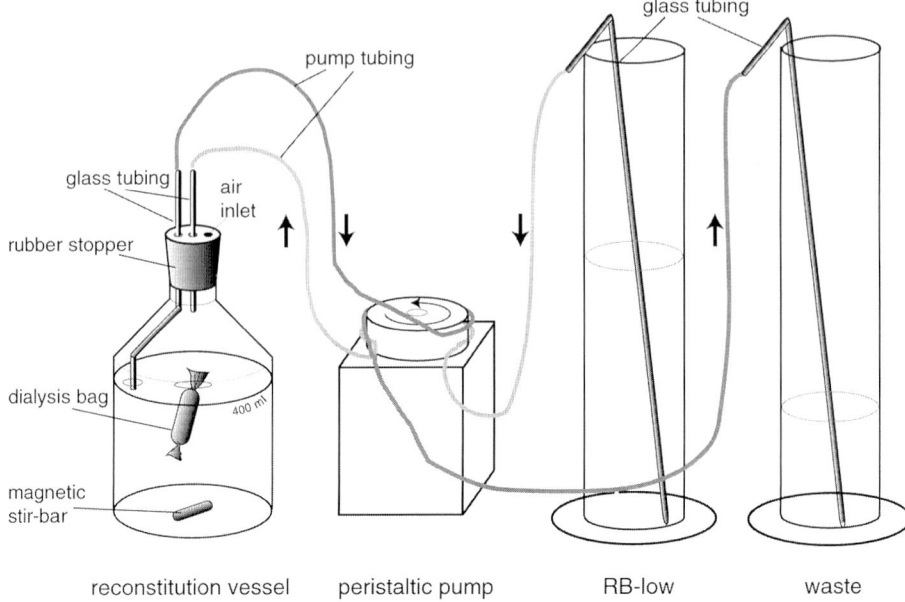

Fig. 1. Schematic drawing of the experimental apparatus for reconstitution. We use a 500-mL glass flask as a reconstitution vessel, and a peristaltic pump with a four-channel head. Standard glass tubes are bent in the appropriate manner and are connected by silicone tubing. The reconstitution vessel contains RB-high to start.

3. Acrylamide stock solution: 29.5% acrylamide, 0.5% *bis*-acrylamide in water. Deionized by stirring with Amberlite MB3, and stored at 4°C.
4. Dialysis membrane (molecular weight cutoff: 6–8 kDa), cut to a circle with a radius of 3 cm. Prepare according to the supplier and rinse thoroughly with distilled water before use.
5. Concentration device as specified in **Subheading 2.3.5.**
6. Storage buffers: TCS buffer:
   20 mM Tris-HCl, pH 7.5, 1 mM EDTA, 1 mM DTT
   CCS buffer: 20 mM K-Cacodylate, pH 6.0, 1 mM EDTA.

3. Methods
3.1. Histone Expression

Expression plasmids for the individual histone proteins and their N-terminally truncated versions, based on the T7-expression system *(9)*, have been described previously *(2)*. High level expression of the *Xenopus laevis* histone genes for H2A, H2B, and H3 does not necessitate adaptation of the codon usage, despite the presence of several codons with low usage in *Escherichia coli*. These proteins can be expressed with similar efficiencies as N-terminal fusion proteins or with the coding region fused directly to the promoter. In contrast, the *Xenopus laevis* gene for histone H4 could only be expressed after redesigning the coding region of H4 to optimize for codon usage in *E. coli* (*2*).

Expression levels of histones H2A and H2B appear to be insensitive to the sequence variant that we have expressed, but histone H3 expression levels do vary between different sequence variants or mutated genes. H4 expression is sensitive to amino acid substitutions and can drop to an undetectable level for certain point mutations. Typical yields for H2A, H2B, and H3 are 50–80 mg of pure protein per liter of cell culture, while yields for H4 are 4–5 times lower (*see* **Note 1**).

1. Transform BL21 (DE3) pLysS cells with 0.1–1 µg of the pET-histone expression plasmid and plate on AC agar plates. Incubate the plates at 37°C overnight.
2. First, perform a test expression by incubating five 5-mL aliquots of 2X TY-AC, each inoculated with one single colony from the agar plate. Shake at 37°C for approx 4 h, or until the OD_{600} is between 0.3 and 0.6. Transfer 0.5 mL of the culture into a sterile Eppendorf tube, add 0.2 mL of sterile glycerol, mix well and store at –80°C (this will serve as the glycerol stock for large-scale expressions). Induce all but one culture by addition of IPTG to a final concentration of 0.2 mM. Leave one sample uninduced as a negative control. Incubate for another 2–3 h at 37°C, harvest by centrifugation, and boil the cell pellets in 100 µL of protein gel loading buffer. Load 20 µL per sample on an 18% SDS-polyacrylamide gel (SDS-PAGE), and determine the culture with the maximum expression.
3. The evening before performing the large scale expression, restreak the glycerol stock for this culture on an AC agar plate, and incubate at 37°C over night (*see* **Note 1**).
4. The next morning, inoculate five aliquots of 4 mL 2X TY-AC media with one colony from this plate and shake at 37°C for approx 4 h ($OD_{600} \sim 0.3$). Use the combined precultures to inoculate 100 mL of 2X TY-AC media. Shake at 37°C for about 2 h, or until the culture is slightly turbid ($OD_{600} \sim 0.4$) (*see* **Note 2**).
5. Inoculate 12 2-L Erlenmeyer flasks containing 500 mL 2X TY-AC media with 8 mL of the 100-mL starter culture. Shake at 200 rpm and 37°C until the OD_{600} has reached 0.6 (this takes about 3 h, *see* **Note 2**). Induce by addition of 0.2 mM IPTG (final concentration) and shake for another 2 h (H3 and H4) or 3 h (H2A and H2B).
6. Harvest the cells by centrifugation at room temperature. Resuspend homogeneously in 100 mL wash buffer and flash-freeze in liquid nitrogen (*see* **Note 3**).

3.2. Histone Purification

The purification protocol involves three steps: preparation of inclusion bodies, gel filtration under denaturing conditions, and HPLC-ion exchange chromatography under denaturing conditions.

Analysis of the pure proteins by gel electrophoresis (SDS- and triton-urea-acid PAGE), mass spectroscopy, amino acid analysis, and by sequencing the

N-terminus using Edman degradation shows that all preparations were highly homogenous and free of modifications. Whereas the terminally truncated histone proteins retain the N-terminal methionine residue, the full-length histones are completely free of this methionine residue and begin with the native sequence *(2)*.

3.2.1. Inclusion Body Preparation

1. Start the equilibration of the Sephacryl S-200 gel filtration column in the morning: prepare 2 L of $9M$ urea in water. Heat to dissolve and deionize with Amberlite MB3 (*see* **Note 4**). Prepare 2 L of SAU-1000 buffer. Equilibrate the column with 2 L of filtered and degassed SAU-1000 buffer at a flow rate of 3 mL/min.
2. To avoid overloading of the gel filtration column, no more than the equivalent of 6 L of cell culture should be processed at one time. For histones with moderate expression (e.g., all H4-variants), up to 12 L of culture can be used. Thaw the cell suspension in a warm water bath. The cell suspension will become extremely viscous as lysis occurs. Stir occasionally until completely thawed (20–30 min). Transfer into a wide, short measuring cylinder and adjust the volume with wash buffer to 150 mL. Reduce viscosity by shearing with a Turrax stirrer. Check whether the mixture is still viscous by using a pasteur pipet, and if necessary, repeat shearing step. Centrifuge immediately for 20 min at 4°C and 23,000g. The pellet contains inclusion bodies of the corresponding histone protein.
3. Resuspend the pellet completely in 150 mL TW buffer using a 10-mL plastic pipet. Spin for 15 min at 4°C at 12,000 rpm. Repeat this step twice with TW buffer and twice with wash buffer (*see* **Note 5**). After the last wash, the drained pellet can be stored at –20°C until further processing.
4. With a spatula, transfer the pellet to a 50-mL centrifuge tube. Add 1 mL of DMSO and soak the pellet for 30 min at room temperature. Mince the pellet with a spatula. Slowly add 40 mL of unfolding buffer and stir gently for 1 h at room temperature. The pellet should eventually almost completely dissolve (*see* **Note 4**).
5. Remove cell debris by centrifugation at 20°C and 23,000g. The supernatant contains the unfolded proteins. Reextract the pellet with 10 mL of unfolding buffer, and combine the supernatants.

3.2.2. Gel Filtration

1. Load a maximum of 60 mL of the sample on the equilibrated S-200 column, at a flow rate of 3 mL/min. Record the elution profile at a wavelength of 280 nm, and collect fractions of an appropriate size.
2. Analyze peak fractions by 18% SDS-PAGE. The first peak will contain DNA and larger proteins but can be merged with the histone peak. Because of the high dilution factor and the small molar absorption of histones, the histone peak is often small and unobtrusive. Pool fractions containing histone proteins (*see* **Note 6**).
3. The protein is dialyzed thoroughly against at least three changes of distilled water containing 2 mM 2-mercaptoethanol, at 4°C. Dialysis bags with a cutoff of 6–8 kDa are sufficient even for the globular domain of H4, but leave enough room for

Table 1
Molecular Weights and Molar Extinction Coefficients (ε) for Full-Length and Trypsin Resistant Globular Domains of Histone Proteins[a]

	Full-length protein			Globular domains	
Histone	Mol. wt	ε (cm/M), 276 nm	Amino acid	Mol. wt	ε (cm/M), 276 nm
H2A	13,960	4050	19–118	11,862	4050
H2B	13,774	6070	27–122	11,288	6070
H3	15,273	4040	27–135	12,653	4040
H4	11,236	5400	20–102	9,521	5040

[a]ε was calculated according to Gill and von Hippel (*14*). The molecular weights were determined by a summation of amino acids and were confirmed by mass spectrometry (*2*).

the volume increase. Determine the concentration of the dialyzed sample, using the molecular extinction coefficients listed in **Table 1**. Lyophilize, and store at –20°C.

3.2.3. Purification by Ion Exchange Chromatography (SP-5PW)

1. Dissolve lyophilized histone protein in SAU-200 and remove insoluble matter by centrifugation. Equilibrate the preparative SP-5PW HPLC column with SAU-200 buffer. Inject a maximum of 15 mg of protein for each run. Using a flow rate of 4 mL/min, elute proteins with the gradients given in **Table 2** (buffer A = SAU–200, buffer B = SAU–600).
2. Analyze the peak-fractions by SDS-PAGE. Pool the fractions containing pure histone protein, dialyze against water as described under **Subheading 3.2.2.3.**, and lyophilize.
3. Dissolve in a small volume of water (determine the concentration using the values given in **Table 1**) and lyophilize in Nunc vials in aliquots suitable for subsequent octamer refolding reactions (*see* **Subheading 3.3.1.**). For example: 4.5 mg H3, 3.5 mg H4, 4.0 mg H2A, 4.0 mg H2B). The purified histones can be stored at –20°C for an unlimited length of time.

3.3. Refolding of the Histone Octamer

The protocol below is valid for the refolding of histone octamers from lyophilized recombinant histone proteins. (All combinations of recombinant *Xenopus laevis* full-length and globular domain histone proteins have been refolded to functional histone octamers.) The method works best for 6–15 mg of total protein. For smaller amounts of protein, scale down the gel filtration column (*see* **Note 7**).

1. Dissolve each histone aliquot to a concentration of approximately 2 mg/mL in unfolding buffer. Use a Pasteur pipet to ensure protein sticking to the sides of the tube is dissolved. Do not vortex. Unfolding should be allowed to proceed for at least 30 min and for no more than 3 h (*see* **Note 4**).

Table 2
Salt Gradients for Elution of Histone Proteins from an SP-Column[a]

H2A, H2B		H3		H4	
t (min)	B (%)	t (min)	B (%)	t (min)	B (%)
0	0	0	0	0	0
10	0	3	30	5	50
11	40	8	30	10	59
46	100	40	100	40	100
60	100	50	100	50	100
61	0	51	0	51	0

[a]Salt gradients for elution of histone proteins from an SP-5PW HPLC column (2.15 × 15.0 cm): the flow rate is 4 mL/min, buffer A is SAUDE 200, buffer B is SAUDE 600.

2. Determine the concentration of the unfolded histone proteins by measuring OD_{276} (**Table 1**) of the undiluted solution against unfolding buffer (remove solid matter by centrifugation, if necessary).
3. Mix the four histone proteins to equimolar ratios and adjust to a total final protein concentration of 1 mg/mL using unfolding buffer.
4. Dialyze at 4°C against at least three changes of 2 L of refolding buffer (for 15 mg setup, use dialysis bags with a flat width of 2.5 cm). The second or third dialysis step should be performed overnight. Octamer should always be kept at 0–4°C.
5. Remove precipitated protein by centrifugation. There should be almost no precipitate in the "ideal" refolding reaction. Concentrate to a final volume of 1 mL.
6. Gel filtration is performed at 4°C at a flow rate of 1 mL/min. Load a maximum of 1.5 mL or 15 mg of the concentrated histone octamer on the Superdex-200 gel filtration column previously equilibrated with refolding buffer. High-molecular-weight aggregates will elute after about 45 mL, histone octamer at 65–68 mL, and histone (H2A-H2B) dimer at 84 mL (*see* **Note 7**).
7. Check the purity and stoichiometry of the fractions on an 18% SDS-PAGE. Dilute by a factor of at least 2.5 before loading onto the gel to reduce distortion of the bands resulting from the high salt concentration. Pool fractions that contain equimolar amounts of the histone proteins.
8. Determine the concentration (A_{276} = 0.45 for a solution of 1 mg/mL). Use for nucleosome core particle reconstitution with DNA immediately or concentrate to 3–15 mg/mL, adjust to 50% (v/v) glycerol, and store at −20°C.

3.4. Reconstitution of Nucleosome Core Particles

Specific DNA fragments of the desired length or sequence can be obtained by a number of methods (e.g., *11,12*). Reconstitution of histone octamer with DNA is accomplished using a modification of the salt gradient method described by Thomas and Butler *(13)*. Briefly, octamer and DNA is mixed at 2 M KCl, and the salt concentration is reduced by dialysis to 0.25 M KCl over a period of 36 h *(12)*. The procedure works equally well for large (up to 10 mg) and small (0.1 mg) amounts of nucleosome core particles. Multiple setups can be dialyzed in one vessel. If smaller amounts need to be reconstituted, use dialysis buttons (e.g., Hampton Research, Laguna Hills, CA), or the apparatus described in Chapter 4.

1. Histone octamer is added to the DNA to a 0.9 molar ratio of octamer to DNA, with a final DNA concentration of 6 μM. Before adding the histone octamer (*see* **Note 8**), adjust the salt concentration of the DNA solution to 2 M using 4 M (or solid) KCl, and add DTT to a final concentration of 10 mM. Incubate at 4°C for 30 min.
2. Prepare 400 mL RB-high and 1600 mL RB-low buffer and chill these to 4°C. Set up the dialysis apparatus as shown in **Fig. 1**. Calibrate the pump to a flow rate of 0.7–0.8 mL/min.
3. Transfer the sample to a dialysis bag and start dialysis against 400 mL RB-high at 4°C under constant stirring. Using the peristaltic pump, continually remove buffer from the dialysis vessel and replace with RB-low. Over a period of 36 h, an exponential gradient is generated (*see* **Note 9**). After the gradient has finished, dialyze for at least 3 h against RB-low. If the samples are not further processed within the next 24 h, dialyze against CCS buffer. If the core particle will be purified by preparative gel electrophoresis (*see* **Subheading 3.5.2.**), dialyze against TCS buffer.

3.5. Purification of the Nucleosome Core Particle

Two methods are described for the purification of nucleosome core particles from free octamer and/or free DNA. Both methods have been optimized for NCP with 146 bp DNA, but can easily be adjusted for nucleosomes with different length DNA. The first method uses HPLC DEAE-ion exchange chromatography *(12)*. Free DNA elutes from the column at a higher salt concentration than NCP and thus can be easily separated from the complex. The second method uses gel electrophoresis under nondenaturing conditions as a purification principle *(2)*. **Table 3** compares the advantages and disadvantages of the two methods. Both methods alone give rise to highly pure NCP preparations; the choice depends on available equipment and on the problem at hand. Ion exchange chromatography is suitable for large-scale preparations on a routine basis. However, certain modification of histone proteins (such as covalently bound heavy atoms) might completely alter the binding and elution properties of the NCP. If a significant amount of material appears as a high-molecular-weight band in a gel shift assay, or if the particle is prone to salt-dependent dissociation, the second method might be more suitable. The two methods can also be used in combination.

Table 3
Purification of Nucleosome Core Particles: Comparison of Two Methods

	DEAE-ion exchange	Preparative gel electrophoresis
Capacity	Up tp 10 mg NCP per run for a preparative column (2.15 × 15.0 cm)	Maximum of 3 mg NCP per run
Time	Fast, reliable	Slow, labor-intensive
Purification	No purification from higher order aggregates	Purification from both DNA and higher order complexes
Elution	Salt elution might cause dissociation	Free choice of elution buffer

Reconstitution usually leads to a heterogeneous population of NCP with respect to the position of the DNA on the histone octamer. A simple heating step (37–55°C for 20–180 min.) usually results in a uniquely positioned NCP preparation for DNA 145 to 147 bp in length which is suitable for biochemical studies and crystallization (*see* **Fig. 3**, and **Note 10**).

3.5.1. Purification by Ion Exchange Chromatography

1. Equilibrate the DEAE-5PW column with TES-250 at 4°C. Centrifuge the reconstitution mixture in Eppendorf tubes at 4°C and inject the supernatant on the column (a maximum of 10 mg core particle). Samples can either be in RB-low, TCS buffer, or CCS buffer. Using a flow rate of 4 mL/min, develop the column with the gradient given in **Table 4** (buffer A = TES-250, buffer B = TES-600) while monitoring the eluent at 260 nm. Adjust gradient for core particles containing longer DNA fragments or different temperatures and salts (NaCl instead of KCl). A typical chromatogram is shown in **Fig. 2A** (*see* **Note 10**).
2. Analyze the peak-fractions by non-denaturing gel electrophoresis, if necessary (*see* **Subheading 3.6.**, and **Note 10**).
3. Pool the peak fractions and immediately dialyze against three changes of TCS buffer at 4°C. Concentrate and store on ice until use. For prolonged storage, dialyze against CCS buffer, or add 5 mM potassium cacodylate at an appropriate pH to prevent microbial growth. Determine the concentration of NCP preparations by measuring the absorbency at 260 nm of a 200- to 500-fold dilution. The yield of the above reconstitution and purification method may vary between 20 and 80% of the DNA added to the reconstitution mixture.

3.5.2. Purification by Preparative Gel Electrophoresis
(see also **Subheading 3.6.**)

1. Prepare 20 mL of a 5% polyacrylamide gel (ratio of acrylamide to *bis*-acrylamide 60:1), containing 0.2X TBE, and pour a cylindrical gel with an outer radius of 28 mm,

**Table 4
Salt Gradient for Elution
of NCP from a DEAE -5PW
HPLC Column**[a]

	NCP146
t (min)	B (%)
0	0
10	0
11	40
46	100
60	100
61	0

[a]The flow rate is 4 mL/min, buffer A is TES-250, buffer B is TES-600.

an inner radius of 19 mm and a height of 50 mm. Polymerize and assemble according to instructions given in the manual for the Model 491 Prep Cell. Prerun under constant recircularization of the buffer for 90 min in 0.25X TBE at 4°C and 10 W. Record a base line at 260 nm using TCS as elution buffer (*see* **Notes 11** and **12**).

2. After reconstitution, dialyze NCP against TCS buffer and concentrate. A maximum of 600 µL or 3 mg is mixed with sucrose to a final concentration of 5% (v/v) and loaded on the preparative gel. Electrophorese at 10 W, and elute the complex at a flow rate of 0.7 mL/min with TCS buffer. Recirculate buffer. Record elution at an OD of 260 nm, and collect fractions of appropriate size (usually 0.7–1.0 mL). Free DNA will appear first, followed by pure NCP, and finally, by higher molecular weight aggregates (**Fig. 2B**, *see* **Note 11**).

3.6. High-Resolution Gel Shift Assay for Nucleosome Core Particles

The protocol given next routinely allows for the separation of NCP with different translational setting of the DNA by only 10 bp (**Fig. 3**; *see* **refs.** *[2,5]*). We have observed that the ratio between acrylamide and bis-acrylamide can completely alter the relative mobilities of different NCP species with respect to the DNA size marker (*2*). The choice of gel buffer also has minor effects on the resolution of the different NCP species.

1. Prepare a native gel with the dimensions $20 \times 20 \times 0.1$ cm, using a 10- to 16-well comb. The gel material is 5% acrylamide with 60:1 acrylamide to *bis*-acrylamide, with 0.2X TBE.
2. Prerun gels for at least 3 h at 4°C and 200 V, while constantly recycling the running buffer (*see* **Note 12**).

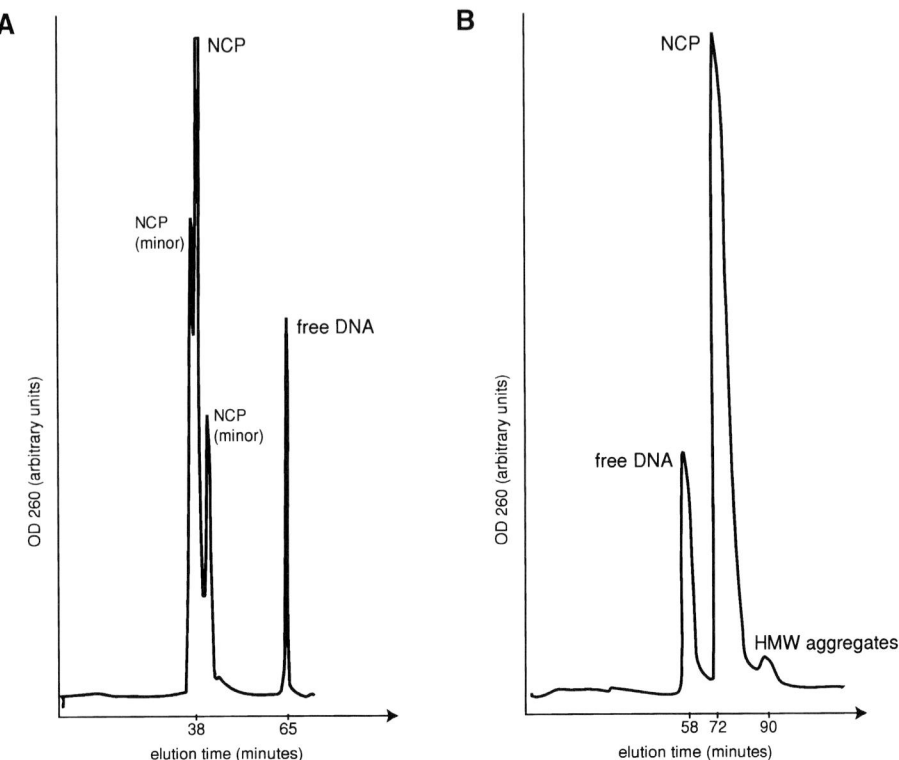

Fig. 2. Purification of reconstituted nucleosome core particle by (**A**) ion exchange chromatography or (**B**) preparative gel electrophoresis. Elution is monitored by the absorbance at 260 nm. Separation conditions are as given in the text. The approximate elution times are shown for each peak, but they are dependent on the flow rate and the geometry of the set-up. (**A**) Purification by HPLC ion exchange chromatography. NCPs with different rotational positions elute at different times from the column. The relative ratio between the main and minor peaks can vary. Usually, major and minor peaks are combined; additional peaks that usually exhibit baseline separation (not shown) are not included. In some cases, free octamer can be observed to elute in the beginning of the gradient (not shown). (**B**) Purification by preparative nondenaturing gel electrophoresis. Note that the three bands observed in **Fig. 3** cannot be discerned by this method. Faster elution will improve the separation of the peaks but will also yield a more dilute sample.

3. Rinse slots well with 0.20X TBE shortly before loading samples. Load 1–2 pmol of core particle solution, containing 5% (v/v) sucrose, in no more than 10 µL. Samples can be supplemented with bromophenol blue for easier handling.
4. Run the gel for a suitable length of time or until bromophenol blue has reached the bottom of the gel. Recycle the running buffer at all times (*see* **Note 12**).

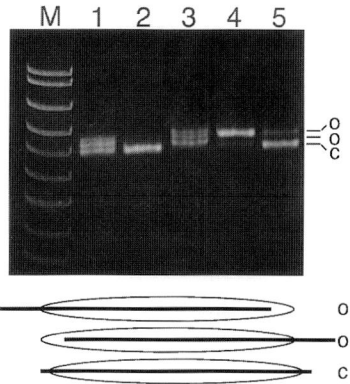

Fig. 3. Nondenaturing gel electrophoresis of NCP reveals multiple positions of the histone octamer on the DNA. Lanes 1–3: purified NCP reconstituted from recombinant full-length histone proteins and a 146-bp fragment derived from the 5S RNA gene of *Lytechinus variegatus*. Before (lanes 1 and 3) and after (lane 2) heating for 1 h at 37°C. Lanes 4 and 5: purified NCP prepared from recombinant full-length histone proteins and a 146-bp palindromic DNA fragment derived from human alpha satellite DNA; before (lane 4) and after (lane 5) heating for 2 h at 37°C. Note that NCP containing the asymmetric 5S RNA DNA fragment reconstitutes in two off-centered and one centered rotational position (marked with o and c, respectively), *(5)*, whereas NCP containing a palindromic sequence reconstitutes mainly to the off-centered positions. Nucleosomes in the two off-centered positions cannot be distinguished on the gel, because the sequence symmetry results in an identical exit angle of the ends of the DNA from the histone octamer. The relative positions of the DNA on the histone octamer in the three bands is shown schematically (*ovals*: histone octamer, *bold line*: DNA).

5. Stain with ethidium bromide. Note that free DNA is stained significantly better by ethidium bromide than DNA bound to the histone octamer. Subsequent staining with Coomasssie brilliant blue is also possible.

4. Notes

1. Glycerol stocks of transformed BL21(DE3)pLys-S can be kept at –80°C for at least 2 mo. If the glycerol stock has not been used for some time, restreak and perform the small-scale expression test again. For some difficult cases (i.e., histone variants), expression of histone proteins in cells lacking the pLysS plasmid proved to be more successful (T. J. Rechsteiner, unpublished results).
2. Values obtained by measurements of the optical density at 600 nm depend on the geometry of the spectrophotometer. We use a Pharmacia Novaspec™ spectrophotometer (Pharmacia). Trial experiments should be performed for each individual histone variant to determine the optimal optical density for induction. We have observed that some point mutants exhibit quite different optima of cell density for induction. The starter culture should never be grown to densities

higher than $OD_{600} \sim 0.6$, since the cells loose the ability to be induced even after dilution.

3. Cells expressing histone proteins (especially H4) are prone to lysis and should be centrifuged at room temperature, and for the same reason, it is not recommended to wash the cell pellet. Resuspend cells well before freezing, as this will improve lysis upon thawing. The cells can be stored at –20°C.
4. Urea in solution is in a slow equilibrium with isocyanate which can irreversibly modify proteins. Do not use urea-containing solutions older than 24 h, and always deionize urea stock solutions before use. Storage of protein in buffers containing guanidinium HCl or urea for more than 24 h is not recommended.
5. If lysis by freezing and thawing is not complete, additional lysis will occur upon the addition of TW buffer and the cell suspension will become viscous again. In that case, repeat shearing. Two consecutive cycles of freezing and thawing of the cell suspension can improve lysis significantly.
6. Unusually early elution of histone proteins from the Sephacryl S-200 gel filtration column (within the large DNA peak), might result from the formation of unspecific complexes between cellular DNA and histones. The presence of 1 M NaCl usually inhibits complex formation between cellular DNA and denatured histones, but this can occasionally remain a problem depending on the shearing of the DNA. For this reason, or if the resolution of the gel filtration column is insufficient, the first fractions of the histone peak might be contaminated with DNA. Before pooling, check the first few histone-containing fractions by UV-spectroscopy for DNA contamination. Discard if the spectrum exhibits a ratio of $OD_{260} : OD_{280} > 1.0$, since DNA and histone will form a precipitate after removal of the salt. Minor precipitate can be removed by centrifugation.
7. Other gel filtration resins of a similar separation range, such as Superose 12 or Sephacryl S-300 (both from Pharmacia) can also be used, but give a lower resolution. Sephadex G-100 does not separate histone octamer from high-molecular-weight aggregates and is therefore not recommended. Separation between histone octamer and excess H2A-H2B dimer is better than from excess $(H3-H4)_2$ tetramer. Yields of pure histone octamer are usually between 50 and 75%. Significant amounts of octamer (or high-molecular-weight aggregates) can remain attached to the column. Clean the column with NaOH as recommended by the supplier.
8. If histone octamer from a glycerol stock is used, dialyze over night against refolding buffer and determine the concentration. The required accuracy in the ratio between histone and DNA ratio cannot be maintained if pipeted directly from the glycerol stock. Histone octamer should always be added last to the reconstitution mixture to avoid premature mixing octamer and DNA at <2M salt concentrations.
9. Ensure that the dialysis bag can circle freely and rapidly to allow constant mixing its contents. Take care to adjust the position of the inlet and outlet tubing as shown in **Fig. 1**. This is important for two reasons:
    a. Rotation of the dialysis bag is inhibited if it gets caught in the tubing.
    b. Uneven pump speed can result in either overflowing or running dry of the dialysis vessel.

Formation of large amounts of precipitate during dialysis, or unexpectedly low yields, may result for two reasons:
   a. An excess of histone octamer (or histone protein) has been added. Any change in the given molar ratio between octamer and DNA will reduce yields significantly.
   b. Stalled motion of the dialysis bag creates an uneven salt gradient within the bag itself. Sometimes, precipitate is formed at salt concentrations of about 400 mM but will dissolve again later.
10. NCPs reconstituted on small DNA fragments are usually heterogeneous with respect to the relative position of the DNA on the histone octamer. This is seen by the appearance of several bands on high-resolution gel shift assays (**Fig. 3**). As a consequence, NCP elutes from the DEAE-column as a major peak with several shoulders (**Fig. 2A**). As a rule, the major peak fractions, including the shoulders, are pooled without analysis by nondenaturing PAGE. The peaks are not distinguished by preparative electrophoresis (**Fig. 2B**). Depending on the sequence and the length of the DNA fragment, the octamer can be moved to one single position by incubation at elevated temperature after reconstitution and purification of NCP (**Fig. 3**). The temperatures and incubation times necessary for this transition have to be checked individually for each sequence and histone octamer. For example, *Xenopus laevis* full-length histone octamer with the 146-bp fragment derived from the 5S RNA gene of *Lytechinus variegatus* is heated for 30 min at 37°C for a complete shift, whereas other sequences might require as long as 2 h at 55°C. Shifting to a unique position occurs completely without dissociation of the DNA; an excess of competitor DNA does not exchange onto the histone octamer during the process.
11. Different conditions for preparative gel electrophoresis can be tested by small-scale native gel electrophoresis experiments, following the guidelines given in the instruction manual for the Model 491 Prep Cell. The ratio between acrylamide and bis-acrylamide, the length of the gel, and the elution speed can greatly alter the relative mobility and the separation of the components. We have also noticed that the choice of elution buffer and electrophoresis buffer greatly influence the relative mobility of the different species. Improved resolution between the different peaks is often a tradeoff with high dilution of the sample. Note that high dilution of NCP during purification might result in a partial dissociation of DNA and octamer.
12. Recycling of the buffer and temperature equilibration are essential for good resolution.

Acknowledgments

We thank our colleagues, especially A. Flaus, A. Mäder, and S. Tan, for support, advice, and discussions. The help of R. Richmond, R. Amherd, and J. Hayek in the preparation of histones is gratefully acknowledged.

References

1. von Holt, C., Brandt, W. F., Greyling, H. J., Lindsey, G. G., Retief, J. D., Rodrigues, J. D., Schwager, S., and Sewell, B. T. (1989) Isolation and characterization of histones. *Meth. Enzymol.* **170,** 431–523.

2. Luger, K., Rechsteiner, T., and Richmond, T. J. (1997) Characterization of nucleosome core particles containing histone proteins made in bacteria. *J. Mol. Biol.* **272,** 301–311.
3. Mellado, R. P. and Murray, K. (1983) Synthesis of yeast histone 3 in an Escherichia coli cell-free system. *J. Mol. Biol.* **168,** 489–503.
4. Luger, K., Maeder, A. W., Richmond, R. K., Sargent, D. F., and Richmond, T. J. (1997) X-ray structure of the nucleosome core particle at 2.8 Å resolution. *Nature* **389,** 251–259.
5. Flaus, A., Luger, K., Tan, S., and Richmond, T. J. (1996) Mapping nucleosome position at single base-pair resolution by using site-directed hydroxyl radicals. *Proc. Natl. Acad. Sci. USA* **93,** 1370–1375.
6. Fukuma, M., Hiraoka, Y., Sakurai, H., and Fukasawa, T. (1994) Purification of yeast histones competent for nucleosome assembly in vitro. *Yeast* **10,** 319–331.
7. Pilon, J., Terrell, A., and Laybourn, P. J. (1997) Yeast chromatin reconstitution system using purified yeast core histones and yeast nucleosome assembly protein-1. *Protein Expr. Purif.* **10,** 132–140.
8. Böhm, L. and Crane-Robinson, C. (1984) Proteases as structural probes for chromatin: the domain structure of histones. *Biosci. Rep.* **4,** 365–386.
9. Studier, F. W., Rosenberg, A. H., Dunn, J. J., and Dubendorff, J. W. (1990) Use of T7 RNA polymerase to direct expression of cloned genes. *Meth. Enzymol.* **185,** 60–89.
10. Lämmli, U. K. (1970) Cleavage of structural proteins during the assembly of the head of bacteriophage T4. *Nature* **227,** 681–685.
11. Simpson, R. T., Thoma, F., and Brubaker, J. M. (1985) Chromatin reconstituted from tandemly repeated cloned DNA fragments and core histones: a model system for study of higher order structure. *Cell* **42,** 799–808.
12. Richmond, T. J., Searles, M. A., and Simpson, R. T. (1988) Crystals of a nucleosome core particle containing defined sequence DNA. *J. Mol. Biol.* **199,** 161–170.
13. Thomas, J. O. and Butler, P. J. G. (1977) Characterization of the octamers of histones free in solution. *J. Mol. Biol.* **116,** 769–781.
14. Gill, S. C. and von Hippel, P. H. (1989) Calculation of protein extinction coefficients from amino acid sequence data. *Anal. Biochem.* **182,** 319–326.

2

Preparation and Analysis of Positioned Nucleosomes

Vasily M. Studitsky

1. Introduction

Short (150–350 bp) DNA fragments containing single nucleosome cores have been employed for investigation of a variety of topics, including binding of regulatory transcription factors to nucleosomes *(1)* and mechanism of transcription of nucleosomal templates *(2)*. The major advantage of this experimental system is relative simplicity of isolation and analysis of uniquely positioned cores.

Typically, nucleosomes are assembled on a DNA fragment containing a nucleosome positioning sequence. There are several methods for reconstitution of nucleosome cores on DNA *(3)*; the one most commonly used involves mixing purified DNA and purified core histones at high salt followed by stepwise reduction in the salt concentration either by dilution or by dialysis (*see* Chapter 1). Initially, H3/H4 bind to DNA as a tetramer and then, at a lower salt concentration, two H2A/H2B dimers bind to each tetramer to form nucleosome cores. The simpler approach is the transfer of histone octamers from "donor" nucleosome cores onto DNA *(1)*; however, it only works for isolation of small (nanogram to microgram) quantities of cores. One alternative method that works at lower salt concentrations involves the use of poly(glutamate) as a carrier for the histones *(3)*, but we have found that the extent of reconstitution is somewhat variable with this technique.

The resulting reconstitute is a mixture of core nucleosomes positioned at different locations on the fragment: The challenge is to isolate the desired uniquely positioned nucleosomes. By far the most powerful method is isolation of cores from a native polyacrylamide gel *(4)*. The method is based on the observation that positional isomers of nucleosome cores have different electrophoretic mobilities (*[5,6]*; **Fig. 1A**). However, cores at symmetrical positions

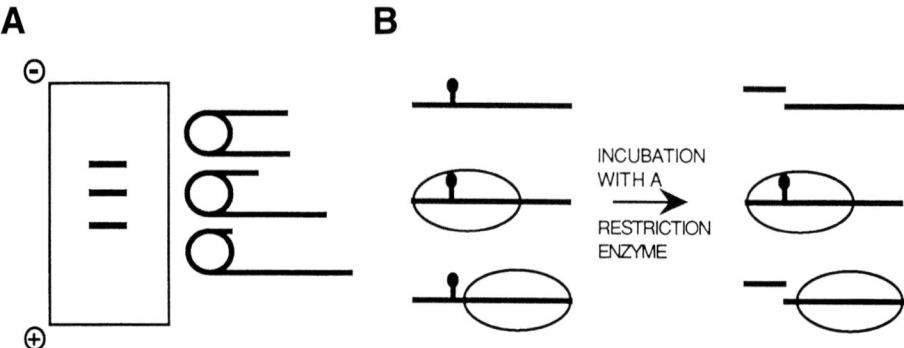

Fig. 1. Two assays for analysis of nucleosome positioning on a short DNA fragment. (**A**) Mobility of nucleosome core in a native gel depends on its position relative to the ends of DNA fragment. (**B**) Restriction enzyme sensitivity assay. Nucleosomes positioned symmetrically (and therefore having the same mobility in a native gel) are differentially sensitive to restriction enzymes and therefore can be resolved in a native gel after incubation with the enzymes.

have the same mobility and cannot be separated in the gel. To overcome this limitation of the method, we developed a new protocol, including predigestion with a restriction enzyme (**Fig. 1B**).

Two methods of mapping of nucleosome core positions along DNA are in routine use in our laboratory: low-resolution mapping with restriction enzymes and high-resolution mapping with micrococcal nuclease (MNase). Both methods are based on the resistance of nucleosomal DNA to endonucleases. In the first method, isolated cores are incubated in the presence of sufficient restriction enzyme for complete digestion of free DNA and then analyzed in a nucleoprotein gel *(7,8)*. The method allows rapid mapping of positions but the resolution is relatively low (approx 10 bp). It is most useful for quantitative information on the fraction of cores in each position. The second method is based on the fact that approx 150 bp of DNA in the nucleosome core are protected from digestion by MNase: core DNA is purified, end-labeled and mapped by digestion with different restriction enzymes *(7,9)*. This method is more involved, but it yields an unambiguous, high-resolution map of all positioned nucleosome cores on the fragment.

2. Materials

2.1. Materials

1. Centricon-10 (Amicon).
2. Pre-wet dialysis tubing (Spectra/Por7; molecular weight cutoff of 8000).
3. G-25 Quick-spin columns (Boehringer Mannheim).

Isolation of Positioned Nucleosomes

4. Siliconized Eppendorf tubes (PGC Scientific).
5. DE-81 paper (Whatman).

2.2. Enzymes

1. T4 polynucleotide kinase (New England Biolabs, [NEB]).
2. Restriction enzymes (NEB).
3. Micrococcal nuclease (MNase, Worthington).
4. Klenow fragment of *E. coli* DNA polymerase I (NEB).

2.3. Reagents

1. Ethidium bromide (10 mg/mL stock).
2. Butyl alcohol.
3. Ethanol.
4. Equilibrated phenol.
5. Chloroform.
6. 5M NaCl.
7. 2-mercaptoethanol.
8. Protease- and nuclease- free BSA (Sigma).
9. Glycerol.
10. Sodium dodecyl sulphate (SDS, 10% [w/w] solution).
11. α-^{32}P-dNTPs; γ-^{32}P-ATP (6000 Ci/mmol; Du Pont NEN).
12. Acrylamide; *N,N'*-methylene-*bis*-acrylamide (Bis) (Bio-Rad).
13. Glycogen, 10 mg/mL solution (Boehringer Mannheim).

2.4. Buffers

1. TAE: 0.04M Tris-acetate, pH 8.0, 1 mM EDTA.
2. HE: 10 mM Na-HEPES, pH 8.0, 1 mM EDTA. Use 99+% pure HEPES (Sigma).
3. Core reconstitution buffers (CRB 1–6): All six buffers contain HE and 10 mM 2-mercaptoethanol; CRB 1–5 contain 5M urea (prepared using an 8M urea solution previously de-ionized by gentle stirring with some BioRad AG501-8X resin (20–50 mesh) and then filtered to remove the resin) and NaCl at the following concentrations: 2M CRB1; 1.2M CRB2; 1M CRB3; 0.8M CRB4; 0.6M CRB5. CRB6 contains 0.6M NaCl but lacks urea.
4. Klenow fragment labeling buffer (KLB): 10 mM Tris-HCl, pH 7.5, 5 mM MgCl$_2$, 7.5 mM dithiothreitol (10X stock solution is available from NEB).
5. Basic transcription buffer (BT): 40 mM HEPES, pH 8.0, 6 mM MgCl$_2$, 2 mM spermidine, 10 mM 2-mercaptoethanol.
6. T4 polynucleotide kinase buffer (KB): 70 mM Tris-HCl, pH 7.6, 10 mM MgCl$_2$, 5 mM dithiothreitol (5X stock solution is available from NEB).

2.5. Proteins

1. H2A/H2B and H3/H4 histone pairs isolated from adult chicken blood by chromatography on hydroxyapatite *(10,11)*.

3. Methods
3.1. Preparation of Short DNA Templates
1. Digest 1 mg of plasmid with restriction enzymes to release the required fragment (*see* **Note 1**).
2. Extract the protein using an equal volume of 1:1 (v/v) phenol:chloroform.
3. Precipitate the DNA with 3 vol of ethanol, wash with 70% (v/v) ethanol, dry, and dissolve in 300 μL of TAE buffer.
4. Resolve the DNA fragments in the digest in a preparative 1.2–1.5% (w/v) agarose gel containing 0.5 μg/mL ethidium bromide and TAE buffer at 4–6 V/cm for 1.5–3 h, depending on the resolution required for clear band separation.
5. Using a long-wavelength UV lamp (to reduce nicking of DNA), identify and excise the required band(s).
6. Put the agarose slice into dialysis tubing containing approx 3 mL TAE buffer (secure using dialysis clips) and electroelute at 4–6 V/cm for 15 min. Reverse the current for 10 s to detach any DNA stuck to the tubing.
7. Transfer the DNA solution to a 6-mL syringe attached to a 0.2- or 0.45-μm filter unit and filter to remove pieces of agarose.
8. Add 3 M Na-acetate (pH 5.5) to a final concentration of 30 mM and extract several times with 2 vol of butanol to reduce the aqueous volume (on the bottom) to ~300 μL.
9. Extract the aqueous phase once with 1 vol of 1:1 (v/v) phenol:chloroform.
10. Precipitate DNA with 3 vol of ethanol, wash with 70% (v/v) ethanol, dry, and dissolve in 100 μL HE.
11. Determine DNA concentration by measuring the A_{260} (using $A_{260} = 20$ for 1 mg/mL DNA) and store at –20°C.

3.2. Labeling DNA Using Klenow Fragment of E. coli DNA Polymerase I
1. Incubate 10 μg DNA with 5 U of the Klenow fragment in 50 μL KLB buffer, supplemented with 100 μCi of α-^{32}P-dNTP at 12°C for 5 min (*see* **Note 2**).
2. Add all four NTP to 100 μM and incubate at 20°C for 10 min. Stop the reaction by adding EDTA to final concentration 20 mM.
3. Extract the DNA with 1:1 (v/v) phenol:chloroform (*see* **Note 3**).
4. Precipitate the DNA with ethanol, wash with 70% (v/v) ethanol, dry, and dissolve in 25 μL HE.
5. Measure the dpm/μL by scintillation counting and its absorbance at 260 μm. Calculate the specific activity of the labeled DNA.

3.3. Reconstitution of Nucleosome Cores Using the Salt/Urea Dialysis Method
1. Cool 500 mL each of CRB1 to CRB5 buffers and 1 L each of CRB6 and HE to 4°C.
2. Add to the DNA solution: 5 M NaCl to a final concentration of 2 M and 2-mercaptoethanol to a final concentration of 10 mM (*see* **Note 4**).
3. Add purified core histones to a molar ratio of 0.9 core histone octamer per DNA molecule (*see* **Note 5**). Adjust the volume to 300–400 μL with HE.

4. Dialyze against CRB1 at 4°C overnight (*see* **Note 6**).
5. Next morning, dialyze successively against CRB2, CRB3, CRB4, and CRB5, each for 1 h. Then dialyze against CRB6 for 3 h and finally against HE for 3 h or overnight (*see* **Note 7**).
6. Transfer the reconstitute to a siliconized Eppendorf tube (*see* **Note 8**) and store at 4°C (do not freeze). Measure the dpm/µL by scintillation counting and the A_{260}. Calculate the specific activity of the labeled DNA in the reconstitute.

3.4. Isolation of Nucleosome Cores from a Preparative Nucleoprotein Gel

1. Pour a 4.5% polyacrylamide gel (acrylamide:bis = 40:1; 17 x 17 x 0.15 cm) containing 20 mM Na-HEPES, pH 8.0, 1 mM EDTA, 5% (v/v) glycerol. Use a comb with teeth approx 2 cm wide.
2. Preelectrophorese the gel for 4.5 h at 12 V/cm. Use 20 mM HEPES, pH 8.0, 1 mM Na-EDTA as running buffer. Change the running buffer and preelectrophorese for another 30 min at 6 V/cm with recirculation of the buffer (*see* **Note 9**).
3. Equilibrate a Centricon-10 unit (Sorvall SS-34 rotor, 4000g, 5 min) with 0.2 mL HE containing 10 µg BSA at 4°C (*see* **Note 10**). Concentrate the reconstitute (SS-34 rotor, 4000g, 1 h) to a final volume of approx 60 µL.
4. Transfer the sample to a siliconized Eppendorf tube, add 1/9 vol 10X TB and incubate in the presence of 30 U of the appropriate restriction enzyme at 37°C for 1 h (*see* **Note 11**).
5. Add EDTA to 20 mM to stop the digestion and sucrose to 10% (*see* **Note 12**). Load the reconstitute on the gel (10 µg reconstituted DNA per 2-cm-wide slot). Electrophorese at room temperature for 5–8 h at 6 V/cm with recirculation of the buffer at approx 300 mL/h.
6. Handle the gel with caution because it is highly radioactive: Cover the wet gel with plastic wrap, apply radioactive or fluorescent markers, and expose it to X-ray film without a screen for 1–10 min.
7. Do not allow the gel to dry out: carefully cut out gel slices containing each band and place them in siliconized, precooled (on ice) 1.5-mL Eppendorf tubes (*see* **Note 13**). Estimate the volume of gel excised by measuring the dimensions of the hole in the gel.
8. Crush the gel in the tube and, for every 100 µL of gel, add 130 µL ice-cold HE containing 150 µg/mL BSA. Rotate overnight at 4°C.
9. Add 100 µl HE containing 150 µg/mL BSA to the tube, and spin the gel pieces down in an Eppendorf centrifuge at maximum speed for 1 min at 4°C. Remove the supernatant immediately, and transfer it to 0.65-mL siliconized Eppendorf tubes.
10. Roughly estimate the fraction of radioactivity eluted from the gel using a Geiger counter; recovery is usually 70–80%. Store isolated cores at 4°C (*see* **Note 14**).

3.5. Mapping the Positions of Nucleosome Cores

3.5.1. Mapping with Restriction Enzymes

1. Digest 0.02 µg (DNA) of labeled cores in 20 µL of BT buffer with 5–10 U of restriction enzyme at 37°C for 20 min (*see* **Note 15**).

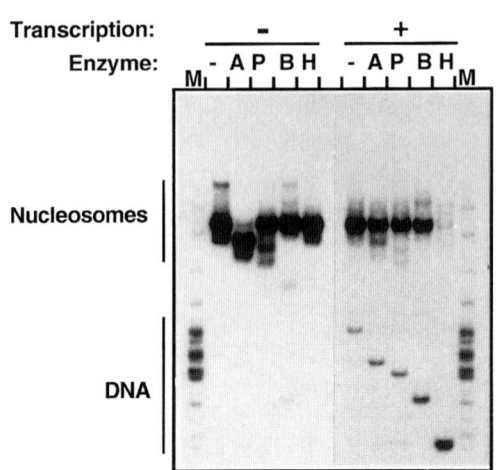

Fig. 2. Mapping of positions of nucleosome cores before and after transcription by SP6 RNA polymerase with restriction enzymes. **(Top)** Restriction map of the 227-bp *Sac*I-*Nco*I template. The template was labeled at the *Nco*I end. The cleavage sites for *Ase*I, *Pac*I, *Bgl*II, and *Hae*II, and a promoter for SP6 RNA polymerase are indicated. Positions of nucleosome core before (in bold) and after transcription are indicated by ovals. **(Bottom)** Analysis of nucleosome cores in a nucleoprotein gel. Mobilities of nucleosomes and histone-free DNA are indicated. Cores were incubated for 20 min in the presence or absence of SP6 RNA polymerase and then digested with *Ase*I, *Pac*I, *Bgl*II, or *Hae*II as indicated. Only the right end of the template was labeled. M, endlabeled *Msp*I digest of pBR322. All enzymes completely digested free templates. At the same time, before transcription the pmajority of cores (90%) are resistant to *Pac*I, *Bgl*II and *Hae*II, and are sensitive to *Ase*I; in contrast, after transcription, the majority of cores are sensitive to *Hae*II and resistant to *Ase*I, *Pac*I, and *Bgl*II. The data suggest that the histone octamer has been transferred on transcription from the labeled to the unlabeled end of 227-bp fragment.

2. Add 1 µL 0.5 M EDTA, pH 8.0, and 5 µL 50% (w/v) sucrose to stop the digestion. Electrophorese the cores in a 4.5% (acrylamide:*bis* = 40:1) gel containing 20 mM HEPES, pH 8.0, 1 mM Na-EDTA, 5% (v/v) glycerol (17 × 17 × 0.15 cm) for 2–3 h at 6 V/cm.
3. Dry the gel on DE-81 paper and expose with an intensifying screen at –80°C overnight or longer if necessary. An example of restriction enzyme mapping of nucleosome core positions before and after transcription is shown in **Fig. 2**.

3.5.2. Mapping with Micrococcal Nuclease

1. Incubate 0.1 µg (DNA) of unlabeled cores in siliconized tubes containing 0.1 mL of BT buffer supplemented with 2 mM CaCl$_2$ and 7 U/µL MNase for 5–25 min at 37°C.
2. Stop the reaction by adding 2.5 µL 0.5 M EDTA, 2 µL 10% (w/v) SDS, and 10 µL 3 M Na-acetate, pH 5.5. Extract DNA twice with 1:1 phenol:chloroform (v/v), precipitate it with ethanol, wash with 70% (v/v) ethanol, and dissolve DNA in 15 µL TAE.
3. Electrophorese DNA in a 6% (acrylamide:*bis* = 40:1) gel containing TAE (17 × 17 × 0.15 cm) for 2 h at 6 V/cm.
4. Stain the gel with ethidium bromide and cut out gel slices containing approx 150 bp DNA. Crush the gel slices in Eppendorf tubes, and elute the DNA overnight in 0.3 mL HE on a rotator at room temperature.
5. Purify DNA from gel pieces by filtration through a 0.45-µm nitrocellulose filter in a 2 mL plastic syringe, precipitate with ethanol and wash with 70% (v/v) ethanol.
6. Incubate DNA in kinase buffer in the presence of 30 µCi γ^{32}P-ATP and T4 kinase for 30 min at 37°C, stop the reaction with EDTA and inactivate the kinase by incubation at 65°C for 15 min. Purify DNA from unincorporated label on G-25 Quick spin columns equilibrated with HE.
7. Divide the DNA sample into aliquots, supplement with appropriate restriction enzyme buffer, and digest overnight with restriction enzymes (*see* **Note 16**).
8. Extract DNA twice with phenol:chloroform (1:1, v/v), then add glycogen as carrier to a final concentration of 0.1 mg/ml, precipitate with ethanol, wash with 70% (v/v) ethanol and dry. Dissolve DNA in TAE.
9. Separate DNA in an 8% (acrylamide:*bis* = 19:1) gel containing TAE (17 × 17 × 0.15 cm) for 3.5 h at 6 V/cm.
10. Dry the gel on DE-81 paper and expose to X-ray film with an intensifying screen at –80°C (*see* **Note 17**). An example of the mapping of nucleosome core positions before and after transcription with micrococcal nuclease is shown in **Fig. 3**.

4. Notes

1. The method is based on the usual protocol for isolation of DNA fragments from agarose gels *(12)*. The DNA solution is then treated with butanol to remove most of the ethidium bromide and to concentrate sample.
2. One of DNA ends must have a recessed 3' end. Use radioactive NTP complementary to first nucleotide (positioned next to the double-stranded portion of DNA fragment) of the 5'-overhang. Incubation at 12°C prevents degradation of 3' ends of DNA by the Klenow fragment.
3. We do not recommend to desalt the DNA solution after labeling because it causes some losses (up to 15% of DNA).
4. We utilize a variant of the salt dialysis method: It involves the addition of urea to the dialysis buffer to prevent nonspecific histone-histone interactions that occurs if histones and DNA are mixed directly at physiologically relevant salt concentrations *(13)*. 2-mercaptoethanol is included throughout the procedure to maintain histone sulphhydryl groups in the reduced state.

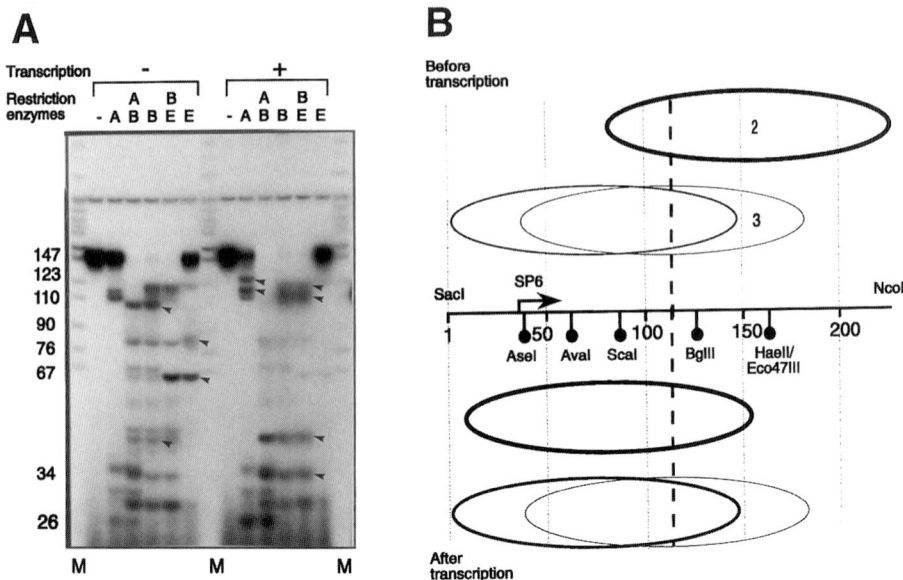

Fig. 3. Mapping of positions of nucleosome cores before and after transcription with micrococcal nuclease. (**A**) Mapping of nucleosome core positions by restriction digestion analysis of purified core DNA. After incubation of the 227-bp nucleosomal template for 20 min in the presence or absence of polymerase, core particles were prepared using micrococcal nuclease. Core DNA (145–155 bp) was gel-purified, labeled at both ends, digested with various restriction enzymes (A: *Ase*I; B: *Bgl*II; E: *Eco*47III) and the products were analyzed in an 8% polyacrylamide gel. A 245-bp-labeled control fragment was included in the digests to correct for losses. The discrete fragments produced on digestion of core DNA indicate the presence of positioned nucleosome cores; these were mapped by measuring the lengths of the bands and using double digests to determine the orientation of fragments with respect to restriction sites. Marker (M): end-labeled *Msp*I digest of pBR322. *Arrowheads* indicate the bands attributed to the major core before or after transcription. Note that patterns of bands before and after transcription are different, indicating transfer of the octamer on transcription. (**B**) Summary of nucleosome core positions before and after transcription. The cores drawn above the restriction map are those observed before transcription; the cores drawn below the map are those observed after transcription. The major positioned core before transcription (complex 2) and the two major cores after transcription are drawn in *bold* outline; minor positioned cores (including complex 3) are outlined with thinner lines. The positions are accurate to +/- 5 bp. (Reproduced from *[7]*).

5. It is very important to measure the core histone concentrations very carefully because if the histone:DNA weight ratio exceeds 1:1, large histone-DNA aggregates are formed. We use A_{230} = 4.2 for 1 mg/mL of either histone pair. For dilutions, pipette histones carefully by "rinsing" the pipette in the solution first.

6. After dialysis, handle CRB1 with caution because it is highly radioactive.
7. The reconstitute obtained after salt/urea dialysis is a mixture of nucleosome cores located at different positions on the DNA fragment, incomplete cores, and free DNA. After proper reconstitution, they are present at the following ratio: approx 40 % of total DNA will be in cores, approx 5% in incomplete cores (mostly [H3/H4]$_2$·H2A/H2B histone hexamers *[7]*), approx 5 % in histone-DNA aggregates, and approx 50% as free DNA. These proportions are very sensitive to the histone:DNA input ratio; if the ratio exceeds 1:1, most of the DNA will be in aggregates. Cores can be separated from free template, aggregates and incomplete cores in a sucrose gradient *(7,14)*, although there is often some cross-contamination of fractions.
8. Use of siliconized tubes is essential to prevent nucleosome disruption during storage at low concentration (1 to 10 mg/mL). Presumably, siliconization prevents irreversible adsorption of histones to the walls of the tube.
9. Preelectrophoresis eliminates unidentified components in the gel (perhaps ammonium persulphate or TEMED) that inhibit subsequent transcription.
10. BSA reduces adsorption of histones to the unit.
11. Recognition site for the restriction enzyme must be present within the region covered by nucleosome in the desired position; the site must not be covered by symmetrically positioned nucleosomes. We found that sites for restriction enzymes are protected even when they are located 5 bp in the nucleosome core *(3)*.
12. Do not add dye to the sample: It can disrupt nucleosomes.
13. We always cut out the free DNA band as well, because it is a useful control—this DNA has the same specific activity as the nucleosome cores have.
14. After gel-isolation, the preparation usually contains approx 90% pure uniquely positioned nucleosome cores and only about 5% free DNA (*see* **Fig. 2**). The cores are stable for at least 6 mo. Concentrations of cores are easily determined by scintillation counting, using the specific activity. Typically, the concentration is 3–10 µg/mL.
15. The majority of restriction enzymes work well in the TB. However some restriction enzymes digest intranucleosomal DNA when incubated in other buffers (i.e., in one of NEB restriction digestion buffers). Therefore, extreme caution is required when it is desirable to use a buffer other than TB.
16. For accurate quantitation of bands, we usually mix some labeled "control" fragment (approx 250 bp long, so that it migrates above intact core DNA in the gel), that does not contain any restriction sites used in analysis of the core DNA. This facilitates accurate comparisons of lanes with the same core DNA, because material losses can be corrected for.
17. Discrete bands in the gel correspond to discrete positions of nucleosomes on the DNA fragment; these positions are deduced from the lengths of the DNA fragments.

References

1. Utley, R. T., Owen-Hughes, T. A., Juan, L.-J., Cote, J., Adams, C. C., and Workman, J. L. (1996) *In vitro* analysis of transcription factor binding to nucleosomes and nucleosome disruption/displacement. *Meth. Enzymol.* **274,** 276–290.

2. Studitsky, V. M., Clark, D. J., and Felsenfeld, G. (1996) Preparation of nucleosomal templates for transcription *in vitro*. *Meth. Enzymol.* **274,** 246–256.
3. Rhodes, D. and Laskey, R. A. (1989) Assembly of nucleosomes and chromatin *in vitro*. *Meth. Enzymol.* **170,** 575–585.
4. Studitsky, V. M., Clark, D. J., and Felsenfeld, G. (1995) Overcoming a nucleosomal barrier to transcription. *Cell* **83,** 19–27.
5. Meersseman, G., Pennings, S., and E.M. Bradbury (1992) Mobile nucleosomes: a general behaviour. *EMBO J.* **11,** 2951–2960.
6. Duband-Goulet, I., Carot, V., Ulyanov, A. V., Douc-Rasy, S., and Prunell, A. (1992) Chromatin reconstitution on small DNA rings. IV. DNA supercoiling and nucleosome sequence preference. *J. Mol. Biol.* **224,** 981–1001.
7. Studitsky, V. M., Clark, D. J., and Felsenfeld, G. (1994) A histone octamer can step around a transcribing polymerase without leaving the template. *Cell* **76,** 371–382.
8. Morse, R. H. (1989) Nucleosomes inhibit both transcriptional initiation and elongation by RNA polymerase III *in vitro*. *The EMBO J.* **8,** 2343–2351.
9. Clark, D. J. and Felsenfeld, G. (1992) A nucleosome core is transferred out of the path of a transcribing polymerase. *Cell* **71,** 11–22.
10. Simon, R. H. and Felsenfeld, G. (1979) A new approach for purifying hictone pairs H2A+H2B and H3+H4 from chromatin using hydroxylapatite. *Nucleic Acids Res.* **6,** 689–696.
11. von Holt, C., Brandt, Greyling, W. F. H. J., Lindsey, G. G., Retief, J. D., Rodrigues, J. de A., Schwager, S., and Sewell, B. T. (1989) Isolation and characterization of histones. *Meth. Enzymol.* **170,** 431–504.
12. Sambrook, J., Fritsch, E. F., and Maniatis, T. (1989) *Molecular Cloning: A Laboratory Manual.* Cold Spring Harbor Laboratory Press, Cold Spring Harbor, NY.
13. Camerini-Otero, D. and Felsenfeld, G. (1977) Supercoiling energy and nucleosome formation: the role of the arginine-rich histone kernel. *Nucleic Acids Res.* **4,** 1159–1181.
14. Noll, N. and Noll, M. (1989) Sucrose gradient techniques and applications to nucleosome structure. *Meth. Enzymol.* **170,** 55–115.

3

Site-Directed Chemical Probing of Histone–DNA Interactions

David R. Chafin, Kyu-Min Lee, and Jeffrey J. Hayes

1. Introduction

The protein–DNA complexes that make up the chromosome serve not only to package the genomic DNA within the confines of the nucleus but to directly participate in the efficient and controlled utilization of the DNA for nuclear processes such as replication, transcription, recombination, and DNA repair *(1)*. Understanding these complex functions will require a molecular description of the multitude of protein-DNA interactions and associations within the chromatin complex. For example, the core histone tail domains make multiple and complex interactions that vary as a function of the condensation state of the chromatin fiber *(2)*. Moreover, the molecular interactions of these flexible domains are undoubtedly modulated by the multiple posttranslational modifications known to occur within these residues *(3)*. Thus, the histone tails represent critical points for signal transduction within the chromatin complex, and these signals are likely to be manifested as subtle structural alterations within the chromatin complex.

Unfortunately, only the simplest components of the chromatin complex are amenable to analysis by standard structural techniques, such as X-ray crystallography or NMR *(4–7)*. Moreover, domains, such as the core histone tails, are not completely defined even in these studies. We therefore have developed a general method to map contacts between histone proteins and nucleosomal DNA. This method allows an unambiguous identification of the points of contact between residues on the histones and the DNA to base-pair resolution. Here we present two variations of this method:

1. A site-directed DNA cleavage method.
2. A site-directed photocrosslinking method.

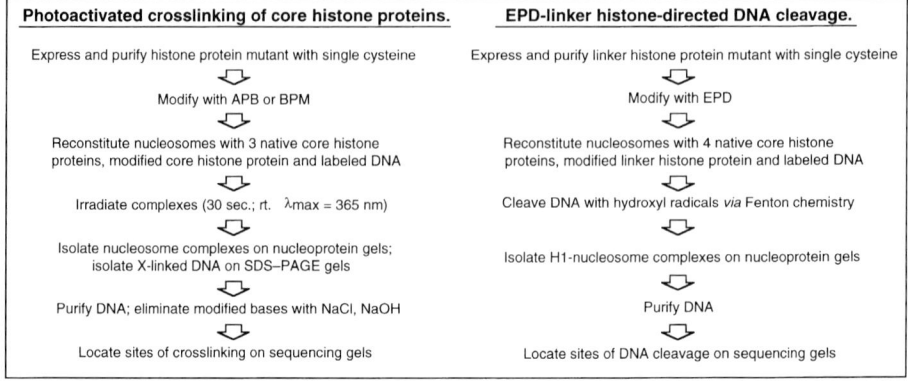

Fig. 1. Overview of the procedures described in this chapter.

Both of these methods involve the production of mutant histone proteins containing single cysteine residues that are substituted for native sequence at rationally selected locations within these proteins. The single sulfhydryl group within these proteins is then modified with a bifunctional reagent, which contains a cysteine-specific moiety at one end and a DNA cleavage or photochemical crosslinking moiety at the other *(8–12)*. These proteins are then assembled into the chromatin complex of interest, and DNA cleavage or crosslinking is initiated. The DNA from such complexes is then prepared, and the location of DNA cleavage or crosslinking events is mapped to single base-pair resolution on DNA sequencing gels. An overview over the procedures is given in **Fig. 1**.

2. Materials

2.1 Construction of Cysteine Substituted Protein

2.1.1. Point Mutation by PCR

1. Oligonucleotide primers: Two primers are complimentary to the 5' and 3' ends of the sequence to be amplified. In addition, if the codon to be altered is located more than approx 10–15 nucleotides from the end of the coding sequence, one additional primer is needed, which must contain the sequence substitutions for the altered codon flanked by 12–15 nucleotides of complementary sequence on each side. Store at –20°C.
2. 10X stock containing all four dNTPs at 10 mM concentration each.
3. Clean, reliable 18 ΩW water, free of chemical contaminants.
4. 10X PCR buffer; can be obtained commercially from the supplier of the PCR enzyme of choice.
5. Vent or *Taq* DNA polymerase; can be obtained from commercial sources.

2.1.2. Ligation and Transformation of PCR Insert into DH5α or BL21 Cells

1. DH5α or BL21 cells can be obtained commercially or prepared in competent form, store at –70°C.
2. Luria broth (LB), sterile.
3. 1000X stock of ampicillin (100 mg/mL).
4. LB-agar plates containing 0.1 mg/mL ampicillin.

2.1.3. Overexpression and Purification of Mutant Histone Proteins

1. 100X (0.2 M) stock of isopropyl β-D-thiogalactopyranoside (IPTG).
2. Luria broth (LB), sterile.
3. 10 mg/mL lysozyme solution.
4. A saturated room temperature solution of PMSF (phenylmethylsulfonylfluoride).
5. Triton-x 100 detergent.
6. 2 M NaCl.
7. A 50 % (V/V) slurry of Bio-Rex 50–100 mesh chromatography resin (Bio-Rad).
8. TE, pH 8.0: 10 mM Tris-HCl, pH 8.0, 1 mM EDTA.
9. TE, pH 8.0 solutions containing 0.5, 0.6, 1.0, and 2.0 M NaCl.

2.2. Reduction and Modification of Cysteine Substituted Proteins with EPD

1. 1 M stock of DTT (Dithiothreitol), made fresh.
2. A 50% slurry of Bio-Rex 100–200 mesh chromatography resin (Bio-Rad).
3. 10 mM Tris-HCl, pH 8.0 solutions containing 0.5, 0.6, 1.0, and 2.0 M NaCl.
4. 0.3 M Stock of EPD (synthesized according to **refs. 8** and **9**) Alternatively, iodoacetamido 1,10 phenanthrolineCu^{+2} can be employed in place of EPD (Molecular Probes, Eugene, OR).
5. Disposable 10-mL plastic chromatography columns (Bio-Rad).
6. Coomassie blue stain: 45% methanol, 10% acetic acid, 2.5 mg/mL Coomassie brilliant blue R250.
7. Destaining solution: 45% methanol, 10% acetic acid.

2.3. Modification of Cysteine-Substituted Proteins with APB

1. 500 µg of cysteine-substituted mutant H2A/H2B dimer.
2. 1 M DTT (Dithiothreitol).
3. 50% slurry of Biorex70 100–200-mesh chromatography resin (Bio-Rad) in 250 mM NaCl, 10 mM Tris-HCl, pH 7.0, 1 mM EDTA.
4. TE, pH 7.0.
5. TE, pH 7.0 containing 1 M NaCl.
6. Disposable 10-mL plastic chromatography column (Bio-Rad).
7. 10 mM APB (4-azidophenacylbromide; Sigma, St. Louis, MO) dissolved in methanol. Prepare a 0.25 M stock of APB dissolved in methanol. Minimize light exposure to APB, because it is extremely light sensitive.
8. ^{14}C- NEM (N-[ethyl-1-^{14}C]-maleimide) in 10 mM NaAcetate, pH 5.0.

2.4. Single 5' Radioactive Modification of Purified DNA

1. Linear DNA fragment with convenient restriction sites on either end, previously phosphatased.
2. 10X T4 polynucleotide kinase buffer (supplied with enzyme).
3. [γ-^{32}P]dATP 6000 Ci/mmol.
4. T4 Polynucleotide Kinase 10,000 U/mL (Promega).
5. 2.5 M ammonium acetate.
6. 95% Ethanol chilled at –20°C.
7. 70% Ethanol chilled at –20°C.
8. 10% SDS.
9. Alkaline phosphatase (Boehringer Mannheim)

2.5. In Vitro Reconstitution of Nucleosomes

1. TE, pH 8.0 (*see* **Subheading 2.3.**)
2. TE, pH 8.0 solutions containing 1.2, 1.0, 0.8, and 0.6 M NaCl.
3. Dialysis tubing molecular weight cutoff: 6000–8000.
4. Sonicated calf thymus DNA, approx 1–2 mg/mL.
5. 5 M NaCl.
6. Purified core histone proteins H2A/H2B and H3/H4. Ours are purified from chicken erythrocyte blood (*see* **Note 7**).

2.6. Maxam-Gilbert G Specific Reaction

1. 10X G specific reaction buffer: 0.5 M NaCacodylate, 10 mM EDTA.
2. Dimethylsulfate (DMS).
3. G reaction stop buffer: 1.5 M NaAcetate, 1 M β-Mercaptoethanol, 0.004 µg/µL sonicated Calf thymus DNA.
4. Piperidine.

2.7. Chemical Mapping of Protein-DNA Interactions with EPD

1. 0.7% agarose made with 0.5X TBE. (Note: Treat all solutions with Chelex-100 resin (Bio-Rad) to remove adventitous redox-active metals.)
2. Histone dilution buffer: 10 mM Tris-HCl, pH 8.0, 50 mM NaCl.
3. 20 mM sodium ascorbate.
4. 1 mM Fe(II)EDTA.
5. 0.15% H_2O_2, freshly made.
6. Stop solution: 50% glycerol, 10 mM EDTA.
7. Series 8000 Microcentrifuge Filtration Devices (e.g., from Lida Manufacturing Corporation).
8. 10 mM Tris-HCl, pH 8.0, 0.1% SDS.
9. Microcentrifuge pestles, (e.g., from Stratagene).
10. Ice-cold 95% and 70% ethanol solutions.
11. 3 M sodium acetate.

2.8. Site-Specific Photocrosslinking of Histone–DNA Interactions Within the Nucleosome Using 4-Azidophenacylbromide (APB)

2.8.1. UV Crosslinking of Nucleosomes Reconstituted with Sequence-Specific (5S) DNA

1. Approx 5 µg of reconstituted nucleosomes containing APB-modified core histones.
2. Spin-X filter tube (Costar).
3. 0.7% agarose gel made with 0.5 X TBE.
4. Microcentrifuge pestles, can be obtained from Stratagene.
5. Light whit mineral oil, can be obtained from Sigma chemicals.
6. UV light source (VWR Scientific LM20E Transilluminator).

2.8.2. UV Crosslinking of Nucleosome Core Particles with Random Sequence DNA

1. Sonicated calf thymus DNA (DNA size 0.5–1 kb; Sigma, St. Louis, MO).
2. Microccocal nuclease, 15 U/µL (Worthington).
3. 0 mM CaCl$_2$.
4. Micrococcal nuclease stop mix: 25% glycerol, 0.25 M EDTA.
5. Buffer E: 25 mM NH$_4$ acetate, 0.1 M EDTA, 2% SDS.
6. [γ-^{32}P]dATP (3000 Ci/mmol).
7. 10 U/µL T4 polynucleotide kinase and 10X T4 polynucleotide buffer.
8. 2% SDS.
9. 15 mg/mL proteinase K.

2.9. Elimination and Cleavage of the Modified Base from Histone–DNA Crosslinked Complexes

1. Buffer E: 25 mM NH$_4$ acetate, 0.1 M EDTA, 2% SDS.
2. 2 M NaOH.
3. 2 M HCl.
4. 20 mM Tris-HCl, pH 8.0.

2.10. Sequencing Gel Analysis

1. Solid Urea.
2. 5X TBE.
3. 40% Acrylamide (19:1 acrylamide:*bis*-acrylamide).
4. 20% APS (ammonium persulfate).
5. TEMED (N,N,N',N'-tetramethylethylenediamine).
6. Formamide loading buffer: formamide, 0.05% bromphenol blue, 0.05% xylene cyanol).

3. Methods

3.1. Overexpression and Purification of Single Cysteine Substituted Proteins

A single amino acid can be substituted in any protein of interest following well-established site-directed mutagenesis protocols that employ PCR with several primers.

1. Standard PCR methods are used to amplify a DNA fragment containing a cysteine codon in place of the wild-type codon. If the codon to be changed is near the end of the amplified coding region, then only two primers are necessary with the change incorporated into one of these "parent" primers. If more central to the sequence, then a 3-primer technique is used with the change incorporated into an internal primer, amplified with one of the parent primers, and then the shorter amplified fragment used as a primer with the remaining parent primer and the original DNA as the template. Finally, if this method fails, two complementary internal primers with the intended change are used to amplify overlapping short fragments using the appropriate parent primers, and then these two fragments are combined with the parent primers and the entire insert amplified without additional template added.
2. Ligate the insert containing the single cysteine substitution into the appropriate expression vector. We typically use the pET expression system (Novagen). Both DNAs must be digested with the same restriction endonucleases. Incubate equimolar amounts of insert DNA and pET3d DNA in 1X T4 ligation buffer. Add 400 U of T4 DNA ligase (Bio-Labs) and incubate at 4°C overnight (*see* **Note 1**).
3. Check the efficiency of the ligation by transforming a small amount of the ligation sample into DH5α cells. Plate the transformation on LB-ampicillin plates and incubate at 37°C overnight.
4. Prepare DNA from several colonies by placing a single colony into 3–5 mL of LB-Ampicillin medium and growing at 37°C. Isolate the DNA from these cultures by standard DNA mini-prep techniques (*see* **Note 2**).
5. Digest part of the isolated plasmid with the original restriction endonucleases used for ligation to liberate the DNA fragment corresponding to the original insert. The plasmids that contain correct inserts can be used to transform BL21 cells for over-expression.
6. Transform the pET plasmid containing the insert into BL21 cells in the same manner as for the DH5α cells *(see above)*.
7. Place one BL21 colony from the LB-ampicillin plate into 200 mL of LB-ampicillin medium.
8. Grow the culture in the absence of IPTG at 37°C to an optical density of 0.6 at 595 nm wavelength light. Add IPTG to a final concentration of 0.2 mg/mL and return the culture to 37°C for 2–4 h (*see* **Note 3**).
9. Pellet the bacteria by centrifugation at 4,000*g*. for 15 min.
10. Decant the supernatant and re-suspended the pellet in 5–10 mL of TE buffer.
11. Add 10 mg/mL lysozyme to a final concentration of 0.2 mg/mL. Then add triton-x 100 to a final concentration of 0.2% and incubate for 30 min at room temperature.
12. Dilute the bacteria twofold with 2 M NaCl to a final concentration of 1 M NaCl. Transfer the bacteria to oakridge centrifuge tubes on ice.
13. Sonicate the bacterial slurry for 6 min total in two-3 min sonications (*see* **Note 4**).
14. Pellet the cell debris by centrifugation at 10,000*g* for 30 min at 4°C.
15. Add PMSF to a final concentration of 1X. Dilute the supernatants twofold with TE buffer to bring the NaCl concentration to 0.5 M.

16. Linker histones and most other proteins will bind directly to the Bio-Rex beads. However, core histone proteins must first be incubated with their partner proteins before they will bind to the chromatography matrix (i.e., H2A with H2B) (*see* **Note 5**).
17. Incubate the diluted supernatant with 12.5 mL of a 50% suspension of Bio-Rex 50–100 mesh beads for 4 h at 4°C with rotation.
18. After 4 h collect the beads in a plastic 10-mL disposable chromatography column. Collect the Flow through fraction in a 50-mL conical tube and freeze.
19. Wash the column with 2–3 column volumes of 10 mM Tris-HCl, pH 8.0 containing 0.6 M NaCl. Collect the first 10 mL of the wash fraction in a 15-mL conical tube and freeze.
20. Elute the bound proteins with two separate single column volume elutions of 10 mM Tris-HCl, pH 8.0 containing 1.0 M NaCl. Collect the 1.0 M elution fractions in separate 15-mL conical tubes and freeze.
21. After elution, wash the column with one column volume of 10 mM Tris-HCl, pH 8.0 containing 2.0 M NaCl. Collect the 2.0 M elution fraction in a 15-mL conical tube and freeze.
22. Check 10 µL of each fraction for protein content by SDS-PAGE.

3.2. Reduction and Modification of Cysteine-Substituted Proteins

3.2.1. Reduction of Cysteine-Substituted Proteins

1. Incubate protein of interest in a 15-mL conical tube with 50 mM DTT final concentration for 1 h on ice.
2. Dilute the protein sample twofold with TE which dilutes the NaCl concentration to 500 mM NaCl.
3. Add 0.8 mL of a 50 % slurry of Bio-Rex (Bio-Rad) 100–200 mesh chromatography resin and incubate with rotation for 2 h at 4°C.
4. Pour slurry into a 10-mL plastic, disposable chromatography column and collect the flow through fraction.
5. Wash the column with 3–5 column volumes of buffer containing 10 mM Tris-HCl, pH 8.0, and 0.5 M NaCl. Immediately remove 20 µL of the freshly eluted sample into a separate eppendorf tube for later analysis on a 12% SDS-protein gel and immediately freeze the 0.5-mL sample to insure that the protein remains reduced. Aliquoting the sample in this manner insures that the sample does not need to be thawed for analysis.
6. An intermediate wash of the column with buffer containing 0.6 M NaCl is performed to remove proteins that are less well bound due to partial degradation. Aliquots of these samples are obtained in the same manner as the previous wash step.
7. Linker histone proteins can be eluted with 0.5-mL aliquots of the same buffer except with 1.0 M NaCl. Typically, five separate 1.0 M NaCl elution steps are performed and collected separately. As previously, 5 µL of the fractions are aliquoted for SDS/PAGE analysis and the samples are frozen immediately. A final elution with buffer containing 2.0 M NaCl buffer will ensure that all of the protein has been eluted from the column.

8. Check the protein content of each aliquot obtained from the elution fractions on a 12% SDS-polyacrylamide gel. After separation, incubate the protein gel in enough Coomassie blue stain to cover it. Stain for approx 1 h at room temperature and destain until the background of the gel is clear.

3.2.2. Modification of Cysteine-Substituted Proteins with EPD

1. The fraction containing the reduced protein to be modified with EPD is thawed on ice. Working as quickly as possible, add a 1.1-fold molar excess of EPD to 60 µL of reduced protein. Incubate for 1 h at room temperature in the dark.
2. Removal of excess cleavage reagent requires one more round of Bio-Rex chromatography, identical to that presented above except that 60 µL of the 50% slurry is added to the protein. The slurry is poured into a column made from a blue 1-mL pipet tip fitted with glass wool at the opening. Wash and elute as in **Subheading 3.2.1.**, but scale all elution volumes to the resin amount. Aliquots for protein analysis are exactly the same size as previously indicated.
3. Post modification labeling with ^{14}C-NEM (N-[ethyl-1-^{14}C]-maleimide) can be used to quantitatively determine the extent of modification with the DNA cleavage reagent. Add 0.25–0.5 µCi of ^{14}C-NEM to each protein aliquot made from the elution fractions of the Bio-Rex column. Ten min later, add two vol of 2X protein loading buffer to the labeling samples and separate the proteins on a 12% polyacrylamide gel. Stain and de-stain the gel as above and dry the gel onto a piece of Whatman filter paper. Visualize the labeled proteins by exposing the dried gel to ultra sensitive Bio-Max autoradiography film.
4. A protein gel at this step performs two functions.
   a. Determine which fractions contain the protein of interest.
   b. Determine the extent of modification with the DNA cleavage reagent.

3.3. Modification with the DNA Crosslinking Reagent APB

1. Place 1 mL of the 50% 100–200 mesh BioRex70 bead slurry equilibrated in TE containing 250 mM NaCl into a transparent 15-mL conical tube.
2. In a separate eppendorf tube, add approx 500 µg of H2A/H2B dimer and reduce in 20 mM DTT final concentration for 30 min at room temperature.
3. Transfer the reduced protein solution into the 15-mL conical tube with the beads.
4. Add 5 mL of TE into the mix and rotate for 1 h at 4°C, reducing the NaCl concentration to 500 mM.
5. Transfer the mix of protein/beads to a small plastic disposable chromatography column. Collect the flow through and wash the column six times with 1 mL of TE.
6. Elute the reduced core histone dimer with 1 ml TE containing 1 M NaCl. Collect the fraction into a clean eppendorf tube.
7. Take about 200 µg of eluted dimer and modify with 100 µM APB in the dark at room temperature for 1 h. Freeze in dry ice and store the modified dimer protein at –80°C.
8. To check the extent of modification, remove a 10-µL (approx 2 µg) aliquot of the APB unmodified and modified H2A/H2B dimer and add 0.5 µL ^{14}C-NEM (N-[ethyl-1-^{14}C]-maleimide).
9. Incubate at room temperature for 30 min and then quench with 1 µL of 0.1 M DTT.

10. Separate the protein samples directly on a 18% SDS/PAGE gel.
11. Stain the gel with Coomassie Blue dye for 1 h and destain as previously overnight (*see* **Subheading 3.2.2.**).
12. Dry the gel onto a piece of Whatman filter paper and expose the dried gel to Kodak Biomax film to determine the efficiency of APB modification by quantitating the amount of ^{14}C-NEM labeling.

3.4. Single 5' Radioactive Modification of Purified DNA

1. Treat approx 5 µg of plasmid DNA or approx 0.5 µg of a purified DNA fragment with the appropriate restriction endonuclease in the manufacturer's buffer.
2. Precipitate the DNA by adjusting the solution to 0.3 *M* sodium acetate and addition of 2.5 vol of cold ethanol.
3. Resuspend the DNA in phosphatase buffer and treat with alkaline phosphatase for 1 h at 37°C.
4. Adjust the solution to 0.1% SDS, phenol extract the solution, and then precipitate the aqueous phase twice with ethanol and sodium acetate.
5. Resuspend the DNA in 10 µL TE and add 2.5 µL of 10X T4 polynucleotide kinase buffer.
6. Add 50 µCi of [γ-^{32}P]dATP and adjust volume to 24 µL with water.
7. Start the reaction by adding 10 U of T4 polynucleotide kinase and incubate for 30 min at 37°C.
8. Stop the kinase with 200 µL of 2.5 *M* ammonium acetate (NH$_4$Oac) and 700 µL of cold 95% ethanol.
9. Pellet the DNA in a microcentrifuge for 30 min at room temperature.
10. Wash the DNA pellet briefly with cold 70% ethanol and dry the DNA in a speedvac concentrator.
11. Dissolve the DNA in 34 µL of TE.
12. Digest the DNA fragment with a second restriction endonuclease that liberates the fragment of interest and yields fragments that can be easily separated by on a gel.
13. Apply the sample to a 6% native polyacrylamide gel.
14. After separation, wrap the gel in saran wrap and expose the gel to film for 1 min, which is sufficient to detect the specific band containing the labeled fragment. The use of fluorescent markers allows alignment of the gel (*see* **Subheading 3.7.2.**).
15. Excise the band of interest from the polyacrylamide gel and place into a clean eppendorf tube. Crush the acrylamide gel slice with a eppendorf pestle and add 700 µL of TE. The labeled fragment will elute overnight with passive diffusion.
16. Split the sample equally into two Series 8000 Microcentrifuge Filtration Devices and spin for 30 min in a microcentrifuge.
17. Precipitate the DNA and dissolve in TE. Add enough TE so that the labeled DNA is approx 1000 CPM/µL (*see* **Note 6**).

3.5. Reconstitution of Nucleosomes by Salt Step Dialysis

The method described here for the reconstitution of nucleosomes allows for large quantities of nearly homogeneous core particles in 12 h *(13)*. These in vitro nucleosomes bind linker histone in a physiologically relevant manner as tested by several assays. Virtually any piece of DNA 147 bp or longer can be used. However to obtain nucleosomes with only one translational position, the DNA sequence should contain nucleosome positioning sequences such as that from the *Xenopus borealis* 5S rRNA gene *(14–16)*. The DNA can be labeled on the 5' or 3' end with commercially available enzymes after phosphatase treatment as described.

1. Add approx 8 µg of unlabeled calf thymus DNA, 200,000–400,000 CPM of *Xenopus borealis* 5S ribosomal DNA labeled at one end, purified chicken erythrocyte core histone protein fractions (H2A/H2B and H3/H4), (*see* **Note 7**), 160 µL of 5 *M* NaCl (2.0 *M* final), and TE to a final of 400 µL.
2. Place the reconstitution mixture into a 6–8 kDa MW cut-off dialysis bag. All subsequent dialysis steps are for 2 h at 4°C against 1 L of dialysis buffers unless specified. The first dialysis buffer is TE containing 1.2 *M* NaCl. Subsequent dialyses are with fresh TE containing 1.0, 0.8, and then 0.6 *M* NaCl. The procedure is completed with a final dialysis against TE overnight. Nucleosomes at this stage can be used for gel shift experiments where EDTA does not interfere.
3. For DNA cleavage experiments with EPD, two additional dialysis steps are required. First dialyze the reconstitutes against 10 m*M* Tris-HCl, pH 8.0 several hours to remove the EDTA. A second dialysis against fresh 10 m*M* Tris-HCl, pH 8.0 removes trace amounts of EDTA and prepares the samples for chemical mapping with EPD.

3.6. Maxam-Gilbert G-Specific Reaction

The G-specific reaction used in the Maxam-Gilbert sequencing method provides a easy and quick method to identify the exact location of bases within any know sequence on sequencing gels. It is used here to determine the sites of DNA to base pair resolution. Because this method is not generally used anymore, the steps are outlined below.

1. Add approx 20,000 CPM of DNA labeled at one end (same DNA used to reconstitute nucleosomes).
2. Add 20 µL of 10X G specific reaction buffer.
3. Add water to a final volume of 200 µL.
4. Start by adding 1 µL of straight dimethylsulfate (DMS) to the tube. Mix immediately and spin briefly in a microfuge (do this in a hood, be careful not to get any DMS on your skin or on gloves. Store DMS in a tightly capped brown glass bottle at 4°C).
5. Add 50 µL of G reaction stop solution and mix immediately.
6. Precipitate the DNA.
7. Dissolve the DNA in 90 µL of H_2O.
8. Add 10 µL of piperidine and incubate at 90°C for 30 min.
9. Dry the DNA solution in a speedvac to completion.

10. Dissolve the DNA in 20 µL of water and repeat the drying step. Repeat this step one more time.
11. Dissolve DNA in 100 µL of TE and store at 4°C.

3.7. Site-Directed Hydroxyl Radical Cleavage of DNA

3.7.1. Binding Single Cysteine Substituted Linker Histone Proteins to Reconstituted Nucleosomes

1. The exact amount of each mutant linker histone protein to add to the hydroxyl radical reactions needs to be determined empirically. This is accomplished by adding increasing amounts of the mutant protein to a fixed volume of reconstituted nucleosomes (typically 5000 CPM) and analysis via a gel shift procedure *(17)*.
2. Add 5% glycerol final to the binding reaction.
3. Add 50 mM NaCl final to the binding reaction (*see* **Note 8**).
4. Incubate the binding reactions for 15 min at room temperature.
5. Separate the complexes on a 0.7% agarose, 0.5X TBE gel. After drying the gel, expose to autoradiography film and determine the amount of protein necessary for good complex formation.
6. Several assays for the correct binding of linker histones to DNA have been performed *(17,18)*. One of the easiest involves a brief digestion with micrococcal nuclease in the chromatosome stop assay *(13)*.

3.7.2. Site-Directed Hydroxyl Radical Mapping of Linker Histone–DNA Interaction

1. Scale up the binding reaction to include 40,000–50,000 cpm of labeled reconstituted nucleosomes, and add enough modified mutant linker histone to form H1-nucleosome complexes.
2. Add glycerol to 0.5% final concentration.
3. Add sodium ascorbate to a final concentration of 1 mM.
4. Add H_2O_2 to a final concentration of 0.0075%.
5. Incubate for 30 min at room temperature in the dark.
6. After 30 min, add 1/10 vol of 50% glycerol, 10 mM EDTA solution.
7. Load samples immediately onto a running (90 V) preparative 0.7% agarose/0.5X TBE gel.
8. Separate the samples so that the H1-nucleosome complexes are well resolved from tetramer and free DNA bands.
9. Next, wrap the gel tightly with Saran Wrap so that the gel can not move within the plastic. Lay fluorescent markers onto various portions of the gel for alignment purposes (can be obtained from Stratagene) or otherwise accurately mark the position of the gel on the film.
10. Expose the wet gel for several hours at 4°C.
11. Next, develop the autoradiograph and overlay onto the wet gel, lining up the fluorescent markers.
12. Cut and remove the agarose containing the H1-nucleosome complexes or bands of interest and place them into Series 8000 Microcentrifuge Filtration Devices.

13. Freeze the filtration tubes containing the agarose plugs on dry ice for 15 min.
14. Spin down the agarose in a microfuge at maximum speed for 30 min at room temperature. The fluid from the agarose matrix will be collected in the 2 mL centrifuge tube surrounding the filtration device.
15. Gently remove the agarose plug from the bottom of the filtration device and place into a clean eppendorf tube. Save the centrifugation devices for use later.
16. Using a microcentrifuge pestle, crush the agarose pellet and add 500 μL of 10 mM Tris-HCl, pH 8.0, 0.1% SDS and continue to crush the agarose.
17. After the agarose is crushed into tiny pieces, place all samples at 4°C overnight or for several hours.
18. Place the crushed agarose into the same centrifugation device and pellet. Spin down the agarose in a microfuge at maximum speed for 30 min at room temperature.
19. Combine identical samples from both spins and precipitate the DNA.
20. Dissolve the DNA in 15 μL of TE.

3.8. Site-Directed Photocrosslinking of Histone–DNA Interactions Within the Nucleosome Using 4-Azidophenacylbromide (APB)

3.8.1. UV Crosslinking of Nucleosomes Reconstituted with Sequence-Specific (5S) DNA

1. Combine 100 μL of nucleosomes reconstituted with a single APB-modified histone and wild-type histones (approx 100,000 cpms) with 10 μL of 25% glycerol. Load the sample on a native 0.7% agarose/0.5X TBE gel and run at 120 V for 3 h.
2. Identify the core histone nucleosome complex by exposing the wet gel to x-ray film (Kodak XR) for 3 h at 4°C.
3. Cut out the agarose containing the octamer histone complex and UV irradiate the gel slice for 20 s with a 365 nm light source. However, the time of irradiation should be first empirically determined (*see* **Note 9**).
4. Place the irradiated gel slice in a Spin-X microcentrifuge filter device and place on dry ice for 30 min.
5. Spin the frozen gel slice at maximum speed in a microcentrifuge for 30 min at room temperature to elute the DNA.
6. Reduce the volume of the eluate in a speedvac concentrator and ethanol precipitate the DNA with 1/10 vol of 3 M sodium acetate and 2.5 vol of cold 95% ethanol. To minimize the loss of sample recovery it possible to go directly from this step to **Subheading 3.9.**, the base elimination step (*see* **Note 10**).
7. Dissolve the DNA pellet in 50 μL of TE and separate the DNA complexes on a 6% SDS/PAGE gel for 4 h at constant 200 V.
8. Identify the nucleosome complex by exposing the wet gel to x-ray film (Kodak XR) for 2 h at 4°C.
9. Cut out the band corresponding to the core histone-DNA complex and crush it with a microcentrifuge pestle.
10. Soak the crushed gel in 500 μL of TE and rotate the tube overnight.
11. Transfer the gel/buffer mixture to a Spin-X filter tubes and pellet for 30 min in a microcentrifuge at 14,000 rpm to elute the DNA from the crushed gel.

12. Reduce the volume of the DNA sample in a speedvac and ethanol precipitate the DNA. Proceed to the base elimination step of the histone-DNA crosslinking (*see* **Subheading 3.9.**).

3.8.2. UV Crosslinking of Nucleosome Core Particles with Random Sequence DNA

1. Mix approximately 40 μg of core histones (20 μg each of H2A/H2B and H3/H4) and 50 μg of calf thymus DNA and perform a salt dialysis nucleosome reconstitution as previously described except in the absence of any labeled DNA (*see* **Note 11**).
2. After reconstitution take 40 μL of the reconstitute (approx 5 μg of nucleosomes) and digest with microccocal nuclease by adding 5 μL of 30 mM $CaCl_2$ and 5 μL of 0.2 U microccocal nuclease. Incubate at 37°C for 10 min.
3. Stop the microccocal digest with 5 μL of stop mix (25% glycerol and 0.25 M EDTA).
4. Load the entire sample on a 0.7% agarose/0.5X TBE gel and separate for 3 h at constant 120 V.
5. After separation, UV irradiate the gel directly for 20 s with a 365 nm light source and stain the gel with ethidium bromide for 1 h.
6. Identify the nucleosome core particle band and cut out the nucleosome complex from the gel (*see* **Note 12**).
7. Place the gel slice in a Spin-X filter tube and freeze on dry ice for 30 min. Elute the protein-DNA complex by centrifugation for 30 min at 14,000 rpm.
8. Reduce the volume of the eluate in a speedvac concentrator and ethanol precipitate the DNA.
9. Dissolve the pellet in 43 μL of H_2O and then 5' end label the DNA by adding 5 μL of 10X polynucleotide kinase buffer, 10 μCi of [γ-^{32}P]ATP and 10 U of polynucleotide kinase.
10. Incubate the sample for 30 min at 37°C and ethanol precipitate the DNA.
11. Dissolve the DNA in 40 μL of TE. **Steps 12–16** are optional (*see* **Note 13**).
12. Load approx 50,000 cpm of the sample on a 6% SDS/PAGE gel and separate for 4 h at constant 200 V.
13. Identify the crosslinked core histone-DNA complex by exposing the wet gel to x-ray film (Kodak XR) for 2 h.
14. Cut out the band corresponding to the histone-DNA complex from the gel and crush the gel slice as before. Resuspend in 500 μL of TE buffer and rotate overnight to elute the DNA.
15. Transfer the gel/buffer mixture to a Spin-X filter tube and pellet for 30 min in a microcentrifuge at 14,000 rpm to elute the DNA from the crushed gel.
16. Reduce the volume of the sample in a speedvac concentrator and ethanol precipitate the DNA.
17. Resuspend the pellet in 48 μL of TE and add 1 μL of 2% SDS and 1 μL of 15 mg/mL proteinaseK and incubate at 37°C for 1 h to digest away the protein crosslinked to the DNA.

18. Load the sample directly onto an 8% nondenaturing polyacrylamide gel (1:19 = bisacrylamide:acrylamide) and run for 2 h at constant 200 V.
19. Identify the labeled DNA band by autoradiography as before for 2 h and cut out the acrylamide slice containing the DNA band.
20. Crush the gel slice as before and resuspend in 500 μL of TE and rotate overnight.
21. Transfer the gel/buffer mixture to a Spin-X filter tube and pellet for 30 min in a microcentrifuge at 14,000 rpm to elute the DNA from the crushed gel.
22. Reduce the volume in a speedvac concentrator and ethanol precipitate the DNA. Wash the DNA pellet with cold 70% ethanol, dry and proceed to the base elimination step (see **Subheading 3.9.**).

3.9. Elimination and Cleavage of the Modified Base from Histone–DNA Crosslinked Complexes

1. Dissolve the DNA pellet in 100 μL of buffer E and layer with 100 μL of mineral oil.
2. Incubate the sample at 90°C for 30 min. Add 5 μL of 2 M NaOH and incubate an additional 60 min at 90°C.
3. Remove the bottom layer (aqueous sample) from the tube and add 6.5 μL 2 M HCl and 100 μL of 20 mM Tris-HCl, pH 8.0 to stop the elimination reaction.
4. Ethanol precipitate the DNA and dry the pellet in a speedvac concentrator.
5. Dissolve the DNA in 4 μL of formamide loading dye (see **Subheading 3.10.** for sequencing gel analysis).

3.10. Sequencing Gel Analysis of H1°aC-EPD Cleavage

1. Add equal numbers of counts from each sample, including the G specific reaction, to clean Eppendorf tubes.
2. Place the sample into a speedvac concentrator and dry to completeness.
3. Dissolve the sample in 4 μL of formamide loading buffer.
4. Heat the samples to 90°C for 2 min to denature.
5. Place sample directly onto ice to prevent re-naturation.
6. Separate samples on a 6% polyacrylamide/8 M Urea sequencing gel running at constant 2000 V.

4. Notes

1. Many ligation procedures are available from primary literature or commercial sources. Ligation of two DNA fragments occurs more rapidly at room temperature or 37°C if the base-pair overlap is sufficiently stable.
2. Many DNA mini-prep procedures are described in detail in ref. *(19)*. The DNA isolated for the techniques described here were from the boiling DNA mini-prep procedure *(19)*.
3. Before proceeding, it is recommended that a small amount of the culture be checked for overexpression of the protein of interest. This can be done by removing 1 mL of the culture before and after induction by IPTG and resolving an aliquot of the total protein by SDS-PAGE.

4. Sonication techniques tend to increase the temperature of the sample quickly which could induce proteolysis of the proteins. The sample must therefore be cooled before and during sonication. Allow several minutes between sonication runs to keep the sample as cold as possible.
5. Histones H2A or H2B do not bind to BioRex beads when purified individually. However, we have found that when allowed to heterodimerize they bind to the column and elute off consistently in 1 M NaCl *(13)*. This characteristic could be because H2A and H2B are completely unfolded when separated from each other *(20)*.
6. Storing labeled DNA in a concentrated form is not advised as autodegredation of the DNA takes place. DNA can be stored for several weeks at approx 5000 cpm/µL.
7. A complication of the in vitro reconstitution procedure is that purified histone proteins are often obtained in two fractions, H2A/H2B and H3/H4 *(21)*. Thus, in addition to total histone mass, the ratio between these two substituents must be empirically adjusted to yield maximum octamer-DNA complexes *(13)*. Often, a small amount of a subnucleosomal band containing $(H3/H4)_2$ tetramer-DNA complexes is observed since many competitor DNAs bind H2A/H2B dimers independent of octamer formation *(13)*. It is possible to assemble tetramers of histones H3 and H4, $(H3/H4)_2$, onto DNA by omitting histones H2A and H2B from the reconstitution *(22)*. This should be done as a control for the identification of subnucleosomal bands observed after nucleoprotein gel electrophoresis and to guard against the misidentification of a di-tetramer-DNA complex [i.e., two $(H3/H4)_2$ tetramers on one DNA] *(23,24)* as a true nucleosome complex. After identification of the tetramer complexes, several reconstitutions containing increasing amounts of H2A/H2B can be prepared, essentially titrating these proteins into the reconstitution mix until the core histone octamer-DNA complex is completely formed *(13)*. Thus, the total input mass of H2A/H2B may be different than that predicted by the theoretical mass ratio between these proteins and the $(H3/H4)_2$ tetramer within the nucleosome core (1:1).
8. Several methods can be used for the incorporation of linker histones into reconstituted mononuclesomes. The method described here involves direct addition of linker histones to mononuclesomes. Linker histones are folded in low-salt solutions in the presence of DNA *(25)*. Indeed, we find that linker histones can be directly mixed to nucleosomes in either 5 or 50 mM NaCl solutions, and these proteins then bind in a physiologically relevant manner *(17)*.
9. Analyze the amount of crosslinked species produced over time by irradiating samples of reconstituted nucleosomes solutions directly in Eppendorf tubes laid directly on the light box and separating crosslinked species on SDS-PAGE as described below. However, instead of exposing the wet gel, dry the gel before exposure. We have found that crosslinking within the tubes approximates crosslinking within the gel slice.
10. In **Subheading 3.8.1.**, **steps 12–16** can be done to enhance the specificity of the crosslinking signal. However, proceeding with these steps entails significant loss of sample and is not necessary when there is a strong enough signal on the sequencing gel *(11)*.

11. Determination of crosslinking with random sequence DNA is done to eliminate any sequence-specific effects to the structure of the nucleosome and to avoid any bias toward a particular sequence in the actual crosslinking reaction.
12. This step requires the use of a UV light source to identify the nucleosome core particle. Therefore it is important to excise the band corresponding to the nucleosomal complex as fast as possible (approx 5 s exposure).
13. In **Subheading 3.8.2.**, we have found that **steps 12–16** are optional. The advantage of these steps is the enhancement of the crosslinking signal. However, because of the significant loss of sample, the benefit of **steps 12–16** has to be determined empirically.

References

1. Wolffe, A. P. (1995) *Chromatin: Structure and Function.* Academic Press London.
2. Hansen, J. (1997) The core histone amino termini: combinatorial interaction domains that link chromatin structure with function. *Chemtracts: Biochem. Mol. Biol.* **10**, 56–69.
3. van Holde, K. E. (1989) Chromatin. Springer-Verlag, New York.
4. Ramakrishnan, V., Finch, J. T., Graziano, V., Lee, P. L., and Sweet, R. M. (1993) Crystal structure of globular domain of histone H5 and its implications for nucleosome binding. *Nature* **362**, 219–224.
5. Cerf, C., Lippens, G., Ramakrishnan, V., Muyldermans, S., Segers, A., Wyns, L., Wodak, S. J., and Hallenga, K. (1994) Homo- and heteronuclear two dimensional NMR studies of the globular domain of histone H1: full assignment, tertiary structure and comparison with the globular domain of histone H5. *Biochemistry* **33**, 11,079–11,086.
6. Arents, G., Burlingame, R. W., Wang, B. W., Love, W., and Moudrianakis, E. N. (1991) The nucleosomal core histone octamer at 3.1? resolution: a tripartite protein assembly and a left-handed superhelix. *Proc. Natl. Acad. Sci. USA* **88**, 10,148–10,152.
7. Luger, K., Mader, A. W., Richmond, R. K., Sargent, D. F., and Richmond, T. J. (1997) Crystal structure of the nucleosome core particle at 2.8 A resolution. *Nature* **389**, 251–260.
8. Ebright, Y. W., Chen, Y., Pendergrast, P. S., and Ebright, R. H. (1992) Incorporation of an EDTA-metal complex at a rationally selected site within a protein: application to EDTA-iron DNA affinity cleaving with catabolite gene activator protein (CAP) and Cro. *Biochemistry* **31**, 10,664–10,670.
9. Ermacora, M. R., Delfino, J. M., Cuenoud, B., Schepartz, A., and Fox, R. O. (1992) Conformation-dependent cleavage fo staphlyococcal nuclease with a disulfide-linked iron chelate. *Proc. Natl. Acad. Sci. USA* **89**, 6383–6387.
10. Chen, Y. and Ebright, R. H. (1993) Phenyl-azide-mediated photocrosslinking analysis of Cro-DNA interaction. *J. Mol. Biol.* **230**, 453–460.
11. Lee, K. -M. and Hayes, J. J. (1997) The *N*–terminal tail of histone H2A binds to two distinct sites within the nucleosome core. *Proc. Natl. Acad. Sci. USA* **94**, 8959–8964.

12. Hayes, J. J. (1996) Site-directed cleavage of DNA by a linker histone–Fe(II)EDTA conjugate: localization of a globular domain binding site within a nucleosome. *Biochemistry* **35**, 11,931–11,937.
13. Hayes, J. J. and Lee, K. -M (1997) In vitro reconstitution and analysis of mononucleosomes containing defined DNAs and proteins. *Methods* **12**, 2–9.
14. Rhodes, D. (1985) Structural analysis of a triple complex between the histone octamer, a Xenopus gene for 5S RNA and transcription factor IIIA. *EMBO J.* **4**, 3473–3482.
15. Hayes, J. J., Tullius, T. D., and Wolffe, A. P. (1990) The structure of DNA in a nucleosome. *Proc. Natl. Acad. Sci. USA* **87**, 7405–7409.
16. Simpson, R. T. (1991) Nucleosome positioning: occurrence, mechanisms and functional consequences. *Prog. Nucleic Acids Res. Mol. Biol.* **40**, 143–184.
17. Hayes, J. J. and Wolffe, A. P. (1993) Preferential and asymmetric interaction of linker histones with 5S DNA in the nucleosome. *Proc. Natl. Acad. Sci. USA* **90**, 6415–6419.
18. Allan, J., Hartman, P. G., Crane-Robinson, C., and Aviles, F. X. (1980) The structure of histone H1 and its location in chromatin. *Nature* **288**, 675–679.
19. Sambrook, J., Fritsch, E. F., and Maniatis, T. (1989) *Molecular Cloning: A Laboratory Manual.* 2nd ed. Cold Spring Harbor Laboratory Press, Cold Spring Harbor, NY.
20. Karantza, V., Baxevanis, A. D., Freire, E., and Moudrianakis, E. N. (1995) Thermodynamic studies of the core histones: Ionic strength and pH dependence of H2A-H2B dimer stability. *Biochemistry* **34**, 5988–5996.
21. Simon, R. H. and Felsenfeld, G. (1979) A new procedure for purifing histone pairs H2A + H2B and H3 + H4 from chromatin using hydroxyapatite. *Nucl Acids Res.* **6**, 689–696.
22. Hayes, J. J., Clark, D., and Wolffe, A. P. (1991) Histone contributions to the structure of DNA in the nucleosome. *Proc. Natl. Acad. Sci. USA* **88**, 6829–6833.
23. Stockley, P. G. and Thomas, J. O. (1979) A nucleosome-like particle containing an octamer of the arginine-rich histones H3 and H4. *FEBS Lett.* **99**, 129–135.
24. Read, C. M., Baldwin, J. P., and Crane-Robinson, C. (1985) Structure of subnucleosomal particles. Tetrameric (H3/H4)2 146 base pair DNA and hexameric (H3/H4)2(H2A/H2B) 146 base pair particles. *Biochemistry* **24**, 4435–4450.
25. Clark, D. J. and Thomas, J. O. (1986) Salt-dependent co-operative interaction of histone H1 with linear DNA. *J. Mol. Biol.* **187**, 569–580.

4

Base-Pair Resolution Mapping of Nucleosomes In Vitro

Andrew Flaus and Timothy J. Richmond

1. Introduction

The position of a nucleosome describes the arrangement of its core 147 base pairs of DNA relative to the core histone octamer *(1)*. The DNA superhelix spiraling around the histone octamer has an outer, solvent exposed face and an inner, histone-associated face *(2)*. Fixing of a particular exposed helical face is known as a rotational phasing of the double helix and controls the accessibility for particular sequences, such as for transcription factor binding sites *(3)*. In general, because DNA on the nucleosome has an approximately 10-bp pitch, bases 10 bp apart have similar accessibility at a first approximation. The rotational phase is easily determined by any nucleolytic digestion pattern of the nucleosomal (e.g., DNaseI, free solution hydroxyl radicals, *[4]*), although the distinction in solvent exposure of DNA shows variability at the edges of the core 147 bp directly associated with the histone proteins. The more rigorous fixing of the exact region of DNA in contact with the histone octamer is known as translational positioning. This not only implies a rotational phase, but also locates the detailed features of nucleosomal DNA relative to the histone octamer structure for a particular sequence *(2)*. For example, the intermediate resolution crystal structure of the nucleosome core particle revealed that the path of the DNA around the histone octamer is not uniform but undergoes a series of sharper bends at specific locations *(5)*. Likewise, biochemical studies have shown that the basic tails of individual histones associate with particular regions of the DNA superhelix *(6)*.

The histone octamer can be rotated by 180° about a *dyad* axis passing through its center to return a symmetrically equivalent structure, the result of the head-to-head association of the histone fold motif formed by histones H3 and H4 *(7)*. Because the histone octamer-associated DNA is rarely itself dyad-symmetric about this dyad axis, the symmetry axis of the nucleosome core is actually a *pseudodyad* because it does not represent a true symmetry at the atomic level *(8)*. We have shown biochemically and in a high-resolution crystal structure that this rotational pseudosymmetry axis passes through the plane of a base pair in the nucleosome *(2,9)*. Hence the assignment of the base pair lying on the histone octamer dyad axis is an exact description of the translational position of DNA in the nucleosome.

Previously, the translational position of a nucleosome has been considerably more difficult to determine experimentally than the rotational phase, since all methods use enzymatic or chemical footprinting and therefore reveal only the outer border of the histone octamer association with the DNA *(10)*. Defining these borders requires subjective assessment of protection endpoints, and relies on the assumption that both ends of the DNA exiting the nucleosome core are equivalently protected. The most popular method has been to use histone octamer-mediated protection of DNA against attack by micrococcal nuclease (MNase *[11]*). The region of protected DNA is then revealed either by restriction enzyme digestion and endlabeling of the resulting fragments or by extending a labeled primer from an internal sequence. The disadvantages of the MNase protection method are that MNase (16.8 kDa) is relatively bulky with respect to the nucleosome *(12)*, has some processivity and sequence bias (K. Luger, unpublished observations, *[13]*) , digests within the core region to create a nonuniform background *(14)*, and requires specific reaction conditions (e.g., temperature; *see* **Note 1**). When performed with care and analysed in combination with rotational phase information, it has been shown that MNase mapping allows the nucleosome position to be estimated with ±2 bp accuracy *(11)*. The method can also be employed on in vivo samples, albeit indirectly *(15,16)*.

A more recent alternative method to determine translational positioning makes use of the high resolution rotational phase signals observed for nucleosomal protection of DNA against hydroxyl radicals generated in free solution by EDTA-chelated Fe^{3+} *(4)*. Hydroxyl radicals are extremely reactive *(17)*, with a typical migration of only 10–15Å in aqueous solution *(18,19)*. When directed at DNA, they lead to strand scission preferentially attacking the C1' and C4' carbons of the deoxyribose ring to generate cleavages, resulting in loss of the base and a 3' phosphate or 3' phosphoglyconate end, respectively *(20,21)*. These two products are indistinguishable for the DNA lengths observed in nucleosomal mapping. Because the free-solution hydroxyl radical footprint of a nucleosome appears to show a change in pitch of the DNA in the 3 turns around the pseudodyad axis, translational positions have been assigned based on this discontinuity *(22)*. However, the high-resolution crystal structure of the nucleosome core particle shows no such change in helical twist at the dyad *(2)*. It is not obvious how the *apparent* twist change comes about, so the validity of the approach remains to be properly justified.

The site-directed hydroxyl radical method presented here *(9)* differs from the free solution hydroxyl radical method in using a disulphide bond to link a small EDTA-derived chemical reagent to histone H4 residue 47 in a modified histone octamer. This residue was identified as close to DNA near the pseudodyad axis of the nucleosome after trials of mutants on the basis of the intermediate resolution nucleosome structure and biochemical information, and is mutated from serine to cysteine to provide an acceptor thiol for the reagent. To avoid potential complications, we also use histone H3 cys110ala, which removes the only native cysteine residue of the higher animal core histones. This mutant behaves identically to the wild-type protein, and histone octamers containing this natural H3 cysteine can be derivatized with the EDTAcyst(NPS) reagent but do not generate any site-specific hydroxyl radical cutting of DNA. The chemical reagent itself (**Fig. 1A**) consists of an EDTA-derived chelating group linked by a peptide bond to cysteamine. The cysteamine thiol is in a disulfide bond with 2-nitrophenylsulfenyl, which is a good leaving group and so activates the EDTA-linked cysteamine of the EDTAcyst(NPS) reagent for exchange with free thiols, such as those of solvent-exposed cysteine residues of histones *(23)*. Once attached to a protein, the disulfide-linked EDTAcyst puts a distance of only approx 7Å between the cysteine thiol and the chelated Fe^{3+} reactive centre. Flexibility is also limited because the planarity of the peptide link leaves only four rotatable bonds in the reagent, and there is potential for hydrogen bonding to anchor the reagent location. The derivatized histone octamer is assembled into a nucleosome containing labeled DNA of the sequence of interest. Simultaneous or subsequent to our development of the EDTAcyst reagent, several other workers synthesized related reagents *(19,24)*, and there are growing numbers of reports of their use in studying nucleic acid interactions. Recently, Hayes has applied this system to probe the interaction of a H1 variant in the chromatosome *(25)*.

To map the DNA translational position, a hydroxyl radical-generating Fenton cycle (**Fig. 1B**) is induced, and the EDTA-derived reagent localizes the generation of hydroxyl radicals. Attached to histone H4 residue 47, the reagent is apparently very tightly directed for attack at a single major site in each DNA strand close to the dyad axis. However, the restricted space between the histone octamer and DNA reduces supply of the hydrogen peroxide and ascorbate substrates for the Fenton cycle, so the site-directed mapping reaction is many times slower than the free-solution hydroxyl radical footprinting method, a reaction which is also carried out at higher Fe^{3+}/EDTA concentrations. Attack by radicals of the local protein *(24)* and of the reagent itself is also likely,

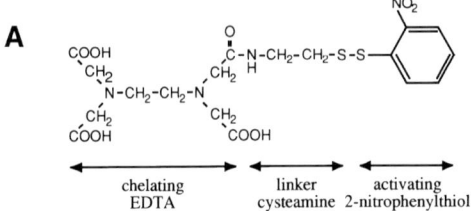
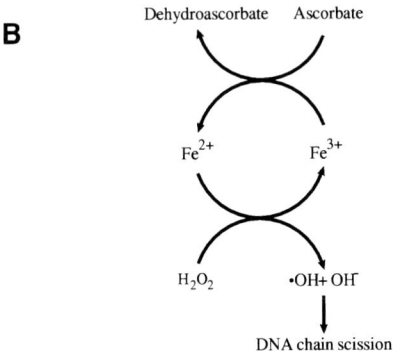

Fig. 1. Chemistry of site-directed hydroxyl radical mapping. (**A**) Structure of the EDTAcyst reagent showing chelating EDTA, cysteamine-derived linker arm, and disulphide exchange-activating 2-nitrophenyl thiol. (**B**) Fenton cycle for production of hydroxyl radicals using ascorbate and hydrogen peroxide, with Fe^{3+} as the localized radical producing center.

especially since sulphur atoms are known to be good sinks for hydroxyl radicals. The decomposition of the disulphide link and the reagent itself therefore suggest why digestion of strands to a maximum extent of approx 10% can be achieved in the mapping reaction.

The hydroxyl radicals generated from the EDTAcyst reagent attached to histone H4 residue 47 create single major cuts in each DNA strand, separated by three base pairs when the complementary strands are aligned (**Fig. 2A**). Therefore, the major cut for each strand must lie two bases 5' of the pseudodyad axis as a consequence of symmetry, and the pseudodyad axis must pass through the plane of the central of these three base pairs (**Fig. 2A** grey-filled ribose, **Fig. 2B** black ribose). Minor cuts also occur five and six bases 3' of the pseudodyad (**Fig. 2B** grey riboses), and the combination of a single strong cut and two weaker ones seven and eight bases in the 3' direction on each strand is diagnostic of a nucleosome position (**Fig. 3**, 2 nucleosome positions with major cuts at –13 and +6). The distance between the C_α of histone H4 residue 47 and

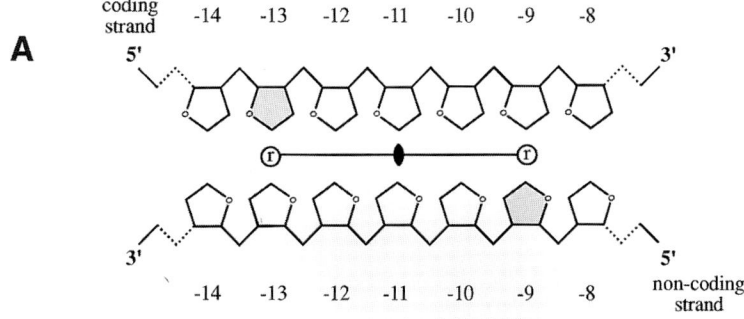

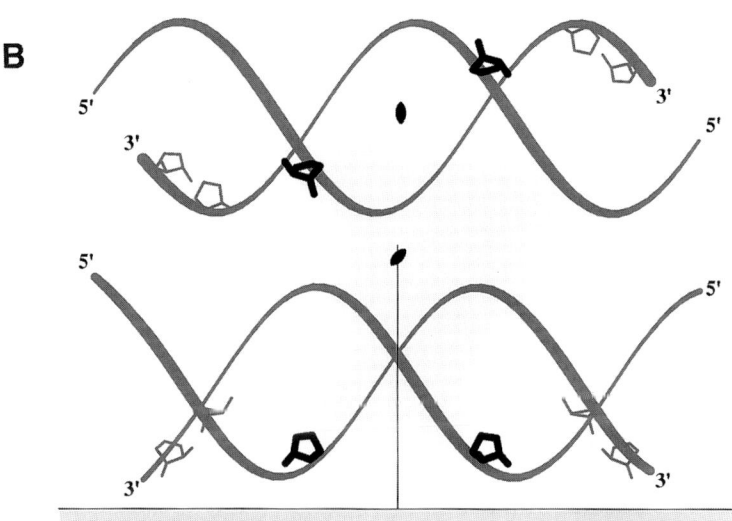

Fig. 2. Schematic diagrams of the site-directed cutting sites. (**A**) Diagram showing the sites of major cutting (grey-filled deoxyribose) on both coding *(upper)* and noncoding *(lower)* strands for the mapping whose coding strand result is in **Fig. 3**. The schematic reagent location (circled "r") and inferred site of the nucleosome pseudodyad axis at –11 are indicated. (**B**) Side and top view of B-form DNA model for the DNA turn across the pseudodyad axis *(marked)* showing the sites of major cutting (black deoxyribose rings) and minor cutting (grey deoxyribose rings).

the C4' of the major cut deoxyribose ring is approx 13Å, with the minor cut bases being those on the opposite strand with the closest approach to the reagent (**Fig. 2B**, grey riboses).

This method is applicable under a wide variety of temperatures and ionic strengths in vitro.

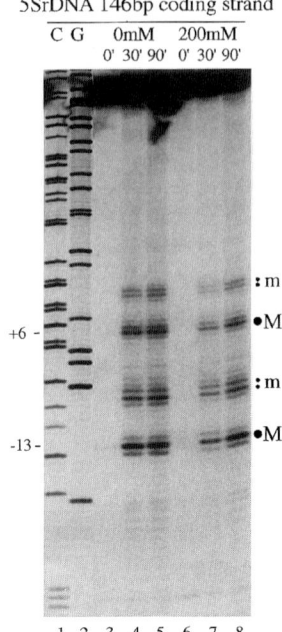

Fig. 3. Denaturing PAGE for mapping of a nucleosome core particle. Reactions were carried out with no (lanes 3–5) or 200 mM NaCl (lanes 6–8) in addition to the standard conditions at 4°C for the indicated times. The major (M) and minor (m) cutting sites are shown by large and small dots, respectively. Lanes 1 and 2 are an arbitrary dideoxy sequencing ladder. The 146-bp DNA fragment is from *Lytechinus variegatus*, uniquely labeled on the coding strand *(9)*.

2. Materials

2.1. Recombinant Core Histone Octamer

Purified core histone octamers (xOct) containing histones H2A, H2B ala7pro, H3 cys110ala, H4 ser47cys are prepared from lyophilisates of each individually expressed and purified protein (*see* Chapter 1) *(26)*.

2.2. EDTAcyst(NPS) Reagent Synthesis

The iron-chelating EDTA-derived reagent S-(2-nitrophenylsulfenyl)-cysteaminyl-EDTA (EDTAcyst(NPS); **Fig. 1A**; **Note 2**) can be synthesized by a simple route from cysteamine, EDTA, *tert*-butanol and 2-nitrophenylsulfenyl chloride *(9)*. The steps in the synthesis are:

1. Protect the thiol of cysteamine using a *tert*-butyl group *(27)*.
2. Recrystallize the *tert*-butyl cysteamine product.

3. Link *tert*-butyl cysteamine via its amino group to the carboxyl of EDTA in a peptide bond, using a large excess of EDTA to promote reaction of only a single carboxyl molecule *(28)*.
4. Purify the product EDTAcyst(tBu) by C_{18} reverse-phase HPLC.
5. Exchange the *tert*-butyl protection for 2-nitrophenylsulfenyl *(27)*.
6. Purify the product EDTAcyst(NPS) by C_{18} reverse-phase HPLC.

2.3. DNA Fragments

We conveniently prepare DNA fragments of 35–450 bp in large quantities using the following general scheme:

1. Multiple direct repeat arrays of up to 32 copies of the desired sequence are cloned in pUC-based plasmids by an array amplification method based on restriction digestions (*see* **Note 3**).
2. Plasmids are isolated using standard alkaline lysis of up to 12 L of *E. coli* culture media.
3. Individual DNA fragments are released from the purified plasmids by large scale restriction digestion.
4. The DNA fragments are purified by standard differential PEG/salt precipitation and DEAE ion-exchange chromatography.
5. DNA strands can be discriminated in the mapping method using radioactive phosphate isotopes by two methods:
   a. 5' endlabeling of both strands followed by digestion of DNA after mapping, using a restriction enzyme with its site close to one DNA terminus to remove the proximal label.
   b. 5' endlabeling of isolated strands separately and then reannealling before nucleosome assembly (*see* **Note 4**).

2.4. Chemicals and Stock Solutions

1. All chemicals we use are from Fluka (Buchs, Switzerland), with the exception of Tris base (Sigma) (*see* **Note 5**).
2. Buffer stock: 50 m*M* Tris-HCl, pH 7.5 prepared by dilution of a metal-free 1*M* stock (*see* **Subheading 2.6.**).
3. Stop solution: 25 m*M* EDTA, pH 8.0, 1 *M* thiourea.
4. CIA: 24 parts chloroform, 1 part isoamyl alcohol (v/v).
5. Phenol/CIA: equivolume mixture of equilibrated phenol and CIA.
6. Precipitation solution: 2.18 *M* sodium acetate, pH 5.2, 91 m*M* $MgCl_2$, supplemented with 0.18 mg/mL tRNA when no subsequent enzyme reactions are to be made on the DNA.
7. Formamide dyes solution: formamide (Fluka) with additional 0.1% xylene cyanol, 0.1% bromophenol blue, 10 m*M* EDTA.
8. TBE electrophoresis buffer: 89 m*M* Tris-borate, pH approx 8.4, 2.5 m*M* EDTA.

2.5. Trace Metal Ion Contaminants

Basic precautions for avoiding trace metal ion contaminants in solutions and equipment are:

1. Chemicals with explicit specifications for low levels of trace metal contaminants should be used (*see* **Note 5**). Only 1 M Tris-HCl, pH 7.5 and 4M potassium chloride (KCl) solutions need to be further purified (*see* **Subheading 2.6.**).
2. All water should be purified to resistance >18MΩ water purified using a MilliQ apparatus (Millipore).
3. Dedicated glass and plasticware is used for all solutions.
4. Trace metal ions are removed from equipment by soaking in 100 mM HCl overnight followed by extensive washing with water *(29)*. It is unnecessary to wash 1.5-mL plastic reaction tubes or pipet tips.

2.6. Removal of Trace Metal Ions using Chelex Resin

1. Prepare a slurry in water of approx 20 g of 200–400 mesh Chelex 100 resin (BioRad, cat. 142-2842).
2. Regenerate this resin by washing well with 1 M NaOH and then 1 M HCl in a Buchner funnel *(29)*.
4. Wash resin thoroughly with 1–2 L of water until the pH is near neutral.
5. Pack Chelex resin slurry into a standard 2.5-cm diameter column by pumping water at approx 4 mL/min using a peristaltic pump.
6. Pass 1–2 L of 1 M Tris-HCl, pH 7.5 solution over this column at 4 mL/min, discarding the first 100 mL.
7. Pass 1–2 L of 4 M KCl solution over this column as in **step 6**.
8. Store both metal-free solutions in acid-washed plastic containers.

2.7. Preparation of Microdialysis Apparatus

A multiwell dialysis apparatus has been designed to facilitate small-scale metal-free nucleosome assembly (*see* **Note 6** and **Fig. 4**).

1. Soak both upper plate and lower block of the apparatus in 100 mM HCl overnight.
2. Rinse both parts of the apparatus well with water.
3. Fill the lower block wells with metal-free 50 mM Tris-HCl, pH 7.5 buffer stock, place the upper plate on top, and leave to soak for >30 min.
4. Temperature-equilibrate the apparatus overnight (*see* **Note 1**).
5. Ensure all wells are dry by blowing air or CO_2 through them.
6. Cut 20–30 mm squares of dialysis membrane from dialysis tubing (6–8 kDa MWCO, prepared by standard methods) using a clean blade.
7. Soak the squares briefly in the starting buffer for the dialysis.
8. Place the dialysis membrane squares over the bottom of the projecting tubes and seal by fitting rubber O-rings which slide to fix into the grooves on the tubes. Ensure there are no folds in the membrane.

3. Methods

3.1. Derivatization of Histone Octamer

1. Add 5 µL 1M DTT to 500 µL ~15 µM xOct and stand on ice for 60 min.
2. Dialyse against three or more 500 mL changes of 5 mM potassium cacodylate buffer (KCac pH 6.0), 2 M NaCl at 4°C for at least 3 h per step (*see* **Note 7**).

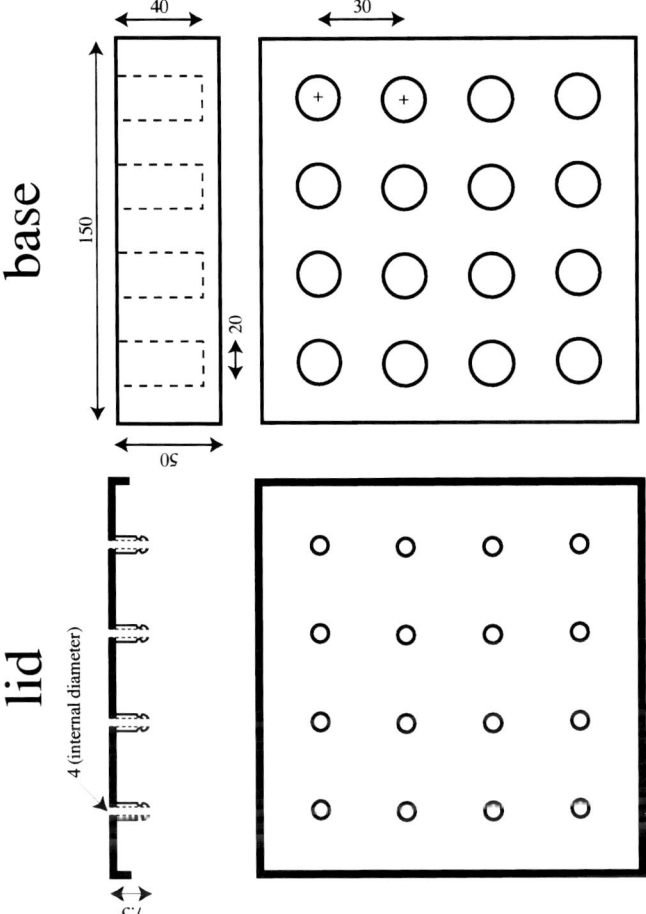

Fig. 4. Apparatus for multiple simultaneous microdialysis. Side and top views of both the lid *(left)* and base *(right)* of the apparatus. All dimensions are in millimetres.

3. Measure the concentration of dialyzed xOct in a spectrophotometer (E[276nm, 1 mg/mL] = 0.44, MW 108000; K. Luger, T. Rechsteiner and T.J. Richmond, unpublished).
4. Prepare fresh EDTACyst(NPS) solution by dissolving 0.5–0.8 mg in 100 mM Tris-HCl, pH 7.5 to a concentration of 10 mM (*see* **Note 2**).
5. Add an equal volume of 4 M NaCl to the dissolved reagent for a final concentration of 5 mM EDTAcyst(NPS) in 50 mM Tris-HCl, 7.5, 2 M NaCl.
6. Add 1 M Tris-HCl, pH 7.5 to make the dialyzed xOct to 50 mM Tris-HCl, pH 7.5 buffer.
7. Immediately add the 5 mM EDTACyst(NPS) solution to a 100-fold molar excess.
8. React overnight at room temperature.

9. Dialyze against three or more changes of 5 mM KCac, pH 6.0, 2 M NaCl at 4°C for at least 6 h per step.
10. Store the derivatized xOct on ice and calculate its final concentration (*see* **Note 8**).
11. Carefully prepare and run 5–20 pmol underivatized and derivatized proteins side-by-side on a fresh standard 18% acrylamide (1:60 *bis*-acrylamide/acrylamide), 1% SDS gel (*see* **Note 9**).

3.2. Assembly of Nucleosomes

1. Prepare the dialysis apparatus (*see* **Subheading 2.7.**), temperature-equilibrated at 4°C (*see* **Note 1**).
2. Prepare a mixture containing 5 µM derivatized xOct with equimolar labeled DNA fragment in a volume of 40 µL and with minimum salt concentration of 1.8 M (*see* **Note 10**).
3. Load 40 µL samples into dialysis well, taking care not to pierce the membranes.
4. Dialyse stepwise for 2 h against 20 mM Tris-HCl, pH 7.5, 1 mM potassium EDTA, pH 6.0–8.0 with consecutively 2 M, 0.85 M, 0.65 M, and 0.5 M potassium chloride per step.
5. Dialyse 2 h against 20 mM Tris-HCl, pH 7.5.
6. Dialyse overnight against 20 mM Tris, pH 7.5 with 0.5–1 mL of buffer-equilibrated Chelex 100 resin slurry.
7. Remove the assembled 5 µM nucleosome samples to prechilled Eppendorf tubes and store on ice (*see* **Note 1**).

3.3. Native PAGE Analysis of Nucleosomes

1. Prepare a 5% acrylamide (1:60 *bis*-acrylamide/acrylamide), 0.2X TBE gel between 16 cm x 18 cm x 1 mm glass plates with 10-mm wide sample wells.
2. Prerun this gel at 250 V for 3 h at 4°C with buffer recirculation.
3. Allow gel to temperature-equilibrate at 4°C overnight.
4. Prepare 1–2 pmol assembled nucleosome samples in 5% sucrose with a total volume per sample of 10–20 µL. Prechill all materials before adding the assembled nucleosome (*see* **Note 1**).
5. Load samples and run samples on gel at 250 V for 3 h at 4°C with buffer recirculation.
6. Dry gel without fixing and expose to X-ray film.

3.4. Mapping Reaction

1. Degas separately two aliquots of 50 mL buffer stock, as well as 50- and 100-mL aliquots of water using acid-washed Buchner flasks and standard water aspiration pumps.
2. Add 39 mg (100 µmol) ammonium ferrous sulphate hexahydrate to the 100 mL H_2O to make a 1 mM solution. Continue degassing while swirling to dissolve this salt.
3. Remove two aliquots of 987 µL buffer from the 50 mL degassed buffer stock to Eppendorf tubes on ice. Label one as "buffer" and the other as "buffer + iron."
4. Add 13 µL degassed 1 mM Fe salt to the buffer tube labeled "buffer + iron."

5. Place 10 μL of 5 μ*M* assembled nucleosome solution into prechilled reaction tubes on ice for each reaction.
6. Add 50 μL "buffer" to each reaction tube.
7. Add 5 μL "buffer + iron" to each reaction tube.
8. Stand on ice for 15 min.
9. While allowing iron to bind in the nucleosome in **step 8**, add 0.10 g (576 μmol) ascorbic acid to the second 50 mL degassed buffer stock to make a 12 m*M* ascorbate solution. Continue degassing while swirling to dissolve.
10. At the same time, add 333 μL 30% H_2O_2 to the 50 mL degassed water to make a 0.2% solution. Continue degassing briefly while swirling to mix.
11. Remove 1 mL of the 12 m*M* ascorbate solution to an Eppendorf tube on ice and label as "ascorbate."
12. Remove 1 mL of the 0.2% H_2O_2 to an Eppendorf tube on ice and label as "H_2O_2 solution."
13. Prepare tubes for zero time point samples by aliquoting 5 μL "stop solution" (*see* **Subheading 2.4.**) and 20 μL "buffer" to a fresh Eppendorf tube for each reaction.
14. After the 15-min equilibration of nucleosome-containing reaction samples (**step 8**), remove 5 μL of each reaction to a separate, prechilled Eppendorf tube for native PAGE analysis (*see* **Subheading 3.3.**).
15. Remove 20 μL of each reaction sample to the Eppendorf tubes containing the zero time point "stop solution"/"buffer" mix prepared in **step 13**.
16. Add 20 μL of "ascorbate" and 20 μL of "H_2O_2 solution" to each reaction in quick succession (*see* **Note 11**).
17. At desired time points, remove samples and add to a fresh Eppendorf tube containing 0.125 vol of stop solution.
18. Immediately after adding samples to stop solution, extract once with phenol/CIA (*see* **Subheading 2.4.**).

3.5. Post-Mapping Reaction Processing

1. If a subsequent restriction enzyme digestion is necessary, extract the samples once with CIA solution (*see* **Subheading 2.4.**).
2. Add 5.5 μL precipitation solution (*see* **Subheading 2.4.**) and 150 μL absolute EtOH to each sample, mix and centrifuge to precipitate.
3. If a restriction enzyme digestion is necessary, resuspend in a suitable buffer and carry out the digestion overnight. Phenol/CIA extract and go to **step 2**.
4. Aspirate off the supernatant and dry the pellet under vacuum for >15 min.
5. Count samples by Cerenkov counting.
6. Resuspend DNA pellets in 1 μL formamide dyes/1000 cpm (*see* **Subheading 2.4.**).

3.6. Denaturing PAGE of DNA Mapping Products

1. Prepare a 8.3 *M* urea, 6% acrylamide (1:20 *bis*-acrylamide/acrylamide ratio), 0.5X TBE gel between 40 cm x 18 cm x 0.1 mm siliconized glass plates and with a 20-well comb (*see* **Note 12**).
2. Assemble the gel in an electrophoresis box with 1X TBE in the upper and lower reservoirs.

3. Pre-run gel for at least 45 min at 45 W. Keep pre-running until you are ready to load the samples.
4. Boil DNA samples for 3–5 min.
5. Stop the prerunning gel and immediately load 2–3 µL of each sample per lane (*see* **Notes 13** and **14**).
6. Electrophorese at 45 W until the bromophenol blue marker, which comigrates with 18–20 base fragments, reaches the bottom (approx 100 min).
7. Fix the gel for 15 min in 10% (v/v) acetic acid, 10% (v/v) methanol.
8. Dry the gel and expose to X-ray film with enhancement screens (e.g., Kodak MR film or Fuji RX film depending on exposure characteristics), or to a freshly erased phosphorimager plate.

3.7. Analysis of DNA Mapping Patterns

1. Identify the single strong cutting sites in each line (e.g., –13 and +6 in **Fig. 3**) by looking for the characteristic single strong cut followed by weaker ones 7 and 8 bases above it in the gel (i.e., in the 3' direction) (*see* **Subheading 1.** and **Notes 14–15**).
2. Mark the location of each strong cutting site for each strand on a linear printout of the DNA sequence text (e.g., circled "r" in **Fig. 2A**).
3. Identify the pseudodyad axis of the nucleosome as the base intermediate between the cuts on complementary strands, which are normally separated by three base pairs (e.g., base-pair –11 in **Fig. 2A**) (*see* **Note 15**).

4. Notes

1. Great care should be taken that nucleosome assembly and all subsequent mapping steps are done strictly at 0–4°C. We have observed a shifting of nucleosome positions upon heating for almost all nucleosomes we have studied, despite the fact that these were "positioned nucleosomes" (A. Maeder, K. Luger, A. Flaus, T. Rechsteiner, and T. J. Richmond, unpublished, *[9,30,31]*). In the most extreme case this can be manifest by incubation for a few minutes at 37°C, by the heat generated in a microcentrifuge, or even during storage of samples on ice for a period of weeks. Some nucleosomes can only be shifted by heating for several hours at 65°C. All possible precautions should be taken to minimise the heating of assembled nucleosomes before and during mapping. The observed shifting has always resulted in nucleosomes moving to positions with the dyad 73 ± 10 bp from the DNA termini *(31)*. We hypothesise that this results from a shift in the equilibrium caused by an extra energetic stability of terminal nucleosomes at low ionic strength (less than approx 400 mM NaCl), combined with the kinetic opportunity to shift/slide without dissociation at elevated temperatures (A. Flaus, A. Maeder, K. Luger, and T. J. Richmond, unpublished). This potential artifact must be considered for any nucleosome observed at a DNA terminus.
2. EDTAcyst(NPS) is stable over a long period when stored as a dried lyophilisate. Some vortexing of the reagent may be necessary as 10-mM aqueous solutions are near the solubility limit.

3. The method depends on having enzyme sites with complementary overhangs flanking the insert, and a third site to one side. Plasmids are digested with either enzyme plus the third; the resulting pieces containing the insert are purified by agarose gel electrophoresis, and finally ligated together to make a "doubled" insert (A. Maeder, T. Rechsteiner, S. Tan, A. Flaus, and T. J. Richmond, unpublished, *[8,9,32]*). The ligation creates a non-recleavable site for the enzymes with complementary overhangs. Unstable dyad symmetric DNA fragments can be cloned as half-fragments and ligated together after isolation *(9,32)*.

4. Isolation of strands can be achieved on a preparative scale by DEAE ion exchange HPLC in 10 mM NaOH using gradients from 400–800 mM NaCl (S. Tan, A. Maeder, A. Flaus, and T. J. Richmond, unpublished). The individual strands are subsequently dephosphorylated, labelled using polynucleotide kinase, and reannealed in a thermal gradient to the complementary unlabeled strand, all according to standard methods.

5. All chemicals are from Fluka (Buchs, Switzerland), who provide explicit specifications for trace metal contaminants, except Tris base (Sigma T1503). Specifically, we use potassium cacodylate, pH 6.0 (cacodylic acid, cat. no. 20835; potassium hydroxide, cat. no. 60370), hydrochloric acid (cat. no. 84420), potassium chloride (cat. no. 60129), ammonium ferrous sulphate hexahydrate (cat. no. 09719), hydrogen peroxide (cat. no. 95300), ascorbic acid (cat. no. 95210), formamide (cat. no. 47671) and thiourea (cat. no. 88810).

6. The dialysis apparatus (**Fig. 4**) consists of a 15-cm square Perspex block with 16 holes (2 cm diameter × 4 cm deep). An overlaying plate has 16 matching projecting tubes (15 mm long × 4 mm internal diameter), each with a rounded lower edge and groove approx 2 mm above. Rubber O-rings fitting in the groove are used to seal squares of dialysis membrane across the open surface of the projecting tube. Samples of 40–50 µL can be conveniently dialyzed against 10-mL buffer in this apparatus by putting small magnetic stirring bars in the large chambers and placing the whole setup on a magnetic stirring plate.

7. Dialysis of the refolded, purified histone octamer in this buffer removes all residual reducing reagents, at the same time protecting the free thiols at low pH. After this point, samples should not be exposed to *any* reducing agents because of the lability of the disulphide bond linking the EDTAcyst reagent to the histone

8. The derivatized xOct is stable for several weeks if stored on ice. The spectrum of the derivatised xOct can show an absorbance maximum at 365 nm if the dialysis has been insufficient to remove all excess reagent or nitrophenylthiol product. (These pass only poorly through dialysis membranes.) Because the nitrophenyl chromophore has extinction coefficients E[365 nm, 1 M] = 3000 and E[280 nm, 1 M] = 4492 (A. Flaus, unpublished), the concentration of the xOct can be calculated from the spectrum. However, if the dialysis bag was turgid, an equally accurate concentration measurement can be obtained by calculation on the assumption that there was no dilution of the material during dialysis.

9. Two additional methods to detect derivatization are using a 40 cm × 18 cm × 0.1 mm triton-urea-acetic acid gel with Coomassie staining or to add ^{63}Ni^{2+} isotope to samples before SDS-PAGE as above, then detect using a fluorogenic

reagent and X-ray film *(9)*. For the simple SDS PAGE observation, use blank samples with equivalent salt concentration in all other lanes on gel, and do not to use *any* reducing agents (e.g., 2-mercaptoethanol) in the loading buffer. The derivatized H4 migrates very slightly but distinctly slower than the underivatized histone in SDS gels of sufficient quality. Incomplete reaction is usually owing to unreduced cysteines in the histone octamer or to insufficient dialysis for removal of reducing reagents from purified xOct.

10. In practice, derivatised xOct of concentration >8 µM and DNA fragments of >20 µM are required. NaCl and KCl are considered to have equivalent water activities.
11. The mapping reaction conditions are therefore 0.05% H_2O_2, 3 mM ascorbate, 37.5 mM Tris-HCl, pH 7.5.
12. We have occasionally observed two species in denaturing PAGE gels of undigested DNA fragments, which we attribute to incomplete denaturation of strands. To overcome these, we employ stronger denaturant conditions in gels (8.3 M urea and/or 25% formamide, *[9]*) and fresh formamide/dye loading buffer where necessary. Sample loading should always be undertaken immediately after the gel pre-run and while both it and the sample remain hot. We note also that lower grades of formamide sometimes contain impurities that affect gel running behavior, and recommend deionizing formamide before use with an ion exchange resin such as Amberlite MB3 (BDH Chemicals).
13. Always include a ladder to allow identification of DNA sizes. Hydroxyl radical attack at a base step leads to loss of the base moiety and a 3' phosphate or 3' phosphoglyconate terminus (*see* **Subheading 1.**). These two products are indistinguishable for the DNA lengths in nucleosomal mapping, and migrate in parallel with terminations generated by the classical Maxam-Gilbert G-specific DMS cleavage (G track). G tracks therefore provide a convenient "ruler" for determining the exact sites of cutting. We have also found empirically that in the gel running conditions described above (**Subheading 3.6.**), dideoxy-terminated ladders from standard enzymatic sequencing with incorporated labels (i.e., no 5' phosphate on primer) migrate two bases slower than the equivalent hydroxyl radical cut fragment in the range approx 90–120 bases, presumably due to the presence in the latter of the 5' phosphate, the charged 3' terminus, and the loss of the terminal base moiety *(31)*.
14. The major difficulty in creating clear mappings is the appearance of a nonspecific background ladder. This can be a consequence of DNA quality, so standard biochemical precautions should be followed such as avoidance of unnecessary oxygenation by shaking, storage of material at –20°C, and use of good quality modification enzymes in their correct buffers. Another significant cause of background laddering is from hydroxyl radicals generated by excess Fe^{3+} ions bound as counter ions to the DNA in the mapping. The protocol here uses a twofold excess of histone-linked reagent to Fe^{3+} ions. However, if materials or buffers contain other di- or multivalent ions, these will compete for reagent chelation and free Fe^{3+} ions, which can then associate with DNA. To avoid this, remove metal ions from all materials and buffers as described in **Subheadings 2.5–2.6.** In our experience, the most important precaution is to acid wash the dialysis apparatus every 2–3 uses.

15. Some variation in the focus of cutting is observed between different sequences (compare cutting at −13 and +6 in **Fig. 3**), although there is always a single major site of attack. The variations presumably reflect subtle differences in the local stereochemistry of the DNA on the nucleosome. Because of this variation, it is not valid to quantitatively infer occupancy from the relative cutting at two different sites. Instead, it is necessary to equate each position mapped with a species in native gels and then compare the quantitation of these species, or to normalize the cutting for each position independently.

Acknowledgments

We are very grateful for the advice, suggestions, and assistance of our colleagues, especially to T. Rechsteiner, A. Maeder, K. Luger, and S. Tan, and to M. Bumke and R. Richmond. We thank S. Halford for supplying EcoRV enzyme.

References

1. Thoma, F. (1992) Nucleosome positioning. *Biochim. Biophys. Acta* **1130,** 1–19.
2. Luger, K., Maeder, A., Richmond, R. K., Sargent, D. F., and Richmond, T. J. (1997) Crystal structure of the nucleosome core particle at 2.8Å resolution. *Nature* **389,** 251–260.
3. Beato, M. and Eisfeld, K. (1997) Transcription factor access to chromatin. *Nucleic Acids Res.* **25,** 3559–3563.
4. Hayes, J. J., Tullius, T. D., and Wolffe, A. P. (1990) The structure of DNA in a nucleosome. *Proc. Natl. Acad. Sci. USA* **87,** 7405–7409.
5. Richmond, T. J., Finch, J. T., Rushton, B., Rhodes, D., and Klug, A. (1984) Structure of the nucleosome core particle at 7Å resolution. *Nature* **311,** 532–537.
6. Lee, K. and Hayes, J. (1997) The N-terminal tail of histone H2A binds to two distinct sites within the nucleosome core. *Proc. Natl. Acad. Sci. USA* **94,** 8959–8964.
7. Arents, G. and Moudrianakis, E. (1993) Topography of the histone octamer surface: repeating structural motifs utilized in the docking of nucleosomal DNA. *Proc. Natl. Acad. Sci. USA* **90,** 10,489–10,493.
8. Richmond, T., Searles, M., and Simpson, R. (1988) Crystals of a nucleosome core particle containing defined sequence DNA. *J. Mol. Biol.* **199,** 161–170.
9. Flaus, A., Luger, K., Tan, S., and Richmond, T. J. (1996) Mapping nucleosome position at single base-pair resolution by using site-directed hydroxyl radicals. *Proc. Natl. Acad. Sci. USA* **93,** 1370–1375.
10. Thoma, F. (1996) Mapping of nucleosome positions. *Methods Enzymol.* **274,** 197–214.
11. Dong, F., Hansen, J. C., and van Holde, K. E. (1990) DNA and protein determinants of nucleosome positioning on sea urchin 5S rRNA gene sequences in vitro. *Proc. Natl. Acad. Sci. USA* **87,** 5724–5728.
12. van Holde, K. E. (1989) *Chromatin.* Springer-Verlag, New York.
13. McGhee, J. D. and Felsenfeld, G. (1983) Another potential artifact in the study of nucleosome phasing by chromatin digestion with micrococcal nuclease. *Cell* **32,** 1205–1215.
14. Cockell, M., Rhodes, D., and Klug, A. (1983) Location of the primary sites of micrococcal nuclease cleavage on the nucleosome core. *J. Mol. Biol.* **170,** 423–446.

15. Mymryk, J. S., Fryer, C. J., Jung, L. A., and Archer, T. K. (1997) Analysis of chromatin structure in vivo. *Methods: A Companion to Methods Enzymol.* **12,** 105–114.
16. Fragoso, G. and Hager, G. L. (1997) Analysis of *in vivo* nucleosome positions by determination of nucleosome-linker boundaries in crosslinked chromatin. *Methods: A Companion to Methods Enzymol.* **11,** 246–252.
17. Halliwell, B. and Gutteridge, J. M. C. (1989) *Free Radicals in Biology and Medicine*. Clarendon Press, Oxford.
18. Halliwell, B. and Gutteridge, J. M. C. (1990) Role of free radicals and catalytic metal ions in human disease: an overview. *Methods Enzymol.* **186,** 1–88.
19. Ebright, Y. W., Chen, Y., Prendergrast, S., and Ebright, R. H. (1992) Incorporation of an EDTA-metal complex at rationally selected sites within a protein: application to EDTA-Iron affinity cleaving with catabolic activator protein (CAP) and Cro. *Biochemistry* **31,** 10,664–10,670.
20. Hertzberg, R. P. and Dervan, P. B. (1984) Cleavage of DNA with methidiumpropyl-EDTA-Iron(II): reaction conditions and product analyses. *Biochemistry* **23,** 3934–3945.
21. Papavassiliou, A. G. (1995) Chemical nucleases as probes for studying DNA-protein interactions. *Biochem. J.* **305,** 345–357.
22. Roberts, M. S., Fragoso, G., and Hager, G. L. (1995) Nucleosomes reconstituted in vitro on mouse mammary tumor virus B region DNA occupy multiple translational and rotational frames. *Biochemistry* **34,** 12,470–12,480.
23. Rydén, L. and Carlsson, J. (1989) Covalent chromatography, in *Protein Purification* (Janson, J.-C. and Rydén, L., ed.), VCH Publishers, New York, pp. 252–274.
24. Ermacora, M. R., Delfino, J. M., Cuenoud, B., Schepartz, A. and Fox, R. O. (1992) Conformation-dependent cleavage of staphylococcal nuclease with disulphide-linked iron chelate. *Proc. Natl. Acad. Sci. USA* **89,** 6383–6387.
25. Hayes, J. J. (1996) Site-directed cleavage of DNA by linker histone-Fe(II) EDTA conjugate: localisation of a globular domain binding site within a nucleosome. *Biochemistry* **35,** 11,931–11,937.
26. Luger, K., Rechsteiner, T. J., Flaus, A. J., Waye, M. M. Y., and Richmond, T. J. (1997) Characterization of nucleosome core particles containing histone proteins made in bacteria. *J. Mol. Biol.* **272,** 301–311.
27. Pastuzak, J. J. and Chimiak, A. (1981) *tert*-Butyl group as thiol protection in peptide synthesis. *J. Org. Chem.* **46,** 1868–1873.
28. Sheehan, J. C. and Hess, G. P. (1955) A new method of forming peptide bonds. *J. Am. Chem. Soc.* **77,** 1067–1068.
29. Schaich, K. M. (1990) Preparation of metal-free solutions for studies of active oxygen species. *Meth. Enz.* **186,** 121–125.
30. Meerseman, G., Pennings, S., and Bradbury, E. M. (1992) Mobile nucleosomes - a general behaviour. *EMBO J.* **11,** 2951–2959.
31. Flaus, A. and Richmond, T. J. (1997) Positioning of nuclesosomes on the MMTV 3'LTR *in vitro*. *J. Mol. Biol.* in press.
32. Palmer, E. L., Gewiess, A., Harp, J. M., York, M. H., and Bunick, G. J. (1995) Large-scale production of palindrome DNA fragments. *Anal. Biochem.* **231,** 109–114.

5

Equilibrium and Dynamic Nucleosome Stability

Jonathan Widom

1. Introduction

Researchers often think of nucleosomes as inert and static structures. But the laws of physical chemistry—which apply to *any* molecular complex—remind us that this picture cannot be correct. Indeed, many early studies revealed that nucleosomes undergo a complex set of assembly/disassembly processes, and more recent studies reveal that nucleosomes also have remarkable dynamic behavior. These properties of nucleosomes have practical consequences for experiments carried out in vitro, and they have important consequences for nucleosome function in vivo. This article will not discuss specific experimental methods; rather, it will summarize several aspects of the equilibrium and dynamic behavior of real nucleosomes that are of particular importance for studies of nucleosome structure and function.

2. Nucleosome Assembly Equilibria

The dissociation/association properties of nucleosomes have been studied intensively for more than two decades and are discussed in a helpful book (*1*). The following discussion will summarize the most important points that have emerged from both earlier and more recent studies, emphasizing the behavior of nucleosomes in approximately physiological solution conditions.

2.1. Nucleosomes Dissociate in the Nanomolar Concentration Range

Nucleosomes are complexes of many subunits: eight histone molecules (nine, counting H1) and a segment of DNA. At equilibrium, molecular complexes obey the law of mass action. Thus, simple dilution should drive nucleosomes to dissociate into subunits, while concentration should favor association (assembly) of subunits into nucleosomes.

The concentration range (of histones and DNA) over which the equilibrium shifts from favoring dissociation to favoring association depends on environmental variables such as temperature, pH, and ionic strength. For any protein-nucleic acid complex, the ionic environment (especially the valences and concentrations of all cations) is a major determinant of stability because of the polyelectrolyte properties of nucleic acids *(2)*. Increasing the concentration or valence of cations decreases the affinities of proteins for nucleic acids. For protein-DNA complexes in which the DNA is bent—such as nucleosomes—the ionic environment is a major determinant of stability also for a second reason, through its effects on the DNA bending force constant *(3)*. But this effect is in the *opposite* direction: Decreasing the concentration and/or valence of cations in solution makes DNA harder to bend, thus destabilizing complexes in which the DNA *is* bent. Consequently for solutions containing only monovalent cations (M^+)—such as Na^+, and $Tris^+$—there is a [M^+] yielding maximal nucleosome stability.

The dissociative behavior of nucleosomes is readily detected in practice. The results of many such earlier studies are summarized in a stability diagram for the [protein and DNA] vs [M^+] concentration plane *(4)*. This picture has been confirmed qualitatively in more recent studies using simple dilution of ^{32}P-labeled nucleosomes (*[5]*; R. Protacio and JW, unpublished) and also in studies using fluorescence polarization of labeled DNA to assess affinities of various histone subunits for DNA *(6)*. In approximately physiological solution conditions, dilution of nucleosomes to the nanomolar or lower concentration range leads to substantial dissociation, yielding free DNA, histones in various states of aggregation *(see below)*, and, potentially, some remaining nucleosomes to which additional histones released from other dissociated nucleosomes have bound *(7,8)*.

2.2. Slow Assembly Kinetics

The behavior of real nucleosomes can be much more complicated than this simple equilibrium picture because of potentially slow kinetics of the assembly reactions. Consider an experiment in which the four core histones plus DNA are mixed together at their native stoichiometries (2 mol of each histone per mol of core particle length DNA) in approximately physiological ionic conditions, at sufficiently high histone and DNA concentrations such that the equilibrium favors nucleosome formation. In actuality, such a reaction does produce some nucleosomes, but many or most of the products are nonnucleosomal histone-DNA complexes and aggregates. These are kinetically trapped "dead-end" complexes, which will not convert to nucleosomes in any reasonable amount of time *(9,10)*. (In vitro nucleosome assembly is carried out using salt-dialysis procedures, which circumvent these kinetics traps. Concentrated [M$^+$] weakens inappropriate histone-DNA contacts, allowing the system to escape from these various nonnucleosomal complexes which, in physiological ionic conditions, would be kinetic dead ends.) A vast set of distinct nonnucleosomal complexes can be formed in such direct mixing experiments. The particular distribution of products that are formed can vary depending on details of the experiment—that is, on the concentrations of histones and DNA, on environmental variables, and even on details of how the components are mixed. It is difficult or impossible to fully characterize these products, so few useful conclusions can be drawn.

Because of this behavior of real nucleosomes, it could prove impossible to assess their true equilibrium stability in physiological conditions: Slow kinetics of reassembly processes could prevent studies of nucleosome dissociation such as those described above from ever reaching true equilibrium. Yet this need not be the case. In dissociation experiments, only very low concentrations of free histones and DNA ever exist, since dissociation occurs in the nanomolar or lower concentration range. It may be that slow forward kinetics arise only from the formation of multimolecular but nonnucleosomal aggregates, and that in sufficiently low concentrations such aggregates essentially do not form. At this time, it is simply not clear whether or for what histone and DNA concentration range association/dissociation processes reach true equilibrium in physiological ionic conditions.

2.3. Slow Disassembly Kinetics and Restricted Subunit Exchange at (Apparent) Equilibrium

True equilibria are dynamic: forward and reverse reactions are constantly and rapidly occurring, with equal rates in both directions. For the case of complexes containing many molecules, a dynamic association/disassociation equilibrium ordinarily implies that subunits should be in free exchange between complexes. Is this the case for nucleosomes?

At sufficiently high (approx 2 *M*) [M$^+$], histone-DNA interactions are sufficiently destabilized *(2)* that all of the components exchange freely and rapidly *(1)*. This fact serves as the basis for nucleosome reassembly procedures that are based on dialysis from concentrated NaCl. For [M$^+$] ≈ 1M, histones and DNA still undergo exchange, although with relatively slow kinetics; this fact serves as the basis for exchange procedures for assembling radiolabeled tracer DNA into nucleosomes.

For studies of nucleosome structure and function, we care particularly about the behavior in approximately physiological conditions. Histone H1 has long been known to be in free exchange between chromatin fragments *(11)* (this free exchange occurs in 50–75 mM [M$^+$], more dilute than physiological; presumably, exchange would occur even more readily at physiological [M$^+$], but the insolubility of H1-containing chromatin in physiological ionic conditions prohibits such experiments). Histones H2A and H2B also undergo exchange between nucleosomes or longer chromatin fragments, although this exchange occurs only slowly, on a time scale of hours *(12)*. H3 and H4 show little or no exchange *(12)*. It is noteworthy that the kinetics of protein subunit exchange between particles correlates directly with the affinity of the differing histone subunits for nucleosomes as measured by the [M$^+$] required to drive their dissociation *(1)*.

Using DNA as the labeled tracer, one finds little exchange of DNA into nucleosomes occurring in physiological solution conditions.

2.4. A Paradox for the Equilibrium Stability of Nucleosomes?

These facts raise an apparent paradox. On one hand, results summarized above establish unequivocally that nucleosomes dissociate when they are diluted into the nanomolar concentration range in physiological solution conditions. Since nucleosomes also continue to exist in these conditions, this suggests that both disassembly and reassembly processes must be facile—that is, nucleosomes act as though they are in rapid dynamic equilibrium. On the other hand, many of the nucleosome components are not in free exchange in those conditions, suggesting that the system cannot in fact be in true dynamic equilibrium.

One formal possibility is that the fraction of nucleosomes that dissociate as [nucleosomes] is decreased to the approx nanomolar or lower concentration range represent a subset of particularly unstable individual nucleosomes. Arguing against this interpretation, however, is that the fraction of nucleosomes that dissociate at low [nucleosomes] exceeds 50% at the lowest [nucleosomes] examined, and could easily have increased further had the experiment been extended to even lower [nucleosomes] *(5)*. Evidently, this represents the behavior of most or all of the nucleosomes in a sample, not just a small subset.

Instead, two alternative interpretations seem likely. In one model, suggested by the structure of the nucleosome itself *(13)*, release of both H2A/H2B heterodimers must preceed the release of the H3/H4 tetramer; exchange of the H3/H4 tetramer is equivalent to and required for exchange of the nucleosomal DNA. This model predicts that exchange of the H3$_2$/H4$_2$ tetramer or DNA would be suppressed at higher [nucleosomes] because the initial equilibria involving stepwise release of both H2A/H2B heterodimers will be driven by high concentration toward the state in which almost all H2A/H2B heterodimers are bound almost all of the time.

Alternatively, free exchange of histone subunits could be hindered by a strongly biased diffusion back to the same nucleosome from which they originated. The large positive histone charge together with the highly negative DNA charge, plausibly provide such a bias. Indeed, one limit of this model is that the histones might often remain held to the DNA by nonspecific ionic interactions through their extended N-terminal tail domains even after their globular domains have dissociated from the nucleosome. Behavior consistent with this view has been reported in nucleosome transcription experiments *(14)* in which it appears that the histone octamer can step around an elongating polymerase without dissociating fully from the DNA. Both of these models could be operative at the same time.

2.5. Dissociation and Association of Histones

The dissociation / association properties of histones on their own are much better defined than when DNA is present, and are now relatively well understood. Classic studies revealed that the histone octamer is the most stable state at sufficiently high [histones] in $2M$ NaCl, and that reduction of the [NaCl] down toward physiological results in dissociation of the octamer into three stable subunits: two H2A/H2B heterodimers and one $H3_2/H4_2$ tetramer (for review, *see [1]*). Recent studies have characterized the dissociation/association properties of these histone subunits in more detail and lead to the following additional conclusions (applicable for approximately physiological solution conditions). H2A/H2B heterodimers have no detectable tendency toward association into larger oligomers. The H2A/H2B heterodimer dissociates into free H2A + H2B with a dissociation constant of approx $10^{-7} M$ *(15)*. In contrast, the $H3_2H4_2$ tetramer has a strong tendency for the formation of higher order aggregates as the protein concentration (or [NaCl]) are increased *(16,17)*. The tetramer dissociates into two dimers with a dissociation constant of approx 15 nM *(15)*. Consistent with the structure of the histone octamer *(18)* and the intact nucleosome core particle *(13)*, these dimers are thought to be heterodimers of H3/H4 *(17)*, rather than homodimers. The equilibrium constant for further dissociation of the H3/H4 heterodimer into free H3 + H4 is not known.

2.6. Competition Experiments
Yield Equilibrium Difference Free Energies

There is considerable interest in measuring free-energy preferences that histone octamers may exhibit for one DNA sequence compared to another. Crothers and colleagues *(19,20)* devised a competition experiment to measure these free-energy preferences. In their approach, radiolabeled tracer DNA competes with a large excess of unlabeled competitor DNA for limiting amounts of histone octamer during nucleosome reconstitution from concentrated NaCl. The products of competitive reconstitutions are separated by native polyacrylamide gel electrophoresis and quantified. The ratio of radiolabeled tracer DNA incorporated into nucleosomes to free tracer DNA defines an equilibrium constant and a corresponding free energy for histone binding of the tracer DNA that are valid for that specific competitive environment. Difference free energies are obtained by subtraction of the free energies for differing radiolabeled tracer molecules measured in parallel experiments having identical competitor DNA and competitive environments.

Results discussed above establish that in concentrated NaCl, histones and DNA have little affinity for each other; they undergo free exchange and readily achieve true equilibrium. A sufficiently gradual decrease in [NaCl] maintains this equilibrium as [NaCl] changes, but gradually "freezes" it in as [NaCl] is decreased into the range where DNA and histone subunit exchange become kinetically slow.

Studies by Crothers and colleagues show that these procedures do indeed establish a true chemical equilibrium which is *subsequently* trapped (frozen) *(19,20)*. Most importantly, the same apparent distributions of labeled tracer molecules are reached regardless of the direction from which the final state is approached—that is, whether equilibrium is approached with the tracer starting out as free DNA or as nucleosomes. This is a standard test for equilibrium. In other tests they show that the results are neither affected by extended incubation times nor by changes in the rate at which the nucleosomes are brought from elevated salt concentrations down to physiological or lower concentrations.

These results establish that the system does reach a true equilibrium, there remains some question as to exactly at what salt concentration this equilibrium applies *(21)*. The partitioning of histones between differing DNA molecules becomes "frozen" in as [NaCl] is decreased much below approx 1 M but, because nucleosomes are mobile in physiological ionic conditions (*see* **Subheading 3.**), the histones remain free to equilibrate along those DNA molecules even in physiological conditions. Thus, nucleosomes resulting from competitive reconstitutions are equilibrated at physiological ionic strengths, on DNA molecules that were chosen at equilibrium at somewhat higher [NaCl].

3. Nucleosome Positioning and Mobility
3.1. Nucleosome Positioning

Nucleosome positioning refers to the fact that, when the DNA is longer than the 147 bp of the nucleosome core particle, the histone octamer can form nucleosomes at many possible locations, yet it may exhibit a preference for a subset of these possible locations. We distinguish translational and rotational positioning. Translational positioning refers to the extent to which a histone octamer selects a particular contiguous stretch of 147 bp of DNA in preference to other stretches of the same length that are translated forwards or backwards along the DNA. Rotational positioning is a degenerate form of translational positioning in which a set of discreet translational positions, differing by integral multiples of the DNA helical repeat, are all occupied in preference to the set of other possible locations. DNA sequences that are intrinsically bent or are anisotropically bendable may lead to rotationally positioned nucleosomes. Other interactions too may contribute to translational or rotational positioning (for a more detailed discussion, see *[22,23]*).

3.2. Nucleosomes Are Mobile

While histone octamers do not exchange freely between differing DNA molecules in physiological solution conditions, they have a surprising ability to translate along DNA, yielding nucleosomes in new positions. This process is often referred to as nucleosome mobility or sliding (although, as will be discussed, the latter seems certain to be a misnomer). Nucleosome mobility is readily detected using a 2-dimensional gel assay that takes advantage of the dependence of the electrophoretic mobility on the position of the octamer between the ends and the middle of a DNA fragment *(24,25)*. The rate of nucleosome mobility is enhanced at elevated temperatures and ionic strengths, but mobility can also be detected in lower temperatures and approximately physiological ionic strengths *(25–27)*. Mobility is strongly suppressed by histone H1 *(28,29)*.

The word "sliding" suggests a mechanism for nucleosome mobility in which the DNA acts as though it is simply pulled along the histone octamer, just as if one pulled on a rope that was coiled around a smooth spool. The structure of the nuclcosomc suggests that such a picture cannot be correct: the histone octamer and DNA interact through myriad bonds and with good steric fits between rough molecular surfaces *(13)*. Such interactions would not allow smooth sliding; sliding intermediates should have prohibitively high energies. Imagine the histone octamer instead as a helical gear, with teeth that mesh with DNA basepairs. Sliding would then correspond to a rotation of the helical gear, each step releasing a gear tooth at one end and engaging a new tooth at the other end. This idea seems unlikely because strong bonds that are lost at one end with each gear-step are replaced with inequivalent or nonexistent bonds at the other end.

An alternative and more plausible model for the mechanism of nucleosome mobility, based on the dynamic property of "site exposure" is discussed in **Subheading 4.**

3.3. Nucleosome Positioning Is Statistical, Not "Precise," and Is Linked to the Free Energy of DNA Binding

Because nucleosomes are mobile in physiological conditions, the positions of histone octamer on the DNA reflect true equilibrium positions. This fact has several important consequences. It implies that nucleosome positioning is not "precise" as often stated, but rather is a statistical property, governed by the laws of chemical equilibrium. Thus, observations of apparent precise positioning actually reflect preferential occupancy of one position together with a general insensitivity of mapping methods to lower levels of occupancy at the set of all other positions. More careful recent studies of positioning in vitro *(30)* reveal occupancy of numerous translational positions that are not related by the DNA helical twist, although even these studies cannot quantify the nonzero occupancies that must exist at all possible positions.

Since positioning is an equilibrium property, there exists a particular mathematical relationship between the free energy of histone-DNA interactions measured in competitive nucleosome reconstitution experiments, and the time- or ensemble-averaged probability of occupancy of the preferred site *(31)*. We consider the simplest case of translational positioning, in which a histone octamer bound to a DNA fragment of length L is transferred from unrestricted exploration of any of the L-147 available nonspecific positions (assumed equivalent for the simplest analysis) to fixed occupancy at a single specific location. We define ΔG_{net} to be the corresponding free energy change and $\Delta\Delta G_{HO}$ as the difference free energy for reconstitution of histone octamer into nucleosomes for two DNA molecules having identical lengths but where one contains a single favored ("specific") position and the other contains no such sites. ($\Delta\Delta G_{HO}$ is thus readily measured by experiment). Then the probability of occupancy of the "specific position" is given by *(31)*:

$$p = (e^{-\Delta Gnet/RT})/(1 + e^{-\Delta Gnet/RT}) = 1/[1 + (L - 147)e^{-\Delta\Delta G_{HO}/RT}] \quad [1]$$

where R is the gas constant and T is the absolute temperature. Related formulas can be derived for rotational positioning, and for cases in which there are multiple distinct translational positions, and so forth.

Since free energies are finite, positioning in vivo will be statistical too, not precise, even though additional energies may contribute to establishing the positional biases *(32)*. It is important to recognize this statistical property of positioning because it has substantial ramifications for mechanisms of gene regulation. When positioning is not precise, essential DNA sequences will sometimes be buried when they need to be accessible or may be accessible when they need to be repressed (buried). Mechanisms proposed for gene regulation must be robust with respect to statistical fluctuations in nucleosome positioning, which are inevitable when free energies are finite.

4. Dynamic Site Exposure

4.1. Nucleosomes Transiently Expose Buried DNA Sites

Recent studies reveal that nucleosomes have a remarkable dynamic property, referred to as "site exposure" *(33–35):* DNA wrapped around the histone octamer is in a constant dynamic equilibrium in which stretches of DNA (less than the entire DNA length) are transiently exposed off the surface of the octamer, allowing other proteins access to buried DNA target sites (**Fig. 1**). This property was initially recognized and quantified in studies of the abilities of restriction endonucleases to cleave DNA at sites throughout the nucleosome *(33)*. Nucleosomes were constructed with DNA containing *patches* of sites, such that sites for multiple restriction enzymes were nested together within roughly one helical turn of DNA. The rate at which buried nucleosomal sites are cleaved by the restriction enzymes, scaled by the rate of cleavage of naked DNA in identical conditions, gives the fraction of time that the nucleosomal DNA "looks like" naked DNA, that is, the equilibrium constant for site exposure, K_{eq}^{conf}. Measuring the apparent K_{eq}^{conf} for the different sites within one patch allows one to discriminate between true site exposure and inadvertent construction of a site that faces "out." Families of constructs were generated that allowed measurement of K_{eq}^{conf} for patches positioned throughout the nucleosome. The DNA sequence was chosen that strongly biased the location of the histone octamer to one specific position, and the positioning was further biased by keeping the DNA length very short so that there would be few fully-wrapped alternative positions. Key control experiments included mapping and assays for the homogeneity of nucleosome positioning; assays for the integrity of the nucleosomes throughout the digestions; and tests for possible direct roles of the restriction enzymes in facilitating site exposure.

4.2. Site Exposure Is Nondissociative and May Occur in a Stepwise Fashion

Several lines of evidence establish that site exposure is nondissociative in nature—that is, buried sites are made transiently accessible without full dissociation of the DNA from the histone octamer. Measured equilibrium constants for site exposure (K_{eq}^{conf}) decrease more-or-less progressively as one moves inward from an end into the middle of the nucleosome, from approx $2-4 \times 10^{-2}$ for sites just inside from an end, to approx $10^{-4}-10^{-5}$ for sites located on or near the nucleosomal dyad. This progressive decrease would not be observed if complete DNA dissociation were required prior to allowing protein access. In a direct binding assay such as illustrated in **Fig. 1** (and *see* **Subheading 4.5.**) histones remain present in a ternary complex comprising histones, DNA, and a bound protein "R" *(36,37)*. And in a restriction enzyme cleavage assay probing sites only a short distance inside the nucleosome, *(33)*, nucleosomes remain intact even after their DNA has been cleaved.

The relatively progressive decrease in K_{eq}^{conf} as one moves inward from an end into the middle of the nucleosome, the fact that binding of multiple pro-

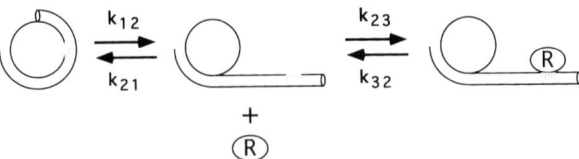

Fig. 1. The site exposure model illustrated for a single nucleosome. The histone octamer is shown from above as a disk with the DNA coiling around it. A particular DNA target sequence (*stippled*) is inaccessible to the regulatory protein (R) that acts on it. k_{12} and k_{21} are position-dependent apparent rate constants for site exposure and recapture, respectively; the equilibrium constant for site exposure, K_{eq}^{conf}, $= k_{12} / k_{21}$. Exposure of sites nearer the middle of the nucleosomal DNA may occur by several successive steps of exposure of shorter segments from an end as illustrated; each smaller step would have its own microscopic rate constants. k_{23} and k_{32} are microscopic rate constants for binding and dissociation of R from its target site and pertain to naked DNA as well as to the exposed state of nucleosomes. Real nucleosomes exist in long chains; but this need not prevent uncoiling. Experiments show that access to buried sites is affected little or not at all when a test nucleosome is embedded in the middle of a long chain of nucleosomes *(39,40)*.

teins to nucleosomal target sites occurs cooperatively (*see* **Subheading 4.6.**), and the structure of the nucleosome itself, are all consistent with a picture in which DNA spontaneously and transiently uncoils inward starting from an end. The structure *(13)* shows the DNA wrapped on the histone surface as making contacts ("bonds") in a small patch, every approx 10bp, each time the phosphodiester backbone (minor groove) faces inward toward the octamer. Thus, uncoiling would naturally proceed stepwise, with an incremental increase in energetic cost (i.e., decreased equilibrium constant for site exposure) associated with each additional 10 bp-long segment uncoiled *(38)*.

4.3. Site Exposure is a Rapid Process

Experiments using coupled enzymatic assays to detect site exposure reveal that site exposure sufficient to allow access to all of the DNA in a nucleosome occurs on a time scale of seconds or faster *(35)*, and possibly *much* faster. (In these studies, as in the earlier studies using restriction enzymes as probes of accessibility (site exposure), the enzymes serve as neutral probes: they do *not* drive the processes that they are being used to detect *[33,35]*). Simple theoretical models for site exposure and recapture processes— assuming either an activated or a diffusive process for recapture—lead to the expectation that $k_{21} \geq 10^5$ s^{-1} (J. Widom, unpublished), in which case the forward process of site exposure (k_{12}) may occur on the millisecond timescale or faster.

4.4. Site Exposure Is Facile Even in the Middle of Long Chains of Nucleosomes

Real nucleosomes exist in long chains; but this need not interfere significantly with site exposure processes. With just modest deformation of the linker DNA, a combined uncoiling coupled to a motion of the uncoiled DNA in a direction parallel to the axis of the nucleosomal disk allows uncoiling beyond the dyad (which is as far as necessary to allow binding anywhere) with no required crossings and with little motion of other nucleosomes. Experiments show that access to buried sites is affected little or not at all when a test nucleosome is embedded in the middle of a long chain of nucleosomes *(39,40)*. Higher levels of chromosome structure may need to be disassembled prior to the site exposure process illustrated here, but they are also believed to possess only marginal stability *(23)*.

4.5. Site Exposure Accounts Semiquantitatively for the Binding of Proteins to Nucleosomal Target Sites

We make the simplifying assumption that site exposure leads to exposure of sufficient nucleosomal DNA, such that the rates and equilibria for binding to an exposed nucleosomal target sequence or to a naked DNA target sequence are identical. Thus

$$N \underset{k_{21}}{\overset{k_{12}}{\rightleftarrows}} S+R \underset{k_{32}}{\overset{k_{23}}{\rightleftarrows}} RS$$

and

$$S + R \underset{k_{32}}{\overset{k_{23}}{\rightleftarrows}} RS \qquad [2]$$

for nucleosomes and naked DNA, respectively, where R is an arbitrary DNA-binding protein, N is a nucleosome, and S is the nucleosome after site exposure or, equivalently, naked DNA.

With this assumption, binding of a regulatory protein to a nucleosomal target sequence will occur with a net free energy change $\Delta G°_{net}$ and an apparent dissociation constant $K_d^{apparent}$ given by:

$$\Delta G°_{net} = \Delta G°_{conf} + \Delta G°_{naked\ DNA}$$

and

$$K_d^{apparent} = K_d^{naked\ DNA} / K_{eq}^{conf} \qquad [3]$$

where K_{eq}^{conf} is the equilibrium constant for site exposure in the nucleosome (k_{12}/k_{21}), $K_d^{naked\ DNA}$ is the dissociation constant for binding to naked DNA (k_{32}/k_{23}), and $\Delta G°_{conf}$ and $\Delta G°_{naked\ DNA}$ are the corresponding free energies.

In this model, K_{eq}^{conf} depends primarily on the translational position of the target sequence within the nucleosome (ΔG_{conf} depends on the length of DNA being uncoiled—that is, on the length of histone-DNA interface that is disrupted). However, the effective K_{eq}^{conf} will also depend on other factors such as the size and shape of the regulatory protein, the rotational setting of the target site around the periphery of the DNA helix, and on DNA bending induced by the protein, since these affect the amount of DNA that must be uncoiled to allow the protein to bind. Because the preceeding **Eq. 3** ignores system-specific details such as these, we expect semiquantitative rather than perfect agreement between the predictions of **Eq. 3** and measured dissociation constants.

The successes of this model are evident *(33,34)*. When experiments reveal that proteins can in fact bind to nucleosomal target sites, this model explains how they got there, and it explains with remarkable accuracy why the observed $K_d^{apparent}$ takes on its particular value, which previously seemed to vary randomly over a 10^5-fold range. The model also clarifies a number of very mysterious observations, in which certain proteins apparently could *not* bind to nucleosomal target sites (including especially remarkable cases where different but very close relatives of those same proteins *could* bind). In these cases, the model reveals that investigators did not carry the binding titration experiments up to the concentrations (approx $K_d^{apparent}$) where binding would have been detected. Binding equilibria are controlled by concentrations of free species, and not, e.g., by ratios of total components present.

4.6. Site Exposure Makes Binding of Multiple Proteins to Real Nucleosomes a Cooperative Process

The site exposure model has within it the potential for important novel cooperative (synergistic) interactions between multiple proteins binding simultaneously to sites within a single nucleosome (**Fig. 2**; *[34]*). This arises from the possibility that, once one protein Y has bound, the binding of a second protein X may take place without having to pay the full energetic penalty for site exposure (here defined as ΔG^0_1), which otherwise would be required. Similarly, the ability of X to bind facilitates the subsequent binding of Y, since at least some of the final free energy penalty for the required conformational change is already paid. These processes are linked in a thermodynamic cycle, hence the same coupling free energy and corresponding binding cooperativity necessarily applies regardless of the order in which the proteins bind. Thus X and Y act cooperatively (synergistically) even if they do not touch: The binding of one protein radically alters the binding ability of the other. No special properties are required of X or Y: they need only bind DNA for this cooperativity to be manifested. X and Y may be two different proteins, or they may be two molecules of the same protein.

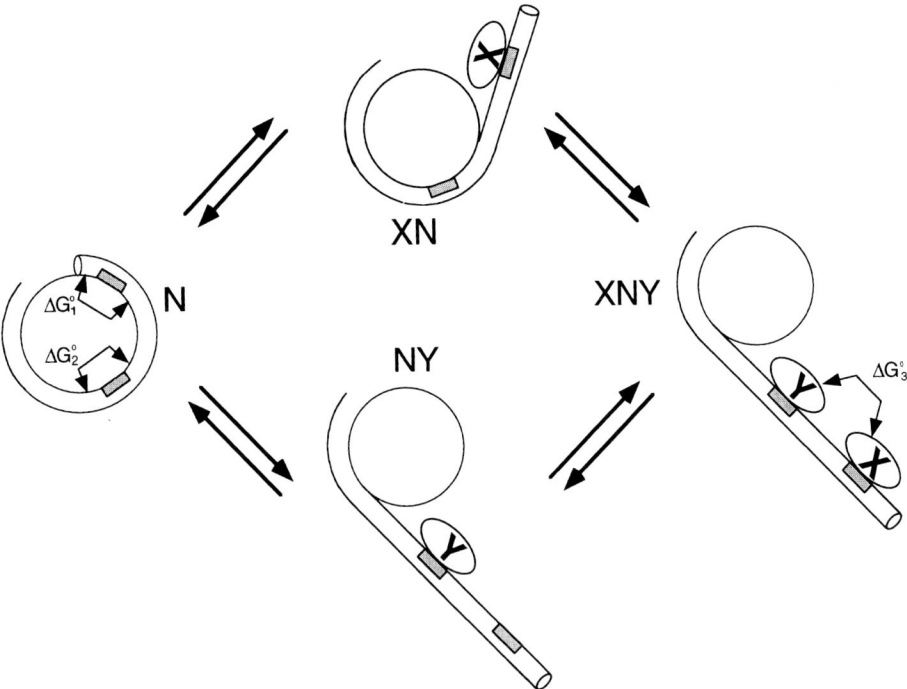

Fig. 2. Cooperativity in the binding of multiple proteins to target sites in a single nucleosome. A nucleosome is shown containing binding sites (*stippled*) for two proteins (X) and (Y). X and Y may be two unrelated proteins or two molecules of the same protein. X is defined as the protein binding to the outermost site, and Y as binding to the innermost site. ΔG_1 is the free energy cost for uncoiling enough DNA so as to expose the site for X. ΔG_2 is the additional free energy cost for uncoiling sufficient additional DNA so as to expose the site for Y. The cooperativity free energy $\delta G_{XY} = -\Delta G_1$. In some cases, X and Y may have "conventional" cooperative interactions, also detectable in their binding to naked DNA (e.g., from favorable protein-protein contacts between X and Y); these are collectively represented as ΔG_3 (which is negative for favorable interactions). The total observed cooperativity will be the sum of $\delta G_{XY} + \Delta G_3$.

The amount of cooperativity between X and Y (the coupling free energy δG_{XY}—the free energy by which the prior binding of X facilitates the binding of Y, or vice versa), is equal to $-\Delta G°_1$—that is, minus one times the energetic cost for exposing the outer more site. In real systems, X and Y may also have direct "conventional" cooperative interactions, represented by $\Delta G°_3$. Then the net cooperativity will be given by the sum of $\delta G_{XY} + \Delta G°_3$. The novel cooperativity free energies inherent in this mechanism (δG_{XY}) are substantially greater than these free energies of previously recognized "conventional" cooperative interactions *(34)*.

Remarkably, this model, *with no adjustable parameters*, accounts quantitatively for a diverse set of experimental results on cooperative binding of various proteins to nucleosomal target sites obtained by another laboratory *(41)*. Moreover, there is good agreement between the predictions of this model and experimental data even using ΔG_{conf} obtained from the restriction enzyme digestion kinetic measurements instead of the δG_{XY} obtained directly from the primary cooperative binding data. These two experiments are completely unrelated except through the site exposure model, and therefore the agreement between the two provides strong evidence for the applicability of the site exposure model to the behavior of real nucleosomes.

An important question for further investigation is whether these cooperative interactions persist across the nucleosomal dyad. The model could allow us to predict the outcome of this experiment only if we had available complete (and reliable) data for the position-dependence of K_{eq}^{conf} (i.e., the detailed pattern of "bonding" energies of all of the histone-DNA contact sites) for sites throughout the nucleosome. At present, sufficiently complete data are not available.

4.7. Site Exposure Can Account for Nucleosome Mobility and for Nucleosome Transcription

The site exposure model provides a physically plausible mechanism for nucleosome mobility, in which a segment of DNA released from the histone surface can loop back and be recaptured at a new position (most likely displaced by a multiple of the DNA helical repeat), creating a nucleosome with a "bulged" loop of DNA. Analogous structures have been postulated as intermediates during nucleosome transcription *(14)*. The bulge could propagate around the nucleosome in a relatively low-energy process; when it runs off the other end, the nucleosome would be found to have moved. The rate of site exposure processes *(35)* is fast compared to rates of spontaneous nucleosome mobility *(24,27)*, suggesting that site exposure could in principle account for nucleosome mobility, and, conversely, that nucleosome mobility itself is unlikely to be the actual mechanism of site exposure.

The site-exposure model also provides a physically plausible mechanism for the progression of RNA or DNA polymerase through a nucleosomal template *(35)*. Successive repetition of two steps—partial uncoiling, followed by elongation up to the next point of steric hindrance—allows full-length elongation. The number of such steps required for elongation through a complete nucleosome in this model depends on the length of DNA released in each step. Site exposure occurs nondissociatively, so the entire DNA chain is not released in just one step. Most likely, as suggested by the structure of the nucleosome, DNA may be exposed in successive approx 10 bp-long steps. Similar suggestions for the mechanism of polymerase progression have been made previously *(42,43)*.

References

1. van Holde, K. E. (1989) *Chromatin*. Springer-Verlag, New York.
2. Manning, G. S. (1978) The molecular theory of polyelectrolyte solutions with applications to the electrostatic properties of polynucleotides. **11,** 179–246.
3. Hagerman, P. J. (1988) Flexibility of DNA. *Ann. Rev. Biophys. Biophys. Chem.* **17,** 265–286.
4. Yager, T. D., McMurray, C. T., and van Holde, K. E. (1989) Salt-induced release of DNA from nucleosome core particles. *Biochemistry* **28,** 2271–2281.
5. Godde, J. S. and Wolffe, A. P. (1995) Disruption of reconstituted nucleosomes: the effect of particle concentration, $MgCl_2$ and KCl concentration, the histone tails and temperature. *J. Biol. Chem.* **270,** 27,399–27,402.
6. Royer, C. A., Ropp, T., and Scarlata, S. F. (1992) Solution studies of the interactions between the histone core proteins and DNA using fluorescence spectroscopy. *Biophys. Chem.* **43,** 197–211.
7. Voordouw, G. and Eisenberg, H. (1978) Binding of additional histones to chromatin core particles. *Nature* **273,** 446–448.
8. Gallego, F., Fernandez-Busquets, X., and Daban, J. R. (1995) Mechanism of nucleosome dissociation produced by transcription elongation in a short chromatin template. *Biochemistry* **34,** 6711–6719.
9. Daban, J.-R. and Cantor, C. R. (1982) Structural and kinetic study of the self-assembly of nucleosome core particles. *J. Mol. Biol.* **156,** 749–769.
10. Daban, J.-R. and Cantor, C. R. (1982) Role of histone pairs H2A, H2B and H3, H4 in the self-assembly of nucleosome core particles. *J. Mol. Biol.* **156,** 771–789.
11. Caron, F. and Thomas, J. O. (1981) Exchange of histone H1 between segments of chromatin. *J. Mol. Biol.* **146,** 513–537.
12. Louters, L. and Chalkley, R. (1984) In vitro exchange of nucleosomal histones H2a and H2b. *Biochem.* **23,** 547–552.
13. Luger, K., Mader, A. W., Richmond, R. K., Sargent, D. F., and Richmond, T. J. (1997) Structure of the nucleosome core particle at 2.8Å pesolution. *Nature* **389,** 251–260.
14. Studitsky, V. M., Clark, D. J., and Felsenfeld, G. (1995) Overcoming a nucleosomal barrier to transcription. *Cell* **83,** 19–27.
15. Scarlata, S. F., Ropp, T., and Royer, C. A. (1989) Histone subunit interactions as investigated by high pressure. *Biochemistry* **28,** 6637–6641.
16. Royer, C. A., Rusch, R. M., and Scarlata, S. F. (1989) Salt effects on histone subunit interactions as studied by fluorescence spectroscopy. *Biochemistry* **28,** 6631–6637.
17. Karantza, V., Freire, E., and Moudrianakis, E. N. (1996) Thermodynamic studies of the core histones: pH and ionic strength effects on the stability of the $(H3-H4)_2$ system. *Biochemistry* **35,** 2037–2046.
18. Arents, G., Burlingame, R. W., Wang, B.-C., Love, W. E., and Moudrianakis, E. N. (1991) The nucleosomal core histone octamer at 3.1Å resolution: a tripartite protein assembly and a left-handed superhelix. *PNAS (USA)* **88,** 10,148–10,152.
19. Shrader, T. E. and Crothers, D. M. (1989) Artificial nucleosome positioning sequences. *Proc. Natl. Acad. Sci. USA* **86,** 7418–7422.

20. Shrader, T. E. and Crothers, D. M. (1990) Effects of DNA sequence and histone-histone interactions on nucleosome placement. *J. Mol. Biol.* **216,** 69–84.
21. Drew, H. R. (1991) Can one measure the free energy of binding of the histone octamer to different DNA sequences by salt-dependent reconstitution? *J. Mol. Biol.* **219,** 391–392.
22. Lowary, P. T. and Widom, J. (1998) New DNA sequence rules for high affinity binding to histone octamer and sequence-directed nucleosome positioning. *J. Mol. Biol.* **276,** 19–42.
23. Widom, J. (1998) Structure, dynamics, and function of chromatin *In Vitro*. *Annu. Rev. Biophys. Biomol. Struc.* **27,** 285–327.
24. Pennings, S., Meersseman, G., and Bradbury, M.E. (1991) Mobility of positioned nucleosomes on 5S rDNA. *J. Mol. Biol.* **220,** 101–110.
25. Meersseman, G., Pennings, S., and Bradbury, E. M. (1992) Mobile nucleosomes: a general behavior. *EMBO J.* **11,** 2951–2959.
26. Varga-Weisz, P. D., Blank, T. A., and Becker, P. B. (1995) Energy-dependent chromatin assembly and nucleosome mobility in a cell-free system. *EMBO J.* **14,** 2209–2216.
27. Ura, K., Hayes, J. J., and Wolffe, A. P. (1995) A positive role for nucleosome mobility in the transcriptional activity of chromatin templates: restriction by linker histones. *EMBO J.* **14,** 3752–3765.
28. Pennings, S., Meersseman, G., and Bradbury, E. M. (1994) Linker histones H1 and H5 prevent the mobility of positioned nucleosomes. *Proc. Natl. Acad. Sci. USA* **91,** 10,275–10,279.
29. Pazin, M. J., Bhargava, P., Geiduschek, E. P., and Kadonaga, J. T. (1997) Nucleosome mobility and the maintenance of nucleosome positioning. *Science* **276,** 809–812.
30. Fragoso, G., John, S., Roberts, M. S., and Hager, G. L. (1995) Nucleosome positioning on the MMTV LTR results from the frequency-biased occupancy of multiple frames. *Genes Dev.* **9,** 1933–1947.
31. Lowary, P. T. and Widom, J. (1997) Nucleosome packaging and nucleosome Popitioning of genomic DNA. *Proc. Natl. Acad. Sci. USA* **94,** 1183–1188.
32. Yao, J., Lowary, P. T., and Widom, J. (1993) Twist constraints on linker DNA in the 30nm chromatin fiber: Implications for nucleosome phasing. *Proc. Natl. Acad. Sci. USA* **90,** 9364–9368.
33. Polach, K. J. and Widom, J. (1995) Mechanism of protein access to specific DNA sequences in chromatin: a dynamic equilibrium model for gene regulation. *J. Mol. Biol.* **254,** 130–149.
34. Polach, K. J. and Widom, J. (1996) A model for the cooperative binding of eukaryotic regulatory proteins to nucleosomal target sites. *J. Mol. Biol.* **258,** 800–812.
35. Protacio, R. U., Polach, K. J., and Widom, J. (1998) Coupled enzymatic assays for the rate and mechanism of DNA site-exposure in a nucleosome. *J. Mol. Biol.* **274,** 708–721.
36. Workman, J. L. and Kingston, R. E. (1992) Nucleosome core displacement in vitro via a metastable transcription factor-nucleosome vomplex. *Science* **258,** 1780–1784.

37. Stager, D. J. and Workman, J. L. (1997) Stable co-occupancy of transcription factors and histones at the HIV-1 enhancer. *EMBO J.* **16,** 2463–2472.
38. Widom, J. (1997) Chromatin: the Nucleosome Unwrapped. *Curr. Biol.* **7,** R653–655.
39. Owen-Hughes, T., Utley, R. T., Cote, J., Peterson, C. L., and Workman, J. L. (1996) Persistent site-specific remodeling of a nucleosome array by transient action of the SWI/SNF complex. *Science* **273,** 513–516.
40. Owen-Hughes, T. and Workman, J. L. (1996) Remodeling the chromatin structure of a nucleosome array by transcription factor-targeted trans-displacement of histones. *EMBO J* **15,** 4702–4712.
41. Adams, C. C. and Workman, J. L. (1995) Binding of disparate transcriptional activators to nucleosomal DNA is inherently cooperative. *Mol. Cell. Biol.* **15,** 1405–1421.
42. Kornberg, R. D. and Lorch, Y. (1992) Chromatin structure and transcription. *Annu. Rev. Cell Biol.* **8,** 563–587.
43. Kornberg, R. D. and Lorch, Y. (1995) Interplay between chromatin structure and transcription. *Curr. Opinion Cell Biol.* **7,** 371–375.

6

Nucleosome Structure and Dynamics

The DNA Minicircle Approach

Ariel Prunell, Mohamed Alilat, and Filomena De Lucia

1. Introduction

DNA supercoiling in eukaryotic viruses was first described by Vinograd and co-workers more than 30 years ago (*see* **ref.** *1* for a review), but its origin was recognized to result from DNA complexation with host histones to form minichromosomes only 10 years later *(2)*. Because they are in contact with endogenous topoisomerases, minichromosomes are relaxed, that is, the linker DNAs that connect nucleosomes to one-another along the minichromosomes are free from torsional constraint. DNA supercoiling therefore results from nucleosome removal during protein extraction. Negative supercoiling of SV40 DNA was found to correspond to a reduction in the DNA linking number, Lk, of approx 26 *(3)*, and the number of nucleosomes in the minichromosomes was approximately the same *(4,5)*. Thus, each nucleosome reduces Lk by one (*see* **Subheading 3.1.**, for Lk definition).

Circularization of DNA fragments into minicircles *(6)* has provided a unique way to investigate the energetics of DNA twisting *(6–8)*. The original purpose in using minicircles instead of plasmid-sized DNAs for chromatin reconstitution *(9)*, was to explore an alternative approach to the so-called chromatin "linking number paradox" *(10)*. This paradox arose from the contradiction between the unit linking number reduction per nucleosome ($\Delta Lk = -1$), and the near 2 turns of a left-handed DNA superhelix wrapped around it *(11,12)*. Such path, as common sense suggested from the two DNA negative crossings which were assumed to be generated, should have instead led to $\Delta Lk = -2$. DNA minicircles, in restricting reconstitution to a single nucleosome, had the advantage over the minichromosome system that DNA supercoiling could now be correlated, through visualization by high-resolution electron microscopy, with the exact path of the DNA around the histone octamer and in the external loop. An unexpected outcome has emerged from the dual study of mononucleosomes on both DNA minicircles and small, linear DNA fragments (linear nucleosomes are not the subject of this chapter). Although these studies confirmed the near two-turn wrapping around the histones, entering and exiting DNAs were found to bend away from each other and the histone surface before they could cross *(13,14)*. This left one negative crossing associated with $\Delta Lk = -1$ *(15)*, and therefore no paradox. In the presence of linker histones, which seal the second turn of the DNA, entering and exiting DNAs were observed to bridge together into a stem, apparently without interwinding, so that the second crossing may not form either *(14)*.

Recently, DNA minicircles were used to reconstitute particles with the histone $(H3-H4)_2$ tetramer, the central part of the histone $(H2A, H2B, H3, H4)_2$ octamer of the nucleosome *(16)*. It was found that particle reconstitution increased with both negative and positive supercoiling of the minicircles. Increase in reconstitution with negative supercoiling is also observed with the octamer in assembling the nucleosome and is known to simply reflect the relief of the free energy of supercoiling upon wrapping into a left-handed superhelix *(9)*. Increase in reconstitution with positive supercoiling, in contrast, was unique to the tetramer and suggested a major structural transition of the particle. This, together with the occurrence of positively supercoiled topoisomers in the particle relaxation equilibria *(16)*, led to a model proposing that a local deformation of the tetramer in the dyad region could result into a rotation of the two crescent-shaped H3-H4 dimers around their H3/H3 interface in the counter-clockwise direction. This rotation transformed the left-handed horseshoe-shaped tetramer into a right-handed superhelical structure. The loss of affinity of the resulting right-handed tetramer for H2A-H2B dimers further suggested that this conformational transition mediated their release during transcription in vivo, presumably as a consequence of the positive supercoiling generated in front of the polymerase *(17)*. When the tetramer returns to its normal left-handed conformation in the wake of the polymerase, H2A-H2B dimers could be recaptured and nucleosomes regenerated *(18)*. In addition to transcriptional elongation, tetramer conformational flexibility may also be involved in chromatin replication, and more generally in processes involving a dynamics of the nucleosome in vivo *(16)*.

This chapter describes the production and purification of DNA minicircles of different supercoiling (termed topological isomers or topoisomers), both at the analytical and preparative levels, and their reconstitution with histone tetramers and octamers. Resulting monomer particles (M_T and M_O, respectively) are visualized by electron microscopy and further characterized by gel electrophoresis. They are relaxed with topoisomerase I, and the data are exploited in terms of DNA supercoiling around them. **Subheading 3.1.** is a reminder of DNA topology basic notions useful to the present analysis.

2. Materials
2.1. Enzymes

1. Alkaline phosphatase from bovine intestine (suspension in $(NH4)_2SO_4$) (Sigma, St. Louis, MO, cat. no. P 5521; or Boehringer Mannheim, Mannheim, Germany, cat. no. 108 138). Centrifuge the suspension for 5 min at 10,000g in the cold, dissolve pellet in water at 0.5 U/µL, and store in aliquots at –20°C. Keep aliquot in current use at 4°C.
2. 10 U/µL T4 polynucleotide kinase (NEB, Beverly, MA, cat. no. 201S).
3. 5 U/µL T4 DNA ligase (Appligene-Oncor, Illkirch, France, cat. no. 120031).
4. 10–12 U/µL calf thymus DNA topoisomerase I (Gibco-BRL, Life Technologies Inc., Gaithersburg, MD, cat. no. 38042-024).
5. Enzymes are used under manufacturer's recommendations, except T4 DNA ligase (*see* **Subheading 3.3.1.**) and DNA topoisomerase I (*see* **Subheading 3.6.**).

2.2. Buffers

1. 50X Tris-acetate buffer: 2 M Tris-acetate, pH 7.8, 1 M sodium acetate, 0.1 M EDTA.
2. 20X TBE buffer: 1 M Tris-base, 1 M boric acid, 20 mM EDTA; pH should be around 8.3.
3. 10X DNA circularization buffer. 0.1 M Tris-HCl, pH 7.5, 0.5 M NaCl, 0.1 M MgCl$_2$, 50 mM dithiothreitol (DTT).
4. 100X TE buffer: 1 M Tris-HCl, pH 7.5, 0.1 M EDTA.
5. 100X TAE buffer: 0.67 M Tris-acetate, pH 7.5, 0.33 M sodium acetate, 0.1 M EDTA.
6. 5X chromatin relaxation buffer: 0.25 M Tris-HCl, pH 7.5, 0.25 M KCl, 25 mM MgCl$_2$, 2.5 mM dithiothreitol.

2.3. Other Materials

1. Histone octamers and tetramers were prepared by hydroxylapatite chromatography *(19,20)* of long soluble chromatin released from duck or chicken erythrocyte nuclei by micrococcal nuclease digestion *(9)*. Pooled fractions were dialyzed against 10 mM Tris-HCl, pH 7.5, 5 mM 2-mercaptoethanol, 0.1–0.25 mM phenylmethylsulfonyl fluoride, and 2 M NaCl, and when necessary, concentrated in Centriplus 30 concentrators (Amicon, Beverly, MA). Aliquots were stored at –80°C until use. Octamer and tetramer concentrations, 1–2 mg/mL, were estimated using A_{230} = 2.75 for a 1 mg/mL solution *(21)* (normal A_{230}/A_{280} ratio approx 7).
2. Acrylamide/bis-acrylamide: 19:1 and 29:1 solutions at 40% (ICN Biochemicals, Aurora, OH, cat. no. 800802 and 800803).
3. [γ-^{32}P] ATP (3000 Ci/mmol, 10 mCi/mL, Amersham, Amersham, Buckinghamshire, UK, cat. no. PB 10168).

4. ATP solution at 10 mM. Store frozen.
5. BSA: albumin bovine, powder (Sigma; cat. no. A 7638). Dissolve at 1 mg/mL in 1X TE buffer. Store at 4°C.
6. Ethidium bromide: powder (Sigma; cat no. E 7637). Dissolve in water at 1 mg/mL, and store frozen in aliquots. Keep in the dark with tube wrapped in aluminum foil.
7. Bromo-phenol Blue and Xylene Cyanole: powders (Bio-Rad, Richmond, CA, cat. no. 161-0404 and 161-0423). Dissolve in water at 1%. When stored at 4°C, part of the bromo-phenol blue precipitates, so incubate at 37°C to redissolve precipitate before use.
8. Netropsin: powder (Merck, Darmstadt, Germany, cat no. 124 791). Dissolve in water at 5 mM, and store frozen in aliquots.
9. Chloroquine: powder (Sigma; cat. no. C 6628). Dissolve in water at 100 mg/mL, and store at 4°C.
10. Phenol/chloroform/isoamyl alcohol 25:24:1.
11. Chloroform/isoamyl alcohol 24:1.
12. Chromatography paper: 3MM Chr paper (Whatman, Maidstone, UK, cat. no. 3030 917).
13. Nensorb 20 Nucleic acid purification cartridges (DuPont NEN, Bad Homburg, Germany, cat. no. NLP 022X).

3. Methods

3.1. Linking Number Difference Associated with Particle Formation

The topoisomer linking number, Lk, is defined by the classical equation *(22–24)*

$$Lk = Tw + Wr \qquad (1)$$

Lk measures the number of times one DNA strand goes around the other when the molecule is unfolded and put flat on a surface. Wr, the topoisomer writhing number, is a function of the 3-D curve followed by the axis of the double helix, and Tw, the twisting number, measures strand rotation about that axis. While Lk is a topological parameter and is always integral, Tw and Wr are geometrical parameters and are generally fractional. In other words, deformations of the DNA that keep phosphodiester bonds intact leave Lk invariant, while Tw and Wr may vary within the constraint of **Eq. 1**.

The linking number difference of the topoisomer, that is its supercoiling, is measured relative to the most probable configuration of the DNA minicircle, and is given by *(8)*

$$\Delta Lk = Lk - Lk_o \qquad (2)$$

Lk_o can be obtained upon relaxation with topoisomerase I, from the mean Lk value of the topoisomer equilibrium population (**Fig. 1**). Lk_o is usually fractional, and so is ΔLk. Alternatively, Lk_o is calculated from the equation *(25)*

$$Lk_o = N / h_o \qquad (3)$$

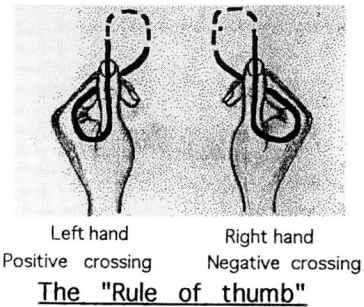
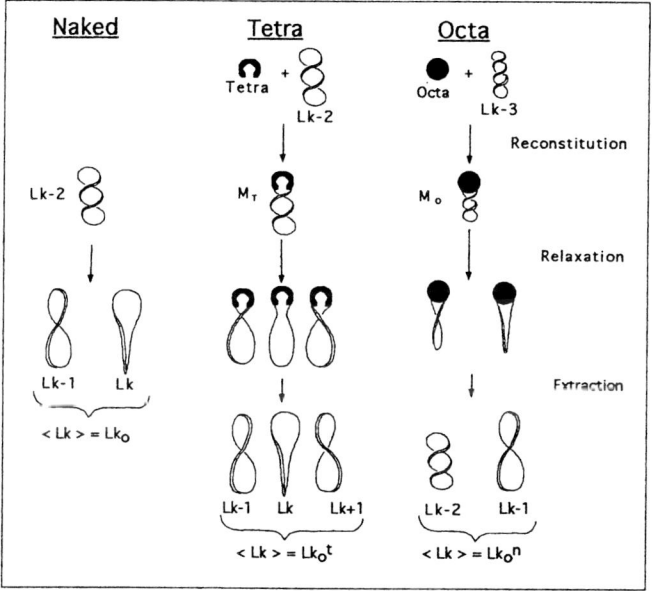

Fig. 1. The "rule of thumb." DNA linking number change (ΔLk) associated with particle formation. The rule offers a convenient way to assess the supercoiling polarity of a crossing, i.e., DNA supercoiling polarity. Naked 350–360 bp minicircles are relatively rigid, and relax into a single topoisomer when their size is close to exact multiples of the helical repeat, or two if their size is in between (*see* **Subheadings 3.3.**). [It is interesting to compare this result with the 5–6 topoisomers obtained upon relaxation of a 4.3 kbp plasmid such as pBR322.] When reconstituted with an octamer (M_O), the same minicircles appear more flexible and relax into two topoisomers regardless of the exact size (*see* example in **Fig. 7B**). Flexibility still increases for tetramer particles (M_T), which always relax into three topoisomers (*see* example in **Fig. 7A**). ΔLk can be calculated from the mean <Lk> values as indicated in **Subheading 3.1.**

when h_o (in bp/turn), the DNA helical repeat under relaxation conditions, is known (the authors use h_o = 10.53 bp/turn in 1X chromatin relaxation buffer at 37°C; **ref. 15**). N is the minicircle size (in bp).

ΔLk identifies the topoisomers. For this purpose, h_o in **Eq. 3** is now taken under chosen reference conditions (the authors use the conditions of topoisomer gel electrophoresis: h_o = 10.56 bp/turn in 1 X Tris-acetate buffer at 27°C; *see* **ref. 26**). A practical topoisomer identification is described below (*see* legend to **Fig. 2**).

ΔLk partitions into twist and writhe according to the differential of **Eq. 1**

$$\Delta Lk = \Delta Tw + Wr \qquad (4)$$

ΔWr = Wr since the reference state has Wr = 0. Wr in this equation governs the topoisomer gel electrophoretic mobility (*see* **Fig. 2**).

Considering now the particle, the most probable configuration of the DNA minicircle partially wrapped around it, Lk_o^p (p = n or t in **Fig. 1** for nucleosomes or tetramer particles), is similarly obtained by relaxation with topoisomerase I. The mean linking number of the equilibrium population of the topoisomers after protein extraction (**Fig. 1**) is

$$<Lk> = \Sigma (Lk_i \cdot r_i) / \Sigma r_i \qquad (5)$$

with Lk_i and r_i being the linking number and the relative amount of topoisomer i *(15)*. The linking number difference associated with formation of the particle, or the DNA supercoiling around the particle, is then

$$\Delta Lk_p = Lk_o^p - Lk_o \qquad (6)$$

3.2. DNA Fragment Purification and ^{32}P Endlabeling

1. Digest plasmid DNA (100 µg) overnight in the recommended restriction buffer supplemented with 100 µg/mL BSA, with 25 U of a restriction endonuclease yielding overhangs of at least 2 nucleotides (*see* **Note 1**). Extract and ethanol-precipitate DNA (*see* **Note 2**), redissolve in 1X TE buffer plus bromo-phenol blue and xylene cyanole, and purify the fragment of interest (approx 250–400 bp long; *see* **Note 3**) by preparative gel electrophoresis in a polyacrylamide gel in 1X Tris-acetate or TBE buffer. After staining with ethidium bromide, the proper gel slice is cut out of the gel, and the fragment recovered by electro-elution, or from the crushed gel *(27)*. Extract, ethanol-precipitate (*see* **Note 2**), and redissolve in 20–50 µL of 1X TE buffer.
2. Dephosphorylate 10 pmoles of the fragment (*see* **Note 4**) (20 pmoles of 5' ends in 100 µL final volume) by incubation with 0.5 U of bovine intestinal phosphatase for 1 h at 37°C. Extract protein immediately (*see* **Note 5**) by shaking energetically with 1 vol. of phenol/chloroform/isoamyl alcohol, remove organic phase from below, and repeat these operations once. Transfer the aqueous phase into another tube. Shake with 5 vol of chloroform/isoamyl alcohol, remove organic

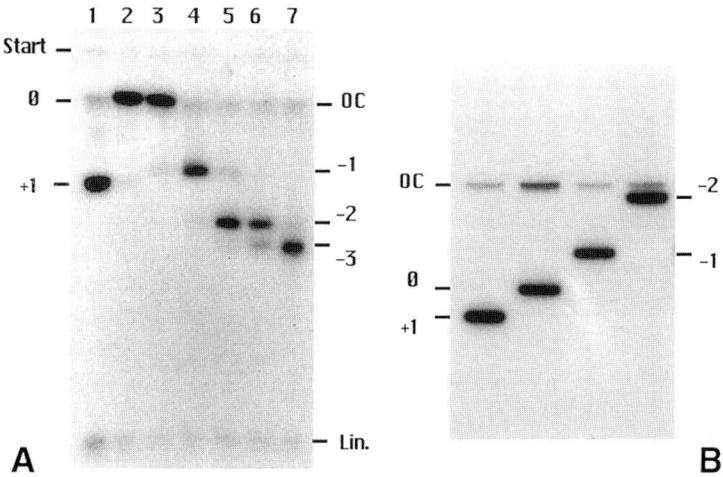

Fig. 2. DNA supercoiling-dependence of topoisomer gel electrophoretic mobility. (**A**) ^{32}P-labeled circularization products of a 359 bp Sau 3A fragment from plasmid pBR322 *(9)*, prepared in the presence of 10 μ*M* netropsin, no ligand, 0.15, 0.3, 0.6, 0.9, and 1.2 μg ethidium bromide per mL (lanes 1–7, respectively), were electrophoresed in a 5% polyacrylamide (acrylamide/bisacrylamide = 19:1, w/w) gel, at room temperature (*see* **Note 9**). 200 V (the gel is 18 cm long) were applied for approx 5 h until the xylene cyanole dye reached the bottom of the gel. Note that little linear DNA remains after the reaction. Topoisomers are identified as follows. Let us consider lane 2 (no ligand). The topoisomer in this lane has the minimal mobility and is therefore the closest to the minicircle most probable configuration under gel conditions. As a consequence, its linking number must be the closest integer to Lk_o = 359/10.56 = 33.99 (*see* **Eq. 3** in **Subheading 3.1.**), that is Lk = 34. **Eq. 2** then gives ΔLk = 34 -33.99 = 0.0 This topoisomer is referred to as topo 0. This figure happens to be an integer, but is generally fractional (*see* **Subheading 3.1.** and **Note 10**). From this, Lk and ΔLk of adjacent topoisomers are immediately assessed. In lane 1, topo +1 with Lk = 35, and in lanes 4–7, topos –1, –2, and –3 with Lk = 33, 32, and 31 (*see* **Note 11**). Note that topos +1, 0, –1 and –2 migrate proportionally to |ΔLk| (and therefore to |Wr|; *see* **Eq. 4** in **Subheading 3.1.** and **ref. 26**), except topo –3 which partially relaxes due to local departure from B-DNA structure induced by high negative supercoiling. (**B**) Purified topoisomers (*see* **Subheading 3.3.1.**) of the same fragment were electrophoresed in a gel identical to that in (**A**) except that 250 μg chloroquine per mL were added to gel and running buffer. Note that the mobility-vs-ΔLk linear relationship remains true in this gel after the ΔLk ~ + 2 introduced by chloroquine (as deduced from the virtual comigration of topo 0 and OC DNA) are taken into account: ΔLk of topos +1, 0, –1 and –2 increases to +3, +2, +1, and 0, respectively. Autoradiograms are shown. OC, open circular DNA. Lin., residual linear DNA.

phase, and repeat these operations twice. Add carrier (*see* **Note 6**), mix, and ethanol-precipitate. Redissolve dephosphorylated fragment in 10 µL of TE 1X buffer.
3. Label 2 pmoles of 5' ends in 5–10 µL final volume by incubation with 2 µL of [γ ^{32}P] ATP and 5 U of polynucleotide kinase for 1.5 h at 37°C in a 1.5-mL microfuge tube. Alternatively, the mixture-containing tube can be placed lying in the middle of the revolving plate of a microwave oven, and the oven switched on for 5 min at the lowest power output. The microwave labelling method is usually slightly less efficient (approx 80 %) than the former.
4. Dilute the kinase reaction mix in 100 µL final volume of 1X TE buffer and purify labeled DNA by filtration in a 1 mL Sephadex G50 spun column (*see* **Note 7**). A complete purification from unreacted ATP usually requires filtration of the first column eluate in a second 1-mL G50 spun column. Usual recoveries are $3–4 \times 10^6$ ^{32}P cpm ($1.5–2 \times 10^6$ ^{3}H cpm; *see* **Note 8**) (in 60–70 µL), all being acid-precipitable.

3.3. DNA Circularization and Topoisomer Purification

Owing to thermal fluctuations in twist and bend, DNA sticky ends within a single fragment may happen to encounter each other in solution, and to be in the proper relative orientation for transient base pairing to form. If a ligase molecule is there, this link may become permanent. Because DNA bending flexibility is limited, fragments approx 170 bp or smaller will not in general circularize. They will only if they are bent, either naturally or as a consequence of binding with another molecule. DNA twisting flexibility is also limited, so that fragment circularizability becomes a periodic function of their size (N) when N is small. Maxima in this function are obtained when the total twist (Tw = N/h_o; Tw of a linear fragment is equivalent to Lk_o in **Eq. 3, Subheading 3.1.**) takes integral values, in which cases the two ends have the most favorable relative orientation for linkage *(28)*. Tw can be decreased by intercalative drugs such as ethidium bromide, or increased through a nonintercalative process by netropsin *(29,30)*, an oligopeptide antibiotics. DNA supercoiling therefore arises upon removal of the ligand, when Tw goes back to its ligand-free value.

3.3.1. Small-Scale Circularization of ^{32}P-Labeled Fragments

1. Incubate $1–2 \times 10^3$ ^{3}H cpm aliquots of the labeled fragment with 1 U of T4 DNA ligase at 15°C for 2 h in 100 µL final volume of 1X DNA circularization buffer supplemented with 0.25 m*M* ATP and 100 µg BSA/mL, and containing increasing concentrations of ethidium bromide (to produce negatively supercoiled topoisomers) or netropsin (to produce positively supercoiled topoisomers). Extract and ethanol-precipitate (*see* **Note 2**) with carrier, dissolve in 10 µL of TE buffer, and submit to electrophoresis in a polyacrylamide gel in 1X Tris-acetate buffer. The circularization of a 359 bp fragment from plasmid pBR322 is shown as an example in **Fig. 2A**. Note that topo 0 (*see* the figure legend for topoisomer identification) comigrates with the open circular DNA (OC) in the higher region of the gel, as expected from its relaxed state. If required, fractionate topo 0 from

OC DNA by adding chloroquine to gel and running buffer. This induces topo 0 to positively supercoil and migrate faster than OC DNA, as shown in **Fig. 2B**. At the same time, topo +1 supercoils more, while topos –1 and –2 partially relax (*see* legend to **Fig. 2B**).

2. Once the appropriate ethidium bromide/netropsin concentrations have been found, circularization is scaled up to a final volume of 400 µL using up to 1×10^6 ^{3}H cpm and 2 U of DNA ligase for 2–3 h at 15°C. Purify topoisomers through electrophoresis as above, place the gel on chromatography paper and cover it with a plastic film. Then expose to an X-ray autoradiographic film at 4°C for 1–3 h. Prior to removing the film, note its position on the gel. For this, mark its contour on the plastic film with a marker, in which case cutting it at one corner prior to exposure helps not to confuse its sides. Alternatively, pierce the X-ray film, the gel and the chromatography paper altogether with a sharp needle at several locations distant from each other. Draw a cross on the exposed X-ray film with a marker at the position of each hole, and align it with the gel on a trans-illuminator, putting the film under the paper. Cut the gel slice out and recover the DNA by proceeding as in **Subheading 3.2.** (*see* **Note 12**).

3.3.2. Large-Scale Circularization of Unlabeled Fragments

Electron microscopy and other physico-chemical assays of the reconstituted particles require ponderable amounts of unlabeled topoisomers to be prepared. A scaling up of the above procedure to approx 10 µg of DNA fragment is unfeasible, since DNA concentration is better kept below 0.2 µg/mL to insure a slow enough intermolecular ligation (oligomerization) compared to the intramolecular ligation (circularization). An original room-temperature procedure was devised in which DNA is progressively delivered into a reaction tube which contains enough DNA ligase for the fragment to be rapidly circularized and therefore removed from the reaction. The procedure requires a peristaltic pump with tubing and a magnetic stirrer, as shown in the scheme of **Fig. 3**.

1. In tube A, mix 0.2 µg of DNA with 1 mL final volume of 1X DNA circularization buffer (containing BSA as in **Subheading 3.3.1.**), add 1 U of T4 DNA ligase/µg of total DNA to be circularized, and introduce a small magnetic bar.
2. In tube B, mix 10–40 µg of fragment with 100–400 µL final volume of 1X BSA-containing DNA circularization buffer.
3. Pump 25–100 µL/h (*see* **Note 13**) of the DNA stock solution in (B) into (A), and switch the magnetic stirrer on at low speed. Circularization is completed in 4 h, although it can be shortened to 2 h or extended overnight by modifying the flow rate accordingly, usually with similar results.
4. Extract DNA (*see* **Note 2**), electrophorese as in **Subheading 3.3.1.**, stain the gel with ethidium bromide and purify the topoisomer of interest as described for linear fragment in **Subheading 3.2.1.**
5. For some applications such as electron microscopy, purify topoisomer from soluble polyacrylamide by chromatography in Du Pont Nensorb 20 nucleic acid

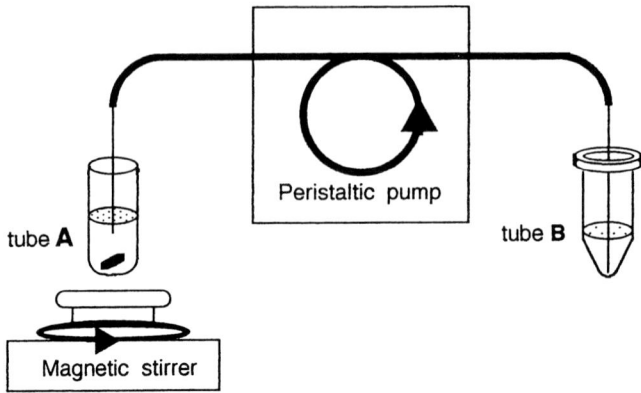

Fig. 3. Scheme of the large-scale circularization set-up. Linear DNA stock solution in tube B is delivered into tube A through glass capillaries and tygon tubing by a peristaltic pump. *See* **Subheading 3.3.2.** for details.

purification cartridges, as specified by the manufacturer. Ethanol-precipitate DNA without carrier, and redissolve in 1X TE buffer. Note that this soluble polyacrylamide does not interfere with nucleosome reconstitution, nor with "chromatin" gel electrophoresis.

To prepare negatively supercoiled topoisomers, chloroquine is preferred to ethidium bromide because it usually gives more reproducible results, presumably because its lower sensitivity to light. For this, the chloroquine concentration to be used in both tubes A and B (500- to 800-fold more than ethidium bromide on a weight basis) is determined by small-scale circularizations, as described in **Subheading 3.3.1.**, except that the assays are performed at room temperature. Netropsin concentration is determined in the same way (*see* **Note 14**).

3.4. Reconstitution of Tetramer (M_T) and Nucleosome (M_O) Particles

Various chromatin reconstitution procedures have been described in the literature, which can be conveniently classified into "low" and "high salt" methods. "Low salt" methods use cellular or nuclear extracts at physiological ionic strength to reconstitute minichromosomes with spaced nucleosomes in an ATP-utilizing reaction (*see* Chapters 1–3). The active component in these extracts usually is an acidic protein, such as nucleoplasmin in *Xenopus* oocytes *(31)* or specific chromatin (nucleosome) assembly factors in somatic cells (*see* **refs. 32** and **33** for reviews). Their role is to complex the histones and transfer them to the DNA. A nonphysiological equivalent of such proteins, poly-glutamic acid, has been successfully used to reconstitute minichromosomes *(34)*, as well as single nucleosomes on DNA minicircles *(9)*. The more primitive "high salt" methods involve a mixing of DNA and pure core histones in 2 M NaCl *(2)*, sometimes in the presence of urea *(35)*, followed by a more or less progressive decrease in the salt concentration by dialysis or dilution (*see* Chapters 1–3). An excess of linker histone-free native nucleosomes can also be used as a histone source in the "histone transfer" method. Early M_O reconstitutions on DNA minicircles used the "salt gradient" dialysis method, in which the dialysis was against an exponential salt gradient *(9,15)*. Subsequently, the so-called "salt-jump" method *(36)*, was used for both M_O *(20)* and M_T particles reconstitutions *(16)*. This method is described below.

3.4.1. The "Salt-Jump" Method

1. Mix DNA (in 1X TE buffer) with 2 µL of 5 *M* NaCl in a 0.5 mL microfuge tube (*see* **Note 15**), and complete to 5 µL with 1X TE buffer. DNA can be either the labeled topoisomer (0.5–1 × 10^3 ^{3}H cpm), supplemented with 0.5 µg of the supercoiled plasmid DNA from which the labeled sequence originates (*see* **Note 16**), or 0.5 µg of the unlabeled topoisomer with the ^{32}P-labeled topoisomer. Similar results are usually obtained at the autoradiographic level, regardless of the carrier.
2. Add the proper amount of histones (usually 0.5 to 1 µL of (H3-H4)$_2$ tetramers or (H2A-H2B-H3-H4)$_2$ octamers at 1–2 mg/mL in 2 *M* NaCl; *see* **Subheading 2.3.**), mix and incubate for 10 min at 37°C. This brings the final volume to 5.5–6 µL. Alternatively, the final volume may be kept at 5 µL, in which case less than 2 µL of 5 *M* NaCl are added in **step 1** in order to keep the final salt concentration at 2 *M*.
3. Add 3 vol (15–18 µL) of 100 µg/mL BSA in 1X TE buffer (TE-BSA), mix and incubate for 20 min at 37°C. Some applications, such as electron microscopy, require to omit BSA at this stage.
4. Dialyze the resulting 20–24 µL against 1X TE buffer for 1.5–2 h at room temperature in a 0.5-mL microfuge tube (*see* **Notes 17** and **18**). Radioactivity recovery is usually 60–70%.

3.4.2. A More Efficient Version of the "Salt-Jump" Method

With octamers, up to 100% of the DNA can be reconstituted at histone/DNA weight ratios (r_w) significantly below 1. With tetramers, in contrast, reconstitution usually reaches a plateau at 50% or less with topos –1 or +1 (*but usually* improves with higher negative supercoiling, *see below*). In that case, increasing r_w further only leads to aggregation. Up to 60–70% reconstitution can be achieved with topos -1 and +1 at a relatively low r_w using a modification of the "salt jump" method, in which the final 0.5 *M* NaCl concentration is reached in several steps.

1. To 5 µL of the above DNA/histone solution in 2 *M* NaCl at 37°C, add 2.5 µL of TE BSA (preequilibrated at 37°C) four times at 15 min intervals. Most importantly, these additions must be made by placing the tip of the pipet at the center of the solution, and not along the tube wall, and *without mixing*. Mixing with the pipet tip, by strokes or by vortexing, brings the reconstitution yield back to that observed in the above single step dilution method.

2. After 15 min, add the final 5 µL (as above and without mixing), and incubate for 10 min more at 37°C.
3. Dialyze against 1X TE buffer as in **Subheading 3.4.1**.

3.5. Analysis of M_T and M_O Reconstitution Products

3.5.1. Electron Microscopy

Direct visualization of reconstituted products can be achieved by classical electron microscopy in the dark-field illumination mode or by scanning transmission electron microscopy as described *(9,20)*. Because metal shadowing gives poor results with such small molecules, samples are rather stained with uranyl acetate. These techniques require the delicate preparation of grids coated with very thin carbon films. Electron micrographs of M_O and M_T particles on topos –2, –1 and +1 of a 359 bp minicircle are shown as examples in **Fig. 4**.

3.5.2. Electrophoresis: The "Chromatin" Gel (see **Note 19**)

1. Prepare a 4% polyacrylamide (acrylamide to bisacrylamide = 29:1; w/w) slab gel in 1X TE or TAE buffer.
2. Rinse the outside of the plates with distilled water and install the gel on your electrophoresis apparatus with 1X TE (or TAE) as running buffer. Recirculate buffer extensively between the two reservoirs (*see* **Note 20**), and ventilate with a fan (this insures a lower and more uniform gel temperature during operation).
3. Pre-electrophorese at 250 V (for an 18-cm long gel) until the intensity drops down significantly and remains stable (at approx 40–50 mA for a 18 cm wide and 0.15 cm thick gel) when persulfate added as a catalyzer for polymerization has diluted into the running buffer (this takes about 1 h).
4. Mix the sample (10–60 µL for 1 cm-wide slots) with 1/5th vol of 60% sucrose in 1X TE buffer containing 0.025% each of xylene cyanole and bromo-phenol blue, and load onto the gel.
5. Electrophorese at 250 V for about 3 h with buffer recirculation and air ventilation until the xylene cyanole dye comes close to the bottom of the gel (*see* **Note 21**).
6. Dry the gel and expose it overnight at –80°C. See **Fig. 5** for examples of M_T particles on topos –1 and +1 of a 359 bp minicircle, and of both M_O and M_T particles on topo –2.6 of a 355 bp minicircle.

3.6. Relaxation of M_T and M_O Particles

As a rule, chromatin is reconstituted on a topoisomer of negative supercoiling sufficient to keep it out of the relaxation equilibrium (*see* **Note 22**, and **Fig. 1**). In practice, this implies to use topoisomers of ΔLk below –1.5 for M_T and below –2.2 for M_O (this criterion is met in **Fig. 6**). Moreover, reconstitution is made at a histone/DNA ratio low enough for the amount of dimer particles to be negligible.

1. Mix 4 µL of chromatin (10^3 and 6×10^3 ^{3}H cpm are typically used for analytical and preparative purposes, respectively) in 1X TE buffer with 1 µL of 5X chroma-

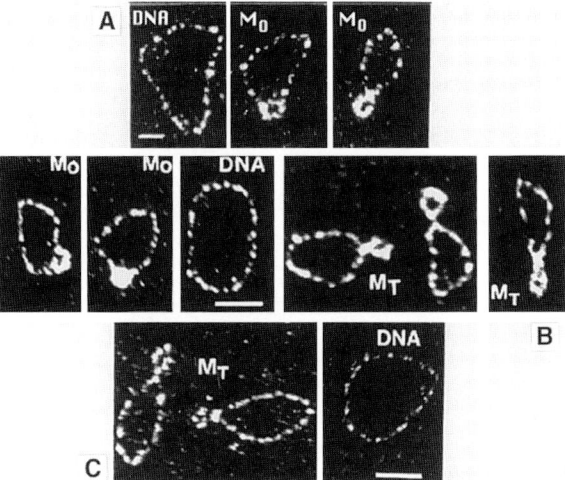

Fig. 4. Electron microscopic visualization of M_O and M_T particles in the dark field mode. M_O and/or M_T on topos −2 (**A**), −1 (**B**), and +1 (**C**) of a minicircle made from a 359 bp Sau3A fragment of pBR322 *(9)*. Unreacted naked topoisomers (DNA) and particles (M_O and M_T) in (**A**), (**B**), and (**C**) originate from the same fields, respectively. Reconstitution products were diluted in 1X TE buffer before adsorption to the grids. As previously shown *(9,15)*, crossed and uncrossed forms of $M_O(-2)$ in (**A**) are in about equal number in 1X TE buffer, whereas in 100 mM NaCl, crossed forms largely predominate. In contrast, $M_O(-1)$ in (**B**) remains uncrossed regardless of the salt concentration *(15)*. Note the low contour length of M_O compared to naked DNA, which results from the hidden DNA turn around that particle (1.75 and 1.4 turns of a superhelix are wrapped in crossed and uncrossed M_O, respectively). In contrast, the contour length of M_T is the same as that of naked DNA, pointing to only approx 0.6 turn of wrapped DNA in that particle. [Galleries were excerpted from published material of the authors' laboratory: (**A**) from **ref.** *9*; (**B**) and (**C**) from **ref.** *16*.] Bars: 30 bp in (**A**) and 60 bp in (**B**) and (**C**).

tin relaxation buffer, add 2–3 U of DNA topoisomerase I, incubate for 30 min at 37°C, and put immediately on ice.
2. Dilute with 3–4 vol of ice-cold TE-BSA, load onto a "chromatin" gel (*see* **Subheading 3.5.2.**), and apply voltage as soon as possible after samples are loaded. Examples of relaxations are displayed in **Fig. 6**.

3.7. Analysis of M_T and M_O Relaxation Products: Data Treatment

This requires to cut the band out of the "chromatin" gel, and to elute DNA. For this, exposure of the gel at 4°C as described in **Subheading 3.3.1.** is not feasible since this would require approx 10-fold as much radioactivity as used in **Subheading 3.6.** Exposure of the wet gel at −80°C would lead to its destruc-

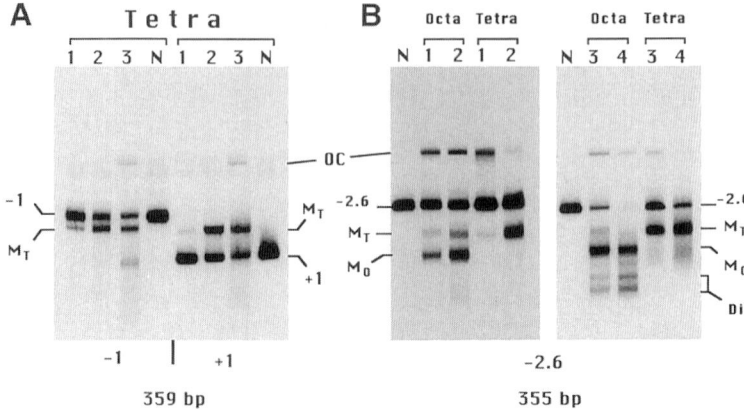

Fig. 5. Electrophoretic patterns of M_T and M_O particles. **(A)** Topos −1 and +1 of a minicircle made from a 359 bp *Bam*HI fragment were reconstituted with tetramers as in **Subheading 3.4.1.** at histones/DNA weight ratios r_w = 0.2, 0.4 and 0.7 (lanes 1–3, respectively). [This 359 bp fragment is identical in sequence to the 357 bp *Bam*HI fragment previously described *(16,52)*, except for a 2 bp addition at its *Taq* I site at position 343 (unpublished data).] Note that naked topo +1 migrates faster than naked topo −1, and that the mobility difference is reduced, but not canceled, upon reconstitution into $M_T(+1)$ and $M_T(−1)$. Such a fast migration of topo +1 probably reflects its crossing in the gel (but not on the electron microscopic grid; *see* **Fig. 4C**) as a potential consequence of the so-called gel "cage effect" *(53)*. **(B)** Topo -2.6 of a 355 bp minicircle made from a Taq I fragment of plasmid pB355 *(40)* was reconstituted with octamers and tetramers as in **Subheading 3.4.1.** at r_w = 0.1, 0.15, 0.2, and 0.4 (lanes 1–4, respectively). [With Lk_o = 33.6 for this 355 bp minicircle, Lk of topo -2.6 is 31; *see* legend to **Fig. 2A**.] Note a slight contamination of M_O particles by M_T particles, and the higher efficiency of octamers to form particles, compared to tetramers, at the same r_w ratio. Di, dinucleosomes. Electrophoreses were at room temperature. Autoradiograms are shown. OC, open circular DNA. N, naked topoisomer.

tion during the freezing-thawing process, while the dry gel would not reswell well enough for elution. To obviate these problems, dry the gel only partially by placing it in the vacuum gel drier for 1 h *at room temperature*, and expose at −80°C with an intensifier screen (autoradiograms in **Fig. 6** were so obtained). The exposed film is then aligned with the gel as in **Subheading 3.3.1.** The DNA is eluted as described in **Subheading 3.2.** and electrophoresed in a chloroquine-containing polyacrylamide gel (**Fig. 7**).

For quantitation, place the gel under a storage phosphor screen overnight and image it in a phosphor imager (Molecular Dynamics). Radioactivity profiles in **Fig. 7** show the topoisomer populations corresponding to M_T and M_O relaxation equilibria shown in **Fig. 6**. Finally, **Table 1** lists the relative amount of each topoisomer in the populations, their mean linking number (<Lk>), and the linking number difference (ΔLk_p) associated with formation of the particles (*see* legend to the table, and **Note 24**).

4. Notes

1. Blunt ends (generated by either one or two restriction endonucleases) and single-nucleotide overhangs do not sustain circularization. Minicircles can still be produced as follows. Oligomerize the fragment with T4 DNA ligase at high concentration (1–5 μg/mL), and digest ligation product with a restriction endonuclease which cleaves only once and generates overhangs of at least 2 nucleotides. Note that oligomerization is poorer with single-nucleotide overhangs. Providing the cleavage position is sufficiently off the middle of the fragment, 3 fragments of different sizes corresponding to the head-to-tail, head-to-head and tail-to-tail arrangements are produced in this secondary digestion, which are fractionated by gel electrophoresis. The recomposed fragment with the original size, derived from the head-to-tail arrangement, generally has a two- to fourfold better yield (in number) than the other two.
2. To extract DNA, dilute it to 200 (or 400) μL final volume with 1X TE buffer in a 1.5-mL microfuge tube. Add 20 (or 40) μL of 10 % SDS, mix, add 50 (or 100) μL of 5 *M* NaCl, mix again and incubate at 37°C for a few min to dissolve precipitated SDS, if any, and vortex with 1 vol of chloroform/isoamyl alcohol at full speed for 1min at room temperature. Spin for 5 min in a microfuge in the cold, and put on ice. Pipet the upper aqueous phase out, while paying attention not to disturb the SDS-protein interface. Add carrier if necessary (*see* **Note 6**), and precipitate with 2–4 vol of 96–100% ethanol. Wash the precipitate with 500 μL 80% ethanol, spin, pour out ethanol, and dry under vacuum. Note that this extraction efficiently removes a number of dyes or drugs, including ethidium bromide, chloroquine and netropsin.
3. The use of minicircles larger than approx 400 bp favors the reconstitution of two or more particles, defeating the purpose of making monomer particles. On the other hand, minicircles smaller than approx 250 bp may be too rigid to sustain a satisfactory reconstitution of M_O particles.
4. The authors use the following equation for the μg-pmoles conversion, where N is the fragment length (in bp):

$$\text{pmoles of DNA fragment} = 1515 \times \mu g/N \qquad (7)$$

5. The authors have observed that freezing the mixture prior to the extraction makes the DNA improper to subsequent use.
6. Recovery of microgram or sub-microgram quantities of DNA through ethanol precipitation is often poor, although it can be improved by spinning longer (30 min or more). Addition of a carrier insures a complete recovery in a 5-min spin. As a carrier, the authors use uncross-linked polyacrylamide, which is prepared by polymerizing acrylamide in the absence of bis-acrylamide *(37)*. Such polyacrylamide is highly soluble and does not interfere with subsequent experiments.

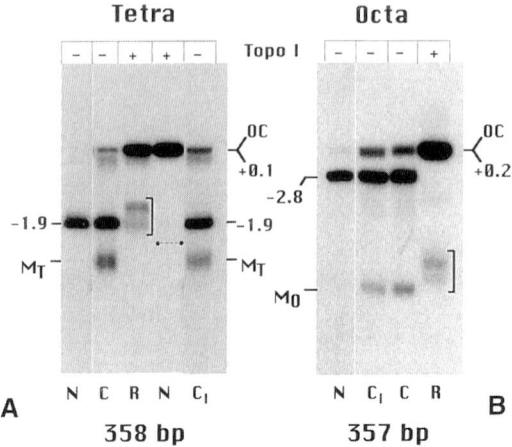

Fig. 6. Relaxation of M_T and M_O particles. M_T (**A**) and M_O (**B**) were reconstituted on topos −1.9 and −2.8 (*see* legends to **Fig. 2A** and **Table 1** for topoisomer identification), respectively, of minicircles made from 358 bp Taq I and 357 bp BamHI fragments. The 358 bp fragment has the same sequence as the 355 bp fragment in **Fig. 5B**, except for the difference in size (*40*), while the 357 bp fragment was referred to in legend to **Fig. 5A**. Electrophoreses were as in **Subheading 3.5.2.**, at room temperature. (A) M_T relaxes into several bands (lane R) corresponding to the three adjacent topoisomers present in the relaxation equilibrium (*see* **Fig. 7A**). The lower band corresponds to $M_T(+1.1)$, and the higher to $M_T(−0.9)$, consistent with the differential migration of $M_T(+1)$ and $M_T(−1)$ in **Fig. 5A**. $M_T(+0.1)$ migrates somewhere in between. A potential difficulty with this fractionation arises from the possible occurrence of a second topoisomer in the relaxation of the unreacted naked topoisomer, depending on the exact minicircle size. Because naked DNA makes the major part of the material, this second topoisomer, even in small relative amount relative to the main topoisomer, will bias the results if comigrating with tetramer relaxation products. To check for this possibility, the naked topoisomer was similarly relaxed (lane N). This relaxation showed a faint band (only visible in the original autoradiogram), whose position is indicated by dots. This band corresponds to topo +1.1 which migrates below $M_T(+1)$ in lane R, consistent again with the fractionation observed in **Fig. 5A**. Bands in lane R were cut (bracket), paying attention not to include topo +1.1. [In the cases of 351 and 360 bp minicircles, for example, this second topoisomer was in larger relative amount, and migrated with the higher and lower regions, respectively, of M_T relaxation products at room temperature. This required "chromatin" gels to be electrophoresed at 6°C and 37°C to shift the contaminating topoisomer upward and downward, respectively (not shown; *see* **Note 9**).] (B) M_O relaxes into multiple bands (lane R). The top one corresponds to $M_O(−0.8)$, and the lower one to $M_O(−1.8)$ (*see* topoisomer composition in **Fig. 7B**), consistent with a nearly relaxed loop in $M_O(−0.8)$ and a supercoiled loop in $M_O(−1.8)$ (*15*). However, $M_O(−0.8)$ itself splits into three sub-bands of unequal intensity which correspond to distinct subsets of nucleosome positions relative to the DNA sequence on the minicircle (*40,52*) (*see* **Note 23**). Note that naked topo −2.8

5 μL of 0.25% uncrosslinked polyacrylamide is used per precipitation. Because soluble polyacrylamide is present when DNA is gel-extracted (*see* **Subheading 3.2.1.**), no additional carrier is needed for ethanol precipitation in this case. It is noteworthy that other such carriers are now commercially available.

7. A 1-mL syringe is secured with a glass-wool plug, and filled with swollen G50 Sephadex in 1X TE buffer, while keeping the outlet open. Prior to use, the column is spun in a swinging-bucket rotor at 1300–1500 g for 5 min. After use, the ^{32}P-containing column is disposed of appropriately.
8. The counts measured directly from the aqueous DNA solution in the ^{3}H channel of a scintillation counter.
9. Topoisomer electrophoretic mobility also depends on the gel internal temperature, through a strong temperature dependence of Tw, and therefore of Wr (*see* **Eq. 1** in **Subheading 3.1.**). For most sequences, Tw decreases by 0.012 degree of angle/bp/°C in temperature elevation *(26,38,39)*. A 10°C elevation in the temperature of the gel in **Fig. 2A**, for example (as resulting from a higher voltage or a lower ventilation), in increasing Wr by +0.12, would lead to significant shifts of topo +1 (downward) and topos –1, –2, and –3 (upward). [Such temperature dependence of minicircle electrophoretic mobility was exploited in the authors' laboratory to measure the helical periodicity of free DNA *(26)* and the thermal flexibility of nucleosomal DNA *(40)*.]
10. Such an identification procedure differs from the common practice to round up ΔLk to the nearest integer. This gives the false impression that topo 0, for example, is fully relaxed, while this is so only when N is an exact multiple of h_o (*see* **Eq. 3**).
11. A proper topoisomer identification requires that sufficiently small increments of ethidium bromide concentration are used during circularization, to prevent any possibility of skipping one topoisomer. This risk does not exist for positive topoisomers because of the poor supercoiling efficiency of netropsin.
12. Purified topoisomers occasionally show extra bands when electrophoresed in "chromatin" gels under low salt conditions (*see* **Subheading 3.5.2.**). This contamination is more frequent in highly negatively supercoiled topoisomers, and presumably originates from comigrating linear or circular oligomers. To prevent this risk, topoisomers may be systematically repurified (with a recovery of about 50%) in "chromatin" gels before use.
13. Such small flow rates were initially achieved owing to an electronic device which alternatively switched the pump, together with the magnetic stirrer, on and off.

---

migrates close to OC DNA (lane N). This reflects a departure from B-DNA structure induced by high negative supercoiling, which tends to relax the molecule. Such a departure is sequence-dependent, and is favored under the low ionic strength conditions of the "chromatin" gel, compared to the "DNA" gel in Tris-acetate (*see* **Fig. 2A**), due to higher intra-molecular DNA-DNA repulsion *(20)*. Relaxed products were cut out of (A) and (B) gels (see brackets). Autoradiograms are shown. OC, open circular DNAs. N, naked topoisomers. C, starting chromatin in 1X TE buffer. C_I, chromatin incubated in 1X relaxation buffer for 30 min at 37°C.

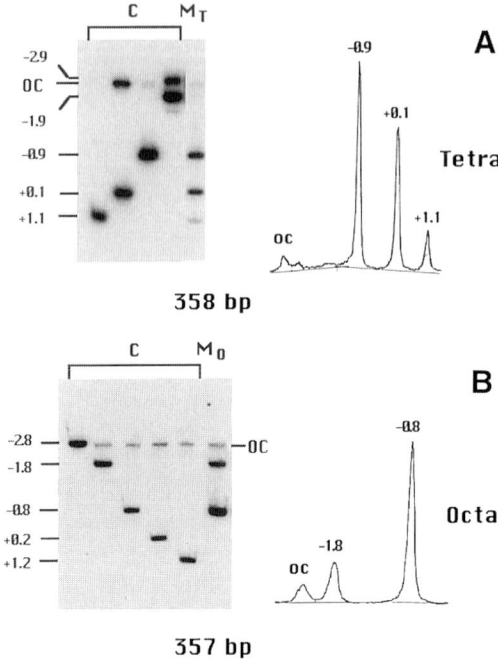

Fig. 7. Analysis of relaxation products. DNA relaxation products eluted from gel slices in **Fig. 6** were electrophoresed in chloroquine-containing gels identical to that shown in **Fig. 2B**, except for the 200 µg chloroquine per mL concentration. Radioactivity profiles in M_T and M_O lanes in (**A**) and (**B**), respectively, were obtained from the reading of the exposed storage phosphor screen, using the Image Quant (Molecular Dynamics) software in an IBM PC computer.

The "on" and "off" periods (2–10 min) were chosen accordingly. More recently, a peristaltic pump able to operate in a step-by-step mode (2232 Microperpex S pump; Pharmacia-LKB, Upsala, Sweden) was used, with similar results.

14. Up to 50% of the desired topoisomer can be obtained in large-scale circularization. In the presence of netropsin, however, this yield may decrease to 20–10%, because of the occurrence of several oligomeric bands. The desired topoisomer is then identified in the preparative gel through the use of the labelled topoisomer prepared through small-scale circularization as a migration marker.
15. Some microfuge tubes, depending on their origin, require a rinse with distilled water for satisfactory reconstitution results.
16. This titrates potential sequence-specific DNA binding proteins copurifying with the histones, which may otherwise generate extra bands in the "chromatin" gel.
17. Dialysis in microfuge tubes secured with a dialysis membrane has been described in the literature *(41,42)*. The authors make an aperture in the tube by cutting the very tip with a razor blade, or better with a "micro tube cutter" (Scienceware, Bel-

Table 1
Analysis of Relaxation Products of 358 bp M_T and 357 bp M_O Particles

	Topoisomer[a]					
Lk	32	33	34	35		
ΔLk (358 bp)	−1.9	−0.9	+0.1	+1.1		
ΔLk (357 bp)	−1.8	−0.8	+0.2	+1.2		
					$<Lk>$[c]	ΔLk_p[d]
M_T	—	53.6	36.3	10.1	33.6	−0.4
M_O	27.0	73.0	—	—	32.7	−1.2
	topoisomer relative amounts (%)[b]					

[a]Topoisomers refer to those present in the relaxation products of M_T and M_O particles in **Fig. 6** (*see* **Fig. 7**). ΔLk of each topoisomer was calculated from **Eq. 2** in **Subheading 3.1.**, using $Lk_o = 358/10.56 = 33.9$ and $357/10.56 = 33.8$.
[b]The relative amount of each topoisomer in M_T and M_O relaxation equilibria was measured from radioactivity profiles in **Fig. 7A,B**, respectively, using the Image Quant (Molecular Dynamics) software in an IBM PC computer.
[c]The mean Lk value of each equilibrium distribution was measured from topoisomer relative amounts using **Eq. 5** in **Subheading 3.1.**
[d]ΔLk_p (index p is for particle) was calculated from **Eq. 6** in **Subheading 3.1.**, with $Lk_o^p = <Lk>$ and $Lk_o = 358/10.53 = 34.0$ and $357/10.53 = 33.9$.

    Art Product, Pequannock, NJ), so that sample layering on the membrane can be made with a pipet; this opening is closed with parafilm (American National Can, Greenwich, CT) during operation. Moreover, such tubes can be easily checked for leakage before use. For this, tubes are over-filled with distilled water so that a convex meniscus is formed over the aperture. The tubes are then left on a paper towel for a minimum of 10 min, so that any leakage becomes apparent through a flattening of that meniscus. Such membrane-secured tubes can be re-used many times when stored immersed in 1X TE buffer in a closed vial at room temperature.

18. A fourfold scaling up of this procedure (2 µg DNA in approx 10 µL initial volume) can be achieved when dialysis of the resulting approx 40 µL is performed in 1.5-mL microfuge tubes. Other operating conditions are identical.
19. The "chromatin" gel often turns out to be a critical step. The way plates are washed is important. In particular, if plates are hand-washed with a detergent, as the authors do, special care must be taken to remove all of it. This is achieved by alternative rinse of the plates and of the hands, many times, with tap water. The plates are finally rinsed with distilled water, placed on paper towel and energetically wiped dry with Kimwipes (Kimberly-Clark). Moreover, unsatisfactory results were often observed with deionized water in the running buffer. The authors use quartz double-distilled water.
20. Recirculation prevents salt exhaustion which would otherwise rapidly occur due to the low ionic strength of 1X TE (or TAE) buffer. Most commercial or home-made electrophoresis apparatus can be efficiently and economically recirculated

with a centrifugal pump ordinarily used for filtering fish water in an aquarium (the authors use model C20 of Rena, Annecy, France). A hole or notch must then be made in the plastic wall of the upper reservoir to insure a safe return of the buffer into the lower reservoir.
21. The running buffer (the authors' apparatus uses 1.8 L) can be used for up to three electrophoreses (within no more than a week interval), when recovered and stored appropriately.
22. Without this precaution, the relaxation equilibrium would show an excess of the topoisomer in question, probably as a consequence of a difficulty of the topoisomerase to be triggered by too low a supercoiling.
23. The reason for $M_O(-0.8)$ fractionation is not clear. If M_O on 359 bp topo −1 of the BamH I fragment fractionates similarly, M_O on 359 bp topo −1 of the Sau3A fragment does not. This fractionation usually occurs with topoisomers of ΔLk close to −1 (when the loop is close to relaxation), and is more pronounced at lower gel temperatures *(40)*. Such a fractionation has been exploited in the authors' laboratory to prepare circular particles with unique nucleosome positioning (unpublished results).
24. The reader may get the impression that relaxations have been made once and for all. In fact, the authors' laboratory is presently involved in some intriguing aspect of these relaxations, that is, a periodic variation of the relaxation equilibria with the minicircle size. This confirms the existence of two states in the M_T particle, but also demonstrates the existence of several states in the M_O particle as well (newly published; *see* **refs. *43*,*44***). More generally, if relaxation experiments have been useful in the past to investigate nucleosome structure and dynamics (*see* **Subheading 1.**), they will undoubtedly help in the future to understand how these properties are affected by the various actors involved in gene regulation at the chromatin level. Among these actors, histone acetylation (*see* **refs. *45–47*** for reviews), nonhistone proteins such as HMGs (*see* **ref. *48*** for a recent review), some transcription factors (**ref. *49***; *see* also **ref. *50*** for a review), and chromatin remodeling complexes (*see* **ref. *51*** for a recent review), have already been identified.

Acknowledgments

The authors are indebted to Dr. A. Sivolob (Kiev), and to Ms. B. Hermier (this laboratory) for valuable contributions to these protocols.

References

1. Bauer, W. R., Crick, F. H. C., and White, J. H. (1980) Supercoiled DNA. *Sci. Amer.* **243,** 100–113.
2. Germond, J. E., Hirt, B., Oudet, P., Gross-Bellard, M., and Chambon, P. (1975) Folding of the DNA double helix in chromatin-like structures from simian virus 40. *Proc. Natl. Acad. Sci. USA* **72,** 1843–1847.
3. Shure, M. and Vinograd, J. (1976) The number of superhelical turns in native virion SV40 DNA and Minicol DNA determined by the band counting method. *Cell* **8,** 215–226.

4. Saragosti, S., Moyne, G., and Yaniv, M. (1980) Absence of nucleosomes in a fraction of SV40 chromatin between the origin of replication and the region coding for the late leader RNA. *Cell* **20,** 65–73.
5. Sogo, J. M., Stahl, H., Koller, Th., and Knippers, R. (1986) Structure of replicating simian virus 40 minichromosomes. The replication fork, core histone segregation and terminal structures *J. Mol. Biol.* **189,** 189–204.
6. Shore, D., Langowski, J., and Baldwin, R. L. (1981) DNA flexibility studied by covalent closure of short fragments into circles. *Proc. Nat. Acad. Sci. USA* **78,** 4833–4837.
7. Shore, D. and Baldwin, R. L. (1983) Energetics of DNA twisting II. Topoisomer analysis.*J. Mol. Biol.* **170,** 983–1007.
8. Horowitz, D. S. and Wang, J. C. (1984) Torsional rigidity of DNA and length dependence of the free energy of DNA supercoiling. *J. Mol. Biol.* **173,** 75–91.
9. Goulet, I., Zivanovic, Y., Prunell, A., and Révet, B. (1988) Chromatin reconstitution on small DNA rings. I. *J. Mol. Biol.* **200,** 253–266.
10. Klug, A., and Lutter, L. (1981) The helical periodicity of DNA on the nucleosome. *Nucleic Acids Res.* **9,** 4267–4283.
11. Finch, J. T., Lutter, L. C., Rhodes, D., Brown, R. S., Rushton, B. M., Levitt, M., and Klug, A. (1977) Structure of nucleosome core particles of chromatin. *Nature* **269,** 29–36.
12. Richmond, T. J., Finch, J. T., Rushton, B. M., Rhodes, D., and Klug, A. (1984) Structure of the nucleosome core particle at 7 Å resolution. *Nature* **311,** 532–537.
13. Furrer, P., Bednar, J., Dubochet, J., Hamiche, A., and Prunell, A. (1995) DNA at the entry-exit of the nucleosome observed by cryoelectron microscopy. *J. Struct. Biol.* **114,** 177–183.
14. Hamiche, A., Schultz, P., Ramakrishnan, V., Oudet, P., and Prunell, A. (1996) Linker histone dependent DNA structure in linear mononucleosomes. *J. Mol. Biol.* **257,** 30–42.
15. Zivanovic, Y., Goulet, I., Révet, B., Le Bret, M., and Prunell, A. (1988) Chromatin reconstitution on small DNA rings. II. DNA supercoiling on the nucleosome. *J. Mol. Biol.* **200,** 267–285.
16. Hamiche, A., Carot, V., Alilat, M., De Lucia, F., O'Donohue, M.-F., Révet, B., and Prunell, A. (1996) Interaction of the histone $(H3-H4)_2$ tetramer of the nucleosome with positively supercoiled DNA minicircles: Potential flipping of the protein from a left- to a right-handed superhelical form. *Proc. Natl. Acad. Sci. USA* **93,** 7588–7593.
17. Liu, L. F. and Wang, J. C. (1987) Supercoiling of the DNA template during transcription. *Proc. Natl. Acad. Sci. USA* **84,** 7024–7027.
18. Jackson, V. (1990) In vivo studies on the dynamics of histone-DNA interaction: evidence for nucleosome dissolution during replication and transcription and a low level of dissolution independent of both. *Biochemistry* **29,** 719–731.
19. Simon, R. and Felsenfeld, G. (1979) A new procedure for purifying histone pairs H2A + H2B and H3 + H4 from chromatin using hydroxylapatite. *Nucleic Acids Res.* **6,** 689–696.

20. Zivanovic, Y., Duband-Goulet, I., Schultz, P., Stofer, E., Oudet, P., and Prunell, A. (1990) Chromatin reconstitution on small DNA rings. III. Histone H5 dependence of DNA supercoiling in the nucleosome. *J . Mol. Biol.* **214,** 479–495.
21. Thomas, J. O. and Butler, P. J. G. (1977) Characterization of the octamer of histones free in solution. *J. Mol. Biol.* **116,** 769–781.
22. White, J. H. (1969) Self-linking and the Gauss integral in higher dimensions. *Am. J. Math.* **91,** 693–728.
23. Fuller, F. B. (1971) The writhing number of a space curve. *Proc. Natl. Acad. Sci. USA* **68,** 815–819.
24. Crick, F. H. C. (1976) Linking numbers and nucleosomes. *Proc. Natl. Acad. Sci. USA* **73,** 2639–2643.
25. Wang, J. C., Peck, L. J., and Becherer, K. (1983). DNA supercoiling and its effects on DNA structure and function. *Cold Spring Harb. Symp. Quant. Biol.* **47,** 85–91.
26. Goulet, I., Zivanovic, Y., and Prunell, A. (1987) Helical repeat of DNA in solution. The V curve method. *Nucleic Acids Res.* **15,** 2803–2821.
27. Maxam, A. M. and Gilbert, W. (1977) A new method for sequencing DNA. *Proc. Natl. Acad. Sci. USA* **74,** 560–564.
28. Shore, D. and Baldwin, R. L. (1983) Energetics of DNA twisting I. Relation between twist and cyclization probability. *J. Mol. Biol.* **170,** 957–981.
29. Snounou, G. and Malcolm, A. D. (1983) Production of positively supercoiled DNA by netropsin. *J. Mol. Biol.* **167,** 211–216.
30. Nunn, C. M., Garman, E., and Neidle, S. (1997) Crystal structure of the DNA decamer d(CGCAATTGCG) complexed with the minor groove binding drug netropsin. *Biochemistry* **36,** 4792–4799.
31. Laskey, R. A., Kearsey, S. E., Mechali, M., Dingwall, C., Mills, A. D., Dilworth, S. M., and Kleinschmidt, J. (1985) Chromosome replication in early Xenopus embryos. *Cold. Spring Harb. Symp. Quant. Biol.* **50,** 657–663.
32. Kaufman, P. D. (1996) Nucleosome assembly: the CAF and the HAT. *Curr. Opin. Cell Biol.* **8,** 369–373.
33. Kaufman, P. D. and Botchan, M. R. (1994) Assembly of nucleosomes: do multiple assembly factors mean multiple mechanisms? *Curr. Opin. Genet. Dev.* **4,** 229–235.
34. Stein, A., Whitlock, J. P. J., and Bina, M. (1979) Acidic polypeptides can assemble both histones and chromatin in vitro at physiological ionic strength. *Proc. Natl. Acad. Sci. USA* **76,** 5000–5004.
35. Camerini-Otero, R. D. and Felsenfeld, G. (1977) Histone H3 disulfide dimers and nucleosome structure. *Proc. Natl. Acad. Sci. USA* **74,** 5519–5523.
36. Stein, A. (1979) DNA folding by histones: the kinetics of chromatin core particle reassembly and the interaction of nucleosomes with histones. *J. Mol. Biol.* **130,** 103–134.
37. Gaillard, C. and Strauss, F. (1990) Ethanol precipitation of DNA with linear polyacrylamide as carrier. *Nucleic Acids Res.* **18,** 378–383.
38. Depew, R. E. and Wang, J. C. (1975) Conformational fluctuations of DNA helix. *Proc. Natl. Acad. Sci. USA* **72,** 4275–4279.

39. Pulleyblank, D. E., Shure, M., Tang, D., Vinograd, J., and Vosberg, H. P. (1975) Action of nicking-closing enzyme on supercoiled and nonsupercoiled closed circular DNA. *Proc.Nat. Acad.Sci. USA* **72,** 4280–4284.
40. Hamiche, A. and Prunell, A. (1992) Chromatin reconstitution on small DNA rings V. DNA thermal flexibility of single nucleosomes. *J. Mol. Biol.* **228,** 327–337.
41. Overall, C. M. (1987) A microtechnique for dialysis of small volume solutions with quantitative recoveries. *Anal. Biochem.* **165,** 208–214.
42. Ceschini, S., Pietroni, P., and Angeletti, M. (1996) A multi-sample microdialysis apparatus for proteins and nucleic acids. *Prep. Biochem. Biotechnol.* **26,** 189–199.
43. Sivolob, A., De Lucia, F., Révet, B., and Prunell, A. (1999) Nucleosome dynamics II. High flexibility of nucleosome entering and exiting DNAs to positive crossing. An ethidium bromide fluorescence study of mononucleosomes on DNA minicircles. *J. Mol. Biol.* **285,** 1081–1099.
44. De Lucia, F., Alilat, M., Sivolob, A., and Prunell, A. (1999) Nucleosome dynamics III. Histone-tail dependent fluctuation of nucleosomes between open and closed DNA conformations: implications for chromatin dynamics and the linking number paradox. A relaxation study of mononucleosomes on DNA minicircles. *J. Mol. Biol.* **285,** 1101–1119.
45. Pazin, M. J. and Kadonaga, J. T. (1997) What's up and down with histone deacetylation and transcription? *Cell* **89,** 325–328.
46. Wade, P. A., Pruss, D., and Wolffe, A. P. (1997) Histone acetylation: chromatin in action. *Trends Bioch. Sci.* **22,** 128–132.
47. Roth, S. Y. and Allis, C. D. (1996) Histone acetylation and chromatin assembly: a single escort, multiple dances? *Cell* **87,** 5–8.
48. Grosschedl, R., Giese, K., and Pagel, J. (1994) HMG domain proteins: architectural elements in the assembly of nucleoprotein structures. *Trends Genet.* **10,** 94–100.
49. Li, R. and Botchan, M. R. (1994) Acidic transcription factors alleviate nucleosome-mediated repression of DNA replication of bovine papillomavirus type 1. *Proc. Natl. Acad. Sci. USA* **91,** 7051–7055.
50. Svaren, J. and Horz, W. (1997) Transcription factors vs nucleosomes: regulation of the PH05 promoter in yeast. *Trends Biochem. Sci.* **22,** 93–97.
51. Tsukiyama, T. and Wu, C. (1997) Chromatin remodeling and transcription. *Curr. Opin. Gen. Dev.* **7,** 182–191.
52. Duband-Goulet, I., Carot, V., Ulyanov, A. V., Douc-Rasy, S., and Prunell, A. (1992) Chromatin reconstitution on small DNA rings IV. DNA supercoiling and nucleosome sequence preference. *J. Mol. Biol.* **224,** 981–1001.
53. Zivanovic, Y., Goulet, I., and Prunell, A. (1986) Properties of supercoiled DNA in gel electrophoresis. The V-like dependence of mobility on topological constraint. DNA-matrix interactions. *J. Mol. Biol.* **192,** 645–660.

7

Analysis of Linker Histone Binding to Mono- and Dinucleosomes

Simon Chandler and Alan P. Wolffe

1. Introduction

The assembly of DNA into the eukaryotic nucleus via an ascending hierarchy of intermediate chromatin structures has two major functional consequences:

1. The intermediate chromatin structures provide a filing system that greatly facilitates the search by RNA polymerase for regulatory DNA and the associated trans-acting factors.
2. The wrapping of DNA around the histones provides a means of compacting a large mass of DNA containing many genes while retaining specific sequences in an exposed and accessible state.

The recognition of eukaryotic genes by RNA polymerase depends on the prior association of specific transcription factors with their cognate DNA sequence elements. This association can be regulated through directed modifications to chromatin structure. Here we discuss experimental methods that reconstruct transcriptionally active and silent chromatin dependent on the presence or absence of linker histones. The aim is to provide an experimental system that allows the functional consequences of structural modifications to be directly assessed on the same chromatin template.

A useful approach to interrelate chromatin structure with transcription has been to make use of short DNA fragments that are long enough to be competent for transcription but short enough to allow aspects of their nucleoprotein organization to be determined *(1–6)*. Our lab has made an extensive analysis of the properties of synthetic mononucleosomes and dinucleosomes containing two 5S rRNA genes *(6,7)*. Each 5S rRNA gene contains intrinsic DNA structure sufficient to direct the rotational and translational positioning of a core histone octamer with respect to DNA sequence *(8–11)*. The properties of dinucleosomal templates differ dramatically from those of mononucleosomes *(6)*. Whereas histone octamers bound to short (<200 bp) DNA fragments are relatively static, histone octamers bound to longer DNA fragments (>400 bp) are mobile *(12–14)*. Nucleosome mobility emerges as a powerful contributory factor for the transcriptional competence of chromatin templates.

2. Materials
2.1. DNA Fragments

A single nucleosome core containing a complete octamer of histones has 180 bp of histone–DNA contacts revealed by hydroxyl radical cleavage, 160 bp of protection revealed by DNase I, and less than 150 bp of strong contacts remaining after limited micrococcal nuclease digestion *(15)*. In view of this information, an appropriately sized DNA fragment for reconstitution with one histone octamer to make a mononucleosome should be not less than 180 bp long, and a fragment used to reconstitute a dinucleosome should be between 360–420 bp long. Longer fragments might lead to additional histone octamers interacting with a proportion of the templates. The use of DNAs of 400–420 bp has the additional advantage that templates of this size can be efficiently transcribed in nuclear extracts of eukaryotic cells *(1)*, whereas DNA become progressively less efficiently transcribed as the length falls below 400 bp (*see* **Note 1**).

DNA is normally radiolabeled at a restriction endonuclease cleavage site using polynucleotide kinase and [α^{32}P] ATP or Klenow DNA polymerase and [γ^{32}P] NTPs. Following this reaction, a second restriction endonuclease is used to generate the uniquely end-labeled DNA of interest. For the examples given here, we make use of either a 214 bp *Eco*RI-*Dde*I fragment from the plasmid pXP-10, which contains the *X. borealis* 5S RNA gene or a 424-bp *Xba*I-*Xho*I fragment derived from plasmid pX5S197-2, which contains two tandemly repeated *Xenopus borealis* 5S rRNA genes.

2.2. Chromatin Purification Reagents

1. Wash buffer: 0.14 *M* NaCl, 15 m*M* sodium citrate, 10 m*M* Tris-HCl, pH 7.5, and 0.25 m*M* phenylmethylsulfonyl fluoride (PMSF).
2. Buffer A: 0.34*M* sucrose, 60 m*M* KCl, 15 m*M* Tris-HCl, pH 8.0, 15 m*M* NaCl, 15 m*M* 2-mercaptoethanol, 0.5% Nonidet P-40, and 0.25 m*M* PMSF.
3. Lysis buffer: 0.2 m*M* Na.EDTA, pH 8.0, 0.25 m*M* PMSF.

2.3. Nucleosome Core Particle Preparation Reagents

1. Digestion buffer I: 0.1 *M* KCl, 50 m*M* Tris-HCl, pH 7.5, and 1 m*M* CaCl$_2$.
2. Digestion buffer II: 25 m*M* NaCl, 10 m*M* Tris-HCl, pH 7.5, 0.1 m*M* EDTA.

2.4. Histone Fractionation Reagent

1. TEP buffer: 10 mM Tris-HCl, pH 8.0, 1 mM Na EDTA, 0.25 mM PMSF.
2. Hydroxylapatite (HAP) column, stored at 4°C in 0.5 M potassium phosphate buffer, pH 7.6

2.5. Chromatin Reconstitution Solutions

1. Buffer CR: 10 mM Tris-HCl, pH 7.5, 1 mM EDTA, 0.1 mM PMSF and 1 mM 2-mercaptoethanol plus either 2 M, 1.5 M, 1 M, 0.75 M, or 0 M NaCl.
2. Binding buffer: 50 mM NaCl, 10 mM Tris-HCl, pH 8.0, 0.1 mM EDTA, 5% glycerol.
3. TE: 10 mM Tris-HCl, pH 7.5, 0.1 mM EDTA.

2.6. Analysis of Nucleosomes

1. Iron (II)/EDTA solution: 1 mM $(NH_4)_2FeSO_4 \cdot 6H_2O$, 2 m$M$ EDTA. This solution may be stored frozen at –20°C.

2.7. Transcription Solutions

1. Priming buffer: 10 mM $MgCl_2$, 5 mM Tris-HCl, pH 7.8, 0.1 mM EDTA, 2.5 mM DTT, 10% glycerol, 10 mM HEPES-KOH, pH 7.4, and 2% polyvinyl pyrrolidone.
2. J buffer: 70 mM NH_4Cl, 7 mM $MgCl_2$, 0.1 mM EDTA, 2 mM DTT, 8% glycerol, and 10 mM HEPES pH 7.4.

2.8. General Solutions and Reagents

1. TE buffer: 10 mM Tris-HCl, pH 7.5, 1 mM EDTA.
2. Linker binding buffer: 10 mM Tris-HCl, pH 8.0, 50 mM NaCl, 0.1 mM EDTA, and 5% glycerol.
3. 1X TBE: 90 mM Tris-HCl, pH 7.5, 90 mM boric acid, 2.5 mM EDTA.
4. Proteinase K buffer: 0.25% (w/v) SDS, 5 mM EDTA.
5. Micrococcal nuclease was purchased as a lyophilized powder from Worthington (Freehold, NJ).
6. Proteinase K, DNaseI, and RNasin were purchased from Gibco-BRL (Gaithersberg, MD).
7. All chemicals were purchased from Sigma-Aldrich (Milwaukee, WI). All enzymes were purchased from New England Biolabs (Beverley, MA).

3. Methods

3.1. Radiolabeling of DNA Fragments

For the examples given, the 214-bp DNA fragment is end-labeled at the *Eco*RI site with T4 polynucleotide kinase and the 424-bp fragment is endlabeled at the *Xba*I site with T4 polynucleotide kinase, electrophoretically purified in a nondenaturing 6% polyacrylamide gel, followed by autoradiography of the wet gel, excision of the gel segment containing the radiolabeled DNA, and electroelution.

3.2. Purification of Donor Chromatin and Histones

1. Chicken erythrocytes or HeLa cells are spun down in 50-mL conical tubes in a clinical centrifuge for 5 min (approx 1100g). The cell pellet is washed twice more in wash buffer (*see* **Note 2**).
2. Nuclei are prepared by resuspending the cell pellet in 40 mL of Buffer A. The detergent (NP-40) in buffer A lyses the cell membrane and helps release nuclei from cytoplasmic contamination. Insoluble cytoplasmic debris is removed by filtering the suspension of broken cells through four to six layers of cheesecloth. Nuclei pass through and are pelleted by centrifugation at 1500g for 5 min.
3. The nuclear pellet is resuspended in Buffer A and repelleted at least twice or until the pellet is white in color. The nuclear pellet can either now be digested with micrococcal nuclease to make nucleosome core particles (*see* **Subheading 3.3.**) or after the final washing the nuclear pellet can be sheared to make "long" chromatin and purified histones. For the latter protocol, nuclei are resuspended in 20 mL lysis buffer. The A_{260} should be measured in 1 M NaOH and the concentration adjusted to 2.5 mg/mL. If storage is required, the nuclei should be frozen rapidly to –70°C in a dry-ice/ethanol bath.
4. Chromatin is prepared from the lysed nuclei by sonication. The extent of sonication necessary to break the chromosome into 3- to 6-kb fragments is determined empirically. The chromatin suspension is centrifuged at 3000g for 5 min to remove insoluble nuclear debris. The long chromatin should be stored at –70°C. Long chromatin is used to prepare purified histones.

3.3. Nucleosome Core Particle Preparation

1. The washed nuclear pellet (*see* **Subheading 3.2.**) is resuspended in digestion buffer I and centrifuged for 5 min at 4°C at 3000g. The nuclear pellet is then resuspended in digestion buffer I to have an A_{260} of approx 120. This nuclear suspension is digested at 37°C for 5 min with 10–20 U of micrococcal nuclease per milliliter of volume. The digestion process is stopped by the addition of 10 mM EDTA, and the suspension is placed on ice.
2. Nuclei are recovered by centrifugation at 12,000g at 4°C for 5 min. The pellet is resuspended in 0.25 mM EDTA, pH 7.5, and the nuclei are allowed to lyse over a 1 h period at 4°C. Nuclear debris is removed by centrifugation at 8000g for 20 min, and the solubilized chromatin in the supernatant is collected.
3. Linker histones are removed by bringing the chromatin to 0.35 M NaCl by the dropwise addition of 4 M NaCl and the addition of approximately 60 mg CM Sephadex C25/mL. The suspension is shaken at 4°C for 2 h before centrifugation at 10,000g to remove the resin. The stripped chromatin is dialyzed overnight against digestion buffer II (*see* **Note 3**).
4. The A_{260} of the dialyzed solution is adjusted to approx 60, $CaCl_2$ is added to 1 mM final concentration, and the stripped chromatin is digested at 37°C with micrococcal nuclease at 90 U/mL. The exact extent of digestion should be empirically determined using a small aliquot. The reaction is stopped after 1–2 h by the addition of EDTA to a final concentration of 10 mM on ice.

5. Solutions are concentrated using a Centriprep-10 U from Amicon (Danvers, MA) up to a final A_{260} of 200. Final purification of mononucleosome core particles is achieved by sucrose gradient centrifugation (5–20%, w/v in 25 mM NaCl, 10 mM Tris-HCl, pH 7.5, 1 mM EDTA). Centrifugation is in a Beckman SW-28 rotor at 82,000g at 4°C for 24 h.

3.4. Histone Fractionation

The most defined and useful source of material for the reconstitution of nucleosome cores is purified core histones. These can be prepared either from long chromatin or nucleosome cores.

1. Chromatin that has been solubilized by sonication or by micrococcal nuclease digestion is dialyzed overnight at 4°C against 0.6 M NaCl in TEP buffer. A hydroxylapatite (HAP) column is equilibrated with 0.6 M NaCl in TEP buffer. Chromatin is applied to the HAP volume (the capacity of hydroxylapatite is 1–5 mg of DNA/ml of chromatin). Fractions (5 mL) are collected as the column is washed with 0.6 M NaCl in TEP buffer until the A_{230} returned to zero. The proteins contained in these fractions are predominatly histones H1 and H5.
2. The column is next washed with 0.93 M NaCl in TEP buffer, and fractions are collected once again until the A_{230} returns to zero. These fractions contain histones H2A and H2B.
3. Finally the column is washed with 2 M NaCl in TEP buffer; these fractions contain histones H3 and H4. Column fractions from the same salt wash are pooled and stored at –20°C. The proteins present in these fractions should be resolved on 18% SDS-PAGE gels to verify that the histones are clearly fractionated and are not degraded. The concentration of histones in each fraction is estimated (A_{230} equals 4.2 for a 1 mg/mL solution) (*see* **Note 4**).

3.5. Reconstitution of Mononucleosomal and Dinucleosomal Templates

Nucleosomal structures can be reconstituted onto radiolabeled DNA fragments either by exchange from chicken erythrocyte core particles or by dialysis from high salt with purified HeLa core histones.

1. In the octamer exchange method, an approx 15-fold mass excess of core particles (7.5 µg) is mixed with radiolabeled DNA (500 ng) in tubes, followed by slow adjustment of NaCl concentration to 1 M.
2. Tubes are incubated at 37°C for 15 min before transfer of the mixture to a dialysis bag (with a molecular size limit of 6–8 kDa). Dialysis is first against buffer CR with 1 M NaCl for 4 h at 4°C; this is followed by dialysis for 4 h against the same buffer except that the NaCl concentration is reduced to 0.75 M. Finally the samples are dialyzed against TE, 1 mM 2-mercaptoethanol overnight (*see* **Note 5**).
3. As an alternative to **steps 1** and **2**, salt dialysis using purified core histones can be performed: Mix radiolabeled DNA (500 ng) and unlabeled DNA (5 µg) with an equal mass of all four core histones in 2 M NaCl. Samples are then dialyzed at

4°C against buffer CR containing the appropriate NaCl concentrations as follows: 2 M NaCl, 1 h; 1.5 M NaCl, 4 h; 1 M NaCl, 4 h; 0.75 M NaCl, 4 h. The final dialysis is overnight against buffer CR at 4°C.
4. After reconstitution, the mononucleosome or oligonucleosomes are loaded on 5–20% sucrose gradients containing TE buffer and 0.1 mM PMSF and then centrifuged for 16 h at 35,000 rpm at 4°C in a Beckman SW41 rotor.
5. Fractions are collected and analyzed by nondenaturing agarose gel (0.7%) electrophoresis in 0.5X TBE.
6. Fractions containing mono- or dinucleosomes can be pooled separately, concentrated to about 2.5 µg/mL using a microcon-30 (Amicon) and dialyzed against TE buffer and 1 mM 2-mercaptoethanol overnight at 4°C. Samples were stored on ice until use.
7. Reconstituted mono- or dinucleosomes (100 ng DNA content) are incubated with various amounts of linker histone (histone H5 in this example) in 10 µL of linker binding buffer at room temperature for 30 min.
8. A simple assay for the binding of linker histones and HMG1 is the resolution of reconstituted nucleoprotein complexes on running 0.7% agarose gels in 0.5X TBE.
9. After electrophoresis the gel is dried and autoradiographed. Titration experiments allow the determination of the binding affinities of histone H1 to mono- or dinucleosomes (*see* **Note 6**).

3.6. Characterization of Reconstituted Mono- and Dinucleosomes

3.6.1. DNaseI Digestion

DNaseI digestion provides information on the rotational positioning of DNA on the histone surface.

1. Reconstituted nucleosomal structures that do or do not contain linker histones (60 ng of DNA) are adjusted to 4 mM $MgCl_2$ concomitantly with the addition of DNaseI (30–60 ng). Digestion is allowed to proceed at 22°C for 1 min, before termination through the addition of EDTA (5 mM).
2. Glycerol (5%, v/v) is added to the sample, and the entire reaction volume is transferred to a running 0.7% agarose gel containing 0.5X TBE.
3. After electrophoresis, the wet gel is exposed to autoradiograph film to allow localisation of the nucleosomes. Complexes are excised from gel and electroeluted. The DNA from these complexes is analyzed on denaturing 6 or 8% polyacrylamide gels containing 7 M urea.

3.6.2. Hydroxyl Radical Cleavage

Since nucleoprotein sample volumes may vary we have given reagent amounts as a ratio of total volume.

1. The cutting reaction (Fenton reaction) is initiated by placing Iron (II)/EDTA solution (1/10 vol), 0.012% H_2O_2 (1/10 vol), and 10 mM L-ascorbic acid (sodium salt) (1/10 vol) on the inner wall of the 1.5-mL Eppendorf tube containing the

nucleoprotein complex (7/10 vol). The reagents are allowed to mix and then added to the solution containing the nucleoprotein complex, which is then mixed thoroughly by pipetting. A naked DNA control can be carried out by digesting 500 ng labeled DNA with the Fenton reaction as above except that 0.1 mM iron (II) EDTA solution is used.
2. The free radical reaction is quenched after 2–4 min at room temperature with the addition of glycerol to 5%, and the entire reaction volume is transferred to an agarose gel as described above.

3.6.3. Micrococcal Nuclease Digestion

This reaction is useful for determining the boundaries of strong histone-DNA interaction

1. Nucleosomal structures (80 ng DNA) are digested with 0.075–0.6 U of micrococcal nuclease for 5 min at 22°C. CaCl$_2$ is added to 0.5 mM concomitantly with addition of micrococcal nuclease.
2. Digestions are terminated by addition of EDTA (5 mM), SDS (0.25% w/v) and proteinase K (1 mg/mL). Protein is extracted via phenol and the DNA is recovered and endlabeled with [γ^{32}P]ATP and T4 polynucleotide kinase.
3. The endlabeled DNA fragments are separated by electrophoresis in nondenaturing 6% polyacrylamide gels. The wet gel is exposed to film to localize digestion products. Kinetic intermediates corresponding to nucleosome and chromatosome should be seen at 146 bp and 167 bp, respectively.
4. DNA cleavage intermediates are excised from the gel and electroeluted. These fragments are recovered and digested with restriction endonucleases to determine micrococcal nuclease cleavage sites (*see* **Note 7**).

3.7. Gel shift and Nucleosome Mobility

The mobility of the nucleoprotein complex in the gel will depend on whether histones bind in the middle of a DNA fragment or at an end. Evidence for the mobility of nucleosomal structures comes from carrying out the two dimensions of electrophoresis. If a nucleosomal structure changes position along the DNA molecule prior to the second dimension, then this will be detected by the appearance of a nucleoprotein complex that migrates at a position away from a simple diagonal in the final autoradiograph of the two dimensional gel (*5,6,13,14*).

1. One and two dimensional gel experiments to show the mobility of nucleosomal structures are performed by loading radioactively-labeled reconstitutes with or without linker histones onto a nondenaturing 5% polyacrylamide (29:1, acrylamide: bisacrylamide) gel at 4°C in 0.5X TBE.
2. The gel is run at a maximum of 10 V/cm and the wet gel is exposed to localise complexes. Each lane of interest is excised, sealed in a plastic bag, and incubated at 37°C for 1 h. Increased temperature increases the mobility of nucleosomal structures.

3. The gel strips are arranged on top of a second nondenaturing gel in the cold, and the second dimension of electrophoresis is performed at 4°C under the same conditions as the first dimension.

3.8. Transcription Reactions

1. *Xenopus laevis* oocyte nuclear extract is prepared by first excising ovaries from adult frogs and allowing fragments of ovary to swell for 3 h in priming buffer.
2. The oocytes are broken using two pairs of forceps. Nuclei are taken up into a pipet tip. Five hundred nuclei are collected into a microfuge tube in a total volume of 1 mL.
3. The nuclei are disrupted by brief vortexing. The lysate is centrifuged at 10,000g for 1 min at 4°C to pellet nuclear debris. The supernatant is used as a transcription extract.
4. Transcription reaction conditions are as follows: 10 ng of radiolabeled reconstituted chromatin template are added to 10 µL of reaction mixture containing 5 µL of oocyte nuclear extract in J buffer supplemented with 0.25 U/µL RNasin and preincubated for 20 min before addition of exogenous triphosphates to 250 µM ATP, CTP, and GTP, 50 µM UTP with 2.5 µCi added of [α^{32}P]UTP. The reaction temperature is 22°C.
5. Radiolabeling of transcripts is allowed to occur for 40 min after the preincubation period. Radiolabeled transcripts are extracted with phenol, precipitated with ethanol, and analyzed by electrophoresis in a 6% denaturing polyacrylamide gel containing 7 M urea.
6. After drying, the accumulation of specific transcripts is determined by autoradiography. The radiolabeled 5S DNA template serves as an internal control for the recovery of nucleic acid. Alternatively, naked DNA encoding a differently sized transcript can be used as an internal transcription control.

4. Notes

1. The ideal range of DNA fragment length for reconstitution of monosomes is between 180 and 230 bp. If the fragment is any larger we have found that reconstitution can give close packed dinucleosomes and hence artificial positioning resulting from mutual exclusion from the native positioning sequences.
2. Vertebrate histones are highly conserved through evolution, and any limited interspecific variation has no known functional significance. The use of chicken, *Xenopus*, or human core histones for the reconstitution of chromatin has not led to variation or lack of reproducibility in experimental results. Chicken erythrocyte nuclei are the favored source of chromatin or histones because of the stability of the nuclei, in particular the virtual absence of proteases and nucleases, and because a large proportion of nuclear proteins are histones. HeLa cells are also a convenient source of nuclei.
3. A quicker method for linker histone removal utilizes ion exchange resin. Dialyze 50 AU (A_{260}) chicken erythrocyte chromatin against TE buffer and 0.7 M NaCl (0.35 M for HeLa chromatin). Equilibrate 3 mL Dowex 50 AG-W X2 resin by the addition of 10 mL 5% NaOH for 10 min. Then spin the resin down, pour off the supernatant, and add 10 mL 5% HCl for 10 min. Neutralize with water and repeat

the NaOH treatment. Neutralize again with water and equilibrate in the chromatin dialysis buffer just described. Add the dialysed chromatin and agitate at 4°C for 1 h. Spin the resin down and remove the supernatant. Run an 18% SDS PAGE gel to check for linker histone removal.

4. When preparing purified histones for salt dialysis reconstitution we have found that preparation of the octamer rather than the separate histones is best, since the stoichiometry of the complex is already set. Dialyze the chromatin against 10 mM K_2PO_4, pH 6.7, 0.73 M NaCl and 0.25 mM PMSF. Equilibrate a 180 mL hydroxylapatite column in the same buffer. Load the dialysed chromatin and run the column at 2.5 mL/min. Collect fractions and wash the column until the A_{260} returns to baseline. This washing step has removed the linker histones. Elute the core histones in 10 mM K_2PO_4, pH 6.7, 2 M KCl, and 0.25 mM PMSF. The histones will be dilute, we have found that concentration on a CM-52 cationic exchange column can overcome this. Pour a 15 mL CM-52 column and equilibrate in TE buffer with 0.2 M NaCl and 0.25 mM PMSF. Dilute the eluted histones in TE buffer with 0.25 mM PMSF until the KCl concentration is reduced to 0.2 M. Load the column and run at 4 mL/min. Wash with 50 mL of equilibration buffer and elute the concentrated histones with 2 M NaCl at 0.5 mL/min. Finally, pass the concentrated octamers through a G-100 filtration column to purify. The octamers are then dialyzed against 2 M NaCl, 0.01 M Na_2PO_4, and 0.25 mM PMSF, and can be stored in this buffer at –70°C.

5. An alternative protocol for salt exchange reconstitution that does away with the need for dialysis is as follows. 500 ng of labeled DNA fragment of the appropriate length is mixed with >2.5µg mononucleosomal chromatin, 1 µL of 100 mM 2-mercaptoethanol, and 2 µL of 5M NaCl with a final reaction volume of 10 µL. Incubate at room temperature for 30 min. Add four aliquots of 5 µL TE buffer at 15 min intervals. After a final 15 min add 170 µL TE buffer. The reconstitute can be stored at 4°C for several days.

6. Gel shift analysis of reconstituted complexes can be useful for quality control purposes as well as binding studies. Acrylamide gel shift is carried out with a 4.5% 37.5:1 monomer ratio gel in 0.25X TBE buffer. Prerun the gel at 4°C for 30 min at 250 V, then load the samples in 5% glycerol (no dye) and run for 2 h at 200 V. We have found that running the gels any longer can result in dissociation of the reconstitute. The acrylamide gel shift allows resolution of different translational positions of nucleosomes on the same fragment. These have slightly different mobilities because of the length of uncomplexed DNA.

7. Mapping translational position using micrococcal nuclease (MNase) requires the reconstitution to contain only the DNA fragment of interest, hence salt dialysis for reconstitution is required. Prepare a water bath at 22°C. Take 70–150 ng salt dialyzed reconstitute (usually approx 5 µL) and add bovine serum albumin to 160 µg/mL. We have found that this stabilises the enzyme. Dilute MNase with 5 mM $CaCl_2$. Digest for 5 min at room temperature and stop the reaction by the addition of 40 µL 2x proteinase K buffer (**Subheading 2.8.**). Phenol extract the samples then re-extract the phenol with 50 µL water to avoid DNA losses. Pre-

cipitate with ethanol in the presence of 10 mM MgCl$_2$ to aid the precipitation of small fragments though a small amount of glycogen works just as well. The products of digestion can then be labeled with [γ-^{32}P]ATP and T4 polynucleotide kinase. Run the labelled products in a 6% acrylamide gel (acrylamide: bisacrylamide 19:1) in 1X TBE buffer. Kinetic intermediates of cleavage can be rescued from the gel and mapped by standard enzyme cleavage reactions. We add a small amount of carrier DNA for this final treatment to avoid overdigestion by restriction enzymes.

References

1. Wolffe, A. P., Jordan, E., and Brown, D. D. (1986) A bacteriophage RNA polymerase transcribes through a *Xenopus* 5S RNA gene transcription complex without disrupting it. *Cell* **44,** 381–389.
2. Losa, R. and Brown, D. D. (1987) A bacteriophage RNA polymerase transcribes *in vitro* through a nucleosome core without displacing it. *Cell* **50,** 801–808.
3. Lorch, Y., La Pointe, J. W., and Kornberg, R. D. (1987) Nucleosomes inhibit the initiation of transcription but allow chain elongation with the displacement of histones. *Cell* **49,** 203–210.
4. Schild, C., Claret, F.-X., Wahli, W., and Wolffe, A. P. (1993) A nucleosome-dependent static loop potentiates estrogen-regulated transcription from the *Xenopus* vitellogenin B1 promoter *in vitro*. *EMBO J.* **12,** 423–433.
5. Studitsky, V. M., Clark, D. J., and Felsenfeld, G. (1994) A histone octamer can step around a transcribing polymerase without leaving the template. *Cell* **76,** 371–382.
6. Ura, K., Hayes, J. J., and Wolffe, A. P. (1995) A positive role for nucleosome mobility in the transcriptional activity of chromatin templates: restriction by linker histones. *EMBO J.* **14,** 3752–3765.
7. Ura, K., Nightingale, K., and Wolffe, A. P. (1996) Differential association of HMG1 and linker histones B4 and H1 with dinucleosomal DNA: structural transitions and transcriptional repression. *EMBO J.* **15,** 4959–4969.
8. Simpson, R. T. and Stafford, D. W. (1983) Structural features of a phased nucleosome core particle. *Proc. Natl. Acad. Sci. USA* **80,** 51–55.
9. Hayes, J. J., Tullius, T. D., and Wolffe, A .P. (1990) The structure of DNA in a nucleosome. *Proc. Natl. Acad. Sci. USA* **87,** 7405–7409.
10. Hayes, J. J., Clark, D. J., and Wolffe, A. P. (1991) Histone contributions to the structure of DNA in the nucleosome *Proc. Natl. Acad. Sci. USA* **88,** 6829–6833.
11. Hayes, J. J. and Wolffe, A. P. (1993) Preferential and asymmetric interaction of linker histones with 5S DNA in the nucleosome. *Proc. Natl. Acad. Sci. USA* **90,** 6415–6419.
12. Meersseman, G., Pennings, S., and Bradbury, E. M. (1991) Chromatosome positioning on assembled long chromatin: Linker histones affect nucleosome placement on 5S DNA. *J. Mol. Biol.* **220,** 89–100.
13. Pennings, S., Meersseman, G., and Bradbury, E. M. (1994) Linker histones H1 and H5 prevent the mobility of positioned nucleosomes. *Proc. Natl. Acad. Sci. USA* **91,** 10,275–10,279.
14. Meersseman, G., Pennings, S., and Bradbury, E. M. (1992) Mobile nucleosomes-a general behavior. *EMBO J.* **11,** 2951–2959.

8

Quantitative Analysis of Chromatin Higher-Order Organization Using Agarose Gel Electrophoresis

Jeffrey C. Hansen, Terace M. Fletcher, and J. Isabelle Kreider

1. Introduction

1.1. Agarose Gel Electrophoresis and Chromatin Higher-Order Organization

In the past five years, there have been numerous molecular and genetic investigations of the mechanisms by which chromatin condenses into interphase chromosomal fibers *(1,2)*. In addition to traditional molecular techniques such as electron microscopy *(3–5)* and analytical ultracentrifugation (*see* Chapter 9; **ref. 6**), it has recently been demonstrated that agarose gel electrophoresis can be used to quantitatively analyze the conformational and configurational changes of chromatin *(7–9)*. The electrophoretic method requires neither large amounts of sample nor a large investment in equipment, making this structural technique potentially more accessible to the typical chromatin researcher. The key to "quantitative" agarose gel electrophoresis is to perform the experiment in a multigel (**Fig. 1**; *see* **Subheading 1.3.**). The multigel approach was developed by Serwer *(10)* and previously has been used to characterize spherical and rod-shaped bacteriophages *(11–12)*. It yields a measure of three important structural parameters: shape, flexibility, and surface charge density. When applied to chromatin, these parameters provide sensitive assays for nucleosomal subunit density, the spacing regularity of the nucleosomal array, and the average extent of chromatin folding *(6,7–9)*. Together, this information can provide unique structural insight into chromatin higher-order organization.

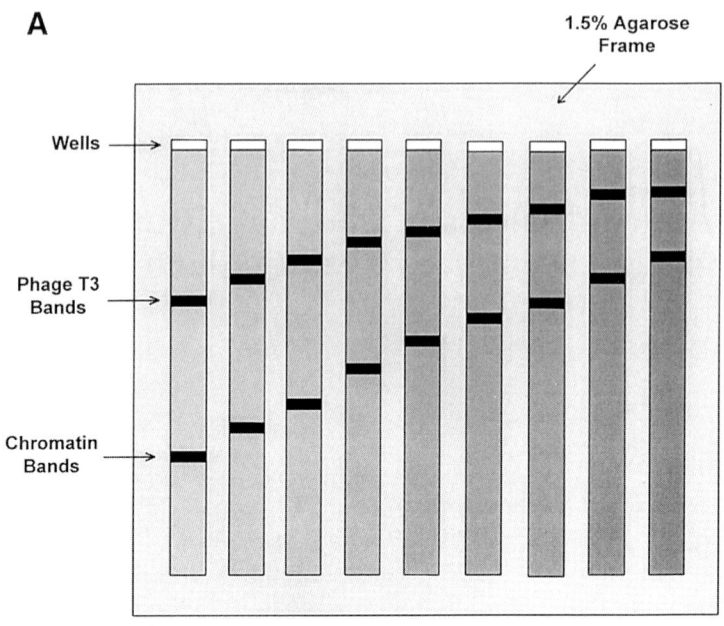

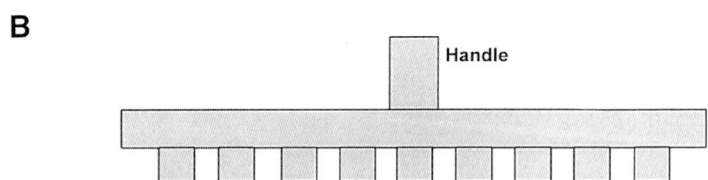

Fig. 1. (**A**) Schematic illustration of a 9-lane multigel. The view is from above after completion of each of the steps described in **Subheadings 3.1.** and **3.2.** Increasing agarose concentrations of the different running gels are represented by increased shading. The position of the chromatin bands relative to the bacteriophage T3 bands are illustrated for a 12-mer nucleosomal array. (**B**) Schematic diagram of a 9-lane slotformer viewed from the side.

1.2. Fundamental Principles of Quantitative Agarose Gel Electrophoresis

Agarose gels are aggregates of double-helical polysaccharide chains. At the microscopic level, they consist of a network of suprafibers that form fairly well-defined pores (*13*). The mobility (μ) of a macromolecule in an agarose gel depends on both its μ in the absence of the gel (μ_o), which is a direct measure of the average electrical surface charge density, and the sieving that occurs during movement of macromolecules through the gel pores *(13,14)*. The extent of sieving is determined by the size, shape, and conformational flexibility of the macromolecule. The equation that best describes the electrophoresis of rigid spheres in agarose gels is:

$$\mu/\mu'_o = (1 - R/P_e)^2 \qquad (1)$$

where P_e is the average gel pore radius, μ is the μ'_o before correction for electroosmosis, and R is the radius of the sphere *(11)*. If the particle is non-spherical, R is replaced by the effective radius (R_e).

1.3. Characterization of Chromatin Higher-Order Organization by Agarose Gel Electrophoresis Requires Use of a Multigel

A multigel is a specialized aggregate agarose gel in which either 9 or 18 different agarose running gels ranging from 0.2–3.0% are physically embedded in a 1.5% agarose frame (**Fig. 1A**; *see* **Subheading 3.1.**). To experimentally determine the μ'_o, P_e, and R_e, one simultaneously electrophoreses both a chromatin sample and a spherical bacteriophage T3 standard in the multigel. The μ'_o is obtained by extrapolating plots of log μ vs agarose percentage to 0% agarose (*see* **Subheading 3.5.**). The experimentally determined μ and μ'_o of bacteriophage T3, together with the known T3 radius (30.1 nm), are used to calculate the P_e for each running gel using **Eq. 1**. The experimentally determined μ and μ'_o of the chromatin sample and the P_e derived from the bacteriophage T3 data are then used to calculate the chromatin R_e at the P_e (*see* **Subheading 3.5.**).

Importantly, in any given experiment μ and μ'_o are influenced by buffer conditions, temperature fluctuations during the run, and voltage potential. All of these parameters tend to differ from one experiment to the next. Thus, an attempt to determine $\mu^{chromatin}$ and μ^{T3} as a function of agarose percentage in separate individual experiments would introduce an unacceptable level of error in the μ'_o and R_e. In direct contrast, the multigel apparatus is designed so that all the samples will be subjected to the same environmental factors during electrophoresis, resulting in μ being dependent solely on agarose concentration. Consequently, in a multigel, the μ'_o and R_e can be determined reproducibly and with analytical precision.

2. Materials

2.1. The Multigel Electrophoresis Apparatus

1. The multigel apparatus consists of a model 850 horizontal submarine electrophoresis unit, a 9- or 18-lane slotformer template (**Fig. 1B**) with the corresponding delrin comb and combholder, and two plastic plates and plate supports. The

9- and 18-lane slotformer templates allow for simultaneous electrophoresis of 9 or 18 different running gels within the multigel, respectively. The entire multigel apparatus is available commercially from Aquebogue Machine and Repair Shop (Aquebogue, NY).

2.2. Other Equipment

1. A high temperature water bath (70–80°C) is required to keep the agarose stock(s) melted prior to pouring the running gels. An oscillating pump setup (e.g., Cole-Parmer oscillating pump together with a steel-cased voltage controller and R3603 connective tubing) is needed to circulate the running buffer between the electrophoresis unit chambers throughout the experiment. In addition to the circulator, a refrigerated water bath is needed for multigel experiments that require >8-h electrophoresis times (*see* **Note 1**). The circulating pump and water bath help control pH and temperature variations during the experiment (*see* **Subheading 3.2., step 3**). Finally, for data analysis purposes one must have access to a digital imaging system.

2.3. Glassware, Running Buffers, and Agarose Source

1. Running gels are prepared in 5 mL and 10 mL Kimble™ borosilicate threaded end tubes; 16 × 125 mm and 20 × 125 mm are used for 18- and 9-lane multigels, respectively. An equivalent tube may be used if it can withstand the 60–70°C water bath temperature.
2. Appropriate running buffers for low-salt multigel experiments are TAE (40 mM Tris-acetate, pH 7.8, 1 mM Na$_2$EDTA) and E (40 mM Tris-HCl, pH 7.8, 1 mM Na$_2$EDTA). When studying chromatin folding, 0.5–2.0 mM *free* MgCl$_2$ (i.e., 1.5–3.0 mM total MgCl$_2$) is included in E buffer to yield EM buffer. TAE is not used for folding experiments since the acetate anion in TAE interferes with interpretation of chromatin folding experiments *(7,8)*. Tris-borate buffers are never utilized for multigel experiments because of anomalous electroosmosis effects.
3. A high-grade, low-electroosmosis- (LE) grade agarose is used for both the multigel frame and running gels. Our laboratory uses molecular biology grade LE agarose purchased from Research Organics, Cleveland, OH (cat. no. 1170A-3). Any equivalent agarose can be used although calculation of the running gel P$_e$ and chromatin R$_e$ is much simpler if the Research Organics LE agarose is used (*see* **Note 2**).

2.4. Chromatin and Bacteriophage T3

1. Bacteriophage T3, a spherical DNA bacteriophage with an R = 30.1 nm *(13)* used to calculate the P$_e$ of each running gel (*see* **Subheading 3.5.**), is purified using the procedure of Serwer *(15)* and stored at approx 50 µg/mL (*see* **Note 3**).
2. The chromatin source needs to be capable of forming discrete bands during electrophoresis and hence must be homogeneous in length. Consequently, the technique is ideally suited for analysis of length-defined nucleosomal arrays reconstituted in vitro from pure core histones and DNA *(6–9)*. High-quality quantitative gel data can be obtained with endogenous chromatin fragments only if

the endogenous chromatin has been extensively fractionated to yield a narrow distribution of oligonucleosome lengths (*see* **Note 4**).

3. Method
3.1. Pouring the Multigel

1. Heat the water bath to 60–70°C. Be sure the water level is sufficient enough to cover the portion of the tubes containing the melted agarose stocks and running gels.
2. Prepare the 1.5% agarose stock solution used to pour the multigel frame. Add 1.8 g agarose to a 250-mL flask and tare the scale to zero. Add 120 g of running buffer to the flask and retare the scale to zero. Melt the agarose in the microwave, reweigh the flask, and add double-distilled water to replace the weight lost during heating.
3. Level the gel bed by adjusting the screws at each end of the electrophoresis unit. Prior to pouring the multigel frame, slide the 13 × 5 cm plastic plates into the grooves at each end of the gel bed using the red plastic supports to achieve the proper plate height. This creates the enclosed area within the submarine electrophoresis unit that will eventually house the multigel. Place the slotformer (**Fig. 1B**) in the gel bed situated approx 5 cm away from the plastic plate nearest the negative electrode. Level the comb and place it against the slotformer. The fit between the comb and slotformer should be snug, and the comb teeth should be aligned with the slots of the slotformer. The bottom of the slotformer should rest directly against the gel bed. Once the 1.5% agarose solution is sufficiently cool to hold, use it together with a Pasteur pipet to seal the interface between the plastic plates and the edges of the gel bed. Allow the agarose to polymerize. To form the multigel frame, subsequently pour the remainder of the 1.5% agarose into the sealed gel bed, making sure it spreads completely around the slotformer. Use forceps or equivalent to remove any visible bubbles. For the 9- and 18-lane slotformers, polymerization of the 1.5% agarose frame takes approx 30 and 60 min, respectively.
4. While the frame is polymerizing, label the required number of borosilicate tubes (9 or 18) with the desired running gel concentrations. At this time, pipet only the appropriate amount of *buffer* into each tube, using the information in **Tables 1** and **2**. Place the tubes in the heated waterbath for 10–15 min.
5. When the frame is completely polymerized, carefully remove the comb. Very carefully remove the slotformer to avoid breaking the narrow portions of the frame that form the running gel slots. Small bits of agarose often remain after the comb is removed. Use a pair of dissecting forceps to remove these agarose bits and burst any large bubbles that may have formed in the running gel slot bottoms during polymerization, as these bubbles will interfere with the multigel image after photography. The thin layer of agarose usually present in the running gel slot bottoms can be ignored once the bubbles are burst. Finally, place the comb back in its original position.
6. Prepare the agarose running gel stock solutions. For both 9- and 18-lane multigels, the 1.0% agarose solution used for 0.2–1.0% agarose running gels consists of 0.6 g agarose + 60 g running buffer, and the 3.0% agarose stock used for

**Table 1
Dilutions for Preparation of Running Gels:
0.2–1.0% Running Gels in 9- and 18-Lane Multigels**

18-Lane Multigel			9-Lane Multigel		
Gel (%)	Buffer (mL)	Stock (mL)[a]	Gel (%)	Buffer (mL)	Stock (mL)[a]
0.2	4.0	1.0	0.2	8.0	2.0
0.3	3.5	1.5	0.3	7.0	3.0
0.4	3.0	2.0	0.4	6.0	4.0
0.5	2.5	2.5	0.5	5.0	5.0
0.6	2.0	3.0	0.6	4.0	6.0
0.7	1.2	3.5	0.7	3.0	7.0
0.8	1.0	4.0	0.8	2.0	8.0
0.9	0.5	4.5	0.9	1.0	9.0
1.0	0.0	5.0	1.0	0.0	10.0

[a]1.0% agarose stock.

**Table 2
Dilutions for Preparation of Running Gels:
1.2–3.0% Running Gels in 18-Lane Multigel**

Gel (%)[a]	Buffer (mL)	Stock (mL)[b]
1.2	3.0	2.0
1.4	2.7	2.3
1.6	2.3	2.7
1.9	1.8	3.2
2.2	1.3	3.7
2.4.	1.0	4.0
2.6	0.7	4.3
2.8	0.3	4.7
3.0	0.0	5.0

[a]Indicated gel concentrations are recommendations and can be altered at the user's discretion.
[b]3.0% agarose stock.

0.9–3.0% agarose running gels consists of 1.8 g agarose + 60 g running buffer. Add the agarose and buffer to a 125-mL Erlenmeyer flask and prepare as described in **Subheading 3.1., step 2**. Place the melted agarose stocks into the heated water bath and anchor the flasks using a donut weight. Let the stock solution equilibrate in the water bath for several minutes.

7. Prepare the desired running gels *one at a time*, following the dilutions indicated in **Tables 1** and **2**. Using a disposable plastic pipet (10-mL and 5-mL pipets for

9- and 18-lane multigels, respectively), add the appropriate volume of 1.0 or 3.0% agarose stock (**Tables 1** and **2**) to the corresponding tube containing the warmed running buffer (*see* **Subheading 3.1., step 4**). Cap the tube, quickly mix well by vortexing, and immediately pipet the agarose into the appropriate running gel slot (usually approx 7 mL fills the slots formed by a 9-lane slotformer and approx 3 mL fills the slots formed by an 18-lane slotformer). For the 0.2–1.0% running gels, a suggested left to right order within a 9 lane multigel frame is 0.8, 0.9, 1.0, 0.2, 0.3, 0.4, 0.5, 0.6, 0.7. The 0.9–3.0% gels can be poured as 0.9, 1.0, 1.3, 1.6, 1.9, 2.2, 2.5, 2.8, 3.0. Analogous ordering should be used for 18-lane multigels.
8. Let the running gels polymerize for approx 1 h. Gently remove the comb, the plastic end plates, and the plastic plate supports. The multigel is now complete and ready for use. If the multigel will not be used the same day, leave it in the gel bed and submerse the entire multigel in running buffer. Seal the entire electrophoresis unit with plastic wrap if the multigel will be stored overnight or longer. Multigels stored at room temperature in this fashion can be used up to 10 d after they have been poured.

3.2. Running the Multigel

1. Prior to loading the gel, place the bottom halves of two 100-mL pipet tips (or something similar) into the plate grooves at the end of the gel bed to keep the multigel from sliding out of the bed during the experiment (because of the vigorous buffer circulation, *see* **Subheading 3.2., step 3**).
2. Each well holds up to 10–15 mL of sample. Our laboratory generally loads approx 0.6 µg of both chromatin and bacteriophage T3, although as little material as is needed to clearly visualize each band by staining can be used. If desired, glycerol (0.5 µL/well) can be used to increase the density of the samples.
3. Multigels should be electrophoresed at 1 V/cm for 8 h or an appropriate V/cm × time equivalent (i.e., 1.15 V/cm for 7 h or 1.33 V/cm for 6 h). These recommendations are based on the µ of linear 2.5 kb DNA and 12-mer nucleosomal arrays, and should be adjusted proportionally for smaller or larger chromatin samples. To start the experiment, electrophorese the samples for approx 5 min without buffer circulation, then turn on the pump and maintain buffer circulation throughout the run. Buffer should be circulated at a flow rate of 350–375 mL/min. In our laboratory, multigels are run at room temperature and temperature variations are kept to within ±3°C. The multigel experiment can be run at other temperatures as long as the temperature of the multigel and running buffer is kept to within ±3°C throughout the experiment.
4. When electrophoresis is complete, carefully remove the multigel from the gel bed by sliding a thin sturdy piece of plastic underneath the multigel. Immediately transfer the multigel to an ethidium-bromide solution (0.5 µg/mL) and stain for 30–45 min at room temperature. Destaining is not usually required; however, if desired, the gel may be destained for 10–20 min in water or running buffer. After staining, photograph the multigel under UV light, placing a fluorescent ruler beside the gel to allow accurate conversions between cm and pixels during later

analysis steps (*see* **Subheading 3.3., step 3**). Keep both the photograph negative and positive. The negative is used to digitize the gel image. If desired, the multigel can be stained for protein by carefully transferring the gel to a glass dish, staining with standard Coomassie blue solutions for 20–30 min, and destaining overnight in 30% methanol, 10% acetic acid.

3.3. Analysis of the Multigel: Digitizing the Gel Image

1. **Steps 2–4** (below) describe the gel imaging method that utilizes public domain NIH Image software for Macintosh *(16)*, in combination with image capture by a digital camera *(6)*. Any equivalent image analysis software/hardware can be used. Gels also can be scanned using a densitometer and analyzed using programs such as ImageQuant (Molecular Dynamics). If an alternative imaging approach is used, analogous manipulations as those outlined in **steps 2–4** will need to be performed.
2. Begin by aligning the film negative on a visible light box so that the image of the top of the gel will be perfectly straight when viewed by the image program. Capture the image by first choosing <Special> in the screen headings and then either choosing <Start Capturing> from the <Special> menu or by using Apple G on the keyboard. Working with the captured image on the computer screen, click on the solid-line icon of the tool bar, press the <Shift> key and draw a straight horizontal line through the middle of the multigel wells. If the gel image is not properly aligned (i.e., is not straight), recapture the image and repeat the above steps.
3. Convert pixels to distance (in centimeters) by clicking on the dotted-line icon from the tool bar, pressing the <Shift> key and drawing a line that corresponds to 1.0 cm on the ruler. Complete the scale conversion by choosing <Analyze> from the headings and <Set Scale> from its menu. Select <Analyze> again and choose <Options>, then select <Length> from the measurement options offered.
4. For each band in each running gel, determine the migration (in centimeters) from the solid line drawn through the wells to the center of each band. Activate the dotted-line icon from the tool bar, press the <Shift> key and vertically drag a dotted line from the well to the center of the desired band, then press the <Shift> key. Record the measurements into a Results Table by selecting <Analyze> followed by <Measure>, or simply press Apple 1. To review the Results Table on the screen, press Apple 2 or choose <Analyze> and <Show Results>. Once all of the band migrations have been measured, save the Results Table.

3.4. Analysis of the Multigel: Determination of μ_o

1. Import the migration measurements stored in the Results Table into an appropriate data processing program (e.g., DeltagraphPro, ImageQuant). Use the data processing software to set up a spreadsheet that converts migration (cm) to μ(cm^2/V·s) using the relationship: μ = [migration (in centimeters)]/[voltage potential (in V/cm) × electrophoresis time (in seconds)].
2. Using the data processing software, create plots of log μ vs agarose percentage (Ferguson plots) for both the chromatin sample and the bacteriophage T3 standard *(7,8,11,12)*. Note that migration and mobility values technically are nega-

tive because DNA, chromatin, and bacteriophage T3 each have a net negative charge and move toward the positive electrode.
3. Obtain the gel-free μ (i.e., μ'_o) by extrapolating the linear region of the Ferguson Plot to the Y-axis (0% agarose). Entire Ferguson Plots rarely are linear. Furthermore, the linear region depends on the size of the macromolecule. For chromatin fragments containing 6–36 nucleosomal subunits, the upper boundary of linearity generally falls between 0.6–1.0% agarose (*see* **Note 5** for a discussion of how this affects the choice of which size multigel template to use).
4. Even small differences in electrophoresis times or the voltage gradient used in different experiments will affect both the $\mu^{chromatin}$ and μ^{T3}. Therefore, to allow accurate comparison of the $\mu'_o^{chromatin}$ obtained from different gels, this term must be converted to the $\mu_o^{chromatin}$ (*see* **Notes 6** and **7**). To accomplish this, one must normalize the $\mu'_o^{chromatin}$ to the μ'_o^{T3}, μ_o^{T3} and μ of electroosmotically driven buffer in the agarose used for the experiment (μ_E). Electroosmosis occurs when counterions present in the agarose preparation move toward the negative electrode causing a current in the opposite direction of the migration of chromatin and T3 bacteriophage and effectively reduces the mobility of the negatively charged samples. The $\mu_o^{chromatin}$ is calculated from the experimentally determined μ'_o^{T3}, and the μ_E and μ_o^{T3} using the following relationship:

$$\mu_o^{chromatin} = [\mu'_o^{chromatin}(\mu_o^{T3} + \mu_E)/\mu'_o^{T3}]\mu_E \qquad (2)$$

The μ_E and μ_o^{T3} in TAE-, E- and EM-buffered Research Organics LE agarose are listed in **Table 3**. *See* **Note 5** for a discussion of how the μ_E and μ_o^{T3} are determined.

3.5. Analysis of the Multigel: Calculation of the Running Gel Pore Size (P_e) and Effective Macromolecular Radius (R_e)

1. **Equation 1** is used for calculation of P_e and R_e. Note that μ and μ'_o terms appear in the calculation. However, since the voltage gradient and electrophoresis time are identical for both the μ and μ'_o, the experimentally determined *migrations* and *gel-free migrations* also can be used to obtain the correct P_e and R_e. Using a spreadsheet in the data analysis software, the P_e for each running gel is calculated from the measured migration, gel-free migration and known R_e (30.1 nm) of bacteriophage T3 using the relationship:

$$P_e = 30.1/[1 - SQRT(T3\ migration/T3\ gel\text{-}free\ migration)] \qquad (3)$$

The calculated P_e value for each running gel is then used to determine the chromatin R_e in that gel using the relationship:

$$R_e = P_e \times [1 - SQRT(chromatin\ migration/chromatin\ gel\text{-}free\ migration)] \qquad (4)$$

4. Notes

1. As alluded to in **Subheading 3.2.**, **step 3**, there are circumstances where the multigel must be electrophoresed at approx 1V/cm for longer than 8 h, for example, for chromatin fragments containing >20 nucleosomal subunits (in these cases longer electrophoresis times are required for the large chromatin fragments

**Table 3
Values of μ_o^{T3} and μ_E in Research Organics LE Agarose Buffered with TAE, E, and EM**

Mobility	TAE	E	EM[a]
$\mu_E{}^{2b}$	1.02×10^{-5}	1.07×10^{-5}	0.97×10^{-5}
μ_o^{T3}	-7.55×10^{-5}	-7.89×10^{-5}	-6.43×10^{-5}

[a]Values are the same in the presence of 0.5–2.0 mM MgCl$_2$.
[b]Indicated mobilities are in units of cm^2/V·s.

to separate from the phage). Under these conditions, a Neslab RTE-Series temperature-controlled water bath/circulator should be employed to help prevent pH and temperature fluctuations. Note that a temperature-controlled water bath generally is not required for multigels run at room temperature for <8 h as long as the fluctuations in room temperature are not significant.

2. As is described in **Subheading 3.4.**, measurement of $\mu_o^{chromatin}$ requires knowledge of both the μ_o^{T3} and μ_E for the particular agarose used in the experiment. To determine μ_E and μ_o^{T3}, bacteriophage T3 is electrophoresed in a single 18-lane multigel in which half of the running gels consist of standard agarose and the other half consist of Seakem Gold™ electroosmosis-free agarose (FMC Bioproducts, Rockport, ME). The μ'_o^{T3} calculated from the Seakem Gold lanes is equal to the μ_o^{T3}. The μ_E equals the μ'_o^{T3} derived from the standard agarose minus the μ'_o^{T3} derived from the Seakem Gold agarose. Importantly, to accurately determine μ_E and μ_o^{T3} in this manner, the multigel must be run under *stringent* temperature control (i.e., ± 0.3°C) with the use of a Peltier-controlled temperature apparatus *(17)*. Both the Seakem Gold agarose, particularly the temperature control system, are potentially prohibitively expensive. We therefore recommend that the Research Organics molecular-biology-grade LE agarose be used for the multigels, because the μ_E and μ_o^{T3} values for this agarose in TAE, E, and EM buffers are already known (**Table 3**) and therefore do not need to be redetermined.

3. A number of complications related to storage, bursting, and staining of the bacteriophage T3 standard can be encountered. The concentrated phage stock should be stored in 1 mM MgCl$_2$ to prevent both bursting (i.e., release of the phage genome as free DNA) and aggregation. During low-salt multigel experiments using TAE or E running buffer, some of the phage inevitably will burst because of the absence of Mg^{2+} *(7)*. When this happens, some free phage DNA will be visualized after ethidium bromide staining *(7)*. This band can be mistaken for the intact phage, but should *not* be analyzed. If there is any question about which band corresponds to intact phage, stain the multigel with Coomassie blue subsequent to staining with ethidium bromide. Only the intact phage will stain with Coomassie. Even staining the intact phage with ethidium bromide sometimes can be difficult. After electrophoresis, incubating the multigel in 1 mM EDTA prior

to staining, or adding 1 m*M* EDTA to the ethidium-bromide solution, will burst the phage within the multigel and yield better band visualization. In these cases, it is not possible to subsequently stain with Coomassie.

4. Chromatin folding experiments are performed in running buffer containing 0.5–2.0 m*M free* MgCl$_2$ *(8,9)*. For these experiments, the chromatin samples can be prepared in two ways. The ideal method is to dialyze the chromatin sample into the same EM running buffer as will be used for the experiment. In this case, the phage should be added to the dialyzed chromatin sample immediately prior to loading. The second method is to add the appropriate amount of concentrated MgCl$_2$ stock to the chromatin/phage mixture to yield the desired salt concentration. Also, for chromatin folding experiments, the multigel frame and running gels may be prepared with or without buffer containing MgCl$_2$. If the multigel is prepared without MgCl$_2$, it should be incubated overnight in the appropriate EM running buffer.

5. As described in **Subheading 3.4.**, **step 3**, the μ'_o is determined by extrapolating the linear region of a plot of log μ vs agarose concentration (Ferguson plot) to 0% agarose. Because the linear region rarely extends above 1.0% agarose, one should use the 9-lane multigel only for determining the $\mu^{chromatin}$ in 0.2–1.0% running gels. When determination of the $\mu^{chromatin}$ in 1.1–3.0% running gels is desired, the 18-lane multigel template should be used and 4–6 of the 18 running gels should be dilute enough to fall into the linear region of the Ferguson plot (thereby yielding sufficient data for accurate determination of the μ'_o in that experiment).

6. The μ_o provides a measure of macromolecular surface charge density, that is, the total charge per surface area. Consequently, for DNA and chromatin, the μ_o is *independent* of length and its absolute value is constant.

7. The $\mu_o^{chromatin}$ is an amalgam of its numerous components, including the DNA, core histones, and any other macromolecules that are associated with the nucleosomal array. In running buffers containing salts, the $\mu_o^{chromatin}$ also is influenced by the charge of the cations that co-migrate with the chromatin during electrophoresis *(9)*. A detailed discussion of interpreting the $\mu_o^{chromatin}$ in the absence and presence of cations is presented in **ref. 9**. The $\mu_o^{chromatin}$ provides a very sensitive assay for the nucleosomal array density *(7)*, the extent of salt-dependent chromatin folding *(8,9)*, and the interaction of highly charged macromolecules, for example, linker histones *(20)*, with nucleosomal arrays.

8. To obtain information about the size, shape and flexibility of chromatin, first plot the $R_e^{chromatin}$ obtained in each running gel as a function of the corresponding P_e *(6–9)*.

9. The value of $R_e^{chromatin}$ should be independent of P_e when the P_e is at least six- to eightfold larger than $R_e^{chromatin}$, that is, in dilute gels *(6–9)*. Importantly, under these conditions, the behavior of the chromatin R_e closely parallels the behavior of the average sedimentation coefficient measured in the analytical ultracentrifuge *(7–9)*. Thus, information about chromatin folding and the nucleosomal density of the chromatin array is obtained from the R_e measured in dilute running gels *(7–9)*. The absolute value of R_e is the true radius only for spherical molecules, but is strongly correlated with the surface area for rod-shaped macromolecules, such as chromatin *(8,9,12)*.

10. In contrast to the situation in dilute running gels, in concentrated gels the value of the R_e approaches that of P_e. Under these conditions, the conformation of a flexible macromolecule becomes deformed during gel electrophoresis and the macromolecule moves end-first through the gel pore network. This phenomenon has been termed *reptation (18,19)*. In a multigel experiment, the reptation exhibited by flexible molecules manifests as a decrease in R_e with decreasing P_e *(6–9,12)*. In contrast, molecules that do not deform during electrophoresis show no P_e-dependence of R_e *(6–8,11)*. Thus, the P_e-dependence of $R_e^{chromatin}$ in concentrated running gels provides an assay for the relative deformability and flexibility of chromatin, which in turn is very sensitive to the nucleosomal density and regularity of nucleosomal spacing of the underlying nucleosomal array *(7–9)*.

References

1. Wolffe, A. P. (1995) *Chromatin: Structure & Function, 2nd edition*. Academic Press, New York.
2. Fletcher, T. M. and Hansen, J.C. (1996) The nucleosomal array: structure/function relationships. *Crit. Rev. Euk. Gene Exp.* **6,** 149–188.
3. Woodcock, C. L. and Horowitz, R. A. (1995) Chromatin organization re-viewed. *Trends Cell Biol.* **5,** 272–277.
4. Belmont, A. S. and Bruce, K. (1994) Visualization of G1 chromosomes: a folded, twisted, supercoiled chromenema model of interphage chromatid structure. *J. Cell Biol.* **127,** 287–302.
5. Leuba, S. H., Yang, G., Robert, C., Samori, B., van Holde, K., Zlatanova, J., and Bustamante, C. (1994) Three-dimensional structure of extended chromatin fibers as revealed by tapping-mode scanning force microscopy. *Proc. Natl. Acad. Sci. USA* **91,** 11,621–11,625.
6. Hansen, J. C., Kreider, J. I., Demeler, B., and Fletcher, T. M. (1997) Analytical ultracentrifugation and agarose gel electrophoresis as tools for studying chromatin folding in solution. *Methods: A Companion to Methods Enzymol.* **12,** 62–72.
7. Fletcher, T., Krishnan, U., Serwer, P., and Hansen, J. C. (1994) Quantitative agarose gel electrophoresis of chromatin: nucleosome-dependent changes in charge, shape and deformability at low ionic strength. *Biochemistry* **33,** 2226–2233.
8. Fletcher, T. M., Serwer, P, and Hansen, J. C. (1994) Quantitative analysis of macromolecular conformational changes using agarose gel electrophoresis: application to chromatin folding. *Biochemistry* **33,** 10,859–10,863.
9. Fletcher, T. M. and Hansen, J. C. (1995) Core histone tail domains mediate oligonucleosome folding and nucleosomal DNA organization through distinct molecular mechanisms. *J. Biol. Chem.* **270,** 25359–25362.
10. Serwer, P. (1986) Use of gel electrophoresis to characterize multimolecular aggregates. *Methods Enzymol.* **130,** 116–132.
11. Griess, G. A., Moreno, E. T., Easom, R., and Serwer, P. (1989) The sieving of spheres during agarose gel electrophoresis: quantitation and modeling. *Biopolymers* **28,** 1475–1484.

12. Griess, G. A., Moreno, E. T., Herrmann, R., and Serwer, P. (1990) The sieving of rod-shaped viruses during agarose gel electrophoresis. I. Comparison with the sieving of spheres. *Biopolymers* **29,** 1277–1287.
13. Serwer, P. (1989) Sieving of double-stranded DNA during agarose gel electrophoresis. *Electrophoresis* **5/6,** 327–331.
14. Rodbard, D. and Chrambach, A. (1970) Unified theory for gel electrophoresis and gel filtration. *Proc. Natl. Sci. Acad. USA* **65,** 970–977.
15. Serwer, P., Watson, R. H., Hayes, S. J., and Allen, J. L. (1983) Comparison of the physical properties and assembly pathways of the related bacteriophages T7, T3 and phi II. *J. Mol. Biol.* **170,** 447–469.
16. O'Neill, R. R., Mitchell, L. G., Merril, C. R., and Rasband, W. S. (1989) Use of image analysis to quantitative changes in form of mitochondrial DNA after ×-irradiation. *Appl. Theor. Electrophor.* **1,** 163–167.
17. Serwer, P. and Dunn, F. J. (1990) Rotating gels: why, how and what. *Methods: A Companion to Methods Enzymol.* **1,** 142–150.
18. Lumpkin, O. J., DeJardin, P., and Zimm, B. H. (1985) Theory of gel electrophoresis of DNA. *Biopolymers* **24,** 1573–1593.
19. Noolandi, J. (1992) Polymer dynamics in electrophoresis of DNA. *Ann. Rev. Phy. Chem.* **43,** 237–256.
20. Carruthers, L. M., Bednar, J., Woodcock, C. L., Hansen, J. C. (1998) Linker histones stabilize the intrinsic salt-dependent folding of nucleosomal arrays: mechanistic ramifications for higher-order chromatin folding. *Biochemistry* **37,** 14,776–14,787.

9

Analytical Ultracentrifugation of Chromatin

Jeffrey C. Hansen and Cynthia L. Turgeon

1. Introduction

1.1. Analytical Ultracentrifugation and the Complexity of Chromatin

The ability of analytical ultracentrifugation to elucidate chromatin structure/function relationships originates directly from its capacity to accurately measure key structural properties of complex macromolecular assemblies in solution. **Figure 1** schematically illustrates the complex nature of chromatin. Newly replicated DNA is wrapped around core histone octamers spaced at approx 200 bp intervals to form nucleosomal arrays, which then interact with linker histones and numerous other nonhistone chromosomal proteins to form "chromatin." Chromatin is conformationally dynamic, undergoing a number of short-range and long-range folding transitions to produce highly condensed interphase chromosomal fibers (**Fig. 1**). For short chromatin fragments studied in vitro, fiber condensation manifests both in the form of intramolecular conformational changes and reversible oligomerization *(1–4)*. In addition, the structure of chromatin fibers and functions such as transcription and replication are irrevocably linked; any given region of a chromosomal fiber can be either functionally active or inactive depending on both its specific complement of chromatin-associated proteins and its overall state of condensation *(1,2)*. Consequently, to biochemically characterize chromatin structure/function relationships in vitro, one must be able to analyze both the intramolecular conformational dynamics and intermolecular interactions of an exceedingly complex macromolecular assembly (e.g., a 12-mer nucleosomal array containing one H1 molecule per nucleosome consists of >100 proteins and 2400 bp of DNA, has a molecular mass of approx 3.5×10^6 D, yet represents only roughly one millionth of an intact eukaryotic chromosome.) The power of analytical

From: *Methods in Molecular Biology, Vol. 119: Chromatin Protocols*
Edited by: P. B. Becker © Humana Press Inc., Totowa, NJ

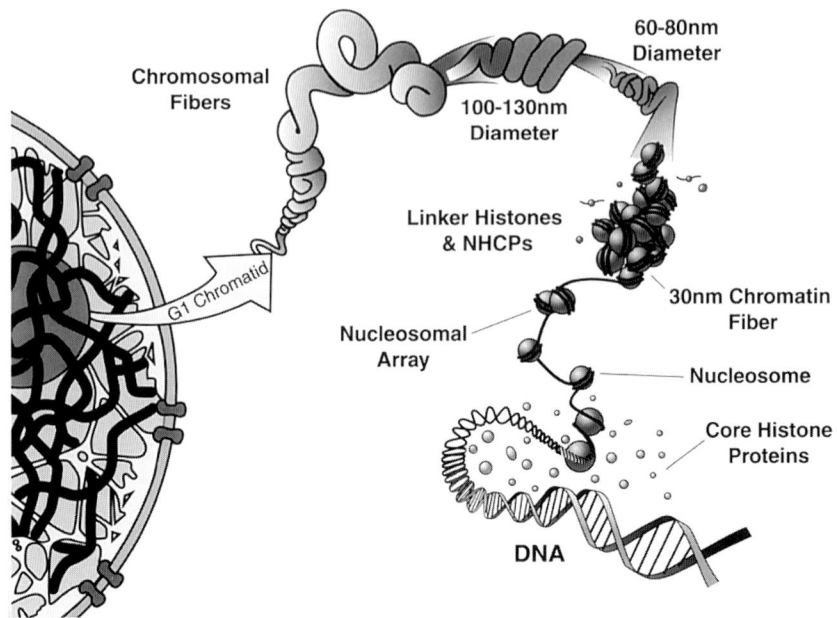

Fig. 1. Schematic illustration of the hierarchical relationships between DNA, chromatin, and interphase chromosomal fibers.

ultracentrifugation ultimately lies in its ability to directly measure both the sedimentation coefficient and molecular weight of complex macromolecular assemblies such as chromatin with analytical precision *(3)*.

1.2. Fundamental Principles of Analytical Ultracentrifugation

The generalized theoretical relationship that describes the sedimentation of macromolecules in the analytical ultracentrifuge and underlies the utility of this technique for studying chromatin is

$$s \propto M/f$$

where s is the sedimentation coefficient, M is the molecular mass, and f is the frictional coefficient. The latter parameter is a direct function of macromolecular shape; for macromolecules having the same M, $f_{sphere} < f_{rod}$. For chromatin samples between 10^4 and 10^8 Dalton, an analytical ultracentrifuge can be used to directly and accurately determine s in a sedimentation velocity experiment and M in a sedimentation equilibrium experiment, thereby also providing an accurate determination of f.

For both sedimentation velocity and sedimentation equilibrium experiments one begins with a uniform solution of macromolecules present in the centrifuge cell when the rotor is at rest. In a sedimentation velocity experiment, suf-

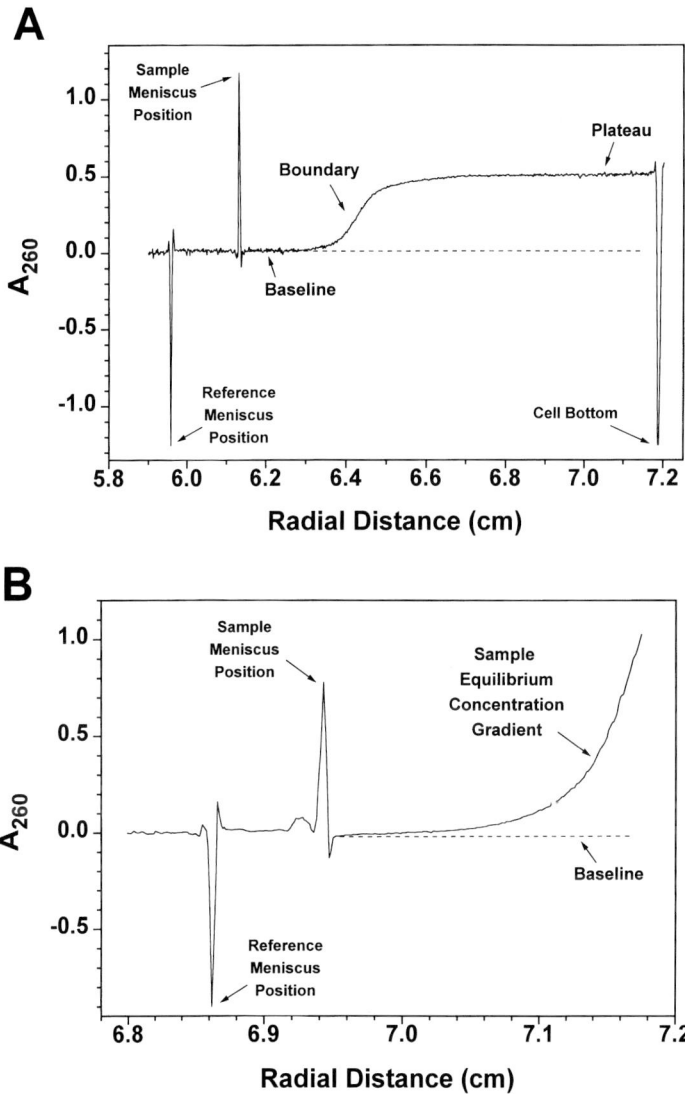

Fig. 2. **(A)** Sedimentation velocity scan. **(B)** Sedimentation equilibrium scan. *See* **Subheading 1.2.** for additional explanation.

ficiently high rotor speeds are used to create a moving boundary that can be visualized by the optical detector of the analytical ultracentrifuge (**Fig. 2A**). The boundary is formed when the macromolecules initially present at and near the meniscus begin to sediment, creating a concentration gradient of macromolecules whose rate of movement is proportional to the macromolecular sedimentation coefficient(s) *(4–8)*. In addition, diffusion of the concentration gradient causes the shape of the boundary to broaden with time. (*See* **Notes 7–12** for a discussion of why this is important.) The analytical ultracentrifuge records both the shape and the radial position of the boundary as a function of time. Data analysis methods exist that allow either an average sedimentation coefficient or a sedimentation coefficient distribution to be derived from this set of boundaries (**refs. 8–10**; *see* **Notes 10–12**). In a sedimentation equilibrium experiment, lower rotor speeds and longer times are used. In this case, macromolecular diffusion balances sedimentation and a static concentration gradient forms within the centrifuge cell at equilibrium (**Fig. 2B**). The slope of the sedimentation equilibrium gradient is a function of macromolecular mass, the presence or absence of macromolecular associations, and the potential presence of thermodynamic nonideality *(4–10)*. Data analysis methods exist that use nonlinear least squares fitting routines to extract all of these macromolecular parameters from sedimentation equilibrium gradients *(8–10)*. Importantly, measurements of s and M in the analytical ultracentrifuge are made in standard aqueous buffers, allowing characterization of macromolecular size and shape under physiologically relevant conditions.

Given this introductory background, it is now apparent that chromatin folding can be documented rigorously in the analytical ultracentrifuge by showing that s increases under conditions where M remains constant *(3,11)*. Macromolecular interactions, be they assembly of DNA with core histone octamers to form nucleosomal arrays or binding of nonhistone chromosomal proteins to nucleosomal arrays, will change both s and M, and may also alter f depending on whether there has been a concomitant interaction-induced change in chromatin conformation. Each of these situations can be dissected experimentally using a combination of sedimentation velocity and sedimentation equilibrium experiments in the analytical ultracentrifuge. Furthermore, when the analytical ultracentrifugation studies are performed with the same chromatin samples under identical solution conditions used for in vitro functional assays (e.g., in vitro transcription), it is possible to directly and rigorously establish the biochemical basis of chromatin structure/function relationships *(12,13)*.

After a lengthy hiatus, Beckman Instruments now manufactures a state-of-the-art analytical ultracentrifuge that acquires data digitally using either absorbance or interference optical systems to detect the sample concentration inside the centrifuge cell *(8–10)*. Combined with the recently available wealth of powerful software programs that allow for the rapid analysis of digital sedimentation velocity and sedimentation equilibrium data *(8–10)*, it is now possible for chromatin researchers at all levels to utilize analytical ultracentrifugation to its fullest. The rest of this article provides a level of technical detail that cannot be found elsewhere in the literature. This information is intended to allow any chromatin researcher with access to one of the new generation analytical ultracentrifuges to immediately be able to successfully perform sedimentation velocity and sedimentation equilibrium analyses of chromatin samples once the requisite criteria are fulfilled.

2. Materials
2.1. Chromatin Samples

As long as the criteria described in **Subheading 3.1.** are fulfilled, there are no limitations regarding the source of the chromatin used for an analytical ultracentrifugation experiment. Consequently, one can just as easily obtain high-quality data using endogenous chromatin fragments purified from isolated nuclei *(14–16)* as with defined nucleosomal arrays reconstituted in vitro from pure core histones and DNA *(3,4,11–13)*. There is, however, a major difference in the interpretability of the data, depending on the chromatin source. Native samples are inherently heterogeneous in terms of both s and M and therefore will yield only highly averaged structural information *(14–16)*, while reconstituted chromatin samples are sufficiently homogeneous to allow structural characterization of the entire distribution of species present under the conditions of the experiment *(3,4,11–13)*. The importance of this point will be discussed further in **Notes 10–12**.

2.2. Analytical Ultracentrifuge and Accessories

The Beckman Optima XL-A analytical ultracentrifuge is equipped with a scanning absorbance optical system capable of directly measuring the absorbance of macromolecules inside a centrifuge cell that is spinning between approx 1000–60,000 rpm. A high-intensity xenon flash lamp together with a removable monochromator produce wavelengths of light ranging from 190–800 nm. The absorbance detector and remainder of the optical system electronics are housed underneath the rotor chamber. Chromatin samples are placed in specialized centrifuge cells consisting of 12-mm epon charcoal-filled or aluminum centerpieces sandwiched between two optical-grade quartz windows. A four-hole titanium An-60 Ti rotor that holds three centrifuge cells and a counterbalance reference generally is the most versatile rotor for chromatin experiments in the XL-A. At different times throughout the experiment the scanning absorbance optical system measures sample absorbance as a function of radial distance inside each of the spinning centrifuge cells, and the data are written to the computer hard drive as ASCII code. These data are referred to as "scans," and are viewed as plots of absorbance vs radial distance (**Fig. 2**).

Recently, Beckman Instruments developed an Optima XL-I analytical ultracentrifuge that contains both the scanning absorption optical system and a Rayleigh interference optical system. The interference optical system measures sample concentration based on the refractive index of the solution *(17)*. Both the XL-A and the XL-I are purchased as an entire package that includes the rotor, monochromator, cell components, and computer system. However, it is important to note that at the time of this writing, interference optics has yet to be used to characterize chromatin. Thus, all information presented in **Subheading 3.** applies to the use of the scanning absorbance optical system, which in most cases is the optical system of choice for experiments involving chromatin anyhow.

2.3. Data Analysis Methods

After the centrifuge run is completed, the digitally acquired data are analyzed on the microcomputer. For both sedimentation velocity and sedimentation equilibrium experiments, numerous analysis methods exist that have been programmed for use with XL-A data. Notes on how to best use the available software for interpreting chromatin experiments are provided in **Notes 10–12**.

3. Method
3.1. Sample Preparation (see Note 1)

1. The sample concentration and volume requirements are as follows: for sedimentation velocity runs, approx 5–15 µg of DNA in a volume of 0.40–0.45 mL. This results in a sample A_{260} of approx 0.25–0.8. For a typical XL-A, the A_{260} range that maximizes the signal:noise ratio of a sedimentation velocity run while staying within the linear region of the XL-A's absorbance detector is 0.6–0.8. For sedimentation equilibrium runs, approx 1.0–1.5 µg of DNA is required in a volume of 0.1–0.13 mL, producing an A_{260} of approx 0.2–0.3. Also required for each type of experiment is a slightly larger volume of the *exact* buffer that the chromatin is dissolved in, which is used as a reference blank during the centrifuge run (*see* **Subheading 3.2.**, **step 3**). It is important to always load more volume in the reference sector so that the reference meniscus position does not fall into the region of the scan that contains the sample boundaries (*see* **Fig. 2A**).
2. Chromatin experiments generally are performed in 10 m*M* Tris-HCl, 0.25–1.0 m*M* EDTA (TE) buffer containing various salts and salt concentrations at a desired pH *(3,11,14–16)*. Other nonabsorbing buffers, e.g., HEPES, have also been used *(12,13)*. The buffers can include any component that does not interfere with the chromatin sample absorbance at 260 nm, e.g., 5% glycerol.

3.2. Assembling and Filling the Centrifuge Cell

1. Before beginning to assemble the cells, check for damaged components. Make sure that quartz windows and centerpieces are scrupulously clean and without scratches or cracks. Use of damaged or dirty components both increases the chance of sample leaks and introduces unnecessary noise in the absorbance scans. *See* **Notes 2–6** for more comments regarding these and related problems.

2. A centrifuge cell is assembled from two window holders, two window gaskets, two window liners, two quartz windows, one centerpiece, one cell housing, and one screw ring with accompanying gasket (**Fig. 3**). For each cell being used, start by building two window assemblies (**Fig. 3A**). First place a new window gasket in each window holder, then insert the window liner so that the approximate middle of the liner aligns opposite the keyway notch on the window holder. With lens tissue, hold the quartz windows by their edges and insert it into the window holder. Adjust the window position so that the scribe mark on the window edge lines up with the keyway notch of the window holder. Be careful not to leave fingerprints, smudges, dust, or lint on the windows during these manipulations (*see* **Notes 3** and **6**).

   Insert one completed window assembly into the cell housing (window facing up) using the keyway on the inside of the cell housing as a guide. Then insert the centerpiece and the other window assembly (window down) in the same manner (**Fig. 3B**). Apply a thin film of Spinkote lubricant to the screw ring gasket and screw ring threads. Lay the screw ring gasket on top of the exposed upper window assembly and manually twist the screw ring into the top of the cell housing (**Fig. 3C**) using the cell-alignment tool. Place the now completely assembled centrifuge cell on the bottom plate of the torque wrench apparatus and apply *exactly* 120 inch-pounds of torque. More torque is unnecessary and will cause the cell housing to distort, which makes it difficult to place the cell in the rotor hole and increases the chances for sample leaks during the centrifuge run.

3. Hold the cell horizontally at eye level with the screw ring facing you. Load the reference buffer in the left cell sector (formed by sandwiching the centerpiece between the two window assemblies—*see* **Fig. 3B**). Loading is best accomplished by using an adjustable pipettor (e.g., P1000 Pipetteman) together with an elongated polypropylene gel loading tip that fits easily into the small exposed centerpiece sector hole on the side of the cell housing (**Fig. 3C**). Repeat the loading procedure, placing the identically buffered solution containing the chromatin sample in the right sector. Loading can often be a frustrating step and helpful strategies are presented **Note 4**. After the loading is complete, place a polyethylene plug gasket over each sector filling hole, and tightening the brass housing screws with a small screwdriver. Tighten screws *no more than $1/8$ turn* beyond where resistance is first met. Even slight overtightening of the housing screws damages the screw threads and may cause the centrifuge cell to distort.

3.3. Preparing the Centrifuge Run

1. Most important, for sedimentation velocity runs, place the rotor and monochromator in the chamber and engage the chamber vacuum prior to assembling and filling the centrifuge cells. Allow the rotor to equilibrate at a temperature 2°C cooler than the desired run temperature before proceeding with the experiment.
2. Once the centrifuge cells are assembled and filled and the rotor temperature is equilibrated, release the chamber vacuum and remove the rotor. The rotor holds three cells and one counterbalance. Set the counterbalance in rotor hole 4 with

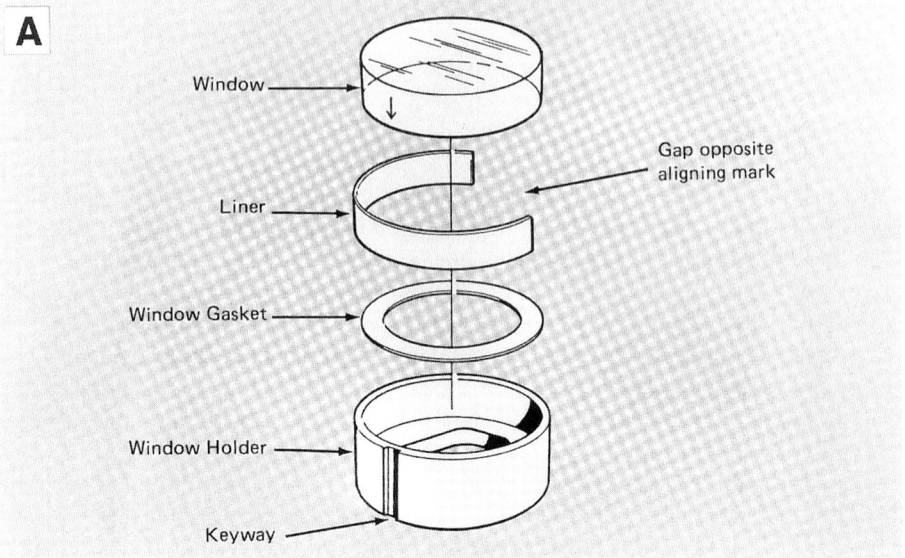
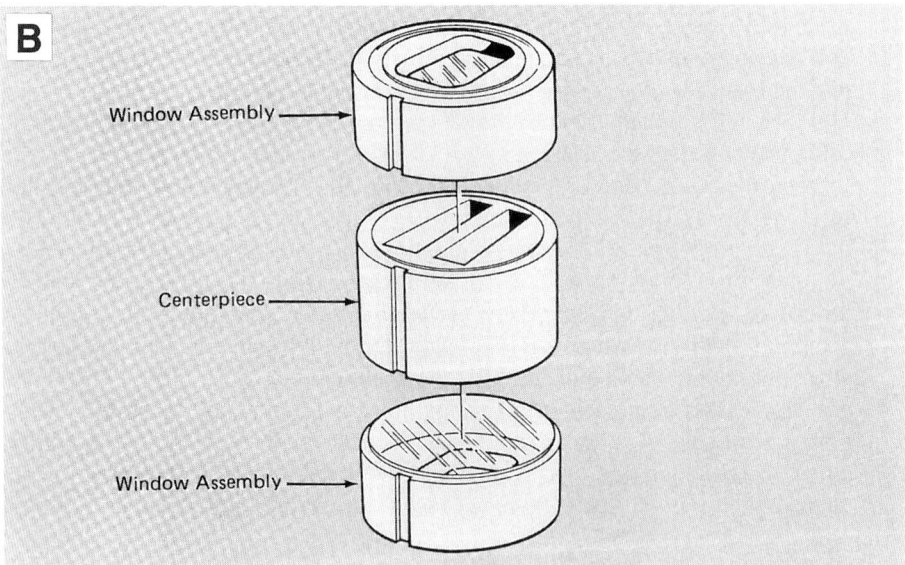

Fig. 3. Components of the centrifuge cell. (**A**) Building of window assemblies as described in **Subheading 3.2.**, **step 2**. (**B**) Stacking of centerpiece and window assemblies. Shown is a representation of how the centerpiece and window assemblies should be stacked inside the cell housing as described in **Subheading 3.2.**, **step 2**. On the opposite side of the centerpiece are the filling holes, which are visible in (**C**). (**C**) Cell housing (containing stacked window assemblies and centerpiece), with screw ring and gasket. (Illustrations are reproduced courtesy of Beckman Instruments).

Analytical Ultracentrifugation

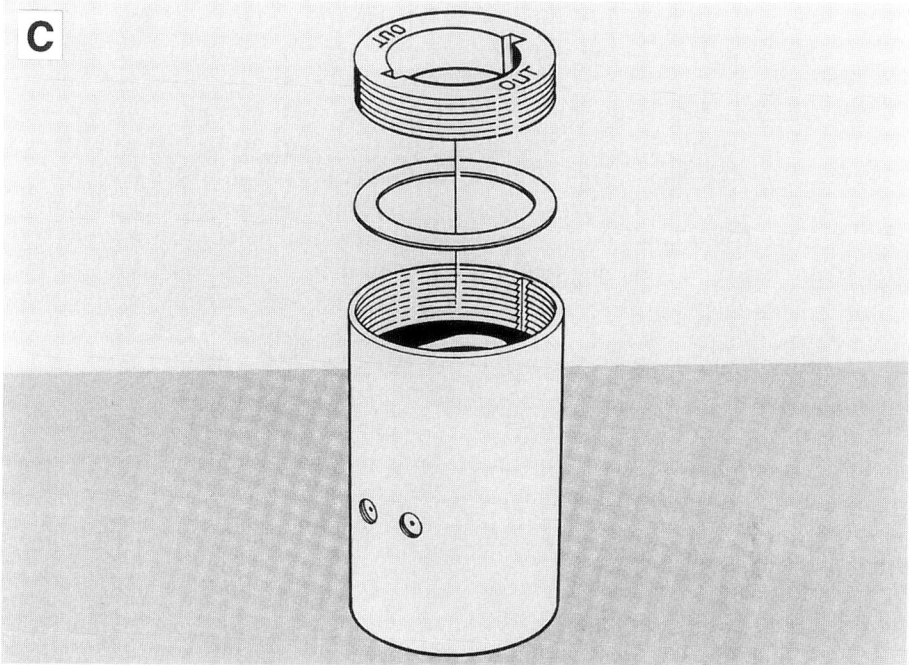

Fig. 3C.

the setscrew facing up and the arrow directed *away* from the center of the rotor. Align the scribe marks on the bottom of the rotor and counterbalance. Tighten the setscrew while maintaining proper alignment. Cell 2 is placed in hole 2 opposite the counterbalance, and must weigh within 0.5 g of the counterbalance. Various-sized weights are screwed into the middle of the counterbalance to adjust the overall weight. The screw weights should not protrude from the top or bottom of the counterbalance. Cells 1 and 3 are placed in holes 1 and 3, respectively, and also must weigh within 0.5 g of each other. (This should not be a problem if the same volume of liquid is loaded in each cell.) The cells are placed in the rotor hole with brass screws facing the center of the rotor and screw rings directed upward. One should be able to push the cell completely into the rotor hole without applying excessive force. When all the cells have been inserted, carefully place the rotor on its side and for each cell align the scribe mark on the bottom of the cell with the scribe mark on the rotor using the cell-alignment tool.

3. Place the rotor on the drive and manually align one of the rotor holes with the absorbance detector opening. Gently insert the monochromator into the chamber, being careful not to bend the guide connectors. Activate the vacuum, and set the run temperature using the keypad on the XL-A console. For a sedimentation velocity run, chromatin samples need to be stable within ±0.2°C of the set run temperature for approx 1 h before the centrifuge run can be started and data can

be acquired. Sedimentation equilibrium runs can be started as soon as the vacuum reaches 50 μm (*see* **Subheading 3.4.**, **step 1**).
4. While the XL-A is pulling vacuum, go to the computer and open a <New File> in the Beckman Data Acquisition User Interface program (version 3.01x) for Windows. For each cell, set the parameters for the sedimentation run and fill in a description for each sample. Create a data directory name to hold the acquired raw data in the Detail dialog box for each cell. Importantly, change the radial step size to 0.001 cm (*see* **Note 7**). Next, in the Options box, choose to overlay approx 20 scans for real-time viewing of data acquisition. In the Method dialog box select the number of scans to be collected, usually the same number that are viewed. Enter the desired temperature and speed of the run into the XL settings box (*see* **Notes 8** and **9** for speed selection).

The last parameters that need to be entered are the scanning mode and number of measurement repetitions at a given radial distance. The respective options are continuous, or step mode, and 1–100 measurements. Sedimentation velocity runs are always performed using a combination of continuous mode and one measurement; step mode and/or multiple repetitions will lead to artificial distortions of the boundary shape because of a scanning (i.e., data collection) speed that is too slow. A completely different set of conditions apply for sedimentation equilibrium experiments because no net movement of the concentration gradient is occurring. In this case, continuous mode is generally used in order to document the close approach to equilibrium. To verify that equilibrium has been reached, and to collect the final scans, use step mode and collect 25–50 absorbance measurements at each radial increment.

3.4. Performing the Centrifuge Run

1. The XL-A will not operate unless the chamber vacuum is ≤50 microns. If the vacuum is >50 μm after **Subheading 3.3.**, **step 4** has been completed, use the mouse to click on <Start Method Scan> in the User Interface program. This activates the diffusion pump without starting the run. In this case, the run is started by pressing the <Start> button on the XL-A keypad after the vacuum has dropped below 50 microns. For a sedimentation velocity run, it is important to allow samples to equilibrate at the set temperature for approx 1 h before starting the run in this manner. Sedimentation equilibrium runs can be started immediately. If the vacuum is ≤50 microns after **Subheading 3.3.**, **step 4** has been completed, a different approach is taken. In this case, after the appropriate time, start the experiment from the computer by using the mouse to click on <Start Method Scan> in the interface program.
2. As the rotor is accelerating to the final set speed, the vacuum should be monitored for a sudden increase in chamber pressure, which is usually indicative of a leak in one of the cells. If a major leak occurs, the rotor may become sufficiently imbalanced for the run to abort. Minor cell leaks come in many forms, but all are identifiable by inappropriate menisci positions in the raw data scans. If cell leaks of any kind have occurred, the run should be aborted.

3. After the run is complete, buffers and samples must be removed from the cells. Use an adjustable pipet with elongated gel-loading tips if you want to reclaim the sample or an analogous vacuum trap setup if the sample is disposable. Disassemble the cells in the reverse order that they were assembled. Because samples can be degraded in the presence of protease and nuclease, one should wear powderless gloves when washing windows and centerpieces. Gently rinse the windows and centerpieces in warm water, then wash by gentle sonication in a dilute, mild liquid detergent solution (e.g., Liqui-nox). Rinse again with deionized water, then with 95% ethanol, and finally again with deionized water. See **Note 5** for additional and more stringent cleaning methods that can be used if necessary.

4. Notes

1. Unlike many other macromolecules, preparation of chromatin samples rarely presents problems. To ensure adequate signal during the experiment, one should always perform a microfuge-based differential centrifugation assay on chromatin sample solutions *(4)* prior to analytical ultracentrifugation in order to determine how much of the total absorbance measured in an ordinary spectrophotometer is coming from high-molecular-weight chromatin oligomers *(4)*. In addition, avoid adding small amounts of concentrated salt stocks to chromatin samples during adjustment to the final salt level; dilute the sample with no more than a 2–3× stock solution or dialyze the sample into the desired solution. Before performing the initial experiments, researchers in all fields are strongly encouraged to download the Sample Preparation Guide prepared by our facility (located at http://bioc02.uthscsa.edu/.biochem/xla.html), which addresses in detail all aspects of sample preparation, including potential problems.
2. Additional useful diagrams relating to assembling the centrifuge cell (*see* **Subheading 3.2.**, **steps 1–3**) can be found in the Beckman XL-A instruction manual.
3. There are various ways that cell components may be damaged when assembling cells, for example, handling windows and centerpieces without lens tissue will likely produce window scratches. Also, dirty cell housing threads, screw ring threads, and/or housing keyways cause resistance during cell assembly and will cause an incorrect amount of torque to be applied.
4. For sedimentation velocity runs, loading the cells can cause much frustration for the beginner (and sometimes even the expert!), because the volumes being used almost completely fill the centerpiece sectors. Holding the cell at eye level offers the best visibility and control. The reference and sample solutions can be added to the cell sectors too quickly, causing sector sides and the fill hole to get splashed. If this occurs, the sample will overflow the loading hole before the sector is full. Overflowing also will occur if the pipet tip is inserted too far into the fill hole, thereby creating a weak vacuum within the sector. It is not easy to salvage these situations. One can try to carefully load more sample with a syringe and needle, if there is sufficient sample to spare. The drawback in this case is that the sample will continue to overflow slightly as the trapped air bubble is displaced with

sample. Another possibility is to tilt the cell and try to load sample into the air pocket. Sometimes nothing works, and the cell will have to be dismantled and the entire cell assembly and sample loading processes repeated.

5. The presence of proteases and nucleases in samples can ruin an analytical ultracentrifuge experiment and may continue to cause problems in subsequent runs because of contaminated cell component surfaces. In most cases, following the cleaning directions in **Subheading 3.4.**, **step 3** will be sufficient to avoid contamination. If needed, soaking the centerpiece in a 100 mM EDTA solution and the windows in 50% nitric acid/50% sulfuric acid (for approx 4 h) will remove remaining contaminants. Also, for experiments involving chromatin and RNA, soaking all cell components in autoclaved diethyl pyrocarbonate-treated water is recommended.

6. There can be many sources of sample or reference solution leaks. Scratched or cracked centerpieces and quartz windows can lead to improper centerpiece-window sealing during cell assembly. Other leaks originate from improperly torqued cells, because of problems with either the polyethylene plug gaskets or brass housing plugs at the fill holes. Whether the leak is slow or fast, the data collected will be of no use.

7. The default radial step size in the User Interface program is 0.003 cm. When performing sedimentation velocity experiments with chromatin samples, which diffuse slowly and produce steep, sharp boundaries, it is important to change the radial step size to 0.001 cm to maximize the number of data points acquired in the boundary region of each scan. To generate the maximum number of data points for the subsequent data analysis, a radial step size of 0.001 cm also should be used when collecting final sedimentation equilibrium datasets.

8. For sedimentation equilibrium experiments, try and collect analyzable data at 3–4 different speeds that produce a ratio of A_{260} (cell bottom): A_{260} (meniscus) of 50–200 without the A_{260} (cell bottom) exceeding 1.2. Because, chromatin samples are large and diffuse very slowly, this will require roughly 1 wk of machine time. For samples being studied for the first time, this process usually involves an educated guess at the correct speed followed by trial and error. Subsequently analyze each dataset to see if the same answer is achieved for each set of conditions.

9. For sedimentation velocity experiments, the speed must be adjusted in a way that best meets several criteria. Optimal boundary resolution is achieved at the fastest possible speed *(18)*. This can be defined as the speed that
    a. allows collection of at least 15 scans,
    b. causes the boundary to move as far toward the bottom of the cell as possible,
    c. maintains a flat plateau region of the scan (*see* **Fig. 2A**) while satisfying criteria a and b.

    It is our experience that very few chromatin samples of any type do not have at least a small amount of high-molecular-weight components in them. In these cases, the speed must be slow enough to prevent the faster material from pelleting during the run, thereby creating a sloping plateau in the latter scans *(18)*.

10. There are numerous methods available for the analysis of both sedimentation velocity and sedimentation equilibrium data in the XL-A *(8–10)*, most of which have been programmed by either Beckman Instruments or other researchers for

use with digital XL-A data. Detailed descriptions of each method fall beyond the scope of this article and can be found in the indicated references. In general, available sedimentation velocity methods fall into three categories:
a. those that yield an average sedimentation coefficient, that is, midpoint or second moment s,
b. one that yields a nondiffusion corrected sedimentation coefficient distribution, g(s*) *(19)*,
c. one that yields a diffusion-corrected integral distribution of s, G(s) *(18,20)*.

The latter method has proven to be the most useful for analyzing chromatin samples because of its ability to detect and characterize the presence of the multiple species that are almost always present in a chromatin sample *(3,4,12,13,21,22)*. Sedimentation equilibrium methods fall into two categories:
a. those that yield a nonquantitative graphical transformation of the data, for example, ln absorbance vs (radial distance)^2 *(8)*,
b. those that use nonlinear fitting routines to derive M and other related parameters *(8,9,23,24)*.

11. When studying chromatin, sedimentation equilibrium experiments tend to be performed less frequently than sedimentation velocity experiments. Endogenous samples are so heterogeneous in M that it is impossible to analyze them quantitatively by sedimentation equilibrium, although comparison of the graphical transformations obtained under different conditions may allow one to determine whether a change in M has occurred *(25)*. Reconstituted chromatin model systems can be assembled in a way that is sufficiently homogeneous to allow nonlinear least squares determination of sample M, which mainly is used during studies of chromatin folding *(3)*. In contrast, sedimentation velocity experiments analyzed by the G(s) method have proven to be essential in elucidating the mechanisms involved in core histone N-termini-dependent chromatin folding and oligomerization *(3,4,11–13,21)*, nucleosome assembly *(22)*, and linker histone interaction with nucleosomal arrays *(26)*.

12. Finally, it is important to realize that although computer programs now perform the rigorous mathematical computations involved in all of the various available data analysis methods, an analytical ultracentrifuge user *must* be familiar with the fundamental principles that underlie the analysis method used. New users in the beginning are strongly encouraged to contact analytical ultracentrifugation specialists for help with analysis and interpretation. This now can be accomplished *easily* using the links found in the University of Texas Health Science Center at San Antonio Department of Biochemistry (http://bioc02.uthscsa.edu) webpage. In addition to digital data acquisition, greatly enhanced "expert accessibility" is one of the most profound differences between the way that analytical ultracentrifugation is practiced today compared with 20 years ago.

References

1. Wolffe, A. P. (1995) *Chromatin: Structure and Function, 2nd edition*. Academic Press, New York.

2. Fletcher, T. M. and Hansen, J. C. (1996) The nucleosomal array: structure/function relationships. *Crit. Rev. Eukaryot. Gene Expr.* **6,** 149–188.
3. Schwarz, P. M. and Hansen, J. C. (1994) Formation and stability of higher order chromatin structures: contributions of the histone octamer. *J. Biol. Chem.* **269,** 16,284–16,289.
4. Schwarz, P. M., Felthauser, A., Fletcher, T. M., and Hansen, J. C. (1996) Reversible oligonucleosome self-association: dependence on divalent cations and core histone tail domains. *Biochemistry* **35,** 4009–4015.
5. Cantor, C. R. and Schimmel, P. R. (1980) *Biophysical Chemistry, Part II.* W.H. Freeman and Co., San Francisco, CA.
6. van Holde, K. E. (1985) *Physical Biochemistry, 2nd Ed.* Prentice-Hall, Englewood Cliffs, NJ.
7. Ralston, G. (1993) *Introduction to Analytical Ultracentrifugation.* Beckman Instruments, Fullerton, CA.
8. Hansen, J. C., Lebowitz, J., and Demeler, B. (1994) Analytical ultracentrifugation of complex macromolecular systems. *Biochemistry* **33,** 13,155–13,163.
9. Hensley, P. (1996) Defining the structure and stability of macromolecular assemblies in solution: the re-emergence of analytical ultracentrifugation as a practical tool. *Structure* **4,** 367–373.
10. Schuster, T. M. and Toedt, J. M. (1996) New revolutions in the evolution of analytical ultracentrifugation. *Curr. Opin. Struct. Biol.* **6,** 650–658.
11. Hansen, J. C., Ausio, J., Stanik, V. H., and van Holde, K. E. (1989) Homogeneous reconstituted oligonucleosomes: evidence for salt-dependent folding in the absence of histone H1. *Biochemistry* **28,** 9129–9136.
12. Hansen, J. C. and Wolffe, A. P. (1992) Influence of chromatin folding on transcription initiation and elongation by RNA polymerase III. *Biochemistry* **31,** 7977–7988.
13. Hansen, J. C. and Wolffe, A. P. (1994) A role for histones H2A/H2B in chromatin folding and transcriptional repression. *Proc. Natl. Acad. Sci. USA* **91,** 2339–2343.
14. Butler, P. J. G. and Thomas, J. O. (1980) Changes in chromatin folding in solution. *J. Mol. Biol.* **140,** 505–529.
15. Walker, I. O. (1984) Differential dissociation of histone tails from core chromatin. *Biochemistry* **23,** 5622–5628.
16. Gale, J. M. and Smerdon, M. J. (1988) Photofootprint of nucleosome core DNA in intact chromatin having different structural states. *J. Mol. Biol.* **204,** 949–958.
17. Laue, T. M., Anderson, A. L., and Demaine, P. D. (1994) An on-line interferometer for the XL-A ultracentrifuge. *Progress in Colloid and Polymer Science* **94,** 74–81.
18. Demeler, B., Saber, H., and Hansen, J. C. (1997) Identification and interpretation of complexity in sedimentation velocity boundaries. *Biophys. J.* **72,** 397–407.
19. Stafford, W. F., III. (1992) Boundary analysis in sedimentation transport experiments: a procedure for obtaining sedimentation coefficient distributions using the time derivative of the concentration profile. *Anal. Biochem.* **203,** 295-301.
20. van Holde, K. E. and Weischet, W. O. (1978) Boundary analysis of sedimentation-velocity experiments with monodisperse and paucidisperse solutes. *Biopolymers,* **17,** 1387–1403.

21. Tse, C. and Hansen, J. C. (1997) Hybrid trypsinized nucleosomal arrays: identification of multiple functional roles of the H2A/H2B and H3/H4 N-termini in chromatin fiber compaction. *Biochemistry* **36,** 11,381–11,388.
22. Hansen, J. C., van Holde, K. E. and Lohr, D. (1991) The mechanism of nucleosome assembly onto oligomers of the sea urchin 5S DNA positioning sequence. *J. Biol. Chem.* **266,** 4276–4282.
23. Johnson, M. L., Correia, J. J., Yphantis, D. A. and Halvorson, H. R. (1981) Analysis of data from the analytical ultracentrifuge by non-linear least squares techniques. *Biophys. J.* **36,** 575–588.
24. McRorie, D. K., and Voelker, P. J. (1993) *Self-Associating Systems in the Analytical Ultracentrifuge.* Beckman Instruments, Inc., Fullerton, CA.
25. Hansen, J. C., Kreider, J. I., Demeler, B. and Fletcher, T. M. (1997) Analytical ultracentrifugation and agarose gel electrophoresis as tools for studying chromatin folding in solution. *Methods: A Companion to Methods Enzymol.* **12,** 62–72.
26. Carruthers, L. M., Bednar, J., Woodcock, C. L., and Hansen, J. C. (1998) Linker histones stabilize the intrinsic salt-dependent folding of nucleosomal arrays: mechanistic ramifications for higher-order chromatin folding. *Biochemistry* **37,** 17,776–14,787.

10

Analysis of Chromatin by Scanning Force Microscopy

Sanford H. Leuba and Carlos Bustamante

1. Introduction

While much is known about the structure of the nucleosome, how a chain of contiguous nucleosomes rearranges to form a fiber is much less understood. Microscopical techniques with nucleosomal resolution (electron microscopy and the newly emerging probe microscopies) have been useful in understanding chromatin fiber structure. The most widely used probe microscopy for biological applications, the scanning force microscope (SFM), also known as the atomic force microscope, is capable of imaging samples under very mild conditions (for recent reviews, see **refs. *1–3***). Samples are not stained or shadowed or subjected to vacuum and are imaged under ambient room temperature conditions in air or in aqueous buffer.

In the SFM, a fine tip is scanned back and forth over an object placed on an atomically flat surface. The tip is deflected by the sample, and this deflection is recorded to provide a topographic map of the sample on the surface. Presently, SFMs operate reliably in ambient conditions in contact and tapping modes. In the contact mode, a constant force is applied from the tip onto the surface so that the tip constantly maintains contact with the surface. In the tapping mode, the cantilever is oscillated at high frequencies (>200 kHz) above the surface. The effect of the tip interacting with the sample is to reduce the amplitude of this oscillation, and this reduction is used to monitor the topography of the sample. In practice, tapping mode minimizes lateral dragging forces and is much more stable than contact mode in air and in liquid for imaging DNA and DNA-protein complexes. These protocols will only describe tapping-mode SFM.

We describe protocols for imaging chromatin fibers with the scanning force microscope for users with little or no prior experience with the microscope. The actual isolation of chromatin fibers from cells and their further manipulation, such as stripping of linker histone and subsequent reconstitution, have been described elsewhere *(4–10)*. In this methods paper, we describe protocols for imaging fixed chromatin fibers on mica in air, unfixed or fixed chromatin fibers on glass in air, and unfixed chromatin fibers on glass in buffer. As these protocols are based on the use of the NanoScope SFM of Digital Instruments (DI), we direct the reader seeking more information to the comprehensive spiral-bound DI Command Reference Manual (1996) and the DI Support Note No. 202 on Tapping Mode in Fluids with the MultiMode SPM *(11)*. Additionally, an excellent Internet directory of other probe microscopy companies, research groups, and journals is maintained by Stephan M. Altmann (www.rzuser.uni-heidelberg.de/~saltmann/spm.html).

2. Materials

1. Triethanolamine (Baker, Phillipsburg, NJ).
2. Electron microscope grade glutaraldehyde (Electron Microscopy Sciences, Fort Washington, PA; cat. no. 16120).
3. Clear ruby mica 0.5-cm diameter disks (New York Mica, NY) or mica (Asheville-Schoonmaker Mica Company, Newport News, VA).
4. Paper cutter board (local office supply; 12 in × 12 in).
5. Magnetic steel sample puck (Digital Instruments, Santa Barbara, CA [www.di.com]).
6. Scotch Magic Tape (3M, St. Paul, MN).
7. Coverglass 12-mm circles (Fisher Scientific, Pittsburgh, PA; cat. no. 12-545-80).
8. SPI Miracle Tip and other tweezers (SPI Supplies, West Chester, PA).
9. Nanopure water system (Barnstead, Dubuque, IA).
10. Coors porcelain combustion boat (cat. no. 22825-103) and Thermolyne type 1300 small benchtop furnace (VWR Scientific Products, S. Plainfield, NJ).
11. Fingernail polish (local drug store; color at the discretion of the researcher).
12. NanoProbe silicon tips (Digital Instruments; or Dr. Olaf Wolter GmbH, Wetzlar-Blankenfeld, Germany [ourworld.compuserve.com/homepages/nanosensors]).
13. Bungy cords (local hardware store). Concrete 10 kg block (local lumber yard store).
14. NanoScope IIIa SFM in the Tapping Mode (Digital Instruments).
15. Parafilm (VWR).
16. Gold coated, sharpened Microlevers (Park Scientific, Sunnyvale, CA [www.park.com]). Liquid cell and silicon O-ring (Digital Instruments).

3. Methods

3.1. Chromatin Dialysis (see Note 1 and Subheading 1.)

1. Dialyze chromatin three times, each time 3 h vs 2 L of 5 mM triethanolamine-HCl, pH 7.0, 0.1 mM EDTA.

3.2. Glutaraldehyde Fixation of Chromatin

1. Carefully break-open an ampoule of 10% glutaraldehyde.
2. To 99 µL of 0.1 mg/mL chromatin (A_{260} = 2.0) in microfuge tube, add 1 µL electron-microscope-grade glutaraldehyde. Upon addition of the heavy, viscous glutaraldehyde, one should see it sink into the chromatin solution.
3. Stir with plastic pipet tip. Place tube on ice and let it sit overnight. Do not shake.

Scanning Force Microscopy

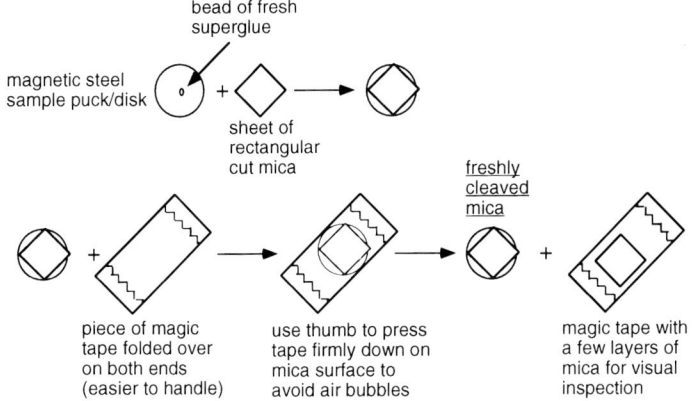

Fig. 1. Schematic of attaching mica with superglue to metal disk and then cleaving the top surface of the mica with Scotch Magic tape.

3.3. Preparation of Freshly Cleaved Mica (Fig. 1)

1. With small bead of superglue, attach mica (*see* **Note 2**) to magnetic steel sample disk/puck.
2. Place strip of $^3/_4$-in wide Scotch Magic Tape over mica.
3. Rub thumb over the tape to make an even seal over the mica.
4. Remove the tape to peel off a few layers of the mica.
5. Examine the peeled-off layers of mica on the tape, if the peeled-off mica makes a shiny smooth surface, then the remaining top layer of mica on the metal disk is also atomically flat and ready for deposition; if not, repeat the operation.

3.4. Preparation of Cleaned Coverglass Circles (see Note 3)

1. Holding coverglass circle carefully with tweezers, rinse it first with a stream of Nanopure water from a squirt bottle, then with 95% ethanol. Allow ethanol on glass to evaporate.
2. With tweezers, place coverglass into center of Bunsen burner flame for about 0.5 s. It is only necessary for the coverglass to briefly turn orange in the blue Bunsen burner flame.
3. Remove coverglass from flame but hold it a few centimeters from the flame to allow it to slowly cool down (*see* **Note 4**).
4. Rinse again with stream of Nanopure water from squirt bottle. Allow to dry.

3.5. Alternative Preparation of Cleaned Coverglass Circles Using 600°C Benchtop Furnace (see Note 5)

1. Place porcelain combustion boat holding coverglass circles into furnace.
2. Carefully watch the temperature of the furnace rise to 600°C.
3. After 5 min at 600°C, turn furnace off, open furnace door, and let it cool off (approx 30 min) until you can safely remove porcelain combustion boat of coverglasses and rinse them with a stream of Nanopure water from squirt bottle.

3.6. Attachment of Cleaned Coverglass Circles to Magnetic Steel Sample Puck/Disk (see Note 6)

1. Place a small bead of fingernail polish on the metal disk and then place the cleaned coverglass circle onto the disk.
2. Press down carefully (to avoid cracking the glass) with tweezers.

3.7. Deposition of Chromatin onto Freshly Cleaved Mica or Cleaned Coverglass Circles

1. Pipet 20 µL of fixed or unfixed 0.1 mg/mL ($A_{260} = 2.0$) chromatin onto center of freshly cleaved mica or cleaned coverglass circle surface. Wait 1 min.
2. Holding metal disk at 45° angle, place ten drops of Nanopure water onto the mica.
3. Without moving your hand holding the metal disk, use your other hand to blot one edge of the mica with No. 4 Whatman filter paper (**Fig. 2A**).
4. Turn on the nitrogen gas and check the flow by directing it towards your upper lip (**Fig. 2B**): the flow should be such that your lip can gently sense it.
5. Flux off (in one direction with respect to the plane of the mica) the remaining visible liquid with the nitrogen gas for up to 15 s (**Fig. 2C**).

3.8. Installation and Adjustment of Silicon-Chip-Holding Cantilever and Tip (see Note 7)

1. Insert the silicon chip by holding cantilever and tip into the chip holder: place the tapping mode chip holder on the corner of the keyboard (*see* **Note 7** and **Fig. 3A**).
2. With your right hand holding the tweezers with the silicon chip, use your left hand to push the chip holder against the keyboard, which releases the spring lever above the groove where the chip is actually mounted (*see* **Note 7** and **Fig. 3A**).
3. After placing the chip into the groove, release your left-hand pressure on the chip holder, and the chip will be secured by the spring lever (*see* **Note 7**).
4. Again, with your left hand, slightly press the back of the chip holder against the keyboard so that the spring holding the chip is released.
5. With the tweezers, carefully nudge the chip over to one side of the groove (*see* **Note 7** and **Fig. 3B**).
6. After the chip is properly positioned, mount the chip holder into the optical head of the microscope.
7. Manual alignment of visible red laser beam (*see* **Note 8**): cut a small piece of stiff paper (Whatman filter paper works nicely) roughly about 0.5 cm × 5 cm and insert it in front of the photodetector (**Fig. 4A**) to catch the red beam of the laser.
8. Use a monocular to determine where the beam is hitting the chip (*see* **Fig. 4B** and **Note 9**).
9. Once you find the beam, adjust it onto the chip, move it down a short incline on the end of the chip onto the cantilever (which looks like a small bee stinger sticking out of the chip), and then move it out to the end of the cantilever using the two knobs obscured by fingers in **Fig. 4B** (*see* **Note 8**).

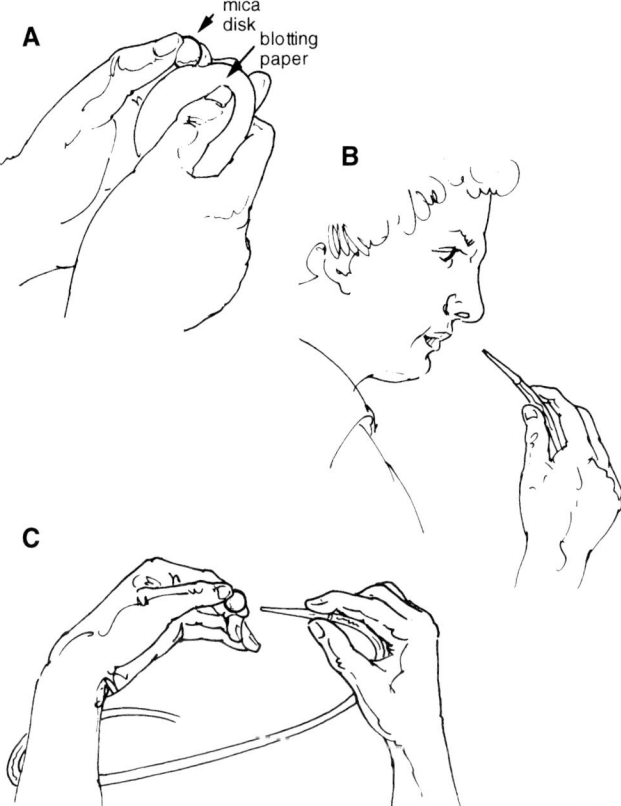

Fig. 2. **(A)** Drawing of the blotting process for imaging in air. The metal disk holding the mica (or coverglass circle) is held edgewise between the thumb and forefinger while the excess liquid is blotted on one edge (so that the flow is in one direction) with Whatman filter paper. **(B)** Blowing of N_2 gas towards ones upper lip to check the flow. One wants the flow to be fast enough that ones upper lip can detect it, but not so fast as to rip through the solution on the surface for samples that are to be imaged in air. **(C)** Removing by nitrogen flux the visible (by eye) liquid for samples to be imaged in air. This step takes at most 15 s.

10. Return to using the short thin piece of paper to adjust the beam on the cantilever on the backside of the tip. Sharply focus the laser beam to a bright red spot on the paper strip. At this point if you move the focus controls in any of three directions, you should immediately lose the spot of the beam on the paper. Remove the paper slip.
11. Adjust the strength of the laser beam signal on the microscope photodetector (*see* **Note 10**): Maximize the inside-perimeter graphical signal of the microscope base oval display by manually adjusting the lever on the backside of the head of the microscope that tilts the mirror back and forth. Often a signal of approx 3.6 can

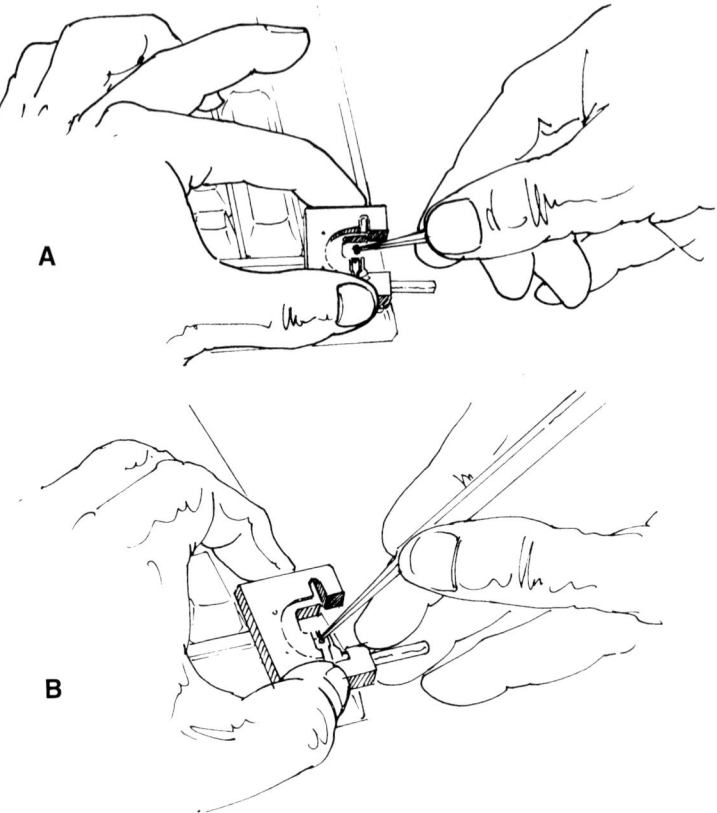

Fig. 3. **(A)** Insertion of a silicon chip holding the cantilever and tip into a tapping mode (in air) chip/cantilever holder. The backside of the holder is juxtapositioned against the slight incline of a computer keyboard to (i) release the spring which holds the actual chip and (ii) to provide an incline for the groove/slot into which the chip slides slightly-down into. **(B)** Tweezer adjustment of the silicon chip in the tapping mode (in air) chip/cantilever holder. Once the chip has been slid into groove **(A)**, then it has to be edged along one side of the groove to create more contact of the chip with the groove and avoid extraneous vibration of the chip. Please note that the angle at which the tweezers are used in **(B)** is different from that in **(A)**. Also notice that here the tweezers are being used not to hold but to nudge the chip over to one side.

be obtained. If you cannot obtain this graphical signal then you have to go back and check the tip, realign the laser beam, reboot the software, replace the tip, and so on.
12. Focus the signal on the center of the four-quadrant photodetector (*see* **Note 11**): rotate the knob (which finely adjusts the photodetector) on the upper left-hand corner on the top of the head of the microscope until the number in the center of the oval display is between –0.2 and 0.2.

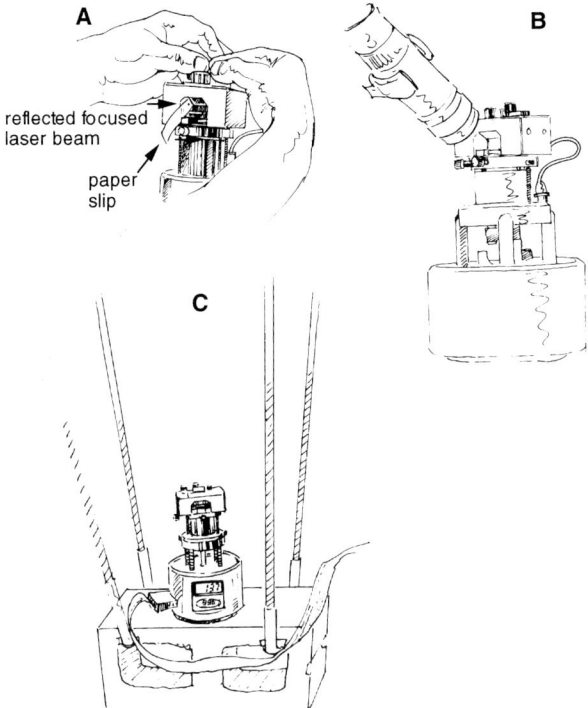

Fig. 4. (**A**) Manual alignment of visible red laser beam. Of the two knobs obscured by fingers, the knob on top right of the head of the microscope adjusts the visible laser beam left and right and the knob on top left back and forth with respect to the sample surface. The point is to adjust the laser beam to the end of the chip, down the beveled face of the chip, onto the bee-sting-look-like cantilever and out to the end of the cantilever on the backside of the tip. At this point the small piece of stiff paper inserted into the head of the microscope will catch the beam of the laser and thus can be used to determine the proper focus of the laser beam using the above mentioned two knobs. (**B**) 45° view of the tip with the monocular stand. This kind of manual setup with the 8 × 30 power monocular can be used to (i) visualize and adjust the laser beam onto the end of the cantilever and (ii) adjust the distance from the tip to the sample surface before placing the microscope on the vibration-free system and starting engagement. (**C**) Placement of the SFM on the vibration-free 10-kg concrete block attached to the ceiling by bungy cords.

13. Manually move the tip toward the surface: place the metal disk holding the substrate (mica or glass disk) with the sample onto the piezo scanner.
14. Place the head of the microscope on top of the piezo scanner and, if need be, secure the two attaching springs of the head to the base of the microscope.
15. Using the monocular to observe the location of the tip with respect to the surface (**Fig. 4B**), manually lower the tip toward the surface by rotating the knobs that manually raise or lower the head to the piezo scanner (*see* **Notes 12** and **13**).

Scan Controls	
Scan size:	0.00 nm
X offset:	0.00 nm
Y offset:	0.00 nm
Scan angle:	0.00 deg
Scan rate:	1.97 Hz
Number of samples:	512
Slow scan axis:	Enabled
Z limit:	440 V

Channel 1	
Datatype:	Height
Z range:	15 nm
Line direction:	Trace
Scan line:	
Real time Planefit:	Line
Offline Planefit:	None
Highpass filter:	Off
Lowpass filter:	Off

Feedback Controls	
Integral gain:	0.50
Proportional gain:	1.00
Look ahead gain:	0.00
Setpoint:	0.00 V
Drive frequency:	300.000 kHz
Drive amplitude:	100 mV
Analog 2:	0.00

Interleave Controls	
Interleave mode:	Disabled
Integral gain:	
Proportional gain:	
Look ahead gain:	
Setpoing:	
Drive frequency:	
Drive amplitude:	
Analog 2:	
Interleave scan:	
Lift start height:	
Lift scan height:	

Other Controls	
Units:	Metric
Color table:	9
AFM Mode:	Tapping
Input attenuation:	1x
Engage Setpoint:	1.0
Min. Engage gain:	
Z modulation:	Disable

Fig. 5. Initial settings for major software controls.

16. Check that the head assembly is on a level plane with respect to the metal disk on the piezo scanner (*see* **Notes 12** and **14**).
17. Place the microscope on a vibration-free platform. In this example the microscope is placed on a 10 kg concrete block (**Fig. 4C**) hanging from the ceiling by bungy cords (*see* **Note 15**).

3.9. Software Controls and Imaging in Air

1. Run the microscope with the software controls (*see* **Fig. 5** for initial settings): place the Scan size, X offset, Y offset, and Scan angle at zero. Set Scan rate to about 2 Hz, Number of Samples to 512, Slow Scan Axis is Enabled, and the Z axis to 440 V.
2. In the next column of numbers, set the Integral gain to 0.5 and the Proportional gain to 1. (The Look ahead gain can be left at zero). Initially set the Setpoint to zero.
3. Set Datatype to Height, Z range to 15 nm, Line direction to either Trace or Retrace.
4. For the other controls, set the Color table at a setting that gives a reasonable (researcher's discretion) range of shades of color. Offline planefit is set to None, Z modulation is Disable, Highpass and Lowpass filters are Off, and Analog 1 is zero.

Scanning Force Microscopy 151

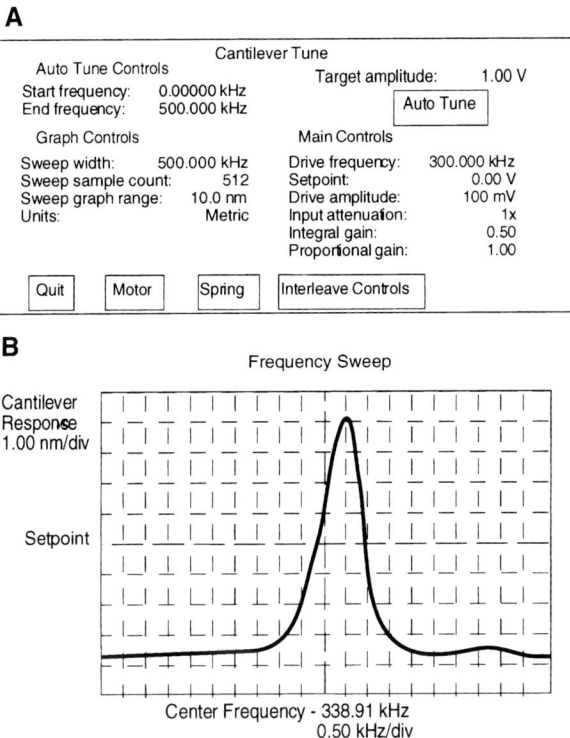

Fig. 6. (**A**) Initial settings for software controls before cantilever tune process. (**B**) Example peak in graph of amplitude of vibration of the cantilever versus frequency of vibration. The graph is centered on an inflection point of this peak before starting engagement.

5. In the Interleave controls, the Interleave scan is Off.
6. Check the cantilever vibration frequency (*see* **Note 16**): select Cantilever Tune from the View menu (*see* **Fig. 6A** for Cantilever Tune startup settings).
7. With the mouse set the Z scan size, Z scan rate, and Graph range to the highest number possible.
8. In the graph of amplitude of the cantilever vibration vs frequency (**Fig. 6B**), find the (usually lone) peak of amplitude. Expand the X-axis from 0 to 500 kHz, locate the peak, and then use the Zoom in and Offset controls to decrease the Sweep width of the X-axis to 10 kHz, and center the peak such that the center vertical line of the plot intersects an inflection point on the left-hand side of the peak (**Fig. 6B**). Some NanoScope users use the inflection point on the right-hand side of the peak.
9. Check that the Setpoint on the microscope base is between 0.5 and 2.0 V (*see* **Note 17**).
10. Engagement of tip onto the sample surface: select the Engage command (under Motor Command window) and the microscope automatically approaches the tip to the sample surface. Once the microscope detects that the tip has engaged the surface, a beginning image will be displayed on the second monitor.

11. Adjust the Scan size to 2000 nm to create a 2000 nm × 2000 nm image.
12. If the image is fuzzy, carefully adjust the Setpoint until the force of the tip on the surface is sufficient enough to maintain stable imaging conditions. In tapping mode, this operation involves *decreasing* the Setpoint setting. We have routinely used a setting between 1 and 2 V for the Setpoint during imaging (*see* **Note 18**).
13. Inspect the gains. It may also be necessary to carefully increase the Integral and Proportional gains to improve the sensitivity of the piezo response to changes in heights of the chromatin fibers on the surface. If the gains are increased too high, it is easy to observe repeated noise in the image (for trouble-shooting *see* **Notes 18–22**).

3.10. Imaging in Liquid (see Note 23)

1. Cover the scanner with a thin film of parafilm to protect the piezo crystal from being destroyed by a short-circuit caused by the liquids used for imaging.
2. Place a Park Scientific Microlever tip in the liquid cell (**Fig. 7A**).
3. As the liquid cell is thicker than the dry cantilever holder, raise the microscope head sufficiently before placing the liquid cell inside it.
4. Focus the laser beam on a cantilever that has a spring constant of 0.5 N/m. You can use the monocular to look through the top of the microscope head to visualize the laser beam and to place it on the end of the chip (**Fig. 7B**).
5. After you have focused the laser beam onto the cantilever (using the piece of paper as described in **Subheading 3.8.**), then you can adjust the distance between the tip and surface: using the monocular, look directly between the liquid cell and the surface (**Fig. 7C**), and you should see a group of spots of reflections (**Fig. 8**) that indicate how close the tip is to the surface. The top and bottom spots are reflections of light from the laser on the tip and the surface, respectively (**Fig. 8**). Use the screws below the scanner to manually move the tip to the surface.
6. Deposit 20 µL of sample onto the cleaned coverglass.
7. Place the silicone O-ring on the surface and place the head of the microscope on top of the O-ring.
8. Now that the laser beam is being reflected through solution, the change resulting from the index of refraction requires that the beam be refocused slightly by tilting the mirror to regain the signal to the photodiode.
9. For tapping mode in liquid, set the Integral gain at 4.0 and the Proportional gain at 1.0. Set the Drive amplitude at 1 or 2 V (which is about 10X this setting for tapping in air).
10. In the Cantilever tune, look for a frequency peak usually approx 7 kHz before starting the engagement of the tip with the surface. Example images of chromatin fibers in buffer are shown in **Fig. 9** (*see* **Note 24**).

3.11. Image Analysis of Chromatin Fibers (see Note 25)

1. Determine (X, Y, Z) coordinates of each nucleosome in a fiber (*see* **Note 25**).
2. Calculate center-to-center distances and angles (*see* **Note 25**).
3. Plot histograms of Frequency vs Height, Center-to-center distance, or Angle.
4. Keep a record of analyzed images by printing a copy (*see* **Note 26**).

Scanning Force Microscopy

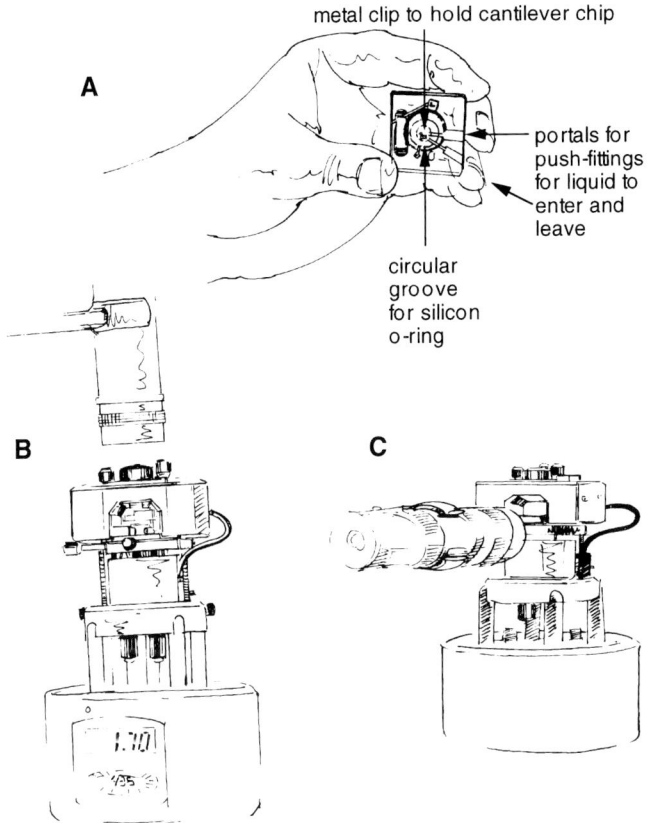

Fig. 7. (A) Carefully handling the glass tapping mode liquid cell from Digital Instruments. The word 'cell' is a slight misnomer as the imaging is done on the surface of the substrate. The liquid cell holds the cantilever chip, provides portals to allow liquid to be flushed through a chamber between the liquid cell and the substrate surface. This chamber is sealed with a silicone O-ring (not shown). (B) Downward view of the tip with the monocular stand. This kind of manual setup with the monocular can be used to visualize and adjust the laser beam onto the end of the cantilever for use with the liquid cell. (C) Horizontal view of the tip and the surface with the monocular stand. This kind of manual setup with the monocular can be used to manually bring the tip close to the surface without crashing it prior to engagement.

4. Notes

1. Dialysis removes small organic contaminants and extraneous salts that may interfere with either the deposition or the imaging steps.
2. If the mica is cut as a rectangle, it is much easier to peel than as a circle. The easiest way to cut the mica is with a paper cutter board.

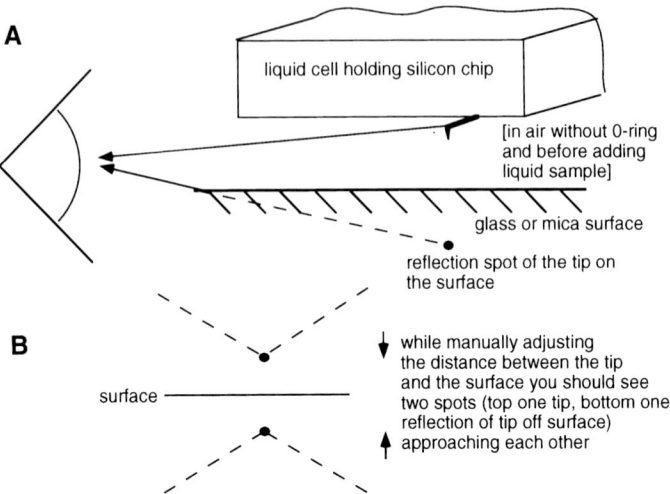

Fig. 8. **(A)** Schematic of the view when looking through the monocular between the liquid cell and the surface before inserting the liquid sample and the silicone O-ring. One should see the spot on the tip and the reflection spot of the tip on the surface. Inner groves for the O-ring which is placed around the tip to seal the chamber are not shown. **(B)** Depiction of the red spots of reflection that one sees through the monocular as one adjusts the distance between the tip and the surface before adding the liquid sample and the silicone O-ring.

3. Coverglasses from the manufacturer have a thin film of oil that must be flamed or baked off and then rinsed before use.
4. This precaution will prevent the glass from cracking as a result of a fast cooling. Typically, however, 50% of the coverglasses are lost to cracking during this early manipulation step.
5. This protocol requires a 600°C bench top furnace. Use proper equipment (i.e., asbestos-like gloves) and care to avoid personal injury.
6. Coverglass circles can be attached to the metal disks either by the use of doublestick tape or by the use of fingernail polish.
7. As there is a push-button on the backside of the chip holder which releases the spring-lever holding the chip, insertion of the chip is easier to perform at a slight incline, such as on a corner of a computer keyboard as indicated in **Fig. 3**. The chip then slides in at an incline and does not fall out. To avoid unnecessary vibration, it works best for one of the long thin edges of the chip to be adjacent to one edge of the groove; thus, once the chip is in place, it is necessary to manually adjust it (**Fig. 3B**) with the tweezers over to one side in its slot or groove, because this groove is slightly wider than the chip.
8. The point of this step is to adjust the laser beam to bounce off the cantilever on the backside of the tip, reflect onto the mirror and then reflect onto the four-

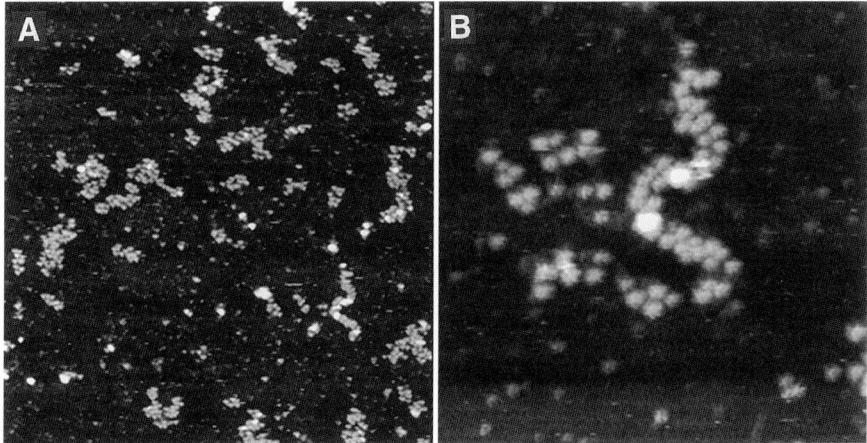

Fig. 9. SFM image from **ref. 10** of unfixed chicken erythrocyte chromatin fibers imaged in 5 mM triethanolamine, pH 7.0, 1 mM sodium butyrate, 0.1 mM EDTA on coverglass. Image sizes are **(A)** 1000 nm × 1000 nm and **(B)** 500 nm × 500 nm. Heights are encoded in shades of gray with low regions depicted in dark gray and higher regions in lighter shades in a scale of 0–15 nm.

quadrant photodetector of the SFM. There are two knobs on the top center of the head of the microscope (see where fingers of both hands are in **Fig. 4A**) that adjust the beam left and right and forward and backward.
9. Digital Instruments supplies an 8 × 30 power monocular with a Bogen stand for viewing tips. However, a high-power clear-focus zoom monocular (1–6 zoom) with stand can be separately and additionally obtained from Nissho Optical Co. Ltd., (Japan, model no. ZMM1) through Labtek (Scotts Valley, CA).
10. On the base of the microscope are two liquid crystal displays. The lower oval display indicates
    a. Graphically the strength of the detected laser beam signal in a range from 0–12.
    b. Numerically how well the beam is centered on the four-quadrant photodetector.
11. Once a maximum graphical signal is obtained, one needs to adjust how well the signal is focused on the center of the four-quadrant photodetector. This digitally displayed number is the difference between the intensities of the signals on opposing sides of the photodetector.
12. The knobs that manually raise or lower the head to the piezo scanner are located above the base of the microscope and directly under the central scanner. In concert with a third motor-driven screw, these three knobs/screws raise or lower the tip relative to the mica surface. Digital Instruments now also makes scanners that are completely motor driven without manual screws.
13. Through the monocular the body of the cantilever, the spot of the laser beam reflecting on the end of the cantilever, and the spot of the laser beam that spills

over the end of the cantilever onto the surface can be seen. A rule of thumb about how close to bring the tip to the surface is to lower the tip until these two laser spots are as close vertically as the horizontal length of the cantilever.

14. This leveling is accomplished by running the motor-driven screw up or down with the switch on the upper right side of the microscope base as well as manually adjusting the two forward screws below the scanner.
15. Some bungy cords do not work well. One has to experiment with different bungy cords or double them up until a sufficient damping of high-frequency vibrations is obtained when using a 10-kg concrete block. Alternatively, one can purchase a tripod/bungy-cord/flat-concrete-block assembly from Digital Instruments. Another possibility are the soundproof imaging cabinets supplied with bungy cords and a heavy base from Molecular Imaging (Phoenix, AZ [www.molec.com]). Once the bungy cords have lost most of their elasticity, they need to be replaced, usually once every year or two.
16. For tapping mode force microscopy, it is necessary to find the correct frequency, generally in the range of 250–350 kHz, to vibrate the cantilever in air.
17. When leaving cantilever tune, the Setpoint will appear automatically on the microscope base in the liquid-crystal display above the oval display. Setpoint values outside the range of 0.5–2.0 V indicate that the cantilever tune needs adjustment such as lowering or increasing the Drive amplitude (usually within the range of 30–250 mV, though generally no greater than 150 mV). Once the cantilever has been tuned, leave these menus, and now the system is ready for engagement (under the Motor Command).
18. Both the Setpoint and the gains should be adjusted slowly with small increments, as too great a change can lead to tip damage, irreversible loss of imaging quality, and the need to replace the damaged tip.
19. An incomprehensible image can be caused from not having a fresh, atomically flat surface on the mica. It may be necessary to peel the mica with Magic tape several times, each time always with a fresh piece of tape. Image the mica surface alone without sample as a control.
20. Sample is imaged as below the surface (e.g., *see* **Fig. 10A**). This contrast inversion problem can likely be attributed to anomalies with tapping mode in air (for extensive discussion *see* **ref. *12***). At this point one has to 'play' with the software controls (i.e., minimize the Setpoint, adjust the Scan angle, reduce Scan speed to 1 Hz, change Scan size to 1000 nm × 1000 nm, carefully adjust Integral and Proportional gains, etc.) in order to obtain an image with the sample above the surface (**Fig. 10B**). It may be necessary to disengage from imaging, adjust tip with tweezers or replace it, and then return to imaging.
21. We have found that it is difficult to get good images of fibers of unfixed chromatin fibers on mica in air. It was only possible to get nucleosome resolution of chromatin fibers deposited from a low salt buffer such as 5 mM triethanolamine-HCl, pH 7.0. Addition of just 10 mM NaCl to this buffer allowed the histones to dissociate from the DNA during the deposition process (deposition, rinsing, blotting, N_2 fluxing). Under these conditions, the images show mostly naked DNA

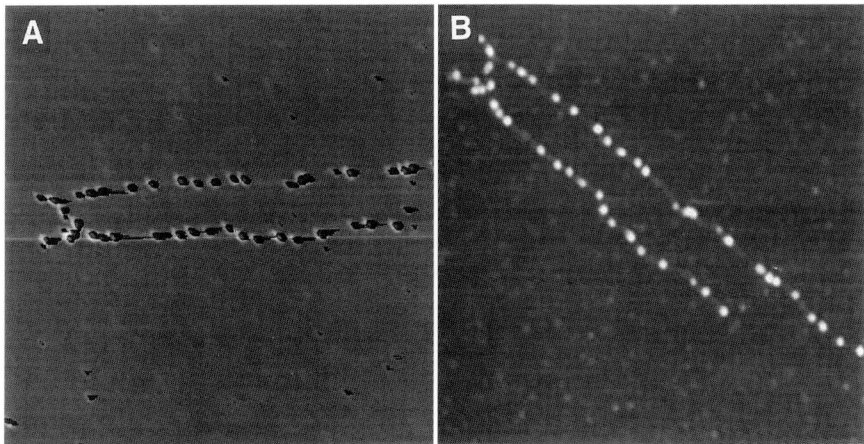

Fig. 10. (**A**) SFM image where the nucleosomes appears to be below the surface of the mica. (**B**) SFM image of exact same fiber upon adjustment of software controls (*see* **Note 20**). Glutaraldehyde-fixed, linker histone-stripped chromatin fibers were deposited from 5 m*M* triethanolamine-HCl, pH 7.0, and imaged in air. Image sizes are 1000 nm × 1000 nm. Heights are encoded in shades of gray with low regions depicted in dark gray and higher regions in lighter shades in a scale of 0–0 nm.

and a few histone octamers. It is much easier to get images of nucleosomal fibers of unfixed chromatin on glass in air.

22. If the images show too few or too many fibers in the field, the easiest way to change the fiber concentration is to adjust the deposition time accordingly. The deposition time can be increased up to typically approx 30 min before the 20 µL drop dries, or can be decreased to a minimal time by rinsing immediately after deposition. Alternatively, one can use slightly more or less concentrated chromatin to begin with. While the 0.1 mg/mL deposition concentration of chromatin is about 100 times the concentrations used for the deposition of DNA restriction fragments, we find that this concentration works reasonably well for chromatin fibers.

23. The set up for imaging in liquids is different from that used in air imaging applications. One needs to set up the microscope completely with the liquid cell (**Fig. 7A**) before the sample is deposited. The steps can be separated into
    a. Protecting the scanner with a piece of Parafilm or thin rubber before starting.
    b. Placing the Park Scientific Microlever tip into the liquid cell, placing the liquid cell into the microscope head, and adjusting the laser beam to bounce off the cantilever on the backside of the tip and focus onto the photodiode.
    c. Adjusting the distance between the tip and the surface.
    d. Depositing the liquid sample onto the surface of freshly cleaved mica or cleaned coverglass and letting it settle for a period of time.

e. Rinsing the sample with the same buffer to remove excess chromatin not on the surface.
   f. Placing the wet sample in the microscope, placing the silicon O-ring onto the surface, and placing the head onto the silicon O-ring surrounding the sample to seal the liquid chamber.
   g. Tuning and engagement of the cantilever.
24. For imaging fibers in buffer, we have observed that the chromatin is more sticky to the glass if the chromatin has been isolated in a procedure in which all buffers contain 1 mM sodium butyrate. The presence of butyrate, a deacetylase inhibitor, maintains the lysines in histone tails in chromatin in higher states of acetylation which appears to result in the chromatin fibers sticking better to the glass surface in buffer. One should keep in mind that the butyrate treatment keeps the histones in a more acetylated state than is found in 'native' chromatin.
25. We have used Alex, a program written in Matlab (Mathworks, Natick, MA) by Mark Young and Claudio Rivetti for the Silicon Graphics Workstation (SGI, Mountain View, CA) for the analysis of our images. The source code for Alex is available through the world wide web at alice.uoregon.edu. In Alex, we have used a homemade routine, "Measure_Chromatin", which records the X, Y, and Z coordinates of each mouseclick on the top of each visible nucleosome in a fiber. (Alternatively, it is possible to use the NanoView program written by Gerry Leatherman in Delphi [Borland, Scotts Valley, CA] for Windows 95/Windows NT. NanoView source code is available from [www.molec.com].) From these coordinates, then it is straightforward to calculate
   a. The center-to-center distance of two adjacent nucleosomes.
   b. The interior angle formed by the intersections of two lines formed by centers of three successive nucleosomes.
   c. The height of each nucleosome observed in the fiber.
   Given coordinates (X_1,Y_1,Z_1) and (X_2,Y_2,Z_2) of two adjacent nucleosomes, the center-to-center distance is calculated by:

$$D_{1,2} = \sqrt{[(X_1 - X_2)^2 + (Y_1 - Y_2)^2 + (Z_1 - Z_2)^2]}$$

Given coordinates (X_1,Y_1,Z_1), (X_2,Y_2,Z_2), (X_3,Y_3,Z_3) of three sequential nucleosomes, the angle ($\varnothing$) is calculated from the law of cosines:

$$\cos(\varnothing_{1,2,3}) = \frac{[(D_{1,2})^2 + (D_{2,3})^2 - (D_{1,3})^2]}{(2\,D_{1,2}\,D_{2,3})}$$

It is simple to assign the coordinates to nearly every nucleosome within images of linker histone-depleted beads-on-a-string fibers. Deciding which nucleosomes are successive in images of three-dimensionally organized, native chromatin fibers is not so straightforward. However, in these three-dimensionally organized fibers, especially at monovalent ionic concentrations of 20 mM and less, it is usually possible to see the repeating structure of the nucleosomes. Typically, the nucleosomes that would take the shortest path to form the fiber are taken to be adjacent, although this choice is somewhat arbitrary. The analysis should be repeated at least once with the same images to ensure against operator bias. The important point to remember with measurements, distributions, and statistics is that they should only be used to quantitatively support morphological differences easily seen by eye.

The present Alex and NanoView software only recognize v. 4 images of NanoScope files; in earlier DI versions, the Z-height is incorrectly read in the image header files. DI v. 3 images can be resaved as version 4 images using DI software. It is also possible to make similar kinds of measurements noted above in the analysis protocol using Digital Instrument software, although it is much more tedious and time-consuming.

26. If the user would like to print or make color slides of images, a simple way is to take photographs of the screen with 100 ASA color print or slide film and a 35 mm camera equipped with either a macro lens or a 70–200 mm zoom lens. The camera is mounted on a tripod approx 3 m in front of the screen with the zoom lens all the way out (Use of the 200-mm lens reduces the problems from the curvature of the computer screen.) After focusing, photographs are taken in a darkened room at 0.5 s with f-stop of 5.6, using a cable release. The user can experiment with exposure times longer than 0.25 s and f-stops of 5.6 or greater (i.e., f-stop of 8, 11, 16). White text will become over exposed and fuzzy in photographs or slides especially during longer exposure times. Alternatively, one can import tiff images into a program such as Adobe Photoshop (www.adobe.com) and then print the images on a color dye-sublimation printer at a cost of approx $5–$10 per page. An economical alternative is to print gray scale images of the screen directly to a laser printer using the DI software.

7. Conclusions

We have described protocols for imaging chromatin with the SFM. With the drawings of the various procedures, it should be possible for the experienced chromatin biochemist/beginning SFM user to obtain similar results. While the tapping mode in air experiments are routine, we should caution the reader that the imaging in buffer experiments are not, and they require extra time, care, and experimentation.

Acknowledgments

We acknowledge the use of the ASU CSSS Scanning Probe Microscopy Facility and thank J. Zlatanova for critical reading and D. Chen for illustrations. This work has been supported by NIH Grant GM32543, NSF Grants BIR 9318945 and MCB 9118482 to C. B., NIH Grant GM16600 to S. H. L., NSF Grant BIR 9513233 to S. M. Lindsay, and NIH Grants CA70274 and GM53517 to R. E. Harrington.

References

1. Bustamante, C. and Rivetti, C. (1996) Visualizing protein-nucleic acid interactions on a large scale with the scanning force microscope. *Annu. Rev. Biophys. Biomol. Struct.* **25,** 395–429.

2. Schaper, A. and Jovin, T. M. (1996) Striving for atomic resolution in biomolecular topography: the scanning force microscope (SFM). *Bioessays* **18,** 925–935.
3. Engel, A., Schoenenberger, C. A., and Müller, D. J. (1997) High resolution imaging of native biological sample surfaces using scanning probe microscopy. *Curr. Opin. Struct. Biol.* **7,** 279–284.
4. Leuba, S. H., Zlatanova, J., and van Holde, K. (1993) On the location of histones H1 and H5 in the chromatin fiber. Studies with immobilized trypsin and chymotrypsin. *J. Mol. Biol.* **229,** 917–929.
5. Leuba, S. H., Zlatanova, J., and van Holde, K. (1994) On the location of linker DNA in the chromatin fiber. Studies with immobilized and soluble micrococcal nuclease. *J. Mol. Biol.* **235,** 871–880.
6. Leuba, S. H., Yang, G., Robert, C., Samori, B., van Holde, K., Zlatanova, J., and Bustamante, C. (1994) Three-dimensional structure of extended chromatin fibers as revealed by tapping-mode scanning force microscopy. *Proc. Natl. Acad. Sci. USA* **91,** 11,621–11,625.
7. Leuba, S. H., Bustamante, C., van Holde, K., and Zlatanova, J. (1998) Linker histone tails and the N-tails of histone H3 are redundant: scanning force microscopy studies of reconstituted fibers. *Biophys. J.* **74,** 2830–2839.
8. Zlatanova, J., Leuba, S. H., Yang, G., Bustamante, C., and van Holde, K. (1994) Linker DNA accessibility in chromatin fibers of different conformations: a reevaluation. *Proc. Natl. Acad. Sci. USA* **91,** 5277–5280.
9. Yang, G., Leuba, S. H., Bustamante, C., Zlatanova, J., and van Holde, K. (1994) Role of linker histones in extended chromatin fibre structure. *Nat. Struct. Biol.* **1,** 761–763.
10. Bustamante, C., Zuccheri, G., Leuba, S. H., Yang, G., and Samori, B. (1997) Visualization and analysis of chromatin by scanning force microscopy. *Methods* **12,** 73–83.
11. Prater, C. B. (1996) TappingMode in Fluids with the MultiMode SPM. *Digital Instruments Support Note Number 202.*
12. Van Noort, S. J., Van der Werf, K., O., De Grooth, B. G., van Holst, N. F., and Greve, J. (1997) Height anomalies in tapping mode atomic force microscopy in air caused by adhesion. *Ultramicroscopy* **69,** 117–127.

11

In Vivo Mapping of Nucleosomes Using Psoralen-DNA Crosslinking and Primer Extension

Ralf Erik Wellinger, Renzo Lucchini, Reinhard Dammann, and José M. Sogo

1. Introduction

In the nucleus of eukaryotic cells, DNA is packaged into a nucleoprotein complex known as chromatin *(1)*. This complex provides the compaction and structural organization of the DNA for processes such as replication, transcription, recombination, and repair. The highly compact structure of chromatin does not only restrict the access of DNA to enzymes and regulating factors, it can even act as an activating principle bringing distant DNA-protein binding sites into close proximity.

The repeating unit of chromatin is the nucleosome, 146 bp of DNA wrapped around a histone octamer. DNA between two adjacent nucleosomes is called linker DNA. Numerous techniques were developed during the past decades to characterize nucleosomes. The position of nucleosomes with respect to the underlying DNA were defined by enzymatic (e.g., MNase or DNase I) and/or chemical treatment of chromatin. The differential accessibility of DNA wrapped over the surface of a histone octamer or in the adjacent linker results in diagnostic cleavage patters or "footprints" that allow the establishment of a nucleosome position.

In order to use nucleases for nucleosome mapping, nuclei have to be isolated following procedures that may distort the native chromatin structure. To overcome such limitations, direct in vivo nucleosome footprinting methods, such as, *dam*-methyltransferases mapping can be used *(2)*. This technique is limited to the analysis of chromatin in organisms that lack such an enzyme (e.g., yeast). Furthermore, since methylation is sequence-specific, methylases only find a target in one out of 35–45 bp, limiting the resolution of these assays.

The psoralen crosslinking technique in combination with electron microscopy *(3,4)* and by band retardation assay *(5)* provides a useful tool for chromatin structure analysis. Psoralen derivatives efficiently intercalate in helical DNA and when irradiated with ultraviolet (UV)-light (366 nm) form covalent crosslinks between pyrimidines of opposite strands. However, in chromatin psoralen crosslinking occurs preferentially within the linker DNA, whereas the nucleosomal DNA is protected against crosslinking. When chromatin is crosslinked with psoralen, the crosslinked DNA is deproteinized and analyzed under denaturing conditions by electron microscopy, arrays of single-stranded (ss) bubbles are visualized. These bubbles correspond to DNA that originally was organized in nucleosomes and therefore was not crosslinked *(6–8)*.

Trimethylpsoralen (TMP) is the most commonly used psoralen for chromatin studies. Even extensive TMP crosslinking of chromosomal DNA, performed under nearly saturating conditions, neither destroys nor disturbs the nucleosomal *(8)* and higher order structure of chromatin *(7)*, at least not at the level seen by micrococcal nuclease digestion and electron microscopy (*see* references above).

To visualize the ss-bubbles in covalently closed circular DNA molecules (e.g., for psoralen crosslinked minichromosomes see **Fig. 1**), it is necessary to introduce at least one nick per molecule, to allow the unwinding of the DNA strands during denaturation (**Fig. 1C,D**, arrowheads; for details see *[8]*). The size distribution of the ss-bubbles and the r-value (ratio of the sum of the length of all ss-DNA sections to the total DNA contour length), which is a measure of the nucleosome density of chromatin *(8)*, is the criterion for quantitative analysis of the structure of a selection of chromatin fragments *(9,10)*.

So far, the psoralen technique has contributed significantly to the characterisation of the chromatin structure during transcription and replication (for reviews *see* **ref. 5,11**). For the first time, it allowed reproducible distinguishing, characterizing, and quantifying of the proportion of transcriptionally active and inactive ribosomal gene copies *(12)*. The crosslinking of nascent RNA strands to the DNA template facilitated the characterization of the chromatin transcribed by polymerase II *(13,14)*. Important information has also been obtained concerning the chromatin structure around the replication fork. The nucleosomal organization in front of and behind the replication machinery has been reported in *(8,10,15)*. More recently, the combination of the psoralen technique with two-dimensional gel electrophoresis and EM allowed the investigation of the replication of transcriptionally active chromatin *(5)*. However, the resolution achieved when nucleosomes or boundaries between the active and inactive chromatin regions were mapped was only ±50 bp, and

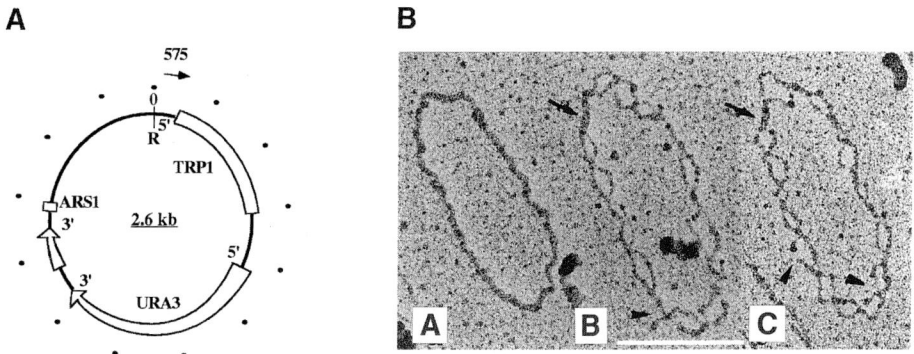

Fig. 1. Physical map and electron micrographs of DNA from psoralen crosslinked YRpTRURAP minichromosomes. (**A**) The physical map of the YRpTRURAP minichromosome is shown *(19)*. Indicated are: genes (open arrows) and their 5' and 3' ends, origin of replication (ARS1), *Eco*RI restriction site (R), primer used for nucleosome mapping (arrows), dots mark 200 bp intervals. Prior to formamide denaturation and spreading *(8)*, the DNA was (**B**) either mock treated or (**C** and **D**) DNase I digested in the presence of 0.3 mg/mL EtBr. *Left:* Example of a covalently closed molecule which appears as double-stranded DNA. *Middle* and *right:* Examples of nicked molecules (single-strand nicks are marked by arrowheads). The relaxed molecules show bubbles consisting of single-strand stretches with a size of 135 ± 39 bp, corresponding to nucleosomal DNA and an r value of 0.79 ± 0.06. Nucleosome-free regions are visible as double-stranded DNA stretches *(arrows)*. The white bar represents 500 nucleotides.

psoralen crosslinking was preferentially performed on high-copy-number sequences (rRNA genes, SV40).

In order to overcome these limitations and to enable a more accurate mapping of nucleosomes in single-copy regions of the genome, we developed a primer-extension based assay. This assay takes advantage of the fact that psoralen acts as a roadblock for DNA modifying enzymes such as exonucleases *(9,16–18)*. As a model system we have used the yeast minichromosome YRpTRURAP (**Fig. 1A**), in which nucleosomes have been mapped with high precision using well established micrococcal nuclease and DNase I footprinting assays *(19,20)*.

Finally, we would like to emphasize two important aspects of the technique. First, the psoralen crosslinking can be done in living cells, an important advantage with respect to the well established assays for nucleosome mapping. A slight drawback of the method is that the intercalation of psoralen is sequence-specific, which could limit the application of the technique for particular sites. However, for the TRURAP minichromosome, on average, one crosslink per 6 bp is theoretically possible.

1.1. Nucleosome Mapping at Low Resolution by Psoralen Crosslinking

A schematical outline of the biochemical approach we used to map nucleosomes at low resolution by psoralen crosslinking of chromatin DNA is shown in **Fig. 2**. After preincubation of the cells with psoralen and UV irradiation, the DNA is isolated and used as substrate for further analysis. In order to detect crosslinked sites, the DNA is first linearized with a restriction enzyme(s) that creates a 5'-prime overhang. This overhang is filled-in by Klenow DNA polymerase (*see* **Note 3**). This blunt-ended DNA then serves as the preferred substrate for λ-exonuclease. The λ-exonuclease degrades the DNA starting from the 5'-end, but the exonuclease reaction is blocked at a psoralen crosslink, leaving an ss-DNA with a free 3'-end. Subsequently, primer extension is performed using a specific oligonucleotide and *Taq* DNA polymerase, until the psoralen adduct acts as a road block for the *Taq* polymerse, and the extension reaction is terminated. The location of psoralen crosslinks can be deduced by the length of the extension products, which are separated by electrophoresis on an alkaline agarose gel and blotted onto a nylon membrane. Indirect endlabeling is done with a specific, labeled single-stranded probe (made by primer extension) in order to only detect nucleosome positioning in the region of interest.

1.2. Examples and Interpretation

We wanted to compare the nucleosome positioning obtained by in vivo psoralen crosslinking to the positioning determined by MNase digestion of isolated yeast nuclei. We therefore used the yeast minichromosome YRpTRURAP chromatin structure which has previously been studied in great detail by MNase and DNase I digestion of isolated nuclei and analysis of the DNA at low and high resolution (*see* **ref. 20** and Chapter 27). In the following we would like to summarize the results of this study which demonstrate the usefulness of our approach.

1. **The psoralen crosslinking efficiency is very high and crosslinkable sites are evenly distributed.** In naked DNA, crosslinking appears as a rather uniform smear (**Fig. 3A**, lanes 3, 5, 7, 9). However, regions rich in 5'TA and 5'AT (5'TA > 5'TA) are overrepresented over regions that do not contain crosslinkable sites (such as the URA3 promoter; **Fig. 3B**, start). Further analysis of regions that, because of their sequence composition, yield less information by complementary techniques (e.g., nuclease digestions) is recommended.
2. **Sites of psoralen intercalation coincide with linker and nonnucleosomal DNA.** In chromosomal DNA, regularly spaced regions appear to be protected from psoralen intercalation. The length of these protected regions is between 140 and 200 bp, which is interpreted as nucleosomal regions (**Fig. 3A**, lanes 2, 4, 6, 8, 10). Linkers between nucleosomes are represented by clearly visible, sharp bands.

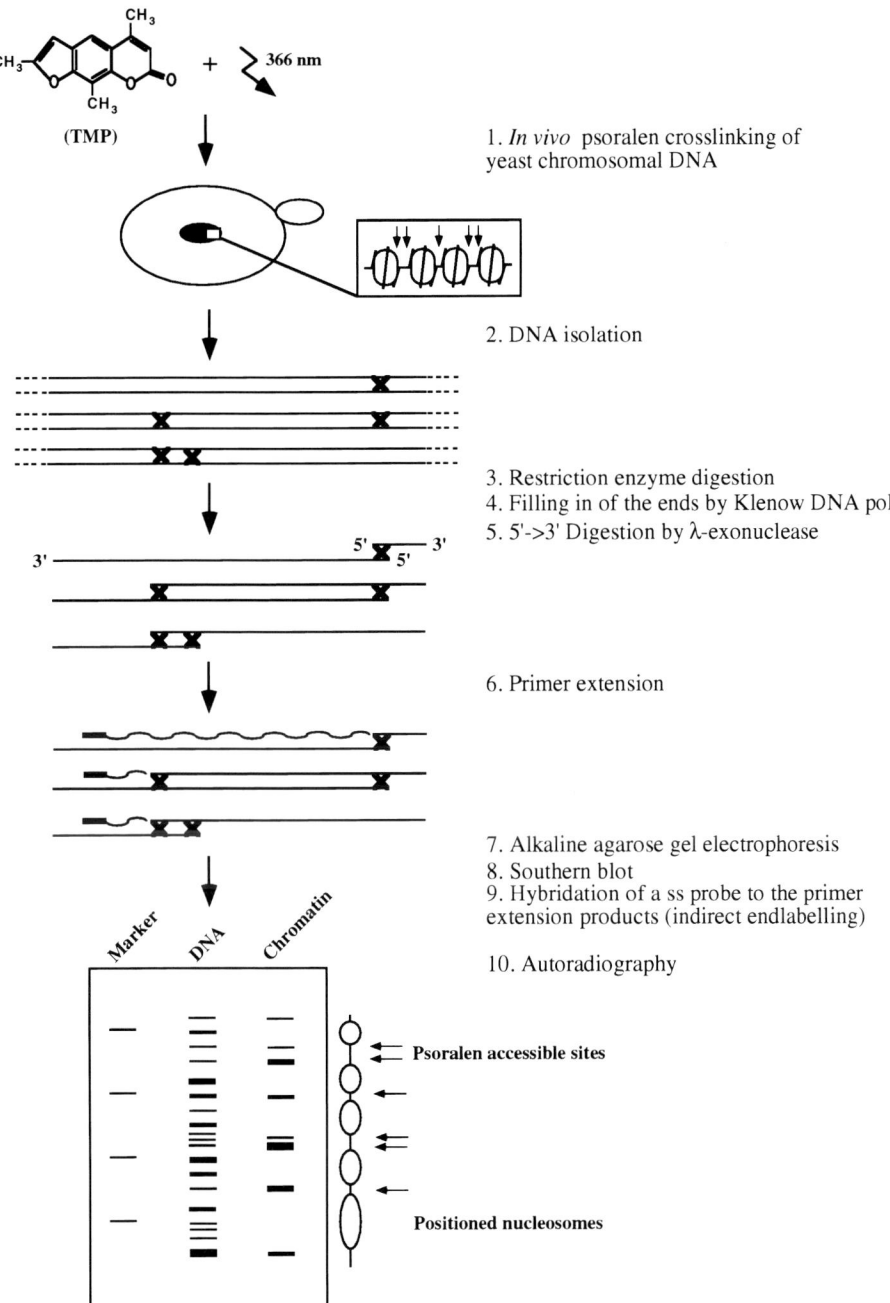

Fig. 2. Schematic representation of nucleosome mapping by psoralen crosslinking and primer extension.

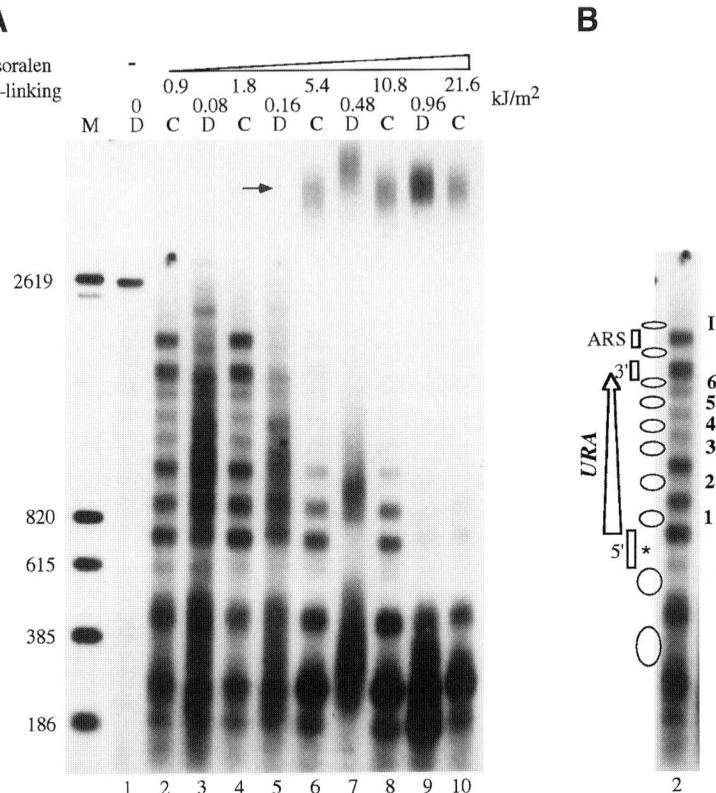

Fig. 3. Analysis of the YRpTRURAP nucleosome structure on the bottom strand by primer extension analysis of λ-exonuclease-treated DNA. (**A**) Mock treated or λ-exonuclease digested DNA was subjected to multiple rounds of primer extension using primer 575. The reaction products were separated by gel electrophoresis on a 1.5% alkaline agarose gel. Southern blot analysis shows the location of psoralen crosslinks in naked DNA (**D**) and chromosomal DNA (**C**). Size marker is shown (**M**). Noncrosslinked DNA was used as a control (lane 1). Crosslinking specifically obstructs the extension reaction (lanes 2–10), and increased crosslinking causes a premature blockage of the polymerase (D, odd lanes, and C, even lanes). The arrow indicates highly crosslinked DNA that remains double-stranded. (**B**) Lane 2 of A is shown. Indicated are: nucleosome positions *(ovals)*, functional elements *(rectangles)*, and the location of the URA3 gene *(empty arrow)*.

3. **The chromatin structure of TRURAP in isolated nuclei and the in vivo chromatin structure as detected by psoralen are very similar.** The nucleosomal pattern obtained by psoralen crosslinking resembles the pattern found by MNase digestion of yeast nuclei (compare **Fig. 3B** and Chapter 26).
4. **All nucleosomes in the URA3 gene are easily detectable.** Note that psoralen can access the linker DNA between tightly packaged nucleosomes (**Fig. 3B**, nucleosomes 4–6).

5. **More than 10 nucleosomes can be mapped at low crosslinking levels.** We find that it is important to perform a time course of psoralen crosslinking. At low crosslinking levels, more than 10 nucleosomes per track can be detected, whereas at high crosslinking levels, the digestion pattern corresponds to not more than two nucleosomes (**Fig. 3A**, even lanes 2, 4, 6, 8, 10).
6. **Primer extension allows the increase of the signal intensity by linear amplification.** We performed 30 cycles of linear amplification in order to increase the sensitivity of the method. The signal intensity could be further increased by more cycles. This allows the use of less material, and therefore mapping of nucleosomes in single copy regions is possible. To analyze the nucleosome structure at high resolution, the primer extension products can also be separated on sequencing gels (unpublished results).

2. Materials

Note that the primers were PAGE-purified and diluted in water to the appropriate working solutions. The primers (Intron, Kaltbrunn, Switzerland) used for theses studies were (map units [MU] are nucleotides on YRpTRURAP; Tm is the annealing temperature):

No. 575 (5'-GAGGGCCAAGAGGGAGGGCATTGGTGAC-3'; 25 to 52 MU; Tm = 65°C), No. 569 (5'-GCAAGCCGCAAACTTTCACCAATGG ACCAG-3'; 160 to 131 MU; Tm = 65°C).

2.1. Psoralen Crosslinking of Yeast Cells

1. Yeast strain FTY23 (*19*).
2. SD (his/trp) medium: 0.67% yeast nitrogen base without amino acids (Difco, Fluka, Buchs, Switzerland), 2% dextrose, 20 mg/L histidine (Fluka, Fluka, Buchs, Switzerland), 20 mg/mL tryptophane (Fluka).
3. Petri dish (Falcon 1008, 35 × 10 mm).
4. A dark corner with yellow light (Sylvania GE "Gold" fluorescent light, Sylvania SA, Meyerin, Switzerland).
5. UV crosslinker with four Sylvania G15T8 black light blue bulbs.
6. TMP (200 µg/mL): 4,5',8-trimethylpsoralen (Sigma, Chemie, Buchs, Switzerland) in ethanol.
7. UVX radiometer: UV Inc., CA.
8. Quiagen Genomic tip G/20 columns and solutions (Quiagen, Basel, Switzerland, cat. no. 10223).
9. TE: 10 mM Tris-HCl, pH 8.0, 1 mM EDTA pH 8.0.

2.2. Psoralen Crosslinking of DNA

1. FTY23; SD (his/trp) medium (*see* **Subheading 2.1.**).
2. Qiagen Genomic Tip G/20 columns and solutions (*see* **Subheading 2.1.**).
3. TE (see **Subheading 2.1.**).
4. Multiwell tray (Nunclon, Gibco BRL, Life Technologies AG, Basel, Switzerland; 24 slots each 15 × 20 mm).

5. TMP (200 µg/mL) (see **Subheading 2.1.**).
6. A dark corner with yellow light (Sylvania GE "Gold" fluorescent light).
7. UV crosslinker with four Sylvania G15T8 black light blue bulbs.
8. Dichloromethane/isoamyalcohol (24/1).
9. 5 M NaCl.
10. 100% Ethanol.
11. 70% Ethanol.

2.3. Preparation of the Primer Extension Template

1. *Eco*RI, 40 U/µL.
2. 10x buffer H: 500 mM Tris-HCl, pH 7.5, 100 mM MgCl$_2$, 1 M NaCl, 10 mM Dithioerythritol.
3. Klenow DNA polymerase: (Boehringer, Mannheim, Rotkreuz, Switzerland) 2 U/µL.
4. 5 mM dNTP solution: Pharmacia Ultrapure dNTP Set (Pharmacia Biotech, Duebendorf, Switzerland) cat. no. 27-2035-01; dilute 100 mM stock solution in water.
5. High Pure PCR Product Purification Kit (Boehringer Mannheim, cat. no. 1732668).
6. Elution buffer: 10 mM Tris-HCl, pH 8.0.
7. λ-Exonuclease: Gibco BRL; 2.9 U/µL.
8. 10X λ-Exonuclease buffer: 670 mM glycine-KOH, pH 9.4, 25 mM MgCl$_2$, 500 µg/mL BSA (Boehringer).
9. StrataClean Resin (Stratagene, Zurich, Switzerland, cat. no. 400714).
10. PCR equilibration buffer: 20 mM Tris-HCl, pH 6.1, 55 mM KCl, 4.3 mM MgCl$_2$.

2.4. Primer Extension and Analysis of the Extension Products

1. PCR-tubes (0.65 mL).
2. 10X PCR buffer: 100 mM Tris-HCl, pH 8.3, 500 mM KCl, 30 mM MgCl$_2$.
3. Primer (0.1 pmol/µL).
4. 5 mM dNTP solution (*see* **Subheading 2.3.**).
5. *Taq* polymerase: Perkin Elmer; 5 U/µL.
6. Highly liquid paraffin.
7. Perkin Elmer thermocycler (PE Applied Biosystems, Rotkreuz, Switzerland).
8. 5X alkaline loading buffer: 250 mM NaOH, 5 mM EDTA, 12.5% Ficoll Type 400 (Sigma), 0.125% bromocresol green (Sigma).
9. Peristaltic pump: (Ismatec, Glattsbrugg-Zurich, Switzerland).
10. 20 cm × 25 cm electrophoresis chamber: BRL.
11. Agarose: Gibco BRL ultrapure.
12. Electrophoresis buffer: 50 mM NaOH, 1 mM EDTA.
13. Nylon membrane: Pall B.
14. X-ray film.

2.5. Oligo Directed Labeling

1. PCR tubes (*see* **Subheading 2.4.**).
2. 10X PCR buffer (*see* **Subheading 2.4.**).

3. Primer (50 pmol/μlL).
4. Primer (1 pmol/μl).
5. 5 mM dNTP solution (*see* **Subheading 2.3.**).
6. 100 μM dATP, dGTP, dTTP solution: Pharmacia; 100 mM stock solution diluted in water.
7. Quarz bidistilled water.
8. *Taq* polymerase: Perkin Elmer; 5 U/μL.
9. Highly liquid paraffin.
10. Perkin Elmer thermocycler.
11. High Pure PCR Product Purification Kit (Boehringer Mannheim).
12. Elution buffer (*see* **Subheading 2.3.**).
13. [α-^{32}P]CTP: Amersham 3000 Ci/mmol.
14. Quick Spin™ Column (Boehringer, cat. no. 1273965).

3. Methods

3.1. Psoralen Crosslinking of Yeast Cells

1. Grow FTY23 cells in SD (his/trp) to exponential phase (1–3 × 10^7 cells/mL).
2. Collect the cells by centrifugation (6000g, 4°C, 10 min) and suspend them at 7 × 10^8 cells/mL in TE.
3. Transfer 1.5 mL cells (1 × 10^9 cells) to a petri dish (35 × 10 mm) to form a thin layer (approx 2 mm). Add 75 μL TMP, mix well and incubate the cells for 5 min under yellow light (*see* **Note 1**).
4. Put the Petri dish on ice-water. Irradiate the cell suspension using four Sylvania G15T8 black light blue lamps (at 2 cm distance with 0.9 (30 s), 1.8 (1 min), 5.4 (3 min), 10.8 (6 min) and 21.6 (12 min) kJ/m^2) (*see* **Note 2**).
5. Pellet the cells by centrifugation (6000g, 4°C, 10 min).
6. Isolate the DNA using Qiagen Genomic Tip G/20 Columns according to the manufactures manual (*see* **Note 4**).
7. Finally, dissolve the DNA in 100 μL TE (approx 150 ng/μL DNA).

3.2. Psoralen Crosslinking of DNA

1. Purify DNA by following **steps 1, 5,** and **6** of **Subheading 3.1.**
2. Dissolve the DNA in 300 μL TE (about 50 ng/μL DNA).
3. Transfer the DNA into a multiwell tray (15 × 20 mm) and add 15 μL TMP to form a thin layer (2.5 mm). Mix well and incubate for 5 min under yellow light.
4. Put the multiwell dish on ice-water. Irradiate the DNA using four Sylvania G15T8 black light blue lamps (at 18 cm distance with 0.08 (5 s), 0.16 (10 s), 0.48 (30 s), and 0.96 (1 min) kJ/m^2) (*see* **Note 3**).
5. Transfer the DNA into an 1.5-mL tube. Add 12 μL NaCl (final conc. 150 mM). Extract the DNA three times with 1 vol of dichloromethane/isoamylalcohol (24:1) (*see* **Note 5**).
6. Precipitate the DNA by addition of 2.5 vol of ethanol. Pellet the DNA by centrifugation (maximal speed in microfuge, 4°C, 30 min). Wash the pellet with 70% ethanol.
7. Finally, dissolve the pellet in 100 μL TE (about 150 ng/μL DNA).

3.3. Preparation of the Primer Extension Template

1. Set up the restriction digest in a 100-μL containing 20 μL DNA (3μg), 5 μL *Eco*RI (200 U), 10 μL 10X buffer H. Incubate at 37°C for 12 h (*see* **Note 6**).
2. Add 1 μL Klenow (2 U) and 2 μL dNTP solution (final conc. 100 μ*M*) and incubate the sample further at 37°C for 1 h.
3. Purify the DNA using PCR purification columns according to the manufactures protocol. Elute the DNA in 100 μL of elution buffer.
4. To 44 μL DNA add 5 μL of 10X λ-exonuclease buffer and 1 μL λ-exonuclease (2.9 U). Incubate the sample at 37°C for 30 min (*see* **Note 7**).
5. Add 0.5–1 μL StrataClean Resin and vortex the sample (*see* **Note 8**).
6. Pellet the resin in a microfuge for 30 s. Transfer the supernatant into a new tube.
7. Add 50 μL (1 vol) equilibration buffer. Store at –20°C.

3.4. Primer Extension and Analysis of the Extension Products

1. Sequentially mix: 10 μL DNA (150 ng total DNA), 1 μL 10X PCR buffer, 1 μL dNTP solution (final conc. 1.25 m*M*), 1 μL primer bottom strand (575; 0.1 pmol/μL), 7 μL diluted *Taq* polymerase (1 U in water) and overlay with 25 μL paraffin.
2. Subject the samples to 30 cycles of repeated denaturation at 95°C for 45 s, annealing at primer specific temperatures (Tm = 65°C) for 5 min and extension for 3 min at 72°C (*see* **Note 9**).
3. Transfer the aqueous phase into a 1.5-mL tube. Store at 4°C.
4. Add 5 μL 5X alkaline loading buffer shortly before electrophoresis. Electrophorese the DNA in an 1.5% alkaline agarose gel (20 cm × 25 cm) at 4°C at 350 mA for 9–10 h with circulation of electrophoresis buffer (2 L) (*see* **Note 10**).
5. Transfer the DNA onto a nylon membranes (Pall B, Pall) using an alkaline blotting protocol (Bio-Rad bulletin 234 890 90-0891).
6. Hybridize to a radioactively labeled DNA fragment (*see* **Subheading 3.5.**).
7. Detect radioactive bands by exposure to X-ray films.

3.5. Probe Synthesis

In order to create a specific ss-probe for the indirect endlabeling, we first amplify a short (approx 150 bp) fragment by PCR. This is done from one end with the primer used for primer extension in combination with a oligonucleotide which primes opposite to the other stand. This PCR fragment is then purified from free primers and nucleotides. Next, using the opposite primer and [α-^{32}P]CTP, the radioactively labeled, single-stranded probe is produced.

1. Sequentially mix: 1 μlL plasmid DNA (0.6 fmol), 3 μL 10X PCR buffer, 2 μL dNTP solution (final conc. 330 μ*M*), 3 μL primer top strand (569; 50 pmol/μL), 3 μL primer bottom strand (575; 50 pmol/μL), 18 μL diluted *Taq* polymerase (1 U in water) and overlay with 25 μL paraffin.
2. Subject the sample to 15 cycles of repeated denaturation at 95°C for 45 s, annealing at primer specific temperatures (Tm = 65°C) for 5 min and extension for 3 min at 72°C.

3. Transfer the aqueous phase into a 1.5-mL tube. Add 70 μL of water to a final volume of 100 μL.
4. Purify the PCR fragment using PCR purification columns according to the manufactures protocol. Elute the DNA in 100 μL elution buffer.
5. Sequentially mix: 12 μL purified PCR fragment (0.6 pmol), 4 μL 10X PCR, 10 μL dATP/dGTP/dTTP solution (final conc. 100 μM), 5 μL primer top strand (569; 0.1 pmol/μL), 3 μL [α-^{32}P]CTP (6 pmol), 6 μL diluted *Taq* polymerase (1 U in water) and overlay with 25 μL paraffin.
6. Subject the sample to 10 cycles of repeated denaturation at 95°C for 45 s, annealing at primer specific temperature (Tm = 65°C) for 5 min and extension for 3 min at 72°C.
7. Transfer the sample onto a Sephadex G50 Quick Spin™ Column. Purify the labelled DNA according to the manufactures manual.
8. Measure the activity using a scintillation counter (approx 3×10^9 dpm/μg DNA are expected).

4. Notes

1. The cells should be preincubated in the dark to ensure the intercalation of TM^P in chromosomal DNA.
2. Since UV-light with a wavelength of 254 nm reverses psoralen crosslinks and damages the DNA, crosslinking should be done using UV-lamps which preferentially emitt 366 nm. The UV-dose is measured using a UVX radiometer.
3. To get a comparable crosslinking of naked DNA and chromosomal DNA, the naked DNA was irradiated with an approx 10-fold lower dose (*see* **Fig. 3**, upper panel).
4. The DNA must be clean and undegraded for further treatment. We find that the Qiagen purified DNA fulfills these criteria.
5. Multiple extractions of the DNA with dichloromethane/isoamylalcohol efficiently removes unbound psoralen.
6. In order to detect as many nucleosomes as possible it is recommended to linearise the DNA close to the sequence which hybridises to the specific primer.
7. Ensure that the DNA is blunt-ended. Only blunt ends serve as a good substrate for the λ-exonuclease. We also tried to digest 2'-overhanging ends with various nucleases (Mung Bean, S1-nuclease) in order to created blunt ended DNA. However, these nucleases introduce some nicks into the dsDNA. Since these nicks block the λ-exonuclease activity, they mimic psoralen crosslinks. Thereby the fidelity of the technique could be affected. Other 5'->3' exonucleases which are blocked by psoralen crosslinks could also be used in order to create a substrate for the primer extension reaction. Again, it is recommended to first test these nucleases for nuclease activity.
8. We find that the nuclease is efficiently removed by StrataClean resin. After purification we did not observe an residual compound or activity which interfered with the primer extension reaction. Since the compositon of the λ-exonuclease buffer is very similar to that of the PCR-buffer, the buffer can be adjusted in order to omit an intermediate DNA-precipitation step.

9. Routinely, we perfom 30 cycles of primer extension. However, depending on the substrate concentration, more or less cycles can be applied.
10. Since hot alkali treatment destroys the agarose, we first dissolve the agarose in water. The polymerized gel is then soaked for 30 min in 1–2 L electrophoresis buffer.

Acknowlegments

We thank F. Thoma for critical discussion. This work was supported by grants from the Swiss National Science Foundation and by the Swiss Federal Institute of Technology, Zürich (ETH) to (J. M. S., NF 31.41827.94).

References

1. Wolffe, A. (1992) In *Chromatin: Structure and Function*. Academic Press, London and New York.
2. Kladde, M. P., Xu, M., and Simpson, R. T. (1997) Direct study of DNA-protein interactions in repressed and active chromatin in living cells. *EMBO J.* **15,** 6290–6300.
3. Thoma, F. and Sogo, J. M. (1988) In *Chromosomes and Chromatin*. CRC Press, Boca Raton, FL, pp. 85–107.
4. Sogo, J. M. and Thoma, F. (1989) In *Methods in Enzymology* (Wassermann, P. M. and Kornberg, R. D., eds.), **170,** pp. 142–165.
5. Lucchini, R. and Sogo, J. M. (1998) Transcription of ribosomal RNA genes by eukaryotic RNA polymerase I (Paule, M. R., ed.) Landes Bioscience, pp. 255–276.
6. Hanson, C., Shen, C., and Hearst, J. (1976) Cross-linking of DNA in situ as a probe for chromatin structure. *Science* **193,** 62–64.
7. Conconi, A., Losa, R., Koller, T., and Sogo, J. (1984) Psoralen-crosslinking of soluble and of H1-depleted soluble rat liver chromatin. *J. Mol. Biol.* **178,** 920–928.
8. Sogo, J., Stahl, H., Koller, T., and Knippers, R. (1986) Structure of replicating simian virus 40 minichromosomes. The replication fork, core histone segregation and terminal structures. *J. Mol. Biol.* **189,** 189–204.
9. Widmer, R. M., Koller, T., and Sogo, J. M. (1988) Analysis of the psoralen-crosslinking pattern in chromatin DNA by exonuclease digestion. *Nucleic Acids Res.* **16,** 7013–7024.
10. Gasser, R., Koller, T., and Sogo, J. M. (1996) The stability of nucleosomes at the replication fork. *J. Mol. Biol.*, **258,** 224-239.
11. Sogo, J. M. and Lasky, R. A. (1995) In *Chromatin Structure and Gene Expression* (Elgin, S., ed.), Oxford University Press, pp. 49-70.
12. Conconi, A., Widmer, R., Koller, T., and Sogo, J. (1989) Two different chromatin structures coexist in ribosomal RNA genes throughout the cell cycle. *Cell* **57,** 753–761.
13. Sargan, D. and Butterworth, P. (1982) Eukaryotic ternary transcription complexes. II. An approach to the determination of chromatin conformation at the site of transcription. *Nucleic Acids Res.* **10,** 4655–4669.
14. De Bernardin, W., Koller, T., and Sogo, J. (1986) Structure of in-vivo transcribing chromatin as studied in simian virus 40 minichromosomes. *J. Mol. Biol.* **191,** 469–482.

15. Gruss, C., Wu, J., Koller, T., and Sogo, J. M. (1993) Disruption of the nucleosomes at the replication fork. *EMBO J.* **12,** 4533–4545.
16. Zhen, W.-P., Buchardt, O., Nielsen, H., and Nielsen, P. E. (1986) Site-specificity of psoralen-DNA interstrand cross-linking determined by nuclease *Bal*31 digestion. *Biochemistry* **25,** 6598–6603.
17. Ostrander, E. A., Karty, R. A., and Hallick, L. M. (1988) High resolution psoralen mapping reveals an altered DNA helical structure in the SV 40 regulatory region. *Nucleic Acids Res.* **16,** 212–227.
18. Kochel, T. J. and Sinden, R. R. (1989) Hyperreactivity of *B-Z* junctions to 4,5',8-trimethylpsoralen photobinding assayed by an exonuclease III/ photoreversal mapping procedure. *J. Mol. Biol.* **205,** 91–102.
19. Thoma, F. (1986) Protein-DNA interactions and NSRs determine nucleosome positions on yeast plasmid chromatin. *J. Mol. Biol.* **190,** 177–190.
20. Tanaka, S., Livingstone, M., and Thoma, F. (1996) Chromatin structure of the yeast URA3 gene at high resolution provides insight into structure and positioning of nucleosomes in the chromosomal context. *J. Mol. Biol.* **257,** 919–934.

12

Preparation of Chromatin Assembly Extracts from *Xenopus* Oocytes

David John Tremethick

1. Introduction

The majority of DNA in eukaryotic cells is packaged by histones and many poorly characterized nonhistone proteins to form a dynamic structure known as chromatin. Chromatin is a periodic structure made up of repeating, regularly spaced subunits, the nucleosomes. Elegant genetic experiments have clearly demonstrated that histones play a central role in transcriptional control *(1)*. Moreover, histones, via protein-protein interactions or by playing an architectural role, can facilitate or inhibit the transcriptional activation process *(1,2)*. It also appears that the function of histones themselves may be regulated by protein modifications and therefore may be targets for cell signaling pathways *(3,4)*.

In addition to providing descriptive details, these genetic studies have provided some mechanistic information concerning the role of histones in regulating transcription. However, to determine the precise molecular details of how chromatin regulates transcription (and how the structure of chromatin itself is regulated), it is essential to mimic the in vivo transcriptional activation process in vitro within a chromatin context. Indeed, numerous studies that have examined transcription factor-nucleosome interactions in vitro have supported, and in some cases extended, the findings from genetic studies *(3–9)*.

A critical feature of chromatin with regard to structure and function is the regular spacing of nucleosomes. Previously, many attempts had been made to reconstitute this periodic structure in vitro using pure histones and DNA and often in the presence of counterions (*see* **ref. 10** for a discussion). Although nucleosomes were assembled, the native periodic structure was not formed. Clearly, other factors besides histones and DNA are required for the regular spacing of nucleosomes. On the other hand, cellular extracts prepared from

Xenopus laevis oocytes *(10,11)*, *Xenopus laevis* eggs *(12)*, and *Drosophila* embryos (**refs. *13,14***; *see* chapter 13) can assemble authentic, regularly spaced chromatin.

Xenopus laevis has been used extensively for studying many different aspects of vertebrate development. These studies include the use of the oocyte as an in vivo test tube for the investigation of transcription *(15)*. Another advantage of using *Xenopus* toads is that oocytes provide a rich source of nuclear proteins, including chromatin assembly components and transcription factors. The oocyte accumulates such a large pool of proteins, because such a store is required by rapidly dividing nuclei in the pre-blastula embryo. The protocol presented here describes the preparation of an efficient extract from oocytes that can assemble the large amounts of chromatin necessary for biochemical analysis. In addition, protocols are described for the isolation of key nucleosome assembly components from the oocyte extract.

After surgically removing the ovary, the oocytes are partially dispersed by collagenase treatment. Following this treatment, and extensive washing, the oocytes are washed in a low ionic-strength extraction buffer. The oocytes are then lysed in the same buffer by a high speed centrifugation step. Unlike homogenization, this method of lysis does not release proteinases and other degrading enzymes *(16)*. The resultant supernatant is basically a cytoplasmic extract (3–6 mg/mL). The oocyte extract can essentially assemble any piece of DNA (of appropriate size) into chromatin with closed circular DNA being assembled more efficiently than linear DNA. The chromatin assembly reaction itself is a time-dependent process in which nucleosomes load gradually onto the DNA template. The majority of nucleosomes load onto template DNA within 30 min and the reaction is allowed to proceed for 5 h to ensure completion of the process. A critical component of the chromatin assembly reaction is the ATP regeneration system. ATP is required for the ordering of nucleosomes into a regular array, and recently several ATP-dependent nucleosome spacing activities have been identified *(10,17,18)*. As described here, all new oocyte extract preparations are first tested for nucleosome formation by employing a DNA supercoiling assay (each nucleosome introduces one negative supercoil into covalently closed circular DNA; *see* Chapter 6) *(11)*. Most important, the authenticity of the assembled chromatin (i.e., whether assembled nucleosomes are organized into a regular array) must then be examined by digesting the chromatin template with micrococcal nuclease *(10,11)*. The nucleosomal repeat length of chromatin assembled using the oocyte extract under the conditions described here is approx 180 bp (**Fig. 1**, lanes 1–5). A major advantage of this assembly system, using template DNA of interest, is its ability to be scaled up, which in turn enables a detailed structural analysis of the chromatin template to be carried out *(11)*.

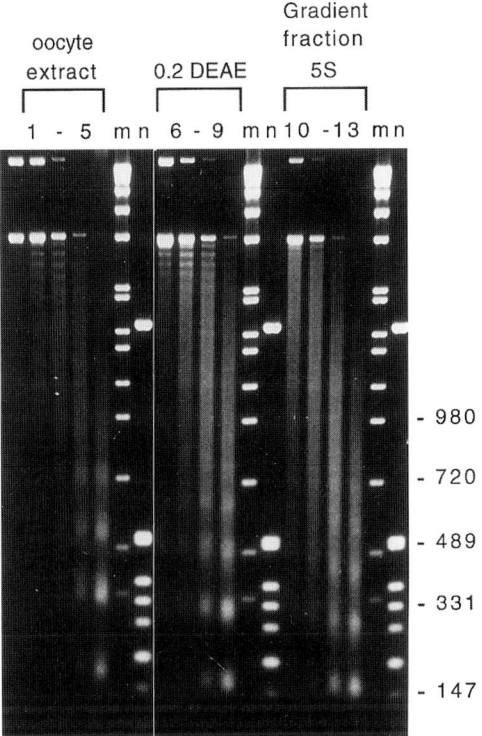

Fig. 1. Micrococcal nuclease digestion analysis of in vitro assembled chromatin. Chromatin was assembled using the unfractionated *Xenopus* oocyte extract (lanes 1–5), a 0.2 *M* NaCl DEAE column fraction (lanes 6–9), and the N1, N2-(H3,H4) complex (lanes 10–13). The assembled chromatin was digested with micrococcal nuclease and the purified DNA was run on a 1.5% agarose gel *(10)*.

A potential problem with regard to using the crude oocyte extract for chromatin/transcription studies is that the oocyte extract is not only enriched with chromatin assembly components but also with transcription factors. The oocyte extract can be fractionated to isolate key nucleosome assembly components and remove many proteins involved in the transcription process (but not all, *see* **ref. *19***) to yield a more defined in vitro system. (In addition, this fractionation work has provided important information concerning the mechanism by which DNA is assembled into chromatin in vitro, *see* **ref. *10*.**) In *Xenopus* oocytes, the two histone pairs exist in two distinct forms. Histones H3 /H4 (which are in the diacetylated form) are complexed with carrier proteins known as N1 and N2 to produce a distinct complex *(20)*. On the other hand, histones H2A/H2B appear to be mostly in the free form, with a subpopulation being complexed with nucleoplasmin *(10)*. In the first fractionation step, the oocyte extract is passed over a weak anion exchanger (DEAE Sephacel). Under the conditions used (buffer containing 100 mM NaCl), noncomplexed histones H2A/H2B flow through the column while the N1, N2-(H3,H4) complex and H2A/H2B associated with nucleoplasmin remain bound. These complexes are eluted off the column with buffer containing 200 mM NaCl. Noncomplexed nucleoplasmin, N1 and N2 can be eluted off the column using buffer containing 800 mM NaCl.

In contrast to the crude oocyte extract, assembly reactions using the 0.2 M DEAE fraction (which contains the N1, N2-(H3,H4) complex), require the addition of histones H2A/H2B. Although histones H2A/H2B purified from the DEAE Sephacel flowthrough can be used, histones H2A/H2B purified from other sources (e.g., chicken red blood cells or Hela cells) are used because the oocyte extract contains three H2A variants *(11)* that could complicate interpretation of results from transcription studies. Although nucleosomes are spaced in a regular array which is dependent on ATP, chromatin assembled using the 0.2M DEAE fraction has a shortened nucleosomal repeat length of 165 bp (**Fig. 1**, lanes 6–9).This indicates that a component(s) responsible for increasing the repeat length from 165–180 bp is missing from the 0.2 M DEAE fraction. However, this fraction is particularly useful for carrying out histone H1 structure/function studies because addition of histone H1 to the reaction restores the repeat length back to approx 180 bp *(10)*.

Finally, the N1, N2-(H3,H4) complex is isolated by running the 0.2 M DEAE fraction on a sucrose gradient—the N1, N2-(H3,H4) complex has a sedimentation coefficient of 5S. The isolated N1, N2-(H3,H4) complex combined with purified histones H2A/H2B provides a well-defined nucleosomal assembly system. It is possible to use this system to assemble single positioned nucleosomes on short DNA fragments *(21)* or assemble plasmid DNA into a nucleosomal template with a wide range of nucleosome densities (by varying the amount of N1, N2-(H3,H4) complex added to the assembly reaction) *(19)*. Assembly reactions receiving these histone components produce a nucleosomal template with a repeat length of 145 base pairs (**Fig. 1**, lanes 10–13). Depending on the experimental aim, longer repeat lengths can be restored by the addition of an ATP-dependent nucleosome spacing activity *(10)*. Finally, this defined system assembles nucleosomes under physiological conditions and therefore provides a valuable system for transcription studies.

2.Materials

2.1. Preparation of Xenopus *Oocyte Extracts*

1. Large female *Xenopus laevis* toads (Nasio, Wisconsin).
2. 10X stock OR2A buffer: 25 mM KCl, 825 mM NaCl, 50 mM HEPES, 10 mM Na$_2$ HPO$_4$, pH 7.6.

Chromatin Assembly Extracts

3. 100X stock OR2B buffer: 100 mM CaCl$_2$, 100 mM MgCl$_2$.
4. Collagenase (Sigma, C-6885).
5. Extraction buffer: 1 mM EGTA, 5 mM KCl, 1.5 mM MgCl$_2$, 10% (v/v) glycerol, 20 mM HEPES, 10 mM β-glycerol phosphate (di sodium salt), 0.5 mM DTT, 10 mg/mL phenylmethylsulfonyl fluoride, 2 mg/mL leupeptin, 2 mg/mL pepstatin, pH 7.5.
6. 50-mL conical tubes (Falcon).
7. SW-41 ultracentrifuge tubes (Beckman).

2.2. Preparation of DEAE Column Assembly Fraction

1. DEAE-Sephacel (Pharmacia).
2. Poly-Prep Chromatography 10-mL Columns (Bio-Rad)
3. Column buffer: 1 mM EDTA, 1.5 mM MgCl$_2$, 20 mM HEPES, 0.5 mM DTT, 10 mg/mL phenylmethylsulfonyl fluoride, 2 mg/mL leupeptin, 2 mg/mL pepstatin, pH 7.5.
4. Protein assay dye reagent (Bio-Rad).
5. 75 mM NaCl Extraction buffer: 75 mM NaCl, 1 mM EGTA, 5 mM KCl, 1.5 mM MgCl$_2$, 10% (v/v) glycerol, 20 mM HEPES, 10 mM β-glycerol phosphate, 0.5 mM DTT, 10 mg/mL phenylmethylsulfonyl fluoride, 2 mg/mL leupeptin, 2 mg/mL pepstatin, pH 7.5.

2.3. Isolation of the N1, N2-(H3,H4) Fraction

1. SW-41 ultracentrifuge tubes (Beckman).
2. 5% Sucrose buffer: 5% (w/v) sucrose, 23 mM KCl, 17 mM NaCl, 0.1 mM EDTA, 2 mM MgCl$_2$, 20 mM HEPES, 1 mM DTT, 10 mg/mL phenylmethylsulfonyl fluoride, 2 mg/mL leupeptin, 2 mg/mL pepstatin, pH 7.5.
3. 20% Sucrose buffer: 20% (w/v) sucrose, 23 mM KCl, 17 mM NaCl, 0.1 mM EDTA, 2 mM MgCl$_2$, 20 mM HEPES, 1 mM DTT, 10 mg/mL phenylmethylsulfonyl fluoride, 2 mg/mL leupeptin, 2 mg/mL pepstatin, pH 7.5.
4. 18% SDS-polyacrylamide gel.

2.4. Purification of Histones H2A/H2B

1. Homogenization buffer: 0.2 mM EDTA, 20 mM Tris-HCl, 1 mM DTT, 10 mg/mL phenylmethylsulfonyl fluoride, 2 mg/mL leupeptin, 2 mg/mL pepstatin, pH 8.0.
2. Glass Dounce homogenizer.
3. Hydroxylapatite (Bio-Rad, DNA-Grade)
4. Hydroxylapatite buffer: 0.5 M K PO$_4$, 1 mM DTT, 10 mg/mL phenylmethylsulfonyl fluoride, 2 mg/mL leupeptin, 2 mg/mL pepstatin, pH 7.5.
5. Dialysis buffer: 200 mM NaCl, 20 mM HEPES, 0.1 mM EDTA, pH 7.5.
6. SS-34 centrifuge tubes (Sorvall).
7. 18% SDS-polyacrylamide gel.
8. Protein assay dye reagent (Bio-Rad).

2.5. Testing the Oocyte Extract for Chromatin Assembly

1. Circular plasmid DNA of interest.
2. *Xenopus* oocyte extract.

3. Extraction buffer: 1 mM EGTA, 5 mM KCl, 1.5 mM MgCl$_2$, 10% (v/v) glycerol, 20 mM HEPES, 10 mM β-glycerol phosphate, 0.5 mM DTT, 10 mg/mL phenylmethylsulfonyl fluoride, 2 mg/mL leupeptin, 2 mg/mL pepstatin, pH 7.5.
4. Creatine kinase (Boehringer Mannheim).
5. ATP buffer: 10 min MgCl$_2$, 30 mM ATP, pH 7.5, 800 mM creatine phosphate, 10 mM β-glycerol phosphate, 0.5 mM DTT, 1 mM EGTA, 5 mM KCl, 10% (v/v) glycerol, 20 mM HEPES, 10 mg/mL phenylmethylsulfonyl fluoride, 2 mg/mL leupeptin, 2 mg/mL pepstatin, pH 7.5.

2.6. Testing the DEAE Fraction for Chromatin Assembly

1. Circular plasmid DNA of interest.
2. DEAE fraction.
3. Extraction buffer: (*see* **Subheading 2.1.**).
4. Creatine kinase (Boehringer Mannheim).
5. ATP buffer: 5.0 mM MgCl$_2$, 30 mM ATP, pH7.5, 450 mM creatine phosphate, 10 mM β-glycerol phosphate, 0.5 mM DTT, 1 mM EGTA, 5mMKCl, 10% (v/v) glycerol, 20 mM HEPES, 10 mg/mL phenylmethylsulfonyl fluoride, 2 mg/mL leupeptin, 2 mg/mL pepstatin, pH 7.5.
6. Pure histones H2A/H2B.
7. Dialysis buffer: 200 mM NaCl, 20 mM HEPES, 0.1 mM EDTA, pH 7.5.

2.7. Testing the N1, N2-(H3,H4) Complex for Nucleosome Assembly

1. Circular plasmid DNA of interest.
2. The N1, N2-(H3,H4) complex.
3. ATP buffer: 30 mM ATP, pH7.5, 270 mM creatine phosphate, 10 mM β-glycerol phosphate, 0.5 mM DTT, 1 mM EGTA, 5 mM KCl, 10% (v/v) glycerol, 20 mM HEPES, 10 mg/mL phenylmethylsulfonyl fluoride, 2 mg/mL leupeptin, 2 mg/mL pepstatin, pH 7.5.
4. Pure histones H2A/H2B.
5. Dialysis buffer: *see* **Subheading 2.6.**

2.8. Processing of Chromatin Samples for DNA Supercoiling Analysis

1. 2.5% Sarcosyl (v/v), 100 mM EDTA.
2. RNase A (Boehringer Mannheim).
3. 2% SDS (w/v).
4. Proteinase K (Boehringer Mannheim).
5. 7.5 M NH$_4$SO$_4$.
6. 100% Ethanol.
7. 70% Ethanol.

3. Methods

3.1. Preparation of Xenopus Oocyte Extracts

1. Anesthetize seven large *Xenopus* toads by hypothermia by placing the animals in an ice bucket (with lid) and covering them with ice for a suitable length of time (*see* **Note 1**).

2. Prepare 1 L of OR 2 Buffer (1X) by adding 100 mL of 10X stock OR2A buffer and 10 mL of 100X stock OR2B buffer to 890 mL of sterile milliQ water.
3. Operate on the toads to remove the ovaries (*see* **Note 2**). Place ovaries from the seven toads into a beaker containing 200 mL of OR2 buffer. Gently swirl and transfer the ovaries to a second beaker containing another 200 mL of OR2 buffer. Repeat this process until the ovaries are washed with 800 mL of buffer.
4. Following the addition of 0.15 g collagenase (*see* **Note 3**) to 200 mL of OR 2 Buffer in a 500 mL flask, the washed oocytes are added. The oocytes are then placed in a water bath at 25°C and shaken gently for 2 h.
5. The collagenase treated oocytes (*see* **Note 4**) are washed extensively with 5 × 200 mL washes of freshly prepared OR 2 buffer (*see* **Note 5**).
6. Half fill 50-mL Falcon tubes with the washed oocytes and top up with freshly prepared extraction buffer. Invert twice and allow the large oocytes to settle before decanting the buffer. Repeat this washing step three times (*see* **Note 6**).
7. Transfer oocytes to a 500-mL beaker containing 100 mL of extraction buffer and proceed to load the oocytes into six SW41 ultracentrifuge tubes at 4°C (*see* **Note 7**). Fill the centrifuge tubes to 1 cm from the top. Decant buffer and replace with fresh extraction buffer.
8. Centrifuge samples at 36,000 rpm (220,000g) for 2 h at 4°C.
9. After centrifugation, place centrifuge tubes in a suitable test tube rack and draw off the supernatant using a 5-mL syringe connected to a 19-gauge needle (*see* **Note 8**). Combine supernatants in a 50-mL tube on ice, add protease inhibitors and aliquot. Store aliquots at –70°C (*see* **Note 9**). Alternatively, the oocyte extract can be further fractionated (*see* **Subheadings 3.2.** and **3.3.**).

3.2. Preparation of DEAE Column Assembly Fraction

1. During the collagenase treatment of the oocytes, prepare a 1.5-mL DEAE column ensuring that the column is washed extensively with 100 mM NaCl column buffer (containing protease inhibitors).
2. Using a conductivity meter, adjust the oocyte extract (with 2 M NaCl) to a final conductivity of 0.1 M NaCl in column buffer (*see* **Note 10**).
3. Very slowly, overnight, load the DEAE column with the oocyte extract (*see* **Note 11**).
4. After allowing all of the supernatant to run in, the column is washed extensively with 10-column volumes using freshly prepared 100 mM NaCl column buffer. Once the column is washed, elute chromatin assembly fractions using freshly prepared 200 mM NaCl column buffer. Collect 500-µL fractions.
5. Using a standard protein-dye assay, combine the three most concentrated protein fractions (*see* **Note 12**). Dialyze against 75 mM NaCl extraction buffer, aliquot and store at –70°C. Alternatively, use these combined fractions to isolate the N1, N2-(H3,H4) complex (*see* **Subheading 3.3.**).

3.3. Isolation of the N1,N2-(H3,H4) Fraction

1. During the column washing step (*see* **step 4**, **Subheading 3.2.**), pour three 11 mL of 5–20% sucrose gradients (plus a balance) in gradient buffer using SW41 ultracentrifuge tubes.

2. Load 500 µL of the combined 200 mM NaCl DEAE fraction onto each gradient and centrifuge samples at 40,000 rpm (270,000g) for 24 h at 4°C.
3. After centrifugation, fractionate gradients into approximately 23 fractions of 500 µL each.
4. Run 10 µL of appropriate fractions (*see* **Note 13**) on a 18% SDS-polyacrylamide gel with standard histones as a control.
5. After silver staining the gel, combine the three gradient fractions (*see* **Note 14**) with the highest concentration of histones H3 and H4. The N1,N2-(H3,H4) complex is then aliquoted and stored at –70°C.

3.4. Purification of Histones H2A/H2B

1. Isolate nuclei from a convenient source of cells (*see* **Note 15**) using a suitable method, and pellet the nuclei by low-speed centrifugation.
2. After weighing nuclei, resuspend nuclei in 5 vol of homogenization buffer.
3. Homogenize with 10 strokes using a Dounce homogenizer and leave on ice for 30 min.
4. Sonicate twice for 60 s.
5. Pellet nuclei debris by high-speed centrifugation (14,500 rpm at 4°C using SS-34 centrifuge tubes).
6. Remove supernatant and dilute with 9 vol of fresh 0.6 M NaCl Hydroxylapatite column buffer.
7. Load diluted supernatant slowly onto hydroxylapatite column (*see* **Note 16**) overnight.
8. Wash column with 10-column volumes of 0.6 M NaCl Hydroxylapatite column buffer.
9. Elute histones H2A/H2B using 0.93 M NaCl Hydroxylapatite column buffer. Collect 10 fractions, the volume of each fraction being equivalent to ¼ of the column volume.
10. Elute histones H3/H4 using 2.0 M NaCl column buffer.
11. Using a standard protein-dye assay, determine which fractions contain protein. Run these fractions on a 18% SDS-polyacrylamide gel to check purity.
12. Combine pure H2A/H2B, and H3/H4 fractions and dialyse against 0.2 M NaCl dialysis buffer.

3.5. Testing the Oocyte Extract for Chromatin Assembly

1. Add increasing amounts (50–500 ng) of circular supercoiled or relaxed plasmid DNA of interest (*see* **Note 17**) to an appropriate number of eppendorf tubes in a total volume of 10 µL, on ice, using extraction buffer as a dilution buffer.
2. Add 5 µL of Creatine kinase (1 ng/µL) followed by 5 µL of ATP buffer (*see* **Note 18**).
3. Finally, add 30 µL of oocyte extract very slowly to the reaction mix (final volume 50 µL) and incubate for 5 h at 27°C (*see* **Note 19**).

3.6. Testing the DEAE Fraction for Chromatin Assembly

1. Add increasing amounts of DEAE fraction (4–16 µL), in duplicate, to a suitable number of 5-mL tubes (*see* **Note 20**).
2. To the DEAE fraction, add a volume of Creatine kinase (1 ng/µL) followed by the same volume of an ATP buffer equivalent to ¹⁄₁₀ of the final volume (*see* **Note 21**).

3. Add pure histones H2A/H2B, at a protein-to-mass ratio of 0.4, to one set of reaction mixtures (*see* **Note 22**). To the other set, add 0.2 *M* NaCl dialysis buffer.
4. Add slowly 300 ng of relaxed plasmid DNA.
5. Add extraction buffer to increase the volume of the reaction mix to the final volume.
6. Incubate for 5 h at 27°C (*see* **Note 23**).

3.7. Testing the N1, N2-(H3,H4) Complex for Nucleosome Assembly

1. Add increasing amounts of the N1, N2-(H3,H4) complex (8–24 µL), in duplicate, to a suitable number of Eppendorf tubes (*see* **Note 24**).
2. To the N1, N2-(H3,H4) complex, add a volume of ATP buffer equivalent to $\frac{1}{10}$ of the final volume (*see* **Note 25**).
3. Add pure histones H2A/H2B, at a protein-to-mass ratio of 0.4 (*see* **Note 26**), to one set of reaction mixtures. To the second identical set, add 0.2 *M* NaCl dialysis buffer.
4. Add slowly 300 ng of relaxed plasmid DNA.
5. Incubate for 5 h at 27°C (*see* **Note 27**).

3.8. Processing of Chromatin Samples for DNA Supercoiling Analysis

1. On completion of the chromatin assembly reaction, add 12.5 µL of a 2.5% sarcosyl, 100 m*M* EDTA solution.
2. Adjust final volumes of samples to 62.5 µL with sterile water.
3. Add 1 µL of RNase A (10 mg/mL) and incubate samples for 30 min at 37°C.
4. Next, add 8 µL of a 2% SDS solution and 8 µL of Proteinase K (10 mg/mL). Incubate overnight at 37°C.
5. Following the overnight incubation, ethanol precipitate the DNA samples by adding 53 µL of 7.5 *M* NH$_4$SO$_4$ followed by 265 µL of ethanol.
6. Following washing DNA samples with 70% ethanol and drying, run the resuspended DNA samples on a 1% agarose gel (*see* **Note 28**).

4. Notes

1. Usually 60 min is sufficient to anesthetize the toads, but occasionally they may respond to handling. An additional 90 min may be required.
2. Two ventral incisions, in an anterior-posterior direction, are made on both sides of the abdomen. The ovary is gently teased out by using fine-tipped forceps. The toad is then sacrificed by pithing.
3. A critical factor in determining the success of the chromatin assembly oocyte extract is the activity of the collagenase (Sigma), which should not be less than 400 U/mg of protein. It is also recommended that a small amount of a specific collagenase batch is first tested, with regard to yielding an active oocyte extract, before purchasing a large batch.
4. The collagenase treatment does not completely disperse the oocytes. Addition of more collagenase can reduce the chromatin assembly capacity of the oocyte extract.

5. The washes remove collagenase, immature oocytes, and connective tissue. For each wash, the OR2 buffer is poured down the side of the flask and the flask is gently swirled. The buffer is then decanted when the large oocytes have settled, at the same time that the small white immature oocytes are still floating. The success of the assembly extract is also dependent on the presence of large, healthy mature (stage 6) oocytes.
6. This step replaces OR2 buffer with extraction buffer. In addition, this washing step further facilitates the removal of immature oocytes because the immature oocytes settle much more slowly than stage 6 oocytes, because extraction buffer contains glycerol.
7. Gently load dispersed oocytes using the wide end of a Pasteur pipet and larger clumps of oocytes using fine tipped forceps. After the final addition of extraction buffer, the buffer should remain clear as cloudiness is indicative of oocyte lysis.
8. Insert the 19-gauge needle, just above the pellet, gently into the ultracentrifuge tube by using a "twisting motion." Remove the clear supernatant and avoid removing any cell debris and the upper lipid layer.
9. Oocyte extracts that have been stored at –70°C have remained active for up to three years. Assembly activity is lost after freeze-thawing.
10. The conductivity reading of the oocyte extract is usually equivalent to a salt concentration of around 50 mM NaCl.
11. It is important to note that for this fractionation step, the preparation of the oocyte extract is scaled up two fold (*see* **Subheading 3.1.1.**). In other words, 14 toads are used to generate sufficient oocytes to fill 12 (rather than 6) SW-41 ultracentrifuge tubes. This generates around 50 mL of oocyte extract which is loaded onto the DEAE column.
12. High chromatin assembly activity is dependent on a high protein concentration.
13. With fraction 1 being the first fraction collected from the bottom of the SW41 tube, fractions 13–21, from the three gradients, are subjected to SDS-polyacrylamide electrophoresis.
14. Routinely, histones H3 and H4 peak in three gradient fractions (usually fractions 16–18). The N1, N2-(H3,H4) complex sediments as a 5S complex.
15. Hela cells *(22)* or red blood cells *(23)* (*Xenopus* or chicken) provide a good source of histones H2A/H2B.
16. One milliliter of packed resin per 3 mg of chromatin.
17. The oocyte extract contains topoisomerase activities which relaxes supercoiled DNA rapidly prior to the assembly of the plasmid into chromatin. However, it is recommended that supercoiled DNA is first relaxed with purified topoisomerase I prior to being used in assembly reactions. It is also recommended that plasmids less than 4 kilobases are used to ensure efficient assembly.
18. In addition to ATP, the ATP buffer contains the substrates needed to yield an ATP regenerating system. Occasionally, extracts are prepared with low supercoiling activity (or that produce a poor micrococcal nuclease digestion ladder). In most cases, lowering the pH of the reaction from 7.5 to 7.0 or below can overcome this problem.
19. The supercoiling assay provides information on the ability of the oocyte extract to assemble nucleosomes *(11)*. However, it is essential to check the quality of the

chromatin assembled by micrococcal nuclease analysis *(10,11)*. It is also worth noting that the nucleosome density required to fully supercoil a circular plasmid template (as assayed in a one dimensional agarose gel) may not be high enough to generate a regular nucleosomal array.
20. A titration is performed with 2 µL increments.
21. The volume of the DEAE fraction added to an assembly reaction is equivalent to 20% of the final volume.
22. It is important to first determine the functional activity of a new histone H2A/H2B preparation. The activity of the H2A/H2B preparation is determined by employing supercoiling enhancement assays and micrococcal nuclease digestion analysis using the N1, N2-(H3,H4) complex *(10,19,24)*. For some H2A/H2B preparations, a protein-to-DNA mass ratio of more than 0.4 is required to assemble complete nucleosomes. The salt contribution to the reaction mixture produced by the addition of H2A/H2B, for the standard reaction described, is 25 mM NaCl.
23. After determining the optimal DEAE fraction concentration required for maximum supercoiling of plasmid DNA, it is also essential to check the integrity of the assembled chromatin by performing micrococcal nuclease digestions *(10)*.
24. Perform a titration from 8–24 µL with 2 µL increments.
25. The volume of N1, N2-(H3,H4) complex added to an assembly reaction is equivalent to 70% of the final volume.
26. A protein-to-DNA mass ratio of more than 0.4 may be required if not all of the histone H2A/H2B dimers are functionally active with regard to assembling complete nucleosomes. For a standard assembly reaction, a NaCl concentration of 30 mM is contributed to the reaction conditions by the addition of histone H2A/H2B.
27. After determining the concentration of the N1, N2-(H3,H4) complex required to give the appropriate extent of supercoiling, confirm nucleosome formation by micrococcal nuclease digestion analysis.
28. Run appropriate relaxed and supercoiled plasmid DNA control lanes *(19)*.

References

1. Grunstein, M., Hecht, A., Fisher-Adams, G., Wan, J., Mann, R. K., Strahl-Bolsinger S, Laroche, T., and Gasser, S. (1995) The regulation of euchromatin and heterochromatin by histones in yeast. *J. Cell Sci.* **19,** 29–36.
2. Wolffe, A. P. (1994) Architectural transcription factors. *Science* **264,** 1100,1101.
3. Wade, P. and Wolffe, A. P. (1997) Histone acetyltransferases in control. *Curr. Biol.* **7,** 82–84.
4. Roth, S. Y. and Allis, C. D. (1996) Histone acetylation and chromatin assembly: a single escort, multiple dances? *Cell* **87,** 5–8.
5. Workman, J. L. and Buchmann, A. R. (1993) Multiple functions of nucleosomes and regulatory functions in transcription. *Trends Biochem. Sci.* **18,** 90–95.
6. Becker, P. B. (1994) The establishment of active promoters in chromatin. *BioEssays* **16,** 541–547.
7. Lu, Q., Wallrath, L. L., and Elgin, S. C. R. (1994) Nucleosome positioning and gene regulation. *J. Cell. Biochem.* **55,** 83–92.

8. Edmonson, D. G. and Roth, S. Y. (1996) Chromatin and transcription. *FASEB J.* **10,** 1173–1182.
9. Wolffe, A. P. and Pruss, D. (1996) Deviant nucleosomes: the functional specialisation of chromatin. *Trends Genetics* **12,** 58–62.
10. Tremethick, D. J. and Frommer, M. (1992) Partial purification, from Xenopus laevis oocytes, of an ATP-dependent activity required for nucleosome spacing in vitro. *J. Biol. Chem.* **267,** 15,041–15,048.
11. Shimamura, A., Tremethick, D., and Worcel, A. (1988) Characterisation of the repressed 5S DNA minichromosomes assembled in vitro with a high speed supernatant of Xenopus laevis oocytes. *Mol. Cell. Biol.* **8,** 4257–4269.
12. Almouzni, G., Mechali, M., and Wolffe, A. P. (1990) Competition between transcription complex assembly and chromatin assembly on replicating DNA. *EMBO J.* **9,** 573–581.
13. Tsukiyama, T., Becker, P. B., and Wu, C. (1994) ATP-dependent nucleosome disruption at a heat-shock promoter mediated by binding of GAGA transcription factor. *Nature* **367,** 525–532.
14. Kamakaka, R. ., Bulger, M., and Kadonaga, J. T. (1993) Potentiation of RNA polymerase II transcription by Gal 4-VP 16 during but not after DNA replication and chromatin assembly. *Genes Dev.* **7,** 1779–1795.
15. Wolffe, A. P. (1991) Xenopus transcription factors: key molecules in the developmental regulation of differential gene expression. *Biochem. J.* **278,** 313–324.
16. Glikin, G. C., Ruberti, I., and Worcel, A. (1984) Chromatin assembly in Xenopus oocytes: in vitro studies. *Cell* **37,** 33–41.
17. Ito, T., Bulger, M., Pazin, M. J., Kobayashi, R., and Kadonaga, J. T. (1997) ACF, an ISWI-containing and ATP-utilising chromatin assembly and remodelling factor. *Cell* **90,** 145–155.
18. Varga-Weisz, P. D., Wilm, M., Bonte, E., Dumas, K., and Becker, P. B. (1997) Chromatin-remodelling factor CHRAC contains the ATPases ISWI and topoisomerase II. *Nature* **388,** 598–602.
19. Tremethick, D., Zucker, K., and Worcel, A. (1990) The transcription complex of the 5 S RNA gene but not Transcription Factor IIIA alone, prevents nucleosomal repression of transcription. *J. Biol. Chem.* **265,** 5014–5023.
20. Kleinschmidt, J. A. and Seiter, A. (1988) Soluble acidic complexes containing histones H3 and H4 in nuclei in *Xenopus laevis* oocytes. *Cell* **29,** 799–809.
21. Ng, K. W., Ridgway, P., Cohen, D. R., and Tremethick, D. J. (1997) The binding of a fos/jun heterodimer can completely disrupt the structure of a nucleosome. *EMBO J.* **16,** 2072–2085.
22. Dignam, J. D., and Lebovitz, R. M., and Roeder, R. G. (1983) Accurate transcription initiation by RNA polymerase II in a soluble extract from isolated mammalian nuclei. *Nucleic Acids Res.* **11,** 1475–1489.
23. Drew, H. R. and Calladine, C. R. (1987) Sequence-specific positioning of core histones on an 860 base-pair DNA. *J. Mol. Biol.* **195,** 143–173.
24. Tremethick, D. J. (1994) High mobility group proteins 14 and 17 can space nucleosomal particles deficient in histones H2A and H2B creating a template that is transcriptional active. *J. Biol. Chem.* **269,** 28,436–28,442.

13

Preparation of Chromatin Assembly Extracts from Preblastoderm *Drosophila* Embryos

Edgar Bonte and Peter B. Becker

1. Introduction

The preblastoderm embryo of *Drosophila melanogaster* is characterized by a series of extremely rapid nuclear divisions. In the absence of relevant protein synthesis replication of the genomes and their concomitant assembly into chromatin relies entirely on stockpiles of maternal histones and assembly factors. Cytoplasmic extracts from preblastoderm embryos, which contain all factors required for efficient chromatin assembly, form the basis of one of the most powerful cell-free systems for chromatin reconstitution under physiological conditions *(1)*. Under the appropriate experimental conditions, incubation of plasmid DNA in these extracts results in the formation of long arrays of nucleosomes with regular spacing, a hallmark of bulk chromatin in nuclei *(2)*. Added single-stranded DNA will be converted into double-stranded DNA before efficient nucleosome deposition *(1)*. Since the crude assembly extract contains a wealth of other chromatin proteins besides histones, the reconstituted chromatin is highly complex *(3)*. The extract is also rich in a number of nucleosome remodeling factors responsible for the dynamic properties of reconstituted chromatin, features that most likely reflect the inherent flexibility of chromatin structure in vivo *(4–6)*. Preblastoderm embryos contain no linker histone H1 *(7)*, but H1-containing chromatin can be reconstituted simply by adding an appropriate amount of purified H1 to the reconstitution reaction *(1)*.

As embryonic fly development proceeds, the maternal pools of histones are rapidly used up to assemble an exponentially growing number of nuclei. Therefore, extracts made from postblastoderm embryos contain many less histones. Remarkably, they are as proficient in chromatin assembly as extracts from preblastoderm embryos if they are supplemented with purified histones *(8)*.

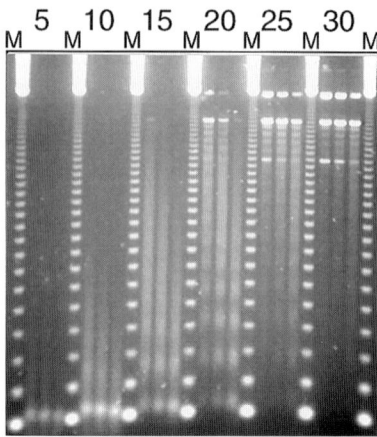

Fig. 1. The regular nucleosomal repeat is visualised with a Micrococcal Nuclease cleavage assay as described in the text. Chromatin in reactions containing increasing amounts (5–30 µL, indicated above the lanes) of *Drosophila* embryo extract. Reconstituted chromatin was digested with MNase for 30, 60, and 120 s as described in the text. The three corresponding DNA samples of each reaction are separated by marker lanes (M, 123 bp ladder, BRL).

This observation can be used to assemble chromatin with variant properties from modified histones (*[2]*, *see* also Chapter 15).

A prerequisite for the preparation of chromatin assembly extract is a healthy *Drosophila* population, cultivated at an appropriate scale *(9)*. Approximately 10 g of preblastoderm embryos (giving rise to approx 10 mL of assembly extract can be collected comfortably during a day from a population of approx 150,000–200,000 flies. In this chapter, we describe the harvesting of these embryos, the preparation of active chromatin assembly extract, and a standard assay for the assessment of reconstituted chromatin (*see* **Fig. 1**). We routinely use this procedure to obtain highly active chromatin assembly extracts from both preblastoderm and postblastoderm embryos. A related procedure specifically for the preparation of extracts from postblastoderm embryos, which differs in several steps from our protocol, has been described by Kadonaga and coworkers *(8)*.

2. Materials
2.1. Harvesting of the Drosophila *Embryos*

1. A healthy population of *Drosophila melanogaster* (*see* **Note 1**).
2. Apple juice agar plates: Recipe for about 200 dishes plates: Melt 500 g agar in 12 L deionized H_2O (bring to boil). When all agar has dissolved turn heater back to 80°C. Add: 3.5 L of apple juice, 750 mL sugar bead syrup, 420 mL 10%

Nipagin (in ethanol) and mix well. Pour approx 80 mL into each large, nonsterile Petri dish. Stack 10 dishes each for cooling and hardening overnight. Pack stacks of 16 in the original plastic bags and store at 4°C.
3. Yeast paste : 220 mL deionized H_2O, 1.4 mL propionic acid (Sigma, Deisenhoten, Germany), 150 g yeast (vacuum packed) (S. I. Lesaffre, 59703 Marcq, France).
4. Stainless steel sieves (Edelstahlprüfsiebe, LINKER; diameter 200 mm, height 50 mm, mesh diameter: 0.125 mm, 0.355 mm, and 0.710 mm) (Kurt Migge, Heidelberg, Germany).
5. Paint brush.
6. Embryo wash (EW): 0.7% NaCl, 0.05% Triton X-100 (Sigma).

2.2. Preparation of Dechorionated Embryos

1. EW (see **Subheading 2.1.**).
2. Bleach: 13% hypochloric acid (Europe) or Chlorox bleach (USA) (see **Note 2**).
3. Test sieve (mesh diameter 0.125 mm).

2.3. Washing and Settling of the Dechorionated Embryos

1. EW (see **Subheading 2.1.**).
2. 0.7% NaCl.
3. EX-10 (Extract buffer): 10 mM N-2-hydroxyethylpiperazine-N'-2-ethanesulfonic acid (HEPES-KOH, pH 7.6), 10 mM KCl, 1.5 mM $MgCl_2$, 0.5 mM ethylene glycol-bis-(ß-aminoethyl ether)-N,N,N',N'-tetraacetic acid (EGTA-KOH, pH 8.0), 10% glycerol, 10 mM ß-glycerolphosphate (added freshly), 1 mM dithiothreitol (DTT) (added freshly), and 0.2 mM phenylmethylsulfonyl fluoride (PMSF) (added fresh) (see **Note 3**).
4. Potter-Elvehjem homogenisator (30 mL) with tight-fitting teflon pestle (Migge, Germany or equivalent).

2.4. Extract Preparation

1. Motor-driven homogenizator (Heidolph).
2. 1 M $MgCl_2$.
3. COREX centrifuge tubes.
4. HB-4 rotor (SORVALL).
5. Polyallomer ultra centrifuge tubes (13 × 51 mm) (Beckman).
6. Highly liquid paraffin (Merck, Darmstadt, Germany).
7. SW 55.1 Ti rotor (Beckman).
8. Disposable 5-cc syringe (Plastipak, Becton Dickinson) and green needle (G21).
9. Liquid nitrogen.

2.5. Testing of the Chromatin Assembly Activity

1. Plasmid DNA (500 ng/µL).
2. 10X McNAP : 30 mM $MgCl_2$, 10 mM DTT, 300 mM creatine phosphate (Sigma, Deisenhofen, Germany), 30 mM ATP, 1.0 µg creatine phosphokinase (Boehringer Mannheim) per 100 µL (see **Note 4**).

3. EX-50 buffer (*see* **Note 3**).
4. *Drosophila* extract.
4. Micrococcal Nuclease (MNase) (50 U/μL, Sigma, *see* **Note 5**).
6. 100 mM CaCl$_2$.
5. MNase stop solution: 2.5% N-lauroylsarcosine (Sigma), 100 mM EDTA-NaOH, pH 8.0 (Merck).

2.6. Purification and Gel Separation of the Nucleosomal "Ladder"

1. 10 mg/mL RNase A (Sigma).
2. 2% Dodecylsulphate, Na salt (SDS) (Boehringer, Mannheim).
3. 10 mg/mL Proteinase-K (Boehringer, Mannheim).
4. 20 mg/mL Glycogen (mussels) (Boehringer, Mannheim).
5. Phenol/chloroform/isoamylalcohol (25:24:1) (Amresco, Solon Ohio).
6. Chloroform/isoamylalcohol (24:1).
5. 7.5 M Ammonium-acetate (*see* **Note 6**).
6. 96% Ethanol.
7. 70% Ethanol.
8. TE, pH 7.5: 10 mM Tris-HCl, pH 7.5, 1 mM EDTA.
9. 5X Orange-G loading dye: 50% glycerol, 5 mM EDTA, 0.3% Orange-G (Sigma).
10. 5X Tris-Glycine running buffer: 0.25 M Tris, 2.0 M Glycine (*see* **Note 7**).
11. 123 bp DNA marker (Gibco-BRL, Life Technologies).
12. Agarose *ultra* PURE (Gibco-BRL).
13. 1.3% agarose gel in 1X Tris/glycine running buffer.
14. 10 mg/mL Ethidium bromide (Sigma) (**Carcinogenic !**).
15. UV-lamp and gel documentation system.

3. Methods

3.1. Harvesting of the Drosophila Embryos

1. Collect embryos laid on apple juice plates with yeast paste during a 90-min time window.
2. Rinse the embryos off the plates with tap water and using the paint brush, collect the embryos in the 0.125 mm mesh sieve (*see* **Note 8**).
3. Using a squeeze bottle with EW buffer rinse the embryos from the sieve into a clean beaker. Store the suspension on ice until extract preparation. Pool the embryos of 4–6 successive collections in EX buffer on ice (*see* **Note 9**).
4. Aspirate the cold EW and replace it for fresh EW at room temperature (RT) and allow embryos to settle (*see* **Note 10**).

3.2. Collection of Dechorionated Embryos

1. Remove the supernatant and adjust the volume with fresh EW to 200 mL.
2. Add 60 mL of 13% hypochloric acid (RT) and stir vigorously for 3 min.
3. Pour the embryos back into the fine collection sieve (mesh diameter 0.125 mm) and rinse **vigorously** with cold tap water (*see* **Note 11**).

Chromatin Assembly Extracts

4. Transfer the dechorionated embryos to a 500-mL cylinder and wash the embryos with 500 mL EW (RT) and let the embryos settle again (*see* **Note 12**).
5. Aspirate off the EW.
6. Wash the embryos with 500 mL 0.7% NaCl (RT) and let them settle again (*see* **Note 12**).
7. Aspirate off the 0.7% NaCl.
8. Wash the embryos with 500 mL EX-buffer (at 4°C) and let them settle again. From now on keep embryos on ice and perform all manipulations at 4°C with precooled buffers (*see* **Note 12**).
9. Aspirate off the EX-buffer as much as possible, without loosing embryos, add additional 100 μL 1 M DTT and 100 μL 0.2 M PMSF to the embryos (*see* **Note 13**).
10. Transfer the embryos to the Potter-Elvehjem homogenisator vessel on ice and allow them to settle for at least 15 min.
11. Aspirate off the EX-buffer as much as possible, leaving behind the packed embryos.

3.3. Extract Preparation

1. Stir up packed embryos to resuspend and homogenise them with one complete stroke at 3000 rpm and 6 strokes at 1500 rpm (*see* **Note 14**).
2. Measure the volume of the homogenate and add 1 M MgCl$_2$ to a final concentration of 6.5 mM. Mix the MgCl$_2$ immediately with an additional homogenisation stroke at 1,500 rpm (*see* **Note 15**).
4. Centrifuge the extract for 5 min at 10.000 rpm (17,000g) in a chilled (4°C) HB4 rotor, using COREX tubes.
5. Collect the turbid cytoplasmic extract (*see* **Note 16**).
6. Transfer the extract to polyallomer ultracentrifugation tubes and fill up, if necessary, with paraffin oil (*see* **Note 17**).
7. Centrifuge for 2 h in a chilled (4°C) SW55.1 Ti rotor at 45,000 rpm (190,000g).
8. Isolate the clear extract with a needle and a syringe (*see* **Note 18**).
9. Flash freeze 300–500-μL aliquots in liquid nitrogen (*see* **Note 19**) and store the aliquots at –80°C.

3.4. Testing of the Chromatin Assembly Activity

1. Set up standard chromatin assembly reactions with increasing volumes of extract as follows :
   1 μL (500 ng) plasmid DNA,
   7 μL 10X McNAP,
   10, 20, 30, 40, or 50 μL *Drosophila* extract,
   52, 42, 32, 22, or 12 μL Ex-50 buffer.
   The final volume for each assembly reaction should be 70 μL.
2. Incubate for 6 h at 26°C.
3. Add to each assembled chromatin sample 100 μL MNase mix: 93 μL Ex-50 buffer, 5 μL 100 mM CaCl$_2$ and 2 μL (50 U/μL) MNase and mix gently.
4. After 30, 60, and 120 s remove a 55-μL aliquot from the reaction and add to 20 μL MNase stop solution.

3.5. Purification and Gel Separation of the Nucleosomal "Ladder"

1. Add to the MNase treated aliquots 1 µL (10 mg/mL) RNase A and incubate for 30 min at 37°C.
2. Add 8 µL 2% SDS and 5 µL (10 mg/mL) proteinase K and incubate over night at 37°C.
3. Add 1 µL (20 mg/mL) glycogen and 10 µL TE pH 7.5.
4. Add 75 µL 7.5 M ammonium acetate.
5. Add 440 µL 96% ethanol and incubate for 10 min in the cold.
6. Centrifuge for 20 min in an table-top centrifuge, remove supernatant.
7. Wash the pellet with 70% ethanol.
8. Air dry the pellet, do not use a SpeedVac (*see* **Note 20**).
9. Resuspend the pellet in 4 µL TE pH 7.5.
10. Add 1 µL 5X Orange-G loading dye.
11. Run the sample in a 1.3% agarose gel with 1X Tris/glycine as running buffer at 5 V/cm (*see* **Note 21**).
12. Stain the gel in running buffer containing 1 µg/100 mL ethidium bromide. A typical reaction can be seen in **Fig. 1**.

4. Notes

1. Reproducible extract preparations mainly depend on a healthy *Drosophila melanogaster* fly population. The conditions for the maintenance of such a fly population have been described in detail (*9*). Fluctuations in temperature and humidity and deviations from a regular day/night rhythm affect the embryo yields adversely. We routinely are using 20,000 flies (on average 20 g) in a 45 dm^3 fly cage. The newly seeded flies are fed with yeast paste on apple juice agar plates. The embryo collection starts at d 3.
2. Incomplete dechorionation is the main reason for bad extracts. Use only fresh bleach. The bleach should be at room temperature.
3. EX-10 buffer is the extract buffer with 10 mM KCl, in a later stage of the protocol we will use EX-50 buffer, this buffer is identical to the EX-10 buffer with exception of the KCl concentration; the EX-50 buffer has 50 mM KCl.
4. The energy-supplier McNAP has to be prepared fresh every time. Some of the components are unstable even when stored at –80°C. Special care should be taken with the creatine phosphokinase (CPK). Resuspend one vial of 20 mg lyophilized CPK in 1000 µL 100 mM imidazol (pH 6.6). Divide the solution in 20-µL aliquots and store at –80°C for a maximum of 2 mo. Prior to use add to one 20-µL aliquot 380 µL 100 mM imidazol (pH 6.6) and use of this only 1 µL in 100 µL of 10X McNAP. Do not freeze/thaw the aliquot.
5. The unit definition of Micrococcal Nuclease differs between the suppliers. We refer to the unit definition of Boehringer Mannheim. The lyophilized MNase is resuspended at 50 U/µL in EX-50 buffer. To obtain a 50 U/µL solution with MNase obtained from Sigma, one should resuspend a lyophilized 500 U vial in 850 µL EX-50 buffer.
6. The 7.5 M ammonium acetate solution should be sterilized by filtering through a 0.2-µM filter.

7. There is no need to adjust the pH of the 5X Tris-glycine stock.
8. Stack the sieves according to decreasing mesh size (0.125 mm down and 0.710 mm top). The tope sieve will retain intact flies, the middle one fly parts and the lower one will retain the embryos.
9. In our hands, 10–15 mL of extract with reproducible high quality is obtained from 80 plates with embryos laid during a 90-min interval.
10. This step is to make sure that the dechorionation will take place at room temperature. Again the main reason for bad extracts is the insufficient dechorionation. Easy aspiration can be performed with a 20-mL Pasteur pipet connected to a water vacuum pump.
11. The embryos in the lowest sieve should be rinsed extremely well (for at least 2 min) with a hard beam of tap water to remove all yeast and residual chorion parts that may still be attached to embryos. Insufficient dechorionation and washes will result in "floating" embryos that do not settle in the following steps, leading to losses and poor quality extracts.
12. Floaters will only become apparent when changing the EW buffer to the 0.7% NaCl buffer and in the later step to Ex-10 buffer. Floating embryos will prevent the tight packing of embryos in the homogenizer, which will dilute the extract.
13. The volume of the embryo suspension is roughly 50 mL. Addition of 100 µL DTT and 100 µL PMSF is sufficient to prevent possible protein degradation during the homogenisation.
14. Depending on the optimal packing of the embryos the first stroke (a complete movement of the pestle up and down) will be a tough one. Take care that the extract doesn't heat up during this first stroke and that you do not suck in air in between the still packed embryos and the pestle. This first complete stroke at 1000–1500 rpm is followed by one stroke at 2000 rpm, one at 3000 rpm and 6 strokes at 1500 rpm.
15. The final concentration (including the $MgCl_2$ in the EX buffer is 6.5 mM *(10)*.
16. After the low speed centrifugation step one obtains a reasonably solid pellet, the cytoplasmic extract and a tight white lipid layer on top. Penetrate the lipid layer along the centrifuge wall with a pipet tip. Avoid the lipid upon collection of the supernatant.
17. The paraffin oil will stabilise the top layer during centrifugation.
18. After the ultracentrifugation step one obtains a solid pellet, a relatively loose white sediment (cloud) on top of the pellet, the clear extract and a floating lipid layer. Collect the clear extract by puncturing the side of the tube with a green needle (G21) on a 5-cc syringe well above the sediment (cloud) but again avoid the lipid layer. Occasionally a white flocculate material is present in the otherwise clear extract. Most of this material can be removed by a 5-min centrifugation in a table-top centrifuge. The presence of the turbid material does not affect the chromatin assembly reaction.
19. The stability of the prepared extract is fairly high. One can freeze/thaw the extract at least three times without obvious losses of the chromatin assembly activity. Storage of longer then 1 yr should be avoided. In general the protein concentration of the extract is roughly 20–30 mg/mL.

20. Drying the pellet in a SpeedVac may denature the Micrococcal Nuclease-treated DNA fragments. Inefficient re-annealing may result in a continuum of fragments rather than a defined pattern representing oligonucleosomal DNA.
21. For optimal resolution and easy determination of the Nucleosome Repeat Length (NRL) one should use combs with narrow teeth (1 mm) and a long agarose gel (20 cm). Run appropriate size markers next to the MNase-digested DNA to determine the repeat length *(2,10)*

References

1. Becker, P. B. and Wu, C. (1992) Cell-free system for assembly of transcriptionally repressed chromatin from *Drosophila* embryos. *Mol. Cell. Biol.* **12,** 2241–2249.
2. Blank, T. A. and Becker, P. B. (1995) Electrostatic mechanism of nucleosome spacing. *J. Mol. Biol.* **252,** 305-313.
3. Sandaltzopoulos, R., Blank, T., and Becker, P. B. (1994) Transcriptional repression by nucleosomes but not H1 in reconstituted preblastoderm *Drosophila* chromatin. *EMBO J.* **13,** 373–379.
4. Tsukiyama, T. and Wu, C. (1995) Purification and properties of an ATP-dependent nucleosome remodeling factor. *Cell* **83,** 1011–1020.
5. Ito, T., Bulger, M., Pazin, M. J., Kobayashi, R., and Kadonaga, J. T. (1997) ACF, an ISWI-containing and ATP-utilizing chromatin assembly and remodeling factor. *Cell* 145–155.
6. Varga-Weisz, P. D., Wilm, M., Bonte, E., Dumas, K., Mann, M., and Becker, P. B. (1997) Chromatin remodelling factor CHRAC contains the ATPases ISWI and toposiomerase II. *Nature* **388,** 598–602.
7. Ner, S. S. and Travers, A. A. (1994) HMG-D, the *Drosophila melanogaster* homologue of HMG 1 protein, is associated with early embryonic chromatin in the absence of histone H1. *EMBO J.* **13,** 1817–1822.
8. Kamakaka, R. T., Bulger, M., and Kadonaga, J. T. (1993) Potentiation of RNA polymerase II transcription by Gal4-VP16 during but not after DNA replication. *Genes Dev.* **7,** 1779–1795.
9. Shaffer, C., Wuller, J. M., and Elgin, S. C. R., ed. (1994) Raising large quantities of Drosophila for biochemical experiments, in *Methods in Cell Biology, Vol. 44.*, (Goldstein, L. S. B. and Fyrberg, E. A., eds.), Academic Press, San Diego, CA, pp. 99–110.
10. Blank, T. A., Sandaltzopoulos, R., and Becker, P. B. (1997) Biochemical Analysis of chromatin structure and function using *Drosophila* embryo extracts. *Methods* **12,** 28–35.
11. Blank, T. A. and Becker, P. B. (1996) The effect of nucleosome phasing sequences and DNA topology on nucleosome spacing. *J. Mol. Biol.* **260,** 1–8.

14

A Solid-Phase Approach for the Analysis of Reconstituted Chromatin

Raphael Sandaltzopoulos and Peter B. Becker

1. Introduction

The value of a solid support was recognized early in history by the ancient Greek engineer Archimedes who, amazed by the power of the leverage machines that he invented, exclaimed that he could even move the entire planet had he only a suitable solid support to rely on. In biochemistry, sophisticated multistep experimental procedures require that a substrate be purified and processed through a sequence of reactions under different optimal conditions. Solid-phase techniques are invaluable because they allow instant and quantitative purification of reaction intermediates and readjustment of new reaction conditions. Here we describe a method for chromatin reconstitution on a solid support and present how solid-phase chromatin can be analyzed or prepared as a substrate in subsequent reactions.

Chromatin reconstitution in crude extracts from *Xenopus* oocytes or eggs or *Drosophila* embryos provides a powerful means to study structure/function relationships in chromatin organization (*see* Chapters 12, 13, and 15) *(1,2)*. For many of those analyses (e.g., the evaluation of the transcriptional potential of a chromatin template), the chromatin must be purified from the complex reconstitution reaction. The most common method for chromatin purification is its centrifugation through a sucrose gradient. Although efficient, this method is time-consuming, does not allow parallel processing of many samples, and many loosely associated chromatin components that may be of pivotal importance (e.g., for chromatin dynamics), may be lost during the long centrifugation.

By contrast, the solid-phase approach enables the rapid, nondisruptive and quantitative purification of chromatin. A linear fragment of DNA that bears the sequences of interest (i.e., enhancer/promoter and gene coding regions) is biotinylated at one end and then immobilized on streptavidin-coated superparamagnetic beads (**Subheading 3.1.** and **Fig. 1**). The bead-coupled tem-

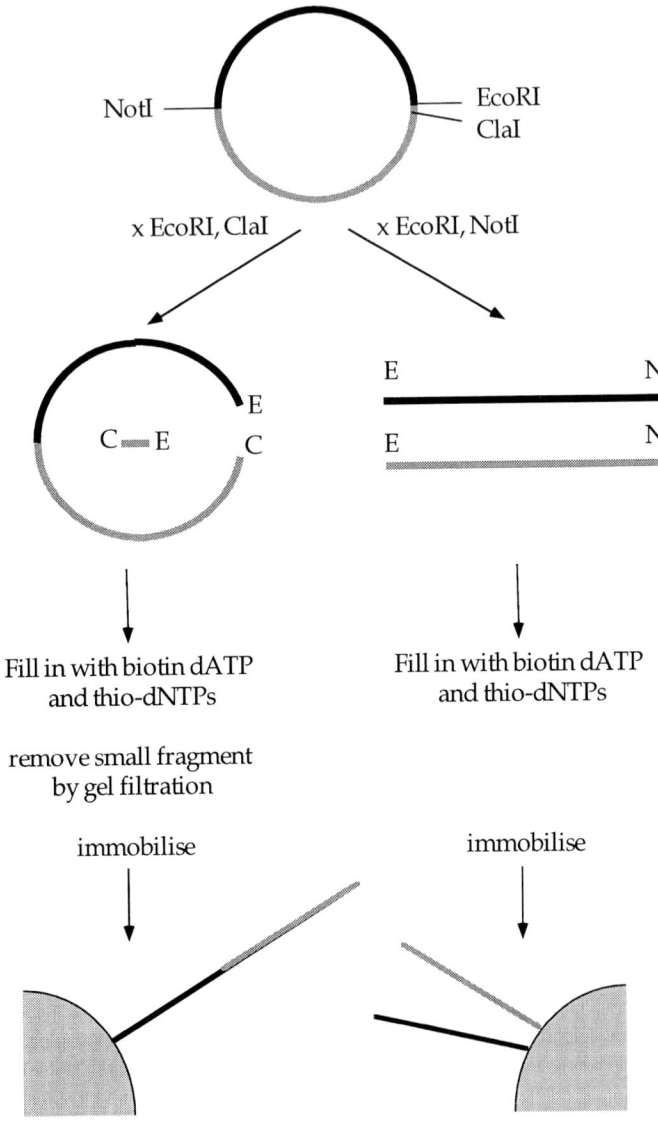

Fig. 1. DNA immobilization strategies (*see* **Note 1**).

plate is then subjected to chromatin assembly (**Subheading 3.2.1.**). Chromatin assembled on immobilized DNA resembles that of soluble DNA with respect to the optimal reconstitution conditions, the kinetics of chromatin assembly, the nucleosomal repeat length, the histone stoichiometry, the association of histone H1, the inhibition of transcription and the association of many nonhistone proteins. The immobilized chromatin can then be purified in a magnetic field, washed as desired, and then used to purify and analyze chromatin-associated proteins (**Subheading 3.3.1.**) or to develop assays for putative chromatin binding proteins. The rapid isolation of immobilized chromatin in a magnetic field facilitates quick buffer exchanges and the efficient removal of soluble components, such as nucleotides or unbound proteins. Therefore, multistep reconstitutions are facilitated; that is, reactions in which the chromatin reconstitution must be separated from other steps, such as the interaction of transcription factors prior to chromatin assembly or subsequent chromatin "remodeling" reactions (for recent applications, *see* **refs. 3–7**).

We also describe how nucleosomes can be reconstituted from pure histones by a salt gradient dialysis procedure on immobilized DNA (**Subheading 3.2.2.**).

2. Materials
2.1. Immobilization of DNA

1. Streptavidin-coated paramagnetic beads (Dynabeads kilobase Binder™ Kit, Dynal, Oslo).
2. Magnetic particle concentrator (Dynal MPC-6, Dynal, Oslo).
3. 0.5 mM or 10 mM biotin-21-dUTP in 50 mM Tris-HCl, pH 7.5, Clontech) or 0.4 mM biotin-14-dATP in 100 mM Tris-HCl, pH 7.5, 0.1 mM EDTA (Gibco-BRL). Biotin-dCTP is available from Gibco-BRL, but we have not tested it yet.
4. 10 mM a-thio-deoxyribonucleotides, pH 8.0 (Pharmacia).
5. Restriction enzymes and 10X digestion buffers (according to the supplier's recommendation).
6. Klenow (Exo⁻) polymerase.
7. 10x Polymerase buffer: 0.1 M Tris-HCl, pH 7.5, 50 mM MgCl$_2$, 75 mM DTT).
8. Vent (Exo⁻) DNA polymerase (New England Biolabs).
9. 10 mg/mL Glycogen (Boehringer Mannheim).
10. Ethanol.
11. Quick Spin columns TE Sephadex G-50 fine (Boehringer) or ChromaSpin column 50-TE (Clontech) or equivalent home-made spin columns.
12. PBS-BSA-NP40: 1.7 mM KH$_2$PO$_4$, 5 mM Na$_2$HPO$_4$, 150 mM NaCl, pH 7.4, 0.05% (w/v) BSA, 0.05% (v/v) Nonidet P40.
13. Wash and binding buffer (WB buffer): 2M NaCl, 50 mM Tris-HCl, pH 8.0, 1 mM EDTA.
14. λDNA (Promega).
15. Kilobase binder reagent (Dynal, Oslo).
16. Shaker with regulated speed and temperature (e.g., thermomixer, Eppendorf).
17. TE: 10 mM Tris-HCl, pH 7.5, 0.1 mM EDTA.
18. Rotating wheel with regulated speed.

2.2. Chromatin Reconstitution on Immobilized DNA
2.2.1. Reconstitution in Drosophila Embryo Extracts

1. Chromatin assembly extract (*see* Chapter 13).
2. 0.5 M MgCl$_2$.

3. McNAP mix (see Chapter 13): 300 mM creatine phosphate, 30 mM ATP, pH 8.0, 30 mM MgCl$_2$, 10 mM DTT, 10 µg/mL creatine phosphokinase.
4. Extract buffer (EX): 10 mM HEPES-KOH, pH 7.6, 1.5 mM MgCl$_2$, 0.5 mM EGTA, 10% (v/v) glycerol, 10 mM β-glycerophosphate, 1 mM dithiothreitol (DTT), 0.2 mM AEBSF (Boehringer).
5. Temperature-regulated chamber with integrated rotating wheel (e.g., hybridization oven).

2.2.2. Nucleosome Reconstitution by Salt Gradient Dialysis

1. Core histones (Boehringer).
2. Two peristaltic pumps.
3. Magnetic stirrer and stirrer bars.
4. DB-1: 2 M NaCl, 10 mM Tris-HCl, pH 7.5, 1 mM EDTA, 0.05% (v/v) Nonidet P40, 1 mM β-mercaptoethanol (freshly added).
5. DB-2: 50 mM NaCl, 10 mM Tris-HCl, pH 7.5, 1 mM EDTA, 0.05% (v/v) Nonidet P 40, 1 mM β-mercaptoethanol (freshly added).
6. Dialysis tubing Spectra/Por 2, MWCO 12–14,000, 2 mL/cm.

2.3. Analysis of Reconstituted Chromatin

2.3.1. Analysis of Chromatin Proteins

1. EX-S, (EX, *see* **Subheading 2.2.1.**), where S represents the concentration of KCl in mM.
2. EX-50-NP40: extract buffer supplemented with 50 mM KCl and 0.05% (v/v) Nonidet P40.
3. 4X SDS-loading buffer: 200 mM Tris-HCl, pH 6.8, 40% (v/v) glycerol, 400 mM β-mercaptoethanol, 4% (w/v) SDS, 0.002% (w/v) bromophenol blue.
4. Equipment for PAGE.

2.3.2. Micrococcus Nuclease Digestion

1. Micrococcus Nuclease (Nuclease from *Staphylococcus aureus* [Boehringer], 50 U/µL in EX buffer).
2. 5X Nuclease stop mix: 2.5% (v/v) sarkosyl, 100 mM EDTA, pH 8.0.
3. TE: 10 mM Tris-HCl, 1 mM EDTA, pH 8.0 .
4. 5X Orange loading buffer: 50% (v/v) glycerol, 5 mM EDTA, pH 8.0, 0.3% (w/v) Orange G.
5. Equipment for agarose gel electrophoresis.

3. Methods

3.1. Immobilization of DNA

3.1.1. Digestion of DNA

1. Cleave plasmid DNA with an appropriate pair of restriction enzymes (example is given for *Cla*I and *Eco*RI; *see* **Note 1** and **Fig. 1**) as follows: Mix 40 µL of super-

A Solid-Phase Approach

coiled plasmid (1 μg/μL = 40 μg), 5 μL 10X digestion buffer, 5 μL Cla I (10 U/μL). Incubate for 3 h at 37°C.
2. Assure complete linearization by analyzing 0.2 μL of the digest by electrophoresis on an 0.8% agarose gel and staining with ethidium bromide (*see* **Note 2**).
3. Add 50 μL H_2O, 11 μL 10X digestion buffer, and 10 μL *Eco*RI (10 u/μL). Adjust the final volume to 160 μL with H_2O. Incubate for 3 h at 37°C.
4. Precipitate DNA: Add 16 μL 3 *M* sodium acetate, pH 5.3. Mix. Add 480 μL ethanol, mix. Incubate for 10 min on ice.
5. Spin 15 min at top speed in a table-top centrifuge. Discard supernatant.
6. Wash pellet with 800 μL 80% ethanol. Dry pellet 2 min in speed vac without heating.
7. Dissolve pellet thoroughly in 40 μL TE.

3.1.2. Biotinylation of DNA

1. Add 7.5 μL 0.4 m*M* biotin-14-dATP, 1.2 μL each of 10 m*M* a-thio-dTTP, 10 m*M* α-thio-dCTP and 10 m*M* a-thio-dGTP (*see* **Note 3**), 6.0 μL 10X polymerase buffer, and 3.5 μL of 5 U/μL Klenow (Exo⁻).
2. Incubate for 1 h at 37°C (*see* **Note 4**).

3.1.3. Removal of Free Biotin (see **Note 5**)

1. Resuspend the matrix of a Quick Spin sephadex G-50 TE spin column. Uncap the top then the bottom of the column.
2. Place in a reaction tube provided (without lid) and let drain in a vertical position (about 5 min).
3. Empty the reaction tube and put the column (together with the reaction tube) in a 15-mL Falcon tube.
4. Spin for 1 min at 1100*g*.
5. Discard flow-through and spin at 1100*g* for 2 min.
6. Replace the collection tube by a fresh one. Apply the biotinylation reaction slowly at the center of the resin without touching the resin.
7. Spin for 2 min at 1100*g* and collect flow-through. The volume of your sample should stay constant (approx 60 μL). Measure optical density at 260 nm to define DNA concentration to account for the losses during gel filtration. Usually the losses are between 10–30%.
8. Add 200 μL 2X WB buffer and 140 μL H_2O. This is the coupling mix that is ready to be added to the beads. The final NaCl concentration must be 2 *M*. Save 1 μL for testing immobilization efficiency.

3.1.4. Coupling of DNA to Dynabeads

1. Resuspend beads well.
2. Remove appropriate amount of bead suspension from the vial. One milligram of beads (100 μL) are required for the immobilization of 1 pmole of DNA (*see* **Note 6**). For example, since 1 pmole of a 5 Kb DNA fragment is 3.3 μg, 1.210 mL (40/3.3 times 100) of bead suspension is needed to immobilize 40 μg of fragment.

3. Place tube on MPC (magnetic particle concentrator) for 1 min.
4. Discard supernatant.
5. Wash beads in 300 µL PBS-BSA-NP40.
6. Wash beads twice with 300 µL WB.
7. Resuspend beads in coupling mix (*see* **Subheading 3.1.3.**, **step 8**).
8. Rotate at room temperature for at least 3 h or overnight.
9. Concentrate beads and remove supernatant.
10. Check 10 µL of supernatant (equivalent to originally 300 ng) on 0.8% agarose gel alongside the uncoupled aliquot (*see* **Note 7**).
11. Resuspend DNA-beads in WB buffer at a concentration of 30 ng of immobilized DNA/µL of buffer and store at 4°C. (Under these conditions they can be stored for several months).

3.1.5. Efficient Immobilization of Very Long DNA (see **Note 6**)

1. Mix: 300 mL λ DNA (100 mg), 40 mL 10X Vent polymerase buffer, 8 µL 10 mM a-thio dGTP, 8 µL a-thio 10 mM dCTP, 8 µL 10 mM a-thio dATP, 4 µL 10 mM biotin-21-dUTP, 5 µL Vent (Exo⁻) DNA Polymerase (2 U/µL) and 27 µL H$_2$O (total volume is 400 µL).
2. Incubate for 30 min at 76°C
3. Add 40 µL 3 M sodium acetate, pH 5.3, and mix gently.
4. Add 1100 µL absolute ethanol. Mix and incubate for 5 min on ice.
5. Spin for 10 min at top speed in a table-top centrifuge.
6. Wash pellet twice with 70% ethanol.
7. Dry and resuspend in 300 µL H$_2$O (approx 1 pmole/100 µL) (*see* **Note 8**).
8. Add an equal volume of 2X WB buffer and transfer to equilibrated beads (**Subheading 3.1.4.**, **steps 1–7**). Then add $1/4$ of this volume of Kilobase binder reagent (*see* **Note 9**). Mix gently.
9. Rotate at room temperature for overnight.
10. Check immobilization efficiency and store DNA beads as in **Subheading 3.1.4.**, **steps 9–11** (*see* **Note 8**).

3.2. Chromatin Reconstitution on Immobilized DNA

3.2.1. Chromatin Reconstitution Using Drosophila Embryo Extracts

1. Resuspend stock of immobilized template. Pipet out appropriate amount of bead-DNA. 900 ng of DNA is sufficient for a MNase assay or analysis of bound histone by silver staining.
2. Concentrate on the MPC. Remove supernatant and wash once with 300 µL of PBS-BSA-NP40.
3. Wash again with 300 µL EX-NP40.
4. Prepare chromatin assembly reaction by mixing 70 µL chromatin assembly extract, 12 µL McNAP, and 38 µL EX buffer for each 900 ng of DNA (*see* **Note 10**).
5. Concentrate bead-DNA, remove supernatant and resuspend beads in complete chromatin assembly reaction.
6. Transfer to 250-µL micro test tubes (*see* **Note 11**).

A Solid-Phase Approach

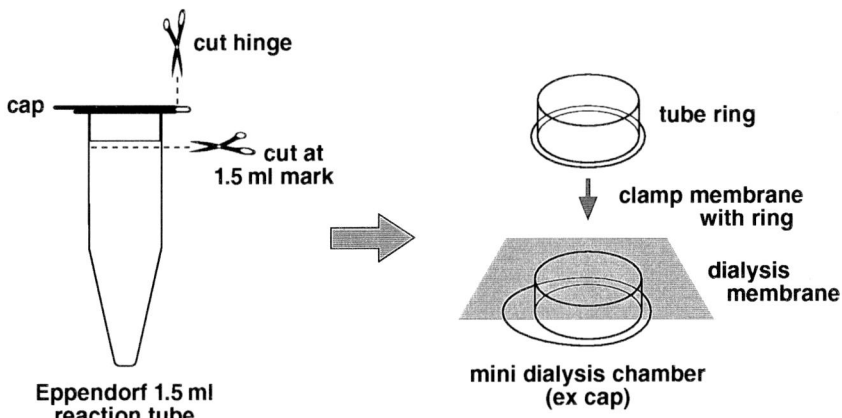

Fig. 2. The preparation of a mini-dialysis chamber (see **Subheading 3.2.2.**). The entire dialysis chamber containing the nucleosome reconstitution is thrown into the beaker with dialysis buffer. The tumbling of the chamber resulting from the vigorous stirring will assure that the beads remain suspended during the 16-h dialysis. We acknowledge the help of Udo Ringeisen in preparing this figure.

7. Rotate at 26°C for up to 6 h the rotation axis being perpendicular to the longitudinal axis of the tube.
8. Check occasionally for aggregation of beads. If necessary, disperse clumps by gently tapping the tube. Some clumping may occur during the first 1–2 h of the assembly reaction. If beads are redispersed once, they usually do not clump again.

3.2.2. Chromatin Reconstitution Using Purified Histones (see **Note 12**)

1. During a salt gradient dialysis reconstitution histones and DNA are first dialyzed into high salt buffer DB-1 (see **Note 12**). During overnight dialysis the salt concentration is reduced by diluting the dialysis buffer with low salt buffer while keeping the volume of the dialysis constant.

   Set up a beaker with 600 mL DB-1 buffer at 4°C and prepare a larger container with 3 L DB-2. Use two peristaltic pumps and appropriate tubing to pump DB-2 at a rate of 3 mL/min into the dialysis beaker containing DB-1 while at the same time pumping the equivalent volume out of the diluted dialysis buffer into a waste container. This set-up assures that the volume of the dialysis reaction remains constant while the salt concentration is reduced.
2. Prepare the samples. Mix 60 µL 5 M NaCl, 82.25 µL TE, 3.75 µL 20 mg/mL BSA, and 4 µL (0.375 mg/mL) purified core histones (see **Note 13**).
3. Prepare the mini-dialysis chamber (**Fig. 2**). Detach the cap of a 1.5-mL reaction tube by cutting the connecting hinge with a pair of scissors. Cut the remaining tube at the 1.25 mL mark. The cut-off ring will serve as membrane clamp. Cut dialysis membrane (12–14,000 MWCO) to 2 × 3 cm pieces. Equilibrate membrane pieces in DB-1 for 30 min.

4. Pipet 50 µL of bead-DNA suspension (30 ng DNA/µL) into a reaction tube. Concentrate the beads on an MPC. Discard supernatant and wash once with 200 µL PBS-BSA-NP40. Wash again with 200 µL of WB buffer (see **Subheading 2.1.**). Concentrate beads, discard supernatant, and resuspend beads in the reaction mix prepared at **step 2**.
5. Transfer the suspension into empty, inverted Eppendorf tube cap. Place a piece of dialysis membrane on top and clamp membrane with the tube ring (see **Fig. 2**). A reaction volume of 150 µL should essentially fill the cap. Avoid trapping air in the cap which will interfere with dialysis.
6. Throw the closed dialysis cap(s) into the dialysis container and start stirring very fast. Vigorous stirring is essential for maintaining beads in suspension.
7. Dialyze for 45 min before you turn on the pumps to dilute the salt concentration. Dialyze until almost all DB-2 has been pumped into the dialysis container (15–16 h).
8. Recover bead suspension by puncturing the membrane with a pipet tip. Transfer to a fresh reaction tube and process as desired.

3.3. Analysis of Reconstituted Chromatin

Reconstituted, immobilized chromatin can be purified from the reaction mix and analyzed in various ways. Chromatin proteins can be separated by SDS-PAGE and visualized by Western blotting, silver staining, or even Coomassie staining, depending on the scale of the reaction. The quality of the reconstituted chromatin can be tested by visualization of the correct histone stoichiometry and a regular nucleosomal array by Micrococcal Nuclease digestion. Chromatin-association of proteins of interest can be tested and the correct stoichiometry of core histones verified.

*3.3.1. Analysis of Chromatin Proteins (see **Note 14**)*

1. If a smaller tube was used for chromatin assembly reaction transfer all liquid to Eppendorf tubes that fit into the MPC. The small tube may be rinsed with 100 µL EX-50-NP40 to ensure complete recovery. Concentrate on an MPC for 1 min (not longer!) and remove supernatant completely. Be careful not to touch the pellet with the tip. This may lead to losses.
2. Wash twice with 200 µL EX-50-NP40. Resuspend well each time by gently tapping the tubes. Do not pipet to resuspend. Remove all supernatant each time. If droplets are dispersed on the tube walls spin for 15 s at 1000 rpm in a bench top minifuge if necessary.
3. Suspend beads in 7.5 µL of EX-Y1-NP40 for elution (see **Note 15**). Concentrate beads and save supernatant. Repeat and pool supernatant for PAGE (total volume 15 µL).
4. Wash beads with 200 µL of EX-Y1-NP40. Discard wash.
5. Proceed to the next salt concentration. Each elution is done by extracting twice in 7.5 µL (save for gel) and a large 200 µL wash (for completeness).
6. Place all samples for PAGE on the MPC for 2–3 min to remove any trapped beads. Recover supernatant into new tubes containing 5 µL 4X SDS loading buffer.

A Solid-Phase Approach

7. Resuspend beads in 20 μL 1X SDS loading buffer. Incubate for 10 min at 37°C. Do not boil (see **Note 16**). Concentrate the beads and save supernatant. This sample represents the proteins that are not eluted even with the most stringent wash applied.
8. Denature all samples for 5 min at 95°C, separate by 15% SDS-PAGE.
9. Stain gel with silver or transfer to membrane for Western blotting.

3.3.2. Micrococcal Nuclease Treatment (see **Note 17**)

1. Assemble 900 ng of immobilized DNA into chromatin as described in **Subheading 3.2.1.** Concentrate chromatin on MPC and remove supernatant.
2. Wash chromatin twice with 100 μL of EX-Y-NP40 (see **Note 15**).
3. Wash beads with 50 μL EX-50-NP40.
4. Resuspend in 120 μL EX-50-NP40 containing 5 mM MgCl$_2$, prewarmed at 26°C.
5. Add 180 μL of MNase premix (168 μL EX-50, 9 μL CaCl$_2$, 3 μL MNase (5 U/μL) prewarmed at 26°C (see **Note 18**).
6. After 30 s, 1 min, and 8 min recover 100 μL into a tube containing 25 μL of nuclease stop mix, and vortex briefly.
7. When all samples are processed, add 1 μL RNase (10 mg/mL), and incubate for 5 min at 37°C.
8. Add 2 μL 20 % SDS and 5 μL proteinase K (10 mg/mL) and digest overnight at 37°C.
9. Concentrate beads on MPC and recover supernatant.
10. Add 90 μL 7.5 M ammonium acetate, pH 5.3, and 0.5 μL glycogen 20 mg/mL. Mix and add 2 vol ethanol.
11. Leave on ice for 5 min and spin for 15 min at top speed in a bench-top centrifuge at 4°C.
12. Wash pellet carefully with 800 μL of 75% ethanol and air dry on the bench. Do not dry pellet in the speed vac as this may cause DNA denaturation!
13. Take pellet up in 8 μL TE and add 2 μL Orange loading buffer (5X).
14. Electrophorese on a 1.3% agarose gel in Tris-glycine buffer (see Chapter 13, see **Note 19**).

4. Notes

1. In order to immobilize a plasmid two restriction enzymes must be selected as follows (see also **Fig. 1**): The plasmid must be linearized with an enzyme leaving a 5' overhang that can be filled-in with biotin-21-dUTP or biotin-14-dATP with Klenow polymerase. In order to prevent the coupling via both ends (which may result in the shearing of the DNA) the linearized DNA must be restricted with a second enzyme leaving a site where no biotin will be incorporated during the fill-in reaction (e.g., blunt ends, 3' overhangs or 5' overhangs with GC-rich sequences). If the secondary cut results in two large fragments, a mixture of both fragments will be immobilized. If the secondary enzyme is chosen such that one large and one very small fragment is produced, this fragment may be removed during the subsequent gel filtration step (**Fig. 1**). Ideally, the biotinylated residue should not be the last nucleotide to be incorporated during the fill-in reaction so

that it can be protected against exonuclease activity by sealing the ends with a-thio-dNTPs (see **Note 3**). Some enzymes that we have used to create an end suitable for biotinylation are *Eco*RI, *Spe*I, *Afl*II, *Hind*III, and *Sal*I. *Not*I and *Cla*I can be used for the other end. These enzymes produce 5' overhang sequences lacking A or T residues which are not filled in with biotin-14-dATP or biotin-21-dUTP.

2. Incomplete restriction enzyme digestion may lead to low coupling efficiency. We routinely check completeness of digestion at each step. Therefore, even when two compatible restriction enzymes are utilized, we prefer to perform the digestions in two steps rather than in one step, in order to monitor digestion efficiency. Digest first with the enzyme that creates the end that will not be biotinylated and assure complete linearization.

3. In order to protect the ends from exonuclease invasion that may occur in some experimental systems, we use a-thio-dNTPs in addition to the biotinylated dNTP to fill in the ends which increase the half-life of the ends in crude exonuclease-containing extracts considerably. Ideally, the biotinylated dNTPs should be shielded by 1-2 α-thio-dNTPs.

4. Poor filling-in by Klenow DNA polymerase affects immobilization. Avoid using ammonium acetate for DNA precipitation as it may inhibit the polymerase. Klenow Exo⁻ is better suited for this application than ordinary Klenow DNA polymerase.

5. Incomplete removal of unincorporated biotin is a common reason for inefficient coupling. Biotin reacts with streptavidin readily and may outcompete the immobilization of DNA. Spin columns from different suppliers have diverse specifications which should be followed precisely.

6. Coupling efficiency drops drastically with increasing length of DNA to be immobilized. For some applications *(3)* long chromatin templates may be particularly useful. We describe here a protocol for efficient immobilization of λ DNA (50 kb) using the Kilobase Binder reagent from Dynal. Approximately 1 pmole of λ DNA can be immobilized per 100 μL of Dynabeads.

7. If coupling was efficient the supernatant from the coupling reaction should be free of DNA (missing band test). In the case of incomplete immobiliszation, comparison of band intensities serves to accurately estimate the percentage of immobilized template. Efficiencies higher than 95% are routinely obtained.

8. It is essential to dissolve the pellet completely at this step. Do not vortex to avoid shear. Allow a long time, if possible overnight, to dissolve DNA pellet. In general, minimize manipulations such as extensive pipeting that may shear the concentrated, viscous lambda DNA. We cut the end of the pipet tips with scissors to widen the tip opening. Avoid pipeting up and down in order to resuspend λ DNA after its precipitation.

9. See Dynal's instructions for up-to date effective concentration.

10. The amount of chromatin assembly extract to be added has to be determined empirically on soluble plasmid DNA. For each amount of extract used, chromatin assembly efficiency is monitored by MNase digestion and agarose gel electrophoresis in order to define the optimum (see Chapter 13). In general 50–90 μL of extract are required for 900 ng of template in a 120 μL reaction. Once the optimal conditions are determined scaling up or down is feasible. If a small

amount of immobilized DNA is to be assembled into chromatin, it is advisable to fill the reaction up with soluble carrier DNA to keep reaction volume conveniently high rather than scaling down.

11. Reaction tubes of different sizes are used in order to match the volume of chromatin assembly reactions. If there is too much empty space in the tube, the reaction mixture spreads all over the surface of the rotating tube. When possible, scale up the chromatin assembly reaction to fill up most of the tube. A small air bubble trapped in the tube will help to maintain the beads dispersed in suspension. Because the magnetic field is much stronger close to the base of the tube, we avoid using relatively big volumes (greater than 600–700 µL) per tube, as this would increase the duration of the concentration (in a viscous milieu this can lead to incomplete recovery). Thus when it is necessary to concentrate a greater volume (e.g., when conditioning a great volume of bead suspension for coupling reaction) split the reaction into aliquots and concentrate them successively. After the first aliquot of beads is concentrated and the supernatant discarded, the second aliquot is added to the tube and so on.

  If the reaction volume is a very small, use small (250 µL), elongated tubes. In this case apply the reaction mixture to the bottom of the tube avoiding contact with its walls. The droplet of the reaction mixture will remain at the bottom of the tube due to surface tension.

12. The nucleosome assembly by salt dialysis is a modification of the one described by Neugebauer and Hörz *(8)*. For further descriptions of salt gradient dialysis procedures, *see* Chapters 1 and 4. Here we concentrate on those modifications to the procedure required when working with immobilized DNA.

13. A ratio of purified core histones to DNA of 1:1 reproducibly results in efficient nucleosome assembly. However, an empirical titration of core histones using soluble DNA may be required. As an internal control in the assembly reactions a short, radioactively labeled and gel-purified PCR fragment may be added in the same dialysis chamber with the immobilized DNA. This will serve to determine the efficiency of nucleosome assembly by a bandshift assay. Complete nucleosome assembly results in a shift of the probe from free to mononucleosome band.

14. A background of proteins sticking nonspecifically to the bead matrix itself is anticipated. Consider the following parameters to optimize the signal-to-noise ratio. First, maximize the amount of DNA per bead by adding an excess of biotinylated DNA in the coupling reaction. Second, preadsorb the beads by washing them a couple of times in a buffer containing 0.01% (w/v) BSA and 0.05 % (v/v) Nonidet P 40. This decreases background and also enables easier handling of the beads by reducing their stickiness. Third, different suppliers provide beads with different matrix characteristics. In our experience, Dynabeads gave a low background when used with *Drosophila* embryo extracts.

15. You have the option to elute proteins sequentially with washes of increasing salt to determine how tightly a protein interacts with chromatin. In general, a buffer containing 400 m*M* KCl strips off most of the chromatin associated proteins while core histones require high salt (2 *M*) for their elution. In the following protocol substitute salt concentration (Y1= salt 1 in m*M*) in the buffers according to your application.

16. Many proteins that interact with the bead matrix *per se* are not eluted in SDS loading buffer unless the beads are boiled. By contrast, chromatin proteins (including histones) are stripped from DNA without boiling. Therefore it is very important to omit boiling of the beads.
17. MNase digestion can be performed with or without prior purification of the template. Here we describe a protocol for MNase digestion of purified chromatin that has been washed. In the case of nuclease treatment without isolation of the DNA from the assembly reaction, approx 10X more MNase units are required. Conversely, the more stringent the washing of chromatin are the less the nuclease is needed.
18. Upon addition the MNase mix, pipet up and down a couple of times to suspend beads. During longer incubation times, resuspend beads once by tapping the tube. Alternatively use the Eppendorf Thermomixer at setting 10.
19. The appearance of the characteristic, ladder-like pattern of DNA fragments generated by MNase analysis and subsequent agarose gel electrophoresis is slightly compromised because only those fragments that are cleaved off the beads (by a double-stranded cut) are recovered for electrophoresis. Because underdigestion may result in only very little DNA on the gel, fine-tuning of the MNase digestion may be required.

    The nucleosome repeat length of immobilized chromatin assembled in *Drosophila* extracts is a bit shorter compared to chromatin assembled on plasmid template under identical conditions. This difference is not owing to the immobilization, but rather reflects a difference between linear and supercoiled DNA *(9)*.

References

1. Almouzni, G. and Méchali, M. (1988) Assembly of spaced chromatin. Involvement of ATP and DNA topoisomerase activity. *EMBO J.* **7,** 4355–4365.
2. Becker, P. B. and Wu, C. (1992) Cell-free system for assembly of transcriptionally repressed chromatin from *Drosophila* embryos. *Mol. Cell. Biol.* **12,** 2241–2249.
3. Heald, H., Tournebize, R., Blank, T. A., Sandaltzopoulos, R., Becker, P. B., Hyman, A., and Karsenti, K. (1996) Self organization of microtubules into bipolar spindles around artificial chromosomes in *Xenopus* egg extracts. *Nature* **382,** 420–425.
4. Sandaltzopoulos, R. and Becker, P. B. (1995) Analysis of protein-DNA interaction by solid-phase footprinting. *Meth. Mol. Cell Biol.* **5,** 176–181.
5. Sandaltzopoulos, R., Blank, T., and Becker, P. B. (1994) Transcriptional repression by nucleosomes but not H1 in reconstituted preblastoderm *Drosophila* chromatin. *EMBO J.* **13,** 373–379.
6. Sandaltzopoulos, R. and Becker, P. B. (1994) Solid phase footprinting: quick and versatile. *Nucleic Acids Res.* **22,** 1511,1512.
7. Sandaltzopoulos, R. and Becker, P. B. (1998). Heat shock factor increases the reinitiation rate from potentiated chromatin templates. *Mol. Cell. Biol.* **18,** 361–367.
8. Neubauer, B. and Hörz, W. (1989) Analysis of nucleosome positioning by *in vitro* reconstitution. *Methods Enzymol.* **170,** 630–644.
9. Blank, T. A. and Becker, P. B. (1996) The effect of nucleosome phasing sequences and DNA topology on nucleosome spacing. *J. Mol. Biol.* **260,** 1–8.

15

Reconstitution and Analysis of Hyperacetylated Chromatin

Wladyslaw A. Krajewski and Peter B. Becker

1. Introduction

Specific acetylation at conserved lysines in the N-terminal tails of histones have been correlated with distinct chromatin structures, association of specific chromatin proteins, accessibility of nucleosomal DNA toward interaction of transcription factors, and unfolded chromatin with increased transcription potential *(1–5)*. Global histone acetylation prevents the folding of the nucleosomal fiber into higher order structures *(6)*. Despite these correlations, the molecular principles governing molecular heterogeneity of chromatin structure and its implications for processes that require a DNA substrate are only poorly understood. But the close correlation between histone acetylation and gene activity suggests a contribution of histone acetylation *(2–4)*.

In order to define the causalities between histone acetylation, specific chromatin structures and gene activity, a biochemical approach is required. Current procedures, such as the chemical acetylation in vitro *(7)* and salt gradient dialysis reconstitution of nucleosomal arrays from hyperacetylated histones *(6)* are limited in their applications because of the nonphysiological reconstitution conditions involved. Here we describe the use of a cell-free system derived from *Drosophila* embryo extracts *(8,9)* for the reconstitution of hyperacetylated chromatin under physiological conditions. This system reconstitutes chromatin of high complexity with respect to the association of nonhistone proteins, with physiological nucleosome spacing and characterised by remarkable dynamic properties *(10,11)*. Hyperacetylated chromatin reconstituted by this procedure exhibits higher conformational flexibility of DNA, elevated DNase I sensitivity and increased potential to be transcribed by RNA polymerase II when compared to control chromatin with similar nucleosome number and den-

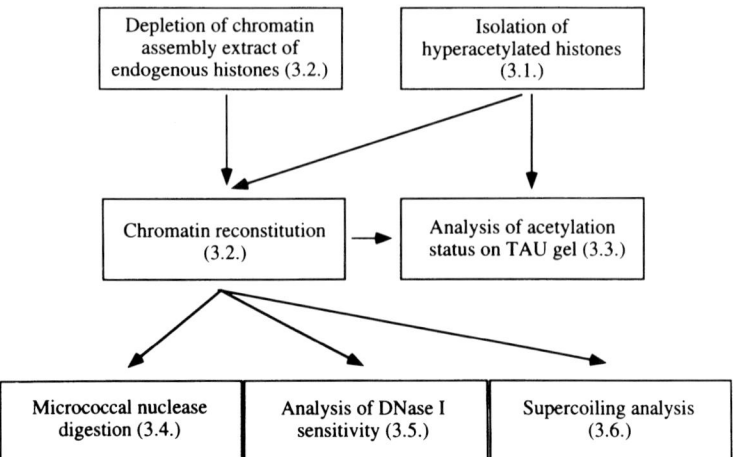

Fig. 1. Overview over the methods presented in this chapter. Numbers in brackets refer to the methods section.

sity *(2,8)*. The recapitulation of important hallmarks of acetylated chromatin *in vitro* opens a new avenue for the analysis of structure/function relationships in acetylated chromatin.

The procedures that will be described are summarized in **Fig. 1**. Chromatin assembly in extracts are derived from *Drosophila* embryos (*see* Chapter 13). Assembly using preblastoderm embryo extracts relies entirely on endogenous pools of histones deposited maternally into the egg *(8)*. These histone pools are used up as the embryo develops, such that extracts from postblastoderm embryos *(13)* need to be supplemented with exogenous histones. Nevertheless, these residual endogenous histones have to be removed before homogeneous chromatin can be reconstituted with exogenous, hyperacetylated histones (*see* **Subheading 3.2.**). These histones are purified from tissue culture cells after treatment with Trichostatin A, an inhibitor of histone deacetylase (**ref. *14***, *see* **Subheading 3.1.**). Inhibition of the deacetylase unballances the equilibrium between histone acetylation and deacetylation leading to accumulation of hyperacetylated histones. A high-resolution discontinuous Trition-Acid-Urea-(TAU) gel system allows to assess histone isoformes differing by a single acetyl group (**ref. *15***, *see* **Subheading 3.3.** and **Fig. 2**). Partial digestion with micrococcal nuclease (*see* **Subheading 3.4.**) and DNase I (*see* **Subheading 3.5.**) provides information about the nucleosomal spacing and the efficiency of reconstitution (**Fig. 3**). The assembly of each nucleosome introduces one superhelical turn into a closed plasmid. Therefore, the number of nucleosomes formed on a plasmid can be evaluated by resolving the topoisomers on chloroquin gels (*see* **Subheading 3.6.** and **Fig. 4**; *see* also Chapter 28).

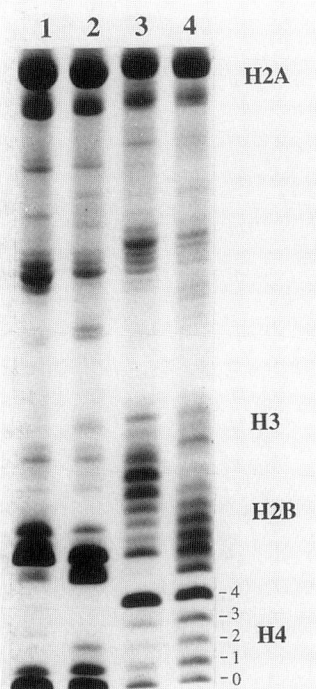

Fig. 2. Analysis of histones in a TAU gel. Histones purified from control (lanes 1, 2) and Trichostatin-treated (lanes 3, 4) mammalian cells, before reconstitution (lanes 1, 3) and after extraction from acetylated chromatin (lanes 2,4) were analyzed. The position of histone H4 isoforms (1, mono-; 4, tetraacetylated) are shown.

2. Materials

2.1. Isolation of Core Histones

1. 0.5 mg/mL Trichostatin A (TSA) in ethanol (Wako Chemicals, Neuss, Germany).
2. 1 M Na-butyrate, pH 8.0.
3. *Complete* protease inhibitor cocktail (Boehringer Mannheim, Germany).
4. Lysis buffer I: 50 mM Tris-HCl, pH 8.0, 25 mM KCl, 25 mM NaCl, 4 mM MgCl$_2$, 10 mM Na-butyrate, 5 mM 2-mercaptoethanol, 0.2 mM PMSF, 0.5 M sucrose, 0.5% Nonidet P40 (v/v) (Sigma, Deisenhofen, Germany), 10 ng/mL of Trichostatin A.
5. Lysis buffer II: 50 mM Tris-HCl, pH 8.0, 50 mM NaCl, 4 mM MgCl$_2$, 10 mM Na-butyrate, 5 mM 2-mercaptoethanol, 0.2 mM PMSF, 0.34 M sucrose, 0.5% Nonidet P40.
6. 200 mM Phenylmethylsulfonyl fluoride (PMSF) in ethanol.
7. 0.25 M H$_2$SO$_4$.
8. 10% Perchloric acid.
9. 0.5 M HCl.

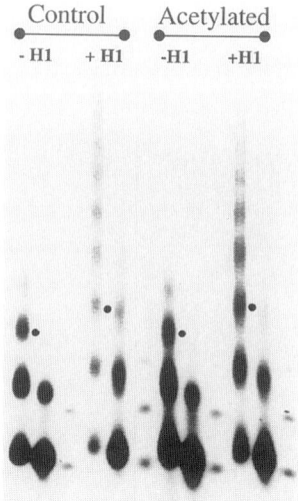

Fig. 3. Analysis of reconstituted chromatin by Micrococcal nuclease digestion. Chromatin was reconstituted using control and acetylated histones in the presence or absence of histone H1. Chromatin was digested with micrococcal nuclease and the resulting DNA fragments were resolved on an agarose gel. The 123 bp DNA ladder (Boehringer) served as size marker. A negative print of the ethidium bromide-stained gel is shown. The trinucleosome-derived fragments are marked with dots to facilitate the comparison of nucleosome repeat lengths.

10. Dialysis tubes with 6000–8000 MW cut-off (Spectrapor Spectrum Medical Industries Lugana Hills, US).
11. Centrifugal concentrators 3K (Filtron Northborough, US).
12. Kollodion bags 10000 MW limit (Sartorius Goettingen, Germany).
13. Sonicator; Branson Sonifier (or equivalent).
14. Sorvall RC-5B refrigerated centrifuge, SA600 or SS34 rotors; (DuPont Instruments).

2.2. Histone Depletion of Assembly Extract and Chromatin Assembly Reaction

1. Extraction buffer (EX): 10 mM HEPES-KOH, pH 7.6, 10 mM KCl, 1.5 mM MgCl$_2$, 0.5 mM EGTA, 10% glycerol (v/v). Add 1 mM DTT and 0.2 mM PMSF prior to use.
2. *Drosophila* embryo extract. For a detailed description of extract preparation, *see* Chapter 13.
3. DNA coupled to paramagnetic particles (*see* **Note 1**).
4. McNAP solution (10X): 30 mM ATP, 300 mM creatine phosphate, 30 mM MgCl$_2$, 10 ng/µL creatine phosphokinase (Sigma), 10 mM DTT, 50 ng/mL Trichostatin A.
5. Bead regeneration buffer: 10 mM Tris-HCl, pH 7.5, 2 M NaCl, 1 mM EDTA, 0.2 mM PMSF.
6. 200 mM Phenylmethylsulfonyl fluoride (PMSF, (Sigma) in ethanol.

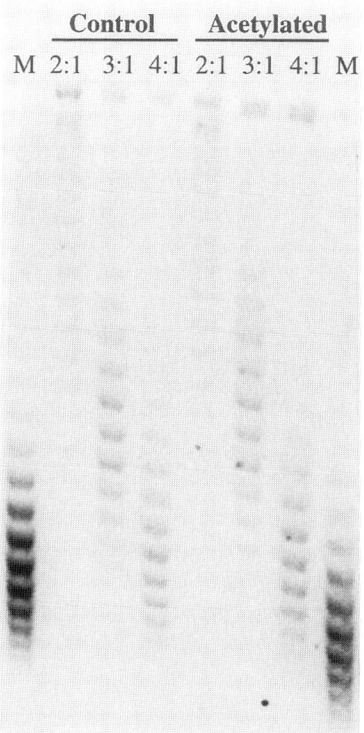

Fig. 4. Acetylated and control chromatin were reconstituted at the indicated histone: DNA ratios. Deproteinization of DNA results in the release of approximately one negative supercoil for each nucleosome. Plasmid topoisomers were resolved on an agarose gel in the presence of chloroquine. As a marker (**M**) the same plasmid isolated from *E. coli* was used (*see* **Note 7**).

7. 1 *M* Dithiothreitol (DTT), (Sigma).
8. Plasmid DNA.
9. Calf thymus histone H1 (Boehringer Mannheim), 0.15 mg/mL in 5X EX buffer. Store in aliquots at –20°C.
10. Magnetic particle concentrator (Dynal Oslo, Norway).

2.3. Analysis of Histone Acetylation in Discontinuous Triton-Acid-Urea Gels

1. Immobilized DNA (*see* **Note 1**).
2. EX buffer: 10 m*M* Hepes-KOH, pH 7.6, 10 m*M* KCl, 1.5 m*M* MgCl$_2$, 0.5 m*M* EGTA, 10% glycerol (v/v). Add 1 m*M* DTT and 0.2 m*M* PMSF prior to use.
3. Elution buffer: 10 m*M* Tris-Cl, pH 7.5, 2 *M* NaCl, 1 m*M* EDTA, 0.2 m*M* PMSF.
4. 50% (w/v) trichloroacetic acid.
5. 0.5 *M* HCl.

6. TAU gel loading buffer: 8 M urea, 5% acetyc acid, 0.1% Pyronin Y (Merck, Darmstadt, Germany).
7. Protogel (National Diagnostics, Hull, UK): 30% acrylamide/ 0.8% bisacrylamide
8. 0.3 M Triton X-100.
9. Thiodiglycol (Sigma).
10. TEMED.
11. 0.05% (w/v) riboflavin (Sigma).
12. 3 M K-acetate, pH 4.0.
13. Staining solution: 25% (v/v) 2-propanol, 10% acetic acid (v/v), 0.2% (w/v) Coomassie blue R-250 (Serva).
14. Destaining solution: 15% (v/v) acetic acid, 10% (v/v) 2-propanol.
14. Yellow photo-filter (Kodak Seattle, WA).

2.4. Chromatin Analysis

1. Micrococcal nuclease (Sigma), 50000 U/mL in EX buffer, 2.6 mM MgCl$_2$. Store in 50-µL aliquots at –20°C.
2. DNase I (Sigma), 10000 U/mL in 1X EX buffer. Store in 50-µL aliquots at –20°C.
3. Wheat germ topoisomerase I, 1 U/ mL (Promega, Mannheim, Germany).
4. EX buffer (*see* **Subheading 2.2.**).
5. 1 M CaCl$_2$.
6. 0.1 M EDTA .
7. 2 mg/mL RNase A (Sigma).
8. 10 mg/mL Proteinase K.
9. Nuclease stop mix: 2.5% (w/v) N-lauroylsarcosine (Sigma), 100 mM EDTA, add 100 ng/mL heat-treated RNase A just before use.
10. Proteinase digestion mix: 1.25 % (w/v) Na-Dodecylsulphate, 1.5 mg/mL Proteinase K.
11. Agarose loading buffer: 7.5% (v/v) glycerol, 0.75 mM EDTA, 0.09 % (w/v) Orange G (Sigma).
12. 123 bp DNA ladder marker (Boehringer-Mannheim)
13. 100 mM Chloroquine diphospate (Sigma).
14. Tris-glycine electrophoresis buffer: 28.8 g/L glycine, 6 g/L Tris base.
15. 0.5 µg/ mL Ethidium bromide .
16. Red filter for photography (Kodak).

3. Methods
3.1. Isolation of Core Histones from Cultured Cells

1. Grow cells of interest under appropriate conditions (*see* **Note 2**).
2. During exponential growth add Trichostatin A to a final concentration of 500 ng/mL medium and incubate cells for further 12 h.
3. Harvest cells on ice (scrape monolayer cultures off the dishes or collect suspension cultures by centrifugation at 1000g for 10 min at 4°C). Cells pellets can be stored at –20°C until used.
4. Thaw cell pellet on ice. The following procedures should be performed on ice or at 4°C, respectively.

5. Resuspend 1 mL of cell pellet in 50 mL of lysis buffer I, containing one tablet of *Complete* protease inhibitor (*see* **Note 3**). Mix well by vortexing for 2 min.
6. Collect nuclei by centrifugation at 1500*g* for 10 min.
7. Resuspend nuclei pellet in 50 mL of lysis buffer II.
8. Collect nuclei by centrifugation at 1500*g* for 10 min.
9. Repeat **step 3** two more times.
10. Resuspend nuclei pellet in 10 mL of 5 mM 2-mercaptoethanol, 0.2 mM PMSF.
11. Add 10 mL of cold 0.25 M H_2SO_4. Mix well and incubate with shaking for 30 min.
12. Pellet nuclei debris by centrifuging at 2500*g* for 10 min.
13. Clear the supernatant by additional centrifugation at 15,000*g* for 30 min.
14. Add an equal volume of cold 10% perchloric acid. Incubate for 1 h on ice.
15. Collect histones by centrifugation at 2500*g* for 15 min.
16. Dissolve histone pellet in 20 mL of 0.5 M HCl, 0.2 mM PMSF, aided by pipeting and sonication for 5 × 10 s at setting "5."
17. Clear solution by centrifugation at 10,000*g* for 30 min.
18. Dialyze overnight against 5 L of 5 mM 2-mercaptoethanol, 0.2 mM PMSF, 10 mM Na-butyrate.
19. Precipitate histones with an equal volume of 10% perchloric acid. Incubate for 1 h on ice.
20. Collect histones by centrifugation at 2500*g* for 15 min.
21. Resuspend pellet in 7.5 mL of 0.5 M HCl.
22. Dialyze overnight against 3 L of 0.5 M HCl, 5 mM 2-mercaptoethanol, 0.2 mM PMSF.
23. Clear solution by centrifugation at 10,000*g* for 30 min.
24. Concentrate to 0.5–0.8 mL using centrifugal concentrators.
25. Dialyze overnight in Kollodion bags against 5 L of 10 mM 2-mercaptoethanol, 0.2 mM PMSF, 10 mM Na-butyrate.
26. Store in aliquots at –70°C until required. For immediate use, add 80% glycerol to a final concentration of 50%, adjust to the desired histone concentration and store at –20°C.

3.2. Chromatin Reconstitution In Vitro

1. In order to deplete the assembly extract of endogenous histones incubate 500 µL *Drosophila* embryo extract with 30 µg of immobilized DNA (*see* **Note 1**), equilibrated with EX buffer (KCl concentration adjusted to assembly conditions).
2. Incubate with agitation for 30 min at 4°C.
3. Remove the immobilized DNA-histone complexes on a magnetic particle concentrator. Histone-depleted extract can be used immediately for the reconstitution reaction. Unused extract can be flash-frozen and stored in liquid nitrogen.
4. Prepare a standard assembly reaction by adding sequentially (final volume of 135 µL (*see* **Note 4**):
    100–105 µL EX buffer, containing the desired KCl concentration (usually 90 mM).
    15–20 µL *Drosophila* embryo extract.
    13.5 µL 10X McNAP energy regeneration buffer.
    1-2.5 µL of core histones (1 mg/mL).
    0.65 µL plasmid DNA (1 mg/mL).

To incorporate histone H1 add 1.0 mL of purified histone H1 (0.15 mg/mL) and reduce the amount of core histones to 1.0 µL (1 mg/mL) (*see* **Note 5**).
5. Incubate reaction mixture at 26°C for 6 h.

3.3. Analysis of Histone Acetylation in Reconstituted Chromatin by Discontinous Triton-Acid-Urea Gel Electrophoresis

1. Scale up a standard chromatin assembly reaction 10-fold (*see* **Subheading 3.2.**), but substitute the plasmid DNA by an equivalent amount of immobilized DNA (*see* **Note 1** and Chapter 14).
2. Incubate reaction mixture at 26°C for 6 h on a rotating wheel.
3. Concentrate beads on a magnetic particle concentrator and remove supernatant.
4. Wash chromatin three times by adding 1 mL of EX buffer, 100 mM KCl followed by concentration on the MPC.
5. Add 0.75 mL of elution buffer. Incubate at 4°C for 30 min with agitation to prevent settling of the beads.
6. Add an equal volume of 50% trichloroacetic acid. Incubate on ice for 1 h.
7. Pellet the histones by centrifugation for 10 min in a table-top centrifuge.
8. Resuspend well in 200 µL of 0.5 M HCl and incubate at 4°C for 30 min with agitation.
9. Clear the solution by centrifugation for 10 min in a table-top centrifuge.
10. To the supernatant add 6 vol of cold acetone.
11. Collect the histones by centrifugation for 10 min in a table-top centrifuge.
12. Remove the supernatant and dry the histone pellet in a speed-vac for 10–15 min.
13. Dissolve the histones in TAU gel loading buffer.
14. Prepare the separating gel as follows (makes 10-mL gel mix):
    Mix 5 mL of Protogel, 0.54 mL of acetic acid, 4.84 g of solid urea, 240 µL of 0.3 M Triton X-100; 100 µL of thiodiglycol, 60 µL of TEMED. Adjust with deionized water to 9 mL. Dissolve urea well (heat at 37°C if necessary). Add 1 mL of 0.05% riboflavin. Mix quickly and pour the separating gel. Polymerize the gel in front of a white lamp.
15. Prepare the stacking gel as follows (makes 5-mL gel mix):
    Mix 1.25 mL of Protogel, 2.4 g of solid urea, 0.625 mL of 3 M K-acetate, pH 4.0, 120 µL of 0.3 M Triton X-100; 50 µL of thiodiglycol; 50 µL of TEMED. Adjust with deionized water to 4.5 mL. Dissolve urea well (heat at 37°C if necessary). Add 0.5 mL of 0.05% riboflavin. Mix quickly and pour the stacking gel. Polymerize the gel in front of a white lamp.
16. Load histone samples.
17. Electrophorese an 8 cm gel for 3.5 h at 200 V using 5% acetic acid as electrophoresis buffer.
18. Stain the gel for 30–60 min and destain until the desired contrast is reached.
19. Photograph the gel using yellow Kodak filter.

3.4. Micrococcal Nuclease Analysis of Reconstituted Chromatin

1. Set up a standard chromatin reconstitution reaction (*see* **Subheading 3.2.**).
2. Add 74 µL of EX buffer containing 50 U of Micrococcal nuclease, 2.6 mM CaCl$_2$ (salt concentration adjusted to assembly conditions).

3. After 1 and 5 min take 100 μL samples and add them to 25 μL of nuclease stop mix.
4. Incubate for 30 min at 37°C.
5. Add 25 μL of proteinase digestion mix and incubate overnight at 37°C.
6. Add 300 μL of ethanol and incubate for 30 min at room temperature.
7. Pellet DNA by centrifugation for 15 min in an table top centrifuge at 4°C (*see* **Note 6**).
8. Remove ALL supernatant and dissolve the pellet in 5–7 μL of loading buffer.
9. Electrophorese on a 1.3% agarose gel in Tris-glycine buffer. For best resolution use a comb with 4 mm wells and a 25 × 16 × 0.4 cm (L × W × H) gel. Run at 120 V const (3–5 h).
10. Stain gel with 2.5-gel volume of water containing 0.5 μg/mL ethidium bromide for 30 min followed by destaining in deionized water for 15 min.
11. Photograph the gel using red Kodak filter.

3.5. Analysis of DNase I Sensitivity

1. Set up a standard chromatin reconstitution reaction (*see* **Subheading 3.2.**).
2. Digest 20 μL of the reaction with 1–12 U DNase I in 10 μL EX buffer, 1.0 mM CaCl$_2$ for 1 min at 26°C (KCl concentration adjusted to assembly conditions).
3. Stop the reactions with 10 μL of nuclease stop mixture. Incubate for 30 min at 37°C.
4. Add 10 μL of proteinase digestion mix and incubate overnight at 37°C.
5. Add 100 μL of ethanol and incubate for 30 min at room temperature (*see* **Note 5**).
6. Pellet DNA by centrifugation for 15 min in a table-top centrifuge at 4°C.
7. Remove the supernatant and dissolve the pellet in 5–7 μL of loading buffer.
8. Electrophorese on an 0.8% agarose gel in Tris-glycine buffer at const 120 V (2–3 h).
9. Stain and destain the gel as in **Subheading 3.4., steps 10** and **11**.

3.6. Analysis of DNA Supercoiling

1. Set up a standard chromatin reconstitution reaction (*see* **Subheading 3.2.**).
2. Stop the reactions and isolate the plasmid DNA (**Subheading 3.4., steps 3–8**).
3. Use plasmid DNA with native negative supercoiling as isolated from bacteria as standard (*see* **Note 7**).
4. Electrophorese on a 1.2% agarose gel in Tris-glycine buffer containing 3–5 mM of chloroquine diphosphate in both gel and running buffer. Use a comb with 4–5-mm wells and a gel of at least 20 cm length gel, not more than 4 mm thick. Run at 80–90 V const (15–20 h).
5. Stain gel with 5 vol of water containing 0.5 μg/mL ethidium bromide for 2 h. Destain in deionized water for 1 h (*see* **Note 8**).
6. Photograph the gel using red Kodak filter.

4. Notes

1. Linear biotinylated plasmid DNA is coupled to magnetic beads (Dynal) according to the manufacturer's instructions (*see* Chapter 14). Immobilized DNA can be re-used up to 5–6 times after washing five times with bead regeneration buffer.

2. We usually grow mammalian cells in standard Dulbecco's media supplemented with 10% fetal calf serum in the atmosphere of 5% CO_2 at 36.5°C.
3. If nonacetylated histones are isolated deacetylase inhibitors Trichostatin A and Na-butyrate may be omitted.
4. The optimal amount of extract has to be determined for every extract preparation. Variation of the KCl concentration (20–120 mM) in the assembly reaction results in differences in the nucleosome repeat length (from 156–180 bp) (for a detailed study see **ref. 16**). Titration of the input histone: DNA ratio from 1.5:1 to 4:1 results in increased nucleosome density (from 20–30 nucleosomes on a 6150 bp plasmid). The exact conditions for desired levels of chromatin assembly and the nucleosome repeat length should be determined empirically.
5. Addition of histone H1 to the reconstitution reaction positively affects the efficiency of chromatin assembly, therefore it is possible to reduce core histone: DNA ratio to 1.5:1.
6. It is important to precipitate DNA by not more then 2–2.5 vol of ethanol in order to avoid coprecipitation of K^+-Dodecylsulphate complexes. For this reason it is also important not to freeze the DNA samples before/during ethanol precipitation.
7. The identical plasmid isolated from bacteria, at a superhelical density of 0.05 *(17)*, serves as a marker (M) which is in this case equivalent to a approx 31 superhelical turns. An equivalent superhelicity on plasmids in chromatin corresponds to an average nucleosme density of about 1 nucleosome/197 bp.
8. Long staining is essential to replace intercalated chloroquine residues by ethidium.

References

1. Mizzen, C. A. and Allis, C. D. (1998) Linking histone acetylation to transcriptional regulation. *Cell. Mol. Life Sci.* **54,** 6–20.
2. Bone, J. R., Lavender, J., Richman, R., Palmer, M. J., Turner, B. M., and Kuroda, M. I. (1994) Acetylated histone H4 on the male X chromosome is associated with dosage compensation in *Drosophila. Genes Dev.* **8,** 96–104.
3. Hebbes, T. R., Clayton, A. L., Thorne, A. W., and Crane-Robinson, C. (1994) Core histone acetylation co-maps with generalized DNase I sensitivity in the chicken b-globin chromosomal domain. *EMBO J.* **13,** 1823–1830.
4. Turner, B. M. and O'Neill, L. P. (1995) Histone acetylation in chromatin and chromosomes. *Sem. Cell Biol.* **6,** 229–236.
5. Vettese-Dadey, M., Grant, P. A., Hebbes, T. R., Crane-Robinsom, C., Allis, C. D., and Workman, J. L. (1996) Acetylation of histone H4 plays a primary role in enhancing transcription factor binding to nucleosomal DNA in vitro. *EMBO J.* **15,** 2508–2518.
6. Garcia-Ramirez, M., Rocchini, C., and Ausio, J. (1995) Modulation of chromatin folding by histone acetylation. *J. Biol. Chem.* **270,** 17,923–17,928.
7. Shewmaker, C. K., Cohen, B. N., and Wagner, T. E. (1978) Chemically induced gene activation: selective increase in DNase I susceptibility in chromatin acetylated with acetyl adenylate. *Biochem. Biophys. Res. Comm.* **84,** 437–445.

8. Becker, P. B. and Wu, C. (1992) Cell-free system for assembly of transcriptionally repressed chromatin from *Drosophila* embryos. *Mol. Cell. Biol.* **12**, 2241–2249.
9. Becker, P. B., Tsukiyama, T., and Wu, C. (1994) Chromatin assembly extracts from *Drosophila* embryos. *Methods Cell Biol.* **44**, 207–223.
10. Sandaltzopoulos, R., Blank, T., and Becker, P. B. (1994) Transcriptional repression by nucleosomes but not H1 in reconstituted preblastoderm *Drosophila* chromatin. *EMBO J.* **13**, 373–379.
11. Varga-Weisz, P. D., Blank, T. A., and Becker, P. B. (1995) Energy-dependent chromatin accessibility and nucleosome mobility in a cell-free system. *EMBO J.* **14**, 2209–2216.
12. Krajewski, W. A. and Becker, P. B. (1998) Reconstitution of hyperacetylated, DNase I-sensitive chromatin characterised by high conformational flexibility of nucleosomal DNA. *Proc. Natl. Acad. Sci. USA* **95**, 1540–1545.
13. Kamakaka, R. T., Bulger, M., and Kadonaga, J. T. (1993) Potentiation of RNA polymerase II transcription by Gal4-VP16 during but not after DNA replication. *Genes Dev.* **7**, 1779–1795.
14. Yoshida, M., Horinouchi, S., and Beppu, T. (1995) Trichostatin A and trapoxin: novel chemicalprobes for the role of histoen acetylation in chromatin structure and function. *BioEssays* **17**, 423–430.
15. Krajewski, W. A. and Luchnik, A. N. (1991) Relationship of histone acetylation to DNA topology and transcription. *Mol. Gen. Genet.* **230**, 442–448.
16. Blank, T. A. and Becker, P. B. (1995) Electrostatic mechanism of nucleosome spacing. *J. Mol. Biol.* **252**, 305–313.
17. Sinden, R.R., J.O., C., and Pettijohn, D.E. (1980) Torsional tension in the DNA double helix measured with trimethylpsoralen in living *E. coli* cells: analogous measurements in insects and human cells. *Cell* **21**, 773–783.
18. Nightingale, K. P., Wellinger, R. E., Sogo, J. M., and Becker, P. B. (1998) Histone acetylation facilitates RNA polymerase II transcription of the *Drosophila* hsp 26 gene in chromatin. *EMBO J.* **17**, 2865–2876.

16

Assembly of Mitotic Chromosomes in *Xenopus* Egg Extract

Anne-Elisabeth de la Barre, Michel Robert-Nicoud, and Stefan Dimitrov

1. Introduction

Cell division in eukaryotes follows an extremely complex plan according to which chromosomes are first duplicated and condensed more than 10,000 times to form the mitotic chromosomes, which are finally separated by the cellular machinery into two new nuclei. Although the fascinating process of assembly of mitotic chromosomes has been observed for more than 100 years, the mechanism of assembly as well as the structural organization of chromosomes are poorly understood. Historically, an important step towards understanding both chromosomal assembly and organization was the development of a methodology for the isolation of "pure" mitotic chromosomes and their biochemical characterization *(1,2)*. The structure of isolated mitotic chromosomes was further studied by microscopic techniques. However, due to the tight compaction of chromatin fibers in chromosomes, their underlying structure could not be viewed by these methods. This problem was overcome by extraction of histone from chromosomes, followed by microscopic visualization of the residual structures *(3–5)*. This led to the suggestion that chromatin fibers were organized into domains, loops attached to a proteinaceous framework called "scaffold." Thus, the scaffolding model of chromosome organization arose from studies where chromosome mitotic structure was initially destroyed. Such an approach, however, has several limitations and may lead to erroneous conclusions *(6)*. Indeed, some authors claim the scaffold to be an artifact, produced during high salt treatment of chromosomes (for details, *see* **ref. 7**).

In the past few years, a very powerful alternative approach for studying both mitotic chromosome organization and nucleus structure has been described and used successfully by different authors *(6,8–12)*. This approach, in contrast to the studies mentioned above, does not destroy chromosomes but rather assembles chromosomes under physiological conditions in mitotic extracts from *Xenopus laevis* eggs.

Xenopus eggs contain a large reservoir of nuclear components, which are used during the very early embryogenesis of this animal. This peculiar property of *Xenopus* eggs led to the development of cell free extracts which can be used for reconstitution of chromosomes and nuclei as well as for the analysis of the dynamic features inherent to these structures *(13–15)*. Activation and inactivation of a protein kinase (mitotis promoting factor [MPF]), present in considerable quantities in the eggs, controls early cell cycles in *Xenopus*. MPF is a multisubunit complex, its two main subunits being kinase p34^{cdc2} and cyclin. Integrity of these subunits is essential for maintenance of the eggs in mitosis. Lysis of *Xenopus* eggs in the absence of chelators of Ca^{2+} or inhibitors of phosphatases leads to cyclin proteolysis and, thus, to destruction of MPF. During extract preparation, eggs are crushed by centrifugation, and EGTA is used together with β-glycerophosphate to preserve the integrity of MPF.

Incubation of *Xenopus* demembraned sperm or somatic nuclei in *Xenopus* mitotic extracts results in mitotic chromosome assembly (*see* **Figs. 1–3**). Immunodepletion of proteins of interest from the extract together with rescue experiments, allow for the elucidation of protein function *(8,11)*. A simple sucrose gradient purification of in vitro assembled mitotic chromosomes permits their purification and subsequent identification of specific mitotic chromosome proteins by electrophoresis. In this way two new polypeptides, called XCAP-C and XCAP-E, were described, which play an essential role in chromosome assembly *(9,10)*. These proteins belong to the structural maintenance of chromosomes (SMC) family and exist in *Xenopus* extract as higher molecular weight complexes called condensins *(10)*. One of these condensins (13S condensin) has a DNA stimulated ATPase activity that induces positive supercoils during interaction with relaxed circular DNA.

The described studies illustrate the power of the *Xenopus* system for dissection of mitotic chromosome assembly. This system will find important applications in future chromosome studies. Here we describe in detail a protocol for the isolation of mitotic *Xenopus* egg extracts and its use in the assembly of mitotic chromosomes. The protocol is relatively simple and produces quality extracts of high reproducibility. Special attention is paid to the individual steps of extract and *Xenopus* sperm preparation, microscopy techniques for chromosome visualization, and immunochemical approaches for specific protein detection and depletion.

Mitotic Chromosome Assembly

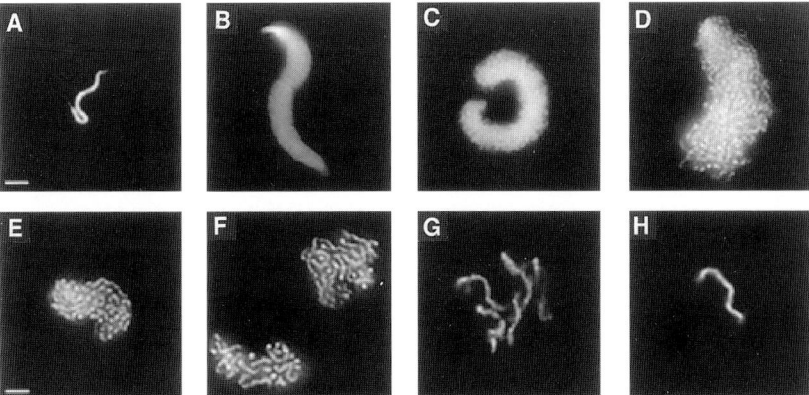

Fig. 1. Structural changes of *Xenopus* sperm chromatin on incubation in mitotic egg extract. (**A**) Control sperm nuclei, (**B**) decondensed sperm after 10 min of incubation, (**C–G**) chromosomal intermediary structures observed after 30, 60, 90, 120, and 150 min, respectively. (**H**) 180 min of incubation results in complete separation of chromosomes. Bar is 5 μm.

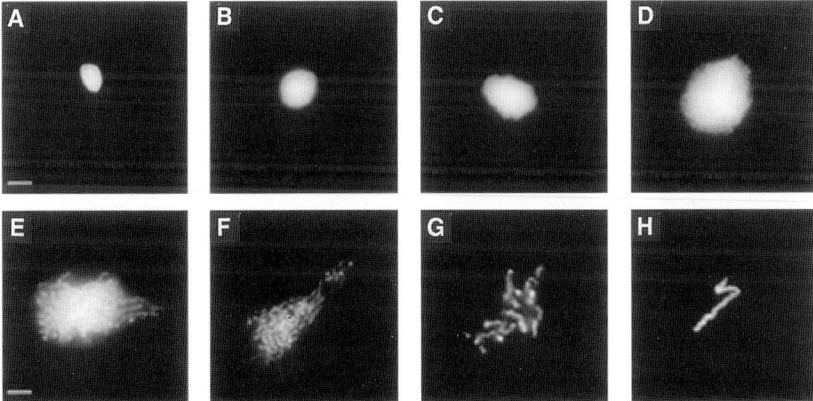

Fig. 2. Structural changes in *Xenopus laevis* erythrocyte nuclei upon incubation in mitotic egg extract. The rearrangements of erytrocyte nuclei are slower and not so well defined as in the case of sperm nuclei. (**A**) Control erythrocyte nuclei, (**B**) decondensed nuclei after 30 min of incubation, changes of nuclei on (**C**) 60, (**D**) 90, (**E**) 120, (**F**) 180, and (**G**) 240 min of incubation. (**H**) Individual chromosomes are observed after 5 h. Bar is 5 μm.

2. Materials
2.1. Collecting Xenopus *Eggs*
1. Human chorionic gonadotropin.
2. 0.1 *M* NaCl.

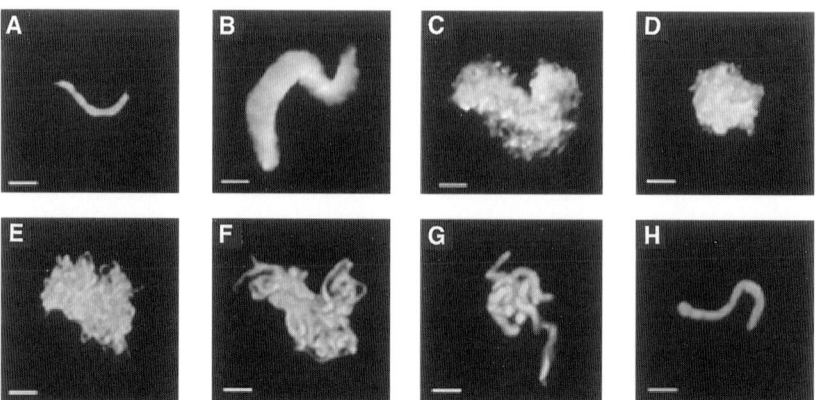

Fig. 3. Three dimensional reconstitution of confocal microscopy data on sperm chromosome assembly. Legend is as in **Fig. 1**. Bar is 5 μm.

2.2. Dejellying of Eggs

1. 2% Cysteine, pH 7.8.
2. MMR: 100 mM NaCl, 2 mM KCl, 1 mM MgSO$_4$, 2 mM CaCl$_2$, 0.1 mM EDTA, 5 mM HEPES, pH 7.8.
3. Stereoscopic microscope.

2.3. Extract Preparation

1. EB: 80 mM β-glycerophosphate (Sigma), pH 7.3, 15 mM MgCl$_2$, 20 mM EGTA, 1 mM DTT.
2. Aprotinin and leupeptin (Boehringer, Mannheim), 5 mg/mL in water.
3. SW41 rotor (Beckman Instruments) or equivalent.
4. Ultracentrifuge
5. Hitachi CS120 ultracentrifuge or Beckman TL-100 table-top ultracentrifuge.
6. TLS-55 (Beckman) swinging bucket rotor or equivalent.

2.4. Preparation of Demembranated Sperm Nuclei

1. Buffer T: 15 mM PIPES, 15 mM NaCl, 5 mM EDTA, 7 mM MgCl$_2$, 80 mM KCl, 0.2 M sucrose, pH 7.4.
2. Buffer R: Buffer T + 3% bovine serum albumin.
3. Buffer S: Buffer T + 20 mM maltose + 0.05% lysolecithin.
4. Fix/stain buffer: Hoechst 33258 at 1 μg/mL in 200 mM sucrose, 10 mM HEPES pH 7.5, 7.4% formaldehyde, 0.23% DAPCO (Sigma), 0.02% NaN$_3$, and 70% glycerol.
5. Clinical centrifuge.
6. Fluorescent microscope.

2.5. Chromosome Assembly Assay

1. Buffer EB (*see* **Subheading 2.3.**).
2. 0.2 M phosphocreatine in 10 mM potassium phosphate buffer, pH 7.0.

3. 0.2 M ATP, pH 7.0.
4. 0.5 mg/mL creatine phosphokinase in 10 mM HEPES, pH 7.5, 50% glycerol.
5. Fix/stain buffer (*see* **Subheading 2.4.**).
6. Fluorescent microscope.

2.6. Immunofluorescence

1. PBS: 125 mM NaCl, 2.7 mM KCl, 1.5 mM KH$_2$PO$_4$, 8.1 mM Na$_2$HPO$_4$, pH 7.5.
2. PFA/PBS: 4% paraformaldehyde in PBS.
3. Antibody buffer: PBS containing 0.1% Triton X-100 and 5% lamb serum (mycoplasma screened, Gibco-BRL).
4. Rinse buffer: PBS containing 0.1% Triton X-100 and 0.5% bovine serum albumin.
5. Fix/stain buffer (*see* **Subheading 2.4.**).
6. Fluorescent microscope.

2.7. Immunodepletion

1. Protein-A sepharose or equivalent.
2. PBS (*see* **Subheading 2.6.**).
3. Bovine serum albumin, 10 mg/mL in PBS.
4. Top-bench centrifuge.
5. Rotary shaker.

3. Methods
3.1. Obtaining and Collecting Xenopus Eggs

1. Induce maturation of oocytes by priming (injecting) female frogs with 100 U of human chorionic gonadotropin (1000 U/mL) 3–5 d before egg collection. Use a 1-mL syringe and a 25-gauge needle.
2. To induce mature egg laying, the evening prior to utilization of the eggs inject each frog (either subcutaneously in the dorsal lymph sac or in the leg frog muscle) with 700–800 U of human gonadotropin (5000 U/mL) (*see* **Note 1**).
3. Transfer each frog to a separate tank, containing about 3 L of 0.1 M NaCl, and allow it to lay eggs overnight (*see* **Note 2**).
4. The following morning, (14–16 h after injection) discard the excess 0.1 M NaCl and collect the eggs in a 75–100-mL glass beaker. Allow the eggs to sediment and carefully remove the aqueous solution by pipeting.

3.2. Dejellying of Eggs (see Note 3)

1. Carry out all operations of egg dejellying at 20–22°C
2. Prepare a fresh solution of 2% cysteine, pH 7.8.
3. To the 100-mL glass beaker containing eggs from one frog, (usually 10–15 mL) add 20–25 mL of 2% cysteine and gently swirl the eggs (swirling helps to remove the jelly coat).
4. Allow the eggs to stand for approx 30–45 s and repeat the swirling. After sedimenting the eggs, discard the solution above them, add again 25 mL of 2%

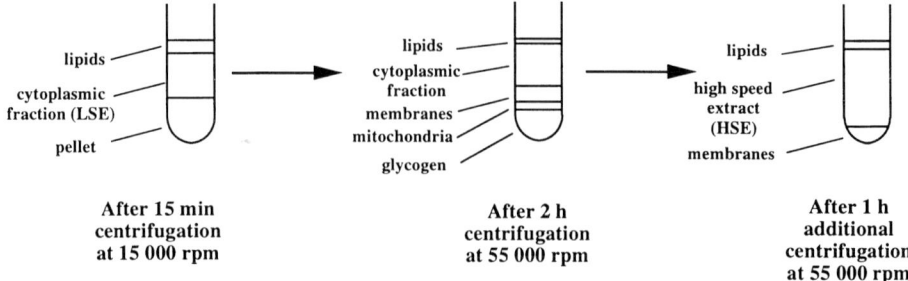

Fig. 4. Schematic illustration of low-speed extract (LSE) and high-speed extract (HSE) preparation.

cysteine, and gently swirl. Remove again the solution above the sedimented eggs and repeat the above operation few more times.
5. Wash eggs several times with 20–25 mL of 0.2 M MMR (see **Note 4**) and transfer to a petri dish. After examination of the eggs under a stereoscopic microscope, discard the bad eggs with the help of a Pasteur pipet (see **Note 5**).

3.3. Mitotic Extract Preparation

The procedure for isolation of mitotic *Xenopus* egg extract is based on the method of Hirano and Mitchison **(6)**.

1. Transfer all the eggs (from different frogs) to a single-glass beaker and rinse them several times with EB (20–22°C) under gentle swirling. Discard as much as possible of the EB.
2. Transfer the eggs to a 12-mL SW41 Beckman rotor centrifuge tube, wait a few minutes until they sediment, remove the EB over the eggs and pack them by centrifugation at 170g (approx 1000 rpm) in a clinical centrifuge.
3. Remove gently the excess EB with a 1-mL Pipetman or with a Pasteur pipet and finally with some absorbent paper (Whatman 3MM paper, for example).
4. Crush eggs by centrifugation for 20 min at 20,000g (15,000 rpm) at 4°C in an SW41 rotor (Beckman Instruments). At the end of centrifugation, three different layers are formed (from top to bottom): lipids, crude cytoplasmic fraction and pellet fraction (see **Fig. 4**).
5. Collect the cytoplasmic fraction slowly by side puncturing the tube with the help of a 1mL syringe and a 20-gage needle (see **Note 6**). Add to the cytoplasmic fraction the protein inhibitors apoprotinin and leupeptin at a concentration of 10 µg/mL.
6. Transfer the cytoplasmic fraction to 2-mL Eppendorf tubes and centrifuge for 15 min at maximal velocity in a bench-top centrifuge to remove the remaining contaminants. The supernatant, termed Low Speed Extract (LSE), can be used for chromosome reconstitution. However, such extracts contain membranes, glycogen, and ribosomes as well as other contaminants. Additionally, it is only stable for a short period of time. To isolate a long-term storage extract, one has to centrifuge at a very high speed (i.e., to isolate high speed extract [HSE]).

7. Transfer the crude extract to 2-mL propylene tubes used for a TLS-55 rotor (Beckman Instruments) and centrifuge for 2 h at 250,000g (52,000 rpm) (*see* **Note 7**). This procedure results in the formation of 5 different layers: lipids, cytoplasm, membranes, mitochondria, and glycogen together with ribosomes (*see* **Fig. 4**).
8. Suck off the lipid layer with the help of a vacuum very carefully and collect the cytoplasmic fraction (the high speed extract, *see* **Fig. 4**). Be careful to include as little as possible of the membrane fraction.
9. Recentrifuge the HSE at 250,000g (52,000 rpm, TLS-55 rotor) for 45 min to remove residual membranes and aliquot the membrane free HSE in 25-µL fractions. Freeze immediately in liquid nitrogen. HSE prepared in this manner maintains its capacity for chromosome assembly for at least 3 mo.

3.4. Preparation of Demembranated Sperm Nuclei

The procedure for preparation of sperm nuclei is based on a modification of the methods of Lohka and Masui *(13)* and Smythe and Newport *(14)*.

1. After anesthetizing a male frog by keeping it for 10 min in a bucket of ice-containing water, sacrifice the animal and open the peritoneal cavity (*see* **Note 8**).
2. Move the intestines carefully to one side and remove the testes by using a small scissors and forceps (the testes are white and oval in shape, each about 1 cm in length; they are located one on each side of the midline and are hard on touching) (*see* **Note 9**).
3. Place the testes in a Petri dish containing cold buffer T and rinse a few times in this buffer.
4. Transfer the testes to an absorbent paper and press them gently rolling them several times to get rid of blood and intestine contamination. To obtain a cleaner testes preparation, remove the fat body contaminants from the testes surface using a scissors.
5. Place the testes in a 20-mL glass beaker containing 2 mL of Buffer T and crush forcefully using a forceps. Centrifuge at 150g (800–1000 rpm in a clinical centrifuge) for approx 10 s.
6. Remove the supernatant and place it in a conical 15-mL Falcon tube. Add 2 additional milliliters of buffer T to the sperm debris (the pellet) and repeat **step 5**. Combine both supernatants in the 15-mL Falcon tube.
7. Centrifuge for 5 min at 1,500g (4000 rpm) and discard the supernatant. The pellet consists of two distinct regions: a white upper one (the sperm) and a lower one containing red blood cells. Remove as carefully as possible the sperm while at the same time avoiding contamination by the red blood cells.
8. Resuspend the sperm in 4 mL of buffer T and repeat **step 7**.
9. Repeat **step 8** two more times.
10. Resuspend the sperm in 100 µL of buffer T, add 300 µL of buffer S and incubate for 5 min at room temperature.
11. Place the sperm suspension on ice, remove a 1-µL aliquot, and check with a fluorescent microscope using fix/stain buffer if the membranes of the sperm nuclei are removed by the lysoleicithin treatment (for details on stain/fix buffer use, *see*

below). On membrane removal, all the sperm nuclei should be uniformly labeled by Hoechst. If this is not the case, incubate the sperm suspension for five additional minutes at 23°C and check again for demembranization. This operation must be repeated until complete demembranization is achieved (*see* **Note 10**).

12. Stop the reaction by adding 1.2 µL of buffer R. After centrifugation at 500*g* (2000 rpm) for 5 min, resuspend the pellet in 1.5 mL of buffer R and recentrifuge under the same conditions.
13. Resuspend the pellet in 50–100 µL buffer T and measure the sperm concentration with the help of a hemocytometer. Aliquot the demembraned sperm in 5 µL fractions and store at –80°C. Sperm can be stored in this form for several years.

3.5. Chromosome Assembly Assay

1. Take a tube with extract (25 µL) and thaw it rapidly. Add 25 µL of EB to the extract.
2. Supplement the diluted extract with components of the ATP regenerating system (20 m*M* phosphocreatine, 2 m*M* ATP, pH 7.0, and 5 µg/mL creatine kinase, final concentration)
3. Add to the diluted extract 1 µL of demembranated sperm (approx 4000 sperm nuclei) and incubate at 20–24 °C.
4. Deposit on a microscopic slide 5-µL aliquots at 10-, 30-, 60-, 90-, 120-, 150-, and 180-min intervals after initial incubation and dilute them immediately with an equal volume of fix/stain buffer.
5. Follow the chromosome assembly by conventional fluorescence microscopy (*see* **Note 11**).

3.6. Immunofluorescence Detection of Proteins During Chromosome Assembly

1. Assemble chromosomes as described in **Subheading 3.5.**
2. Remove aliquots of 10 µL at different times (10, 30, 60 90, 120, 150, and 180 min) after initiation of the assembly reaction and mix with 40 mL of 4% PFA/PBS in ice. Incubate for 15 min on slides in a moist chamber at room temperature.
3. Immerse all slides for 5 min in 50 mL of 4% PFA/PBS at room temperature.
4. Wash twice with PBS at 4°C for 5 min.
5. Incubate each slide for 1 h at 37°C with 200 µL of the specific antibody at a suitable dilution in antibody buffer.
6. Wash three times for 10 min at 4°C with PBS, 0.1% triton X-100, 0,5% BSA.
7. Incubate each slide for 30 min at 37°C with 200 µL of fluoresceine-conjugated goat anti-rabbit IgG (20 µg/mL).
8. Wash three times for 10 min at 4°C with PBS, 0.1% Triton X-100, 0,5% BSA.
9. Counterstain the chromosomes with 8 µL of Hoechst 33258 (per slide) in fix/stain buffer.

3.7. Immunodepletion (see Note 12)

The technique of chromosome assembly in *Xenopus* extracts facilitates elucidation of the role of specific proteins in both the formation and maintenance of mitotic chromosomes. This is achieved by immunodepletion and rescue experiments *(6,11,12)*. The immunodepletion protocol described below is based on the procedure of Dasso et al. *(11)* and Dimitrov et al. *(16)*.

1. Centrifuge 0.8–0.9 mL of protein-A Sepharose beads for 1 min at 1000–1500g on a bench top centrifuge.
2. Wash the pellet three times with 3.5 vol PBS, each time centrifuging as in **step 1**.
3. Block protein-A sepharose beads by two 15-min room-temperature incubations with 3.5 vol bovine serum albumin (10 mg/mL) in PBS. Perform the centrifugation between each blocking step as in **step 1**.
4. Wash the beads three times with 3.5 bead vol of PBS.
5. Add 1.5 mL affinity purified antibody (350 mg/mL) in PBS solution to 250 mL of the packed beads and incubate the mixture for 1 h at room temperature with gentle rotation (*see* **Note 13**). For preparation of mock depleted extract, repeat, using purified immunoglobulin G instead.
6. Wash three times both bead preparations as described in **step 2**.
7. Resuspend the beads in PBS and divide each bead preparation (one containing the antibody, the other containing purified immunoglobulin G) into two equal parts.
8. Pellet one part of each type of beads in a bench top centrifuge at 1500g, discard the supernatant and remove as much as possible of the residual PBS solution by pipeting with a 100-µL Pipetman and microsequencing loading tip (*see* **Note 14**).
9. Add four volumes (approx 400 µL) of extract to each of the "dried" bead preparations (approx 100 µL) and incubate at 4°C for 1.5 h with rotation.
10. Pellet the beads by centrifugation at maximum speed on a bench top centrifuge and remove the supernatants.
11. Repeat **step 8** for the two remaining bead preparations and add to each of them the respective supernatant from **step 10**.
12. Incubate as in **step 9** and centrifuge for 5 min at maximal speed on bench-top centrifuge. The supernatants are the depleted extracts.
13. Check depletion for both extracts by Western blotting, using 0.5, 2.5, and 10 µL of each extract for electrophoresis loading.

4. Notes

1. Sometimes upon injection of the frog in the leg, the site of injection may bleed. To stop the bleeding, press for 1–2 min to the site of injection.
2. It is very important to transfer each frog to an individual tank after injection, because some of the frogs may lay bad eggs. This will save you time and effort when sorting the eggs.
3. A criterium for good dejellying is the tight egg packaging: upon dejellying, the volume of the eggs decreases several times.
4. Egg washing with 0.2 M MMR has to be carried out quickly (for rapid removing of cysteine) and gently: dejellied eggs are fragile, have a tendency to stick not only to each other, but also to the glass and to lyse.
5. Good eggs are 1–1.2 mm in diameter and the vegetal yellow region is well separated from the dark animal. Abnormal eggs are usually larger, white-grey in color

or show nonuniform pigmentation. Egg discarding is performed with a pipet pump and a Pasteur pipet cut at the end to provide a diameter of approx 2 mm. The end of the pipet is passed over a flame to eliminate sharp glass ends obtained on pipet cutting. This is a necessary step in pipette preparation: otherwise sharp ends of the pipet may severely damage the eggs.

6. On removal of the crude cytoplasmic fraction, the syringe needle may sometimes become blocked by some egg debris. An alternative approach to collecting the cytoplasmic fraction, is to use a Pasteur pipet to pierce the lipid layer and to suck the cytoplasmic fraction. However, in this way the LSE will be contaminated with lipids and a second low speed 20,000g (15,000 rpm) centrifugation is recommended.
7. Occasinally, after 2 h centrifugation at 250,000g (52,000 rpm) a cloudy layer in the cytoplasmic quasi-transparent fraction may be observed. To resolve this problem, centrifuge for an additional hour.
8. We recommend the use of 2–3 frogs for sperm isolation: it is easier to work with greater numbers of testes and sperm yield is usually higher.
9. When removing sperm take care not to contaminate with red blood cells, otherwise you may have to repeat the centrifugation and purification procedure many times, which will lead to severe sperm losses.
10. Carefully check the sperm demembranisation by fluorescence microscopy: partially demembranated sperm does not assemble well in mitotic chromosomes.
11. We have observed that Propidium iodide (PI), which is usually used in confocal microscopy for chromosome visualization, is unable to counterstain chromosomes because of the presence of β-glycerophosphate in the egg extracts, which inhibits fluorescence of PI. Instead of PI, we successfully used YOYO-1 DNA fluorescent dye. The fluorescence excitation/emission maxima (in nanometers) of DNA bound YOYO-1 is 491/509. This dye is well-suited to excitation by the 488 nm line of the argon laser. This fluorescent DNA stain is very sensitive and must be conserved at –20°C.
12. The described protocol is for immunodepletion of 400 µL of extract. Depletion of a smaller amount of extract is not recommended, due to potential losses of extract during the immunodepletion procedure.
13. Incubation under rotation is essential for immunodepletion, since it allows good mixing of the extract and beads.
14. The complete removal of the PBS solution at **step 7** ("drying" of beads), is important: incomplete removal will dilute the extract and may destroy its assembly capacity.

References

1. Laemmli, U. K., Cheng, S. M., Adolph, K. W., Paulson, G. R., Brown, J. A., and Baumbach, W. R. (1977). Metaphase chromosome structure: the role of non-histone proteins. *Cold Spring Harbor Symp. Quant. Biol.* **42**, 351–360.
2. Gasser, S. M. and Laemmli, U. K. (1987) Improved methods for the isolation of individual and clustered mitotic chromosomes. *Exp. Cell Res.* **173**, 85–98.

3. Boy de la Tour, E. and Laemmli, U. K. (1988) The metaphase scaffold is helically folded: sister chromatids have predominantly opposite helical handedness. *Cell* **55,** 973–944.
4. Paulson, J. R. and Laemmli, U. K. (1977) The structure of histone-depleted metaphase chromosomes. *Cell* **12,** 817–828.
5. Mirkovitch, J., Mirault, M.-E., and Laemmli, U. K. (1984) Organization of the higher-order chromatin loop: specific DNA attachement sites on nuclear scaffold. *Cell* **29,** 223–232.
6. Hirano, T. and Mitchison, T. J. (1993) Topoisomerase II does not play a scaffolding role in the organization of mitotic chromosomes assembled in *Xenopus* egg extracts. *J. Cell Biol.* **120,** 601–612.
7. Earnshaw, W. C. (1991) Large scale chromosome structure and organization. *Curr. Topics Struct. Biol.* **1,** 237–244.
8. Hirano, T. and Mitchison, T. J. (1991) Cell cycle control of higher-order chromatin assembly around naked DNA in vitro. *J. Cell. Biol.* **115,** 1479–1489.
9. Hirano, T. anf Mitchison, T. J. (1994) A heterodimeric coiled-coil protein required for mitotic chromosome condensation in vitro. *Cell* **79,** 449–458.
10. Hirano, T., Kobayashi, R., and Hirano, M. (1997) Condensins, chromosome condensation protein complexes containing XCAP-C, XCAP-E and a *Xenopus* homolog of the *Drosophila* barren protein. *Cell* **89,** 511–521.
11. Dasso, M., Dimitrov, S. I., and Wolffe, A.P. (1994) "Nuclear assembly and replication is independent of linker histone" *Proc. Natl. Acad. Sci. USA* **91,** 12,477–12,481.
12. Ohsumi, K., Katagiri, C., and Kishimoto, T. (1993) Chromosome condensation in *Xenopus* mitotic extracts without histone H1. *Science* **262,** 2033–2035.
13. Lohka, M. and Masui, Y. (1983) Formation *in vitro* of sperm pronuclei and mitotic chromosomes induced by amphybian ooplasmic components. *Science* **220,** 719–721.
14. Smythe, C. and Newport, J.W. (1991) Systems for the study of nuclear assembly, DNA replication, and nuclear breakdown in *Xenopus* laevis egg extracts. *Methods Cell Biol.* **35,** 449–468.
15. Murray, A. W. (1991) Cell cycle extracts. *Methods Cell Biol.* **36,** 581–605.
16. Dimitrov, S. I., Dasso, M. C., and Wolffe, A. P. (1994) Remodeling sperm chromatin in *Xenopus* laevis egg extract: the role of core histone phosphorylation and linker histone B4 in chromatin assembly, *J. Cell Biol.* **126,** 591–601.

17

Nucleotide Excision Repair Coupled to Chromatin Assembly

Pierre-Henri Gaillard, Danièle Roche, and Geneviève Almouzni

1. Introduction

The packaging of DNA into a chromatin structure within the eukaryotic nucleus can affect processes such as DNA replication, transcription, recombination and repair. During nucleotide excision repair (NER), a major DNA repair pathway, rearrangements of the nucleosomal organisation are observed *(1)*. These rearrangements can be envisioned as the rapid succession of disassembly and reassembly events. A tight co-ordination between the actual DNA repair event and the chromatin assembly process will be critical to fully restore a functional genome. A step toward the dissection of these events was recently accomplished by the development of an assay for both chromatin assembly and NER on the same DNA molecules in cell-free systems competent for the two processes *(2,3)*. Both chromatin assembly and NER have been independently analysed in a variety of cell-free systems. Efficient chromatin assembly can be reproduced in crude extracts derived from *Xenopus* oocytes or eggs *(4–7)*, *Drosophila* embryos *(8–10)*, or from human cells *(11–14)*. Extracts competent for the NER process can be derived from a variety of cultured mammalian cells *(15,16)*, cultured *Drosophila* cells *(17)*, and *Xenopus* eggs *(18)*.

To follow in parallel NER and chromatin assembly, two criteria had to be fulfilled:

1. To have an extract competent for both processes, and
2. To have an assay to follow the two events at the same time on the same molecules.

We have taken advantage of the independent, powerful in vitro approaches for chromatin assembly and NER, and attempted a simultaneous analysis of

From: *Methods in Molecular Biology, Vol. 119: Chromatin Protocols*
Edited by: P. B. Becker © Humana Press Inc., Totowa, NJ

NER and chromatin assembly on the same damaged circular DNA template *(2,3)*. Both *Xenopus* high-speed egg extracts and *Drosophila* embryo extracts could support the reaction. Human cell extracts had to be complemented to allow DNA repair to occur concomitantly with chromatin assembly.

For a successful analysis, a DNA substrate containing defined lesions that can be repaired through NER is a prerequisite. DNA damaged with UVC is a convenient substrate, relatively easy to obtain at low cost. The amount of lesions induced by UVC can be estimated and controlled *(19)*. The two major DNA lesions induced by UVC (cyclobutane pyrimidine dimers and 6-4 photoproducts) are both repaired through the NER pathway. A minor type of lesion, pyrimidine hydrates, repaired through the base excision repair pathway, may be eliminated through additional treatment of the material (*see* **Note 5**, **ref. *19***). Another key aspect is to work with a DNA preparation in which the presence of nicks is limited, because DNA synthesis from these sites generates background signal in the repair assay. Alternatively, DNA substrates containing a unique defined lesion at a specific site in a circular DNA duplex can be prepared *(20)*. These latter substrates are more appropriate to examine the repair process at a single site *(3)*. Furthermore, the presence of lesions on a closed circular molecule offers the advantage of exploiting the changes of its topological properties accompanying chromatin assembly for analysis in a supercoiling assay *(21)*.

Our experimental strategy of superimposing NER and chromatin assembly assays is recapitulated in **Fig. 1**. In these assays, it is essential to contrast the data obtained on the total population of molecules with those obtained on the labeled material. This reveals any preferential assembly on the repaired material, possibly reflecting a link between repair and chromatin assembly.

The protocol described in this chapter corresponds to our standard complementary assays that we designate, respectively, "repair synthesis and supercoiling assay" and "repair synthesis and MNase assay" using *Xenopus* HSE. Examples of experimental data obtained with both assays are shown in **Figs. 2** and **3**, respectively. Our optimised reaction conditions for NER and chromatin

---

Fig. 1. A convenient substrate in a combined NER/Chromatin assembly assay consists of a closed circular plasmid DNA irradiated with UVC. This is incubated in a soluble cell extract in the presence of a radiolabelled nucleotide whose incorporation during the DNA synthesis step of the repair process allows to estimate the NER efficiency. Chromatin assembly can be monitored by two simple complementary assays (i) a supercoiling assay and (ii) a MNase assay. The supercoiling assay makes use of the topological properties of closed circular DNA molecules. These circular molecules, when incubated in a chromatin assembly extract, undergo conformational changes corresponding to the progressive deposition of nucleosomes while a topoisomerase activ-

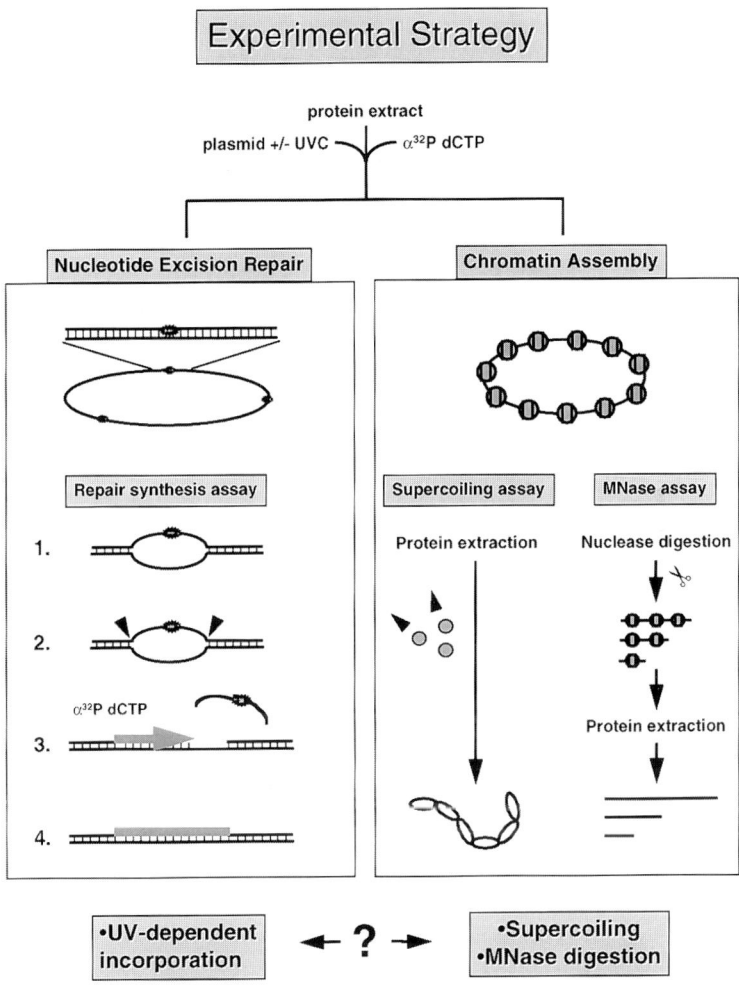

ity allows the absorption of the constraints *(21)*. After deproteinisation, topoisomers with an increasing number of negative supercoils reflect the extent of assembly. Resolution and detection of topoisomers is achieved using gel electrophoresis. The extent of chromatin assembly on the labelled (repaired) molecules and the total population of molecules is then compared (*see* **Fig. 2**). The MNase assay for chromatin assembly makes use of a nuclease, usually microccocal nuclease, to cleave the most accessible regions in a chromatinized DNA. Cleavage occurs preferentially in the linker region between adjacent core nucleosomes generating digestion products whose sizes are multiple of the basic nucleosomal unit. The corresponding DNA fragments, when analysed by gel electrophoresis, give rise to a characteristic profile or « nucleosomal ladder ». The regularity of the pattern and the spacing between adjacent bands provides information on the quality of the reconstituted chromatin on the labelled (repaired) molecules and the total population of molecules (*see* **Fig. 3**).

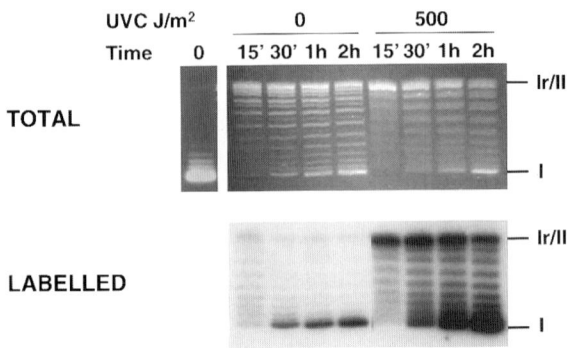

Fig. 2. Reaction was carried out for 15 min, 30 min, 1 h, and 2 h with nontreated (0) or UVC treated DNA (500 J/m2) and samples were processed for gel analysis as indicated in **Subheading 3.1.** "Total": DNA as visualised after ethidium bromide staining. "Labeled": DNA as visualised by autoradiography of the dried gel. Migration position corresponding to relaxed DNA (Ir), nicked DNA (II), and supercoiled DNA (I) are indicated. The slow migrating form or relaxed DNA (Ir) at early time ("Total," 15 min) arises from the action of topoisomerases in the extract that relax the initially negatively supercoiled plasmid ("Total," 0 min). Progressive supercoiling is detected with the appearance of topoisomer intermediates and an accumulation of fast migrating forms corresponding to the position of fully supercoiled DNA (I). UV-specific incorporation is observed on the UV-treated plasmid. A low level of background incorporation on the nontreated plasmid is due to the presence of residual nicks in the plasmid preparation or to non-specific nicking activities ("Labeled"). The time course analysis shows a slow migrating form or relaxed DNA (Ir) at early time ("Labeled," 15 min) and a progressive supercoiling is then observed with the appearance of topoisomer intermediates and an accumulation of fast migrating forms corresponding to the position of fully supercoiled DNA (I). *See* Note on the labeled material, the high proportion of supercoiled molecules indicative of a preferential assembly of repaired molecules (*see* **Note 10**).

assembly are presented and critical parameters discussed. The incubation of the UVC-treated plasmid in cell-free extracts is always paralleled with a control reaction using nontreated plasmids. This can be adapted to the use of other biological systems. Because detailed description for the preparation of the cell-free extracts (*see* Chapters 12 and 13) can be found elsewhere, our focus will be on the presentation of the novel assays.

2. Materials
2.1. Preparation and UV Treatment of DNA Templates

1. A 500 mL culture of bacteria (*E. coli* strain DH5α from BRL) transformed with pBluescript KS+ (Stratagene) grown for 16 h in Luria broth media.

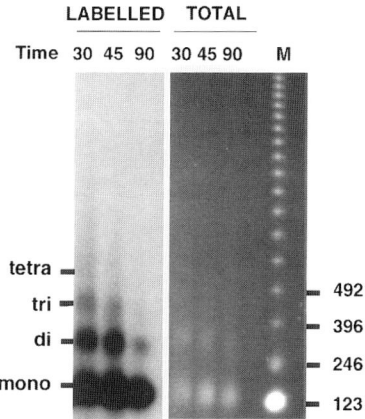

Fig. 3. Reaction was carried out with a DNA substrate irradiated at 500 J/m^2 for 4 h, and digestion with MNase was as described in **Subheading 3.2**. Aliquots were withdrawn at the indicated times (30, 45, and 90 s.) and processed for gel analysis as described in **Subheading 3.2**. "Total": DNA as visualized after ethidium bromide staining. "Labeled": DNA as visualized by autoradiography of the dried gel. Molecular weight marker (123 bp from BRL) was run in parallel. Position of DNA fragments corresponding to mono-, di-, and trinucleosomal forms are indicated. The repetition unit is about 180 bp.

2. Qiagen plasmid purification Mega-kit.
3. TE pH 8.0:10 mM Tris-HCl, pH 8.0, 1 mM EDTA pH 8.0.
4. Spectrophotometer: Ultrospec 2000 (Pharmacia, Biotech).
5. 10X TBE concentrated stock solution: 108 g Tris base, 55 g boric acid, 40 mL 0.5 M EDTA pH 8.0, dissolved in deionized water to a final volume of 1 L.
6. Agarose (Ultra Pure, Sigma) for a minigel: 0.8% agarose 0.25 µg/mL ethidium bromide, running buffer: 1X TBE, 0.25 µg/mL ethidium bromide.
7. Germicidal lamp with a 254 nm peak (Philips).
8. Latarjet Dosimeter.

2.2. Repair Synthesis and Supercoiling Assay

1. Nontreated and UVC-irradiated pBS plasmid at 50 µg/mL.
2. *Xenopus* high-speed egg extract (HSE, *see* **Note 1**).
3. 5X Reaction buffer: 25 mM MgCl$_2$, 200 mM Hepes-KOH, pH 7.8, 2.5 mM dithiothreitol (DTT; Sigma), 200 mM phosphocreatine (di-Tris salt, Sigma). The 5X Reaction buffer can be prepared in advance, aliquoted, and stored at –80°C. Each aliquot is used only once (*see* **Note 2**).
4. 100 mM ATP (Pharmacia) aliquoted and stored at –80°C. Each aliquot is used only once.
5. Creatine phosphokinase (Boehringer) 2.5 mg/mL in H$_2$O aliquoted and stored at –80°C. Each aliquot is used only once.
6. α ^{32}P dCTP, 3000 Ci/mmol (Redivue, Amersham).

7. Stop Mix 1: 30 mM EDTA, 0.7 % SDS.
8. RNase A (Boehringer) 10 mg/mL in 0.01 M sodium acetate pH 5.2, heat for 15 min at 100°C and adjust the pH with 0.1 vol of Tris-HCl pH 7.4. Aliquot and store at –20°C.
9. Proteinase K, 4 mg/mL (Boehringer), aliquots stored at –20°C.
10. Phenol:chloroform:isoamyl alcohol (25:24:1) (Bioprobe).
11. 5 M Ammonium acetate.
12. 100% Ethanol stored at –20°C.
13. 70% Ethanol stored at –20°C.
14. 5X Loading buffer: 0.42% bromophenol blue, 50% glycerol.
15. 50X TAE: 242 g Tris base, 57.1 mL glacial acetic acid, 100 mL 0.5 M EDTA, pH 8.0 diluted in deionized water to a final volume of 1 L.
16. Agarose (Ultra pure, Sigma) for a large gel: 1.2% agarose, Running buffer: 1X TAE.
17. Ethidium bromide (Bio-Rad) 1 µg/mL in distilled water.
18. Intensifying screens for X-ray films.
19. Amersham Hyperfilm MP.
20. Nylon membrane (Hybond N$^+$, Amersham).
21. Vacuum transfer apparatus (Appligene).
22. 0.25 N HCl.
23: Transfer buffer: 0.4 N NaOH ,1.5 M NaCl.
24. Random priming labeling kit (Amersham).
25. Hybridization buffer tablets (Amersham), 2X SCC/0.1% SDS, 1X SSC/0.1% SDS, 0.5X SSC/0.1% SDS.

2.3. Repair Synthesis and Microccocal Nuclease Digestion Assay

Items 1–6 as in **Subheading 2.2.**

7. MNase Digestion buffer: 10 mM HEPES-KOH, pH 7.6, 50 mM KCl, 1.5 mM MgCl$_2$, 0.5 mM EDTA, 10% glycerol, 10 mM β-glycerophosphate, 1 mM DTT.
8. 100 mM CaCl$_2$.
9. Microccocal nuclease (MNase, Boehringer) solution in water at 15 U/µL aliquoted and stored at –20°C.
10. Stop mix 2: 2.5% N. Lauroyl Sarcosine, 100 mM EDTA.
11. RNaseA, DNase free, 2 mg/mL (Boehringer).
12. Proteinase K (Boehringer), 10 mg/mL.
13. 20% SDS.
14. Glycogen (Boehringer), 20 mg/mL.
15. 3 M Sodium acetate, pH 5.2.
16. 100% Ethanol stored at –20°C.
17. 70% Ethanol stored at –20°C.
18. 10X TBE as in **Subheading 2.1.**
19. Agarose (Ultra Pure, Sigma): a large 1.3% agarose gel in TBE buffer; running buffer. 1X TBE.
20. Molecular-weight marker 123 bp ladder (BRL).
21. Loading buffer for MNase digestion samples: 50% glycerol.

22. Loading buffer for 123 bp ladder: *see* **Subheading 2.2.**, **item 14**.
23. *See* **Subheading 2.2.**, **items 17–20**.

3. Methods

3.1. Preparation and UV Treatment of DNA Templates

1. Purify pBS plasmid from bacteria using a Qiagen plasmid purification kit. Resuspend the collected plasmid in TE pH 8.0 and determine the DNA concentration by measuring the optical density at 260 nm (OD_{260} of $1 \approx 50$ µg/mL). Up to 2 mg of supercoiled plasmid can be obtained from a 500-mL culture. While keeping the bulk of the plasmid preparation on ice, briefly check its quality by running a 200-ng aliquot on a 0.8% agarose minigel containing 0.25 µg/mL of ethidium bromide. If nicked form (migrates above the supercoiled form) is detected it is necessary to further purify the supercoiled form through 1 or 2 sucrose gradients as described *(19)*, otherwise directly aliquot the plasmid preparation in 50–100-µL aliquots at 50–100 µg/mL and store at –80°C (*see* **Note 3**).
2. Adjust the distance between the germicidal lamp (254 nm) and the sample to get the desired fluence rate measured with a Latarjet dosimeter. When adjusting the UV-fluence rate make sure that the photosensitive cell is placed at the exactly same position than that of the sample to irradiate. As previously recommended *(19)* it is important to use relatively low fluence rates to allow DNA to mix by diffusion. The height of the lamp is generally adjusted to get a fluence rate of 0.5 $J/m^2/s$. The time of irradiation is determined according to the final dose to apply to the plasmid (ex. irradiate 16 min 40 s for a 500 J/m^2 dose).
3. For the UV treatment, load 50-µL drops of a 50 µg/mL solution of supercoiled plasmid DNA on a piece of parafilm placed on ice or inside an inverted top of a Petri dish placed on ice.
4. Place under the germicidal lamp and irradiate for the appropriate time.
5. Irradiated plasmid is then either aliquoted in 50–100-µL aliquots and stored back at –80°C or is further processed to eliminate pyrimidine hydrates (*see* **Note 4**).

3.2. Repair Synthesis and Supercoiling Assay

1. Set up a 25-µL standard reaction with 300 ng of plasmid, 200–400 µg of protein extract, 5 m*M* $MgCl_2$, 40 m*M* HEPES-KOH, pH 7.8, 0.5 m*M* DTT, 40 m*M* phosphocreatine, 4 m*M* ATP, 2.5 µg creatine phosphokinase, 5 µCi α^{32}PdCTP by adding the following reagents to an Eppendorf tube on ice, in the indicated order (*see* **Note 5**):

    5 µL 5X reaction buffer,
    1 µL ATP,
    1 µL creatine kinase,
    1.5 µL H_2O,
    0.5 µL α^{32}PdCTP,
    10 µL HSE

    homogenize by pipeting up and down several times (never vortex) when adding the extract.

2. To start the reaction, add 6 μL of nontreated or UVC-irradiated plasmid (300 ng), briefly mix by gently pipeting up and down (never vortex), and quickly transfer to a thermostated water bath at 23°C. Incubate for the desired time (*see* **Note 6**).
3. Stop the reaction by adding 25 μL stop mix 1.
4. Add 50 μL H_2O to increase the volume and facilitate extraction procedure.
5. Add 5 μL RNase A (at 2 mg/mL), incubate 30 min at 37°C.
6. Add 5 μL proteinase K (at 4 mg/mL), incubate 30 min at 37°C.
7. Add 110 μL phenol:chloroform:isoamyl alcohol and vortex 10 s minimum each tube. Centrifuge 10 min at 15,000g at room temperature and collect 95 μL of the aqueous upper phase and transfer into a clean tube. Be careful not to take up any of the interphase.
8. Precipitate DNA with 1 vol of ammonium acetate (95 μL) and 2 vol (380 μL) of cold 100% ethanol.
9. Centrifuge 30–45 min at 15,000g at 4°C to collect DNA pellet, wash with 800 μL of cold 70% ethanol, centrifuge 5–10 min at 15,000g at 4°C and carefully remove the supernatant, dry the pellet in the Speed Vac and resuspend in 16 μL H_2O.
10. Add 4 μL of 5X Loading buffer and load on a 1.2% agarose gel in 1X TAE without ethidium bromide (*see* **Note 7**) and run 20 h at 1.5 V/cm (*see* **Note 8**).
11. Soak the gel in the DNA staining solution, rinse for 30 min in water at room temperature, and place it on a UV transilluminator equipped with a camera system to obtain an image for analysis of the migration pattern of the total DNA loaded on the gel. If the intensity of the signal corresponding to the DNA in the gel is sufficient for a good picture continue with **step 12a**, otherwise with **step 12b**.
12a. Dry the gel on Whatman 3MM paper for 2 h at 80°C in a vacuum gel dryer (Bio-Rad) and continue with **step 13** only.
12b. Transfer the DNA onto a nylon membrane and continue with **steps 13–15**.
13. Expose either the dried gel or the membrane (before probing) against an X-ray film for autoradiography to visualize the migration pattern of the radiolabelled DNA (= repaired DNA). Use two intensifying screens and leave at –80°C for at least 4 h. Exposition time may vary according to the repair efficiency of the extract (*see* **Note 9**). The higher proportion of supercoiled molecules in the labelled molecules is indicative of a preferential assembly of repaired molecules (*see* **Note 10**). For common troubleshooting *see* **Note 14**).
14. Prepare a specific ^{32}P radiolabeled probe using purified plasmids as a template in a random priming reaction using a random priming labeling kit (Amersham), and probe the membrane following standard procedures as recommended by manufacturer (*see* **Note 11**).
15. Expose the probed membrane against an X-ray film. Since the radiolabeled probe presents a high specific activity, the time of exposure is reduced to 1 h at room temperature without intensifying screens.

3.3. Repair Synthesis and Microccocal Nuclease Digestion Assay

1. Set up a repair and chromatin assembly reaction as described in **Subheading 3.2.1.**, scaled up threefold to 75 μL (*see* **Note 12**).

2. Incubate the reaction at 23°C for the desired time (4 h are sufficient for both repair and chromatin assembly reactions to be completed).
3. During incubation prepare three stop-tubes containing 30 µL stop mix 2 and prepare a digestion mix containing 200 µL MNase digestion buffer, 10 µL $CaCl_2$ 100 mM, and 3 µL MNase at 15 U/µL.
4. Add to the 75 µL repair and chromatin assembly reaction at the desired time, the 213 µL digestion mix and incubate at room temperature.
5. Remove 90-µL aliquots at 30, 45, and 90 s. Immediately transfer into the corresponding stop-tube containing the 30 µL of stop mix 2.
6. Add 10 µL RNaseA at 2 mg/mL and incubate 30 min at 37°C.
7. Add 5 µL proteinase K at 10 mg/mL and 3.5 µL 20% SDS and incubate 15 h at 37°C.
8. Precipitate the DNA by adding 1 µL glycogen at 20 mg/mL, 15.5 µL sodium acetate 3M, 380 µL 100% ethanol. Vortex and store at –80°C for 30 min.
9. Centrifuge 30 min at 15,000g at 4°C, wash the pellet with 800 µL 70% ethanol. centrifuge 10 min at 15,000g at 4°C before removing the ethanol. Repeat twice. Dry the DNA pellet in the Speed Vac and resuspend in 5 µL H_2O. Add 2 µL of glycerol 50% as loading buffer (*see* **Note 13**).
10. Load on a 1.3% agarose gel in 1X TBE buffer and run at 5 V/cm for 3 h. Load in parallel a size marker using 5X loading buffer. The bromophenol blue in the 5X loading buffer serves as a convenient migration control. Electrophoresis is stopped when the dye is at approx 2 cm from the bottom of the gel.
11. Continue as described in **steps 11–15** in **Subheading 3.2.** (*see* **Note 14**).

4. Notes

1. For preparation of various type of extracts to use in this assay *see* for *Xenopus* HSE Chapters 12,15 for *Drosophila* preblastoderm embryo extracts *(13)*, human whole cell *(19)*, and cytosolic cell extracts *(11,24)*. *Xenopus* HSE and *Drosophila* embryo extracts are crude extracts. Their protein concentration is estimated by a Bradford assay (Bio-Rad) and ranges between 20 mg and 40 mg/mL. They contain an important pool of endogenous dNTPs. The dNTP concentration can be determined by isotopic dilution as previously described *(25)*. This is important to take into account to determine the specific activity of the radiolabelled dNTP in the reaction and to be able to estimate the efficiency of the repair process. The dNTP contribution of *Xenopus* HSE and *Drosophila* embryo extracts is sufficient for the in vitro repair reaction and it is not necessary to add any exogenous dNTPs. The average value for dCTP concentration in the HSE is 50 µM. The salt concentration in the extract can be estimated in equivalent KCl by conductivity. The salt condition under which the reaction is carried out can thus be adjusted. Generally final salt concentrations ranging between 35 and 70 mM equivalent KCl are suitable for both NER and chromatin assembly in both *Xenopus* HSE and *Drosophila* embryo extracts.

    Human cell extracts are dialysed. Thus, the salt concentration of the extract corresponds to the dialysis buffer and exogenous dNTPs need to be added to the reaction mix: 100 µM except for the deoxynucleotide corresponding to the label which is reduced to 40 µM in the 5X reaction buffer.

All extracts are stored in small 50–100-µL aliquots at –80°C to avoid multiple freeze-thawing cycles. An aliquot is generally refrozen only once. Aliquots are frozen by quick freeze in liquid nitrogen before being stored at –80°C.

2. Thawing and refreezing 5X reaction buffer can alter DTT and phosphocreatine concentrations as well as dNTPs when working with human cell extracts.
3. Avoid repetitive thawing–refreezing of the plasmid preparation which may lead to single strand breaks (nicked form) and will result in increased background incorporation which will reduce the apparent UV-specific incorporation.
4. Plasmids carrying pyrimidine hydrates can be eliminated by digesting the irradiated plasmids with *E. coli* Nth protein (prepared by the method of Asahara et al. *[26]*) which cleaves DNA specifically at pyrimidine hydrate sites. Plasmids which do not carry any pyrimidine hydrates remain uncut by the enzyme and are purified as supercoiled form through two sucrose gradients as described *(19)*.
5. Experiments with several different reactions can be carried out. In that case, it is convenient to set up a reaction premix containing the elements common to the various reactions (i.e., 5X reaction buffer, ATP, creatine kinase, radiolabel, and, in some cases, the water) and the extract. It is important to determine for each new extract the ideal reaction conditions by eventually adjusting the protein/DNA ratio, the ionic conditions and the specific activity, to optimise both chromatin assembly and DNA repair.
6. Both repair and chromatin assembly are generally completed after 4 h. Specific UV-dependent incorporation (repair) can be detected already after 15 min.
7. It is absolutely essential to run the gel in the absence of ethidium bromide. The presence of this intercalating agent can induce supercoiling of relaxed molecules during the run and confuse interpretation.
8. The appropriate agarose concentration of the gel is determined according to the size of the plasmid to analyse. For 3- to 4-kb plasmids use 1.2% agarose gels and 0.8% agarose gels for 6- to 7-kb plasmids. Resolution of the topoisomers is optimized by running the gel at about 1.5 V/cm for 20 h. It is important to avoid overheating of the gel. Topoisomers resolution can be further improved by maintaining a homogenous distribution of the electrolytes by recircularizing the running buffer. The absolute number of superhelical turns, corresponding to each assembled nucleosome, can be determined by visualization of the plasmid topoisomers on a two dimentional (2D) agarose gel *(27)*. In addition, closed circular and open circular (nicked) plasmids can only be separated in a 2D gel. In some cases, it can be informative to determine the amount of nicked plasmids, as these may correspond to plasmids assembled into chromatin in which there are remaining nicks due to incomplete repair events *(2)*. 2D gels are set up as classical 1D gels but after migration in the 1st dimension, the gel is rotated by 90° and chloroquine added at 10 µg/mL to the electrophoresis buffer. The gel is then left to equilibrate for 45 min in the dark. The second dimension electrophoresis is then run in the dark under the same conditions than for the first dimension run. During this second run it is important to recirculate the running buffer to maintain an even distribution of chloroquine.

9. The repair efficiency of an extract can be determined by previously described methods *(19)*. Bands corresponding to the DNA substrates can be excised from the dried gel, and the incorporated radioactivity directly measured by Cerenkov counting. The incorporation of dCMP during the repair synthesis reaction is calculated taking into account the specific activity of the label and the background incorporation in the nontreated control DNA. Alternatively, the amount of incorporated radioactivity can be estimated by densitometric measures using a phosphorimager.
10. Quantitation of the distribution of the topoisomers can be carried out on both the Labeled DNA and the Total DNA by densitometric scanning of the images.
11. Probes with a specific activity up to 1.9×10^9 dpm/µg can be obtained using a Megaprime™ DNA labelling system (Amersham) using α^{32}P dCTP as label.
12. Reactions are scaled up several fold to allow one to carry out several types of digestion on the DNA of a same repair and chromatin assembly reaction. The general principle is to add MNase directly to the reaction mix at the end of the repair and chromatin assembly reaction. Volumes equivalent to roughly 300 ng of DNA are then withdrawn at appropriate times and immediately transferred into a tube containing a stop mix to arrest the MNase action. Here, we describe the setting up of reactions scaled up threefold to allow to digest DNA for 15, 45, and 90 s.
13. It is important not to use a loading buffer containing bromophenol blue as this dye will migrate at about the same position than the mononucleosomal DNA and will interfere with the analysis of the migration pattern. However, as an alternative to 50% glycerol alone, it is possible to include in the loading buffer 5 m*M* EDTA, 3 mg/mL orange G (Sigma).
14. Troubleshooting:
    a. Low repair synthesis signals. The extracts may not be efficient. Check protein concentration. Check also the specific activity taking into account the dNTP contribution of the extract. Increase if necessary the radiolabel input. Adjust the ionic conditions in the reaction considering the salt contribution of the extract. Check the energy regenerating system and the quality of the ATP used for the reaction.
    b. Low DNA recovery. For some extract the standard extraction procedure is not appropriate. It may be necessary to add an interphase wash step after phenol extraction with half volume of TE.
    c. No supercoiling. As for the repair synthesis this can be due to a low activity of the extract or to a low protein concentration. Adjust the ionic conditions in the reaction taking into account the salt contribution of the extract. Adjust the protein/DNA ratio.

    Nicking activities may have affected the reaction products during the extraction procedure. Stop mix and phenol:chloroform:isoamyl alcohol should be freshly prepared, and DNA should be handled with care and loaded rapidly on the gel.
    d. Nucleosomal ladders with smeary bands. This can be a result of the remaining RNA or proteins or traces of salt that have not been eliminated correctly in the

samples loaded on the gel. Check also the energy regenerating system and the quality of the ATP used for the reaction.

Acknowlegments

We thank J. G. Moggs, J.-P. Quivy, and S. Holmes for critical comments. This work was supported by an ATIPE grant from the CNRS.

References

1. Friedberg, E. C., Walker, G. C., and Siede, W. (1995) ASM Press, Washington, DC, pp. 291–294.
2. Gaillard, P.-H. L., Martini, E. M. D., Kaufman, P. D., Stillman, B., Moustacchi, E., and Almouzni, G. (1996) Chromatin assembly coupled to DNA repair: a new role for chromatin assembly factor-I. *Cell* **86,** 887–896.
3. Gaillard, P.-H. L., Moggs, J. G., Roche, D. M. J., Quivy, J.-P., Becker, P. B., Wood, R. D., and Almouzni, G. (1997) Initiation and bidirectional propagation of chromatin assembly from a target site for nucleotide excision repair. *EMBO J.* **16,** 6281–6289.
4. Laskey, R. A., Mills, A. D., and Morris, N. R. (1977) Assembly of SV40 chromatin in a cell free system from *Xenopus* eggs. *Cell* **10,** 237–243.
5. Glikin, G. C., Ruberti, I., and Worcel, A. (1984) Chromatin assembly in *Xenopus* oocytes: *in vitro* studies. *Cell* **37,** 33–41.
6. Almouzni, G. and Méchali, M. (1988) Assembly of spaced chromatin involvement of ATP and DNA topoisomerase activity. *EMBO J.* **7,** 4355–4365.
7. Almouzni, G. and Méchali, M. (1988) Assembly of spaced Chromatin promoted by DNA synthesis in extracts from *Xenopus* eggs. *EMBO J.* **7,** 665–672.
8. Nelson, T., Hsieh, T.-S., and Brutlag, D. (1979) Extracts of Drosophila embryos mediate chromatin assembly in vitro. *Proc. Natl. Acad. Sci. USA* **76,** 5510–5514.
9. Becker, P. B. and Wu, C. (1992) Cell-free system for assembly of transcriptionally repressed chromatin from *Drosophila* embryos. *Mol. Cell. Biol.* **12,** 2241–2249.
10. Kamakaka, R. T., Bulger, M., and Kadonaga, J. T. (1993) Potentiation of RNA polymerase II transcription by Gal4-VP16 during but not after DNA replication and chromatin assembly. *Genes Dev.* **7,** 1779–1795.
11. Stillman, B. (1986) Chromatin assembly during SV40 DNA replication in vitro. *Cell* **45,** 555–565.
12. Banerjee, S. and Cantor, C. R. (1990) Nucleosome assembly of simian virus 40 DNA in a mammalian cell extract. *Mol. Cell. Biol.* **10,** 2863–2873.
13. Gruss, C., Gutierrez, C., Burhnans, W. C., DePamphilis, M. L., Koller, T., and Sogo, J. M. (1990) Nucleosome assembly in mammalian cell extracts before and after DNA replication. *EMBO J.* **9,** 2911–2922.
14. Krude, T. and Knippers, R. (1991) Transfer of nucleosomes from parental to replicated chromatin. *Mol. Cell. Biol.* **11,** 6257–6267.
15. Wood, R. D. (1996) DNA repair in eukaryotes. *Annu. Rev. Biochem.* **65,** 135–167.
16. Sancar, A. (1996) DNA excision repair. *Annu. Rev. Biochem.* **65,** 43–81.

17. Shimamoto, T., Tanimura, T., Yoneda, Y., Kobayakawa, Y., Sugasawa, K., Hanaoka, F., Oka, M., Okada, Y., Tanaka, K., and Kohno, K. (1995) Expression and functional analyses of the *DXPA* gene, the *Drosophila* homolog of the human excision repair gene *XPA*. *J. Biol. Chem.* **270,** 22,452–22,459.
18. Shivji, M. K. K., Grey, S. J., Strausfeld, U. P., Wood, R. D., and Blow, J. J. (1994) Cip1 inhibits DNA replication but not PCNA-dependent nucleotide excision repair. *Curr. Biol.* **4,** 1062–1068.
19. Wood, R. D., Biggerstaff, M., and Shivji, M. K. K. (1995) Detection and measurement of nucleotide excision repair synthesis by mammalian cell extracts *in vitro*. *Methods: A Companion to Methods Enzymol.* **7,** 163–175.
20. Moggs, J. G., Yarema, K. J., Essigmann, J. M., and Wood, R. D. (1996) Analysis of incision sites produced by human cell extracts and purified proteins during nucleotide excision repair of a 1,3-intrastrand d(GpTpG)-cisplatin adduct. *J. Biol. Chem.* **271,** 7177–7186.
21. Germond, J. E., Rouvière-Yaniv, J., Yaniv, M., and Brutlag, D. (1979) Nicking-closing enzyme assembles nucleosome-like structures in vitro. *Proc. Natl. Acad. Sci. USA* **76,** 3779–3783.
22. Almouzni, G. (in press) Assembly of chromatin and nuclear structures in *Xenopus* egg extracts, in *Chromatin: A Practical Approach, IRL Practical Approach Series*, ed.H.Gould, Oxford Univ. Press, in press.
23. Blank, A., Sandaltzopoulos, R., and Becker, P. B. (1997) Biochemical analysis of chromatin structure and function using *Drosophila* extracts. *Methods: A Companion to Methods Enzymol.* **12,** 28–35.
24. Li, J. J. and Kelly, T. J. (1984) Simian virus 40 DNA replication in vitro. *Proc. Natl. Acad. Sci. USA* **81,** 6973–6977.
25. Blow, J. J. and Laskey, R. A. (1986) Initiation of DNA replication in nuclei and purified DNA by cell-free extracts of *Xenopus* eggs. *Cell* **47,** 577–587.
26. Asahara, H., Wistort, P. M., Bank, J. F., Bakerian, R. H., and Cunningham, R. P. (1989) Purification and characterization of *Escherichia coli* endonuclease III from the cloned nth gene. *Biochemistry* **28,** 4444–4449.
27. Peck, L. J. and Wang, J. C. (1985) Transcriptional block caused by a negative supercoiling induced structural change in an alternating CG sequence. *Cell* **40,** 129–137.

18

Photolyase

A Molecular Tool to Characterize Chromatin Structure in Yeast

Magdalena Livingstone-Zatchej, Bernhard Suter, and Fritz Thoma

1. Introduction

Folding of DNA into nucleosomes and higher order chromatin structures restricts its accessibility to proteins and drugs. Hence, the location of histone octamers on the DNA sequence (nucleosome positions) as well as structural and dynamic properties of nucleosomes may play important roles in gene regulation, replication and DNA repair. Conventional approaches to characterize chromatin structure include (partial) purification of chromatin and characterization of DNA accessibility to nucleases (micrococcal nuclease, DNaseI) and chemical cleavage reagents (hydroxyl radicals, methidium propyl-EDTA-iron, copper phenanthroline). The cleavage sites are monitored using low- and high-resolution footprinting protocols. These techniques, however, expose the problem that chromatin extraction procedures could alter chromatin composition and structure, including nucleosome positioning. To investigate chromatin structures in vivo, alternative approaches are applied, such as expression of prokaryotic methyltransferases in *Saccharomyces cerevisiae*, the genome of which contains no endogenous detectable methylation *(1,2)*. The sites of methylation can be measured after DNA isolation using methylation-sensitive restriction enzymes. This approach, however, requires expression of a foreign gene, and the resolution is restricted because of the sequence specificity of the methyltransferases.

Here, we present a new approach that makes use of a natural DNA repair enzyme in yeast, DNA photolyase. Cyclobutane pyrimidine dimers (CPDs) and 6-4 photoproducts (6-4PD) are the two major classes of stable DNA lesions generated by ultraviolet light (UV; 254 nm). Pyrimidine dimers are removed by two pathways, nucleotide excision repair (NER) and photoreactivation. NER is a ubiquitous multistep pathway in which more than 30 proteins are involved that sequentially execute damage recognition, excision of an oligonucleotide with the pyrimidine dimer, and gap repair synthesis *(3)*. As an alternative or additional pathway to NER, a wide variety of organisms, including bacteria, yeast, plants, invertebrates, and many vertebrates, can revert CPDs by CPD photolyase in the presence of photoreactivating blue light (350–450 nm), restoring the bases to their native form *(4–6)*. In contrast to the complex NER pathway, in which damage recognition and repair is done by different proteins, photoreactivation depends on a single enzyme, and the reaction can be strictly controlled by presence or absence of photoreactivating light. It was recently demonstrated that photoreactivation is tightly regulated by chromatin structure *(7)*. CPDs in nuclease-sensitive regions (such as "open" promoter regions or the ARS1 origin of replication and in linker DNA between nucleosomes) were repaired very rapidly (within 15 min), whereas CPDs in nucleosomes were repaired slowly (within 2 h) *(see below)*. Hence, by monitoring CPD repair by photoreactivation, one can make direct conclusions about the accessibility of CPDs to photolyase in chromatin in vivo. Additionally, the indirect endlabeling approach permits the measuring of CPDs and their removal in larger regions, whole genes, and, in particular, in the transcribed region of genes. In transcriptionally active genes, the nontranscribed strand is repaired more quickly than the transcribed strand, because RNA polymerases blocked at CPDs prevent photoreactivation. The strand bias of photorepair is relieved when genes are inactivated *(8)*. Hence, the photoreactivation assay permits acquisition of additional information about the transcriptional state of a gene.

The approach is outlined in **Fig. 1** and is compared with the traditional way of mapping nucleosomes by micrococcal nuclease footprinting.

1. The NER pathway of yeast should be inactivated to avoid competitive repair processes. This can be done by deletion of the *Rad1* gene, which is essential for the incision step of NER.
2. DNA damage is induced by irradiation of yeast cultures with ultraviolet light (predominantly 254 nm). This generates CPDs throughout the genome. Polypyrimidine regions, such as T-tracts in promoter regions of yeast genes, are "hot spots" of CPD formation and accumulate high yields of CPDs *(7)*. The CPD yields can be modulated by protein-DNA interactions, chromatin structure and DNA sequence *(9–12)*.
3. Repair of CPDs by photolyase (photoreactivation) is done by limited exposure of the cell suspension to photoreactivating light. Aliquots are taken at different repair times.
4. In order to analyze CPD yields and CPD distribution along the DNA sequence, DNA is purified and treated with T4-endonuclease V (T4-endoV). T4-endoV is a CPD-specific enzyme that introduces single-strand cuts at CPDs *(13)*.

Photolyase: A Molecular Tool

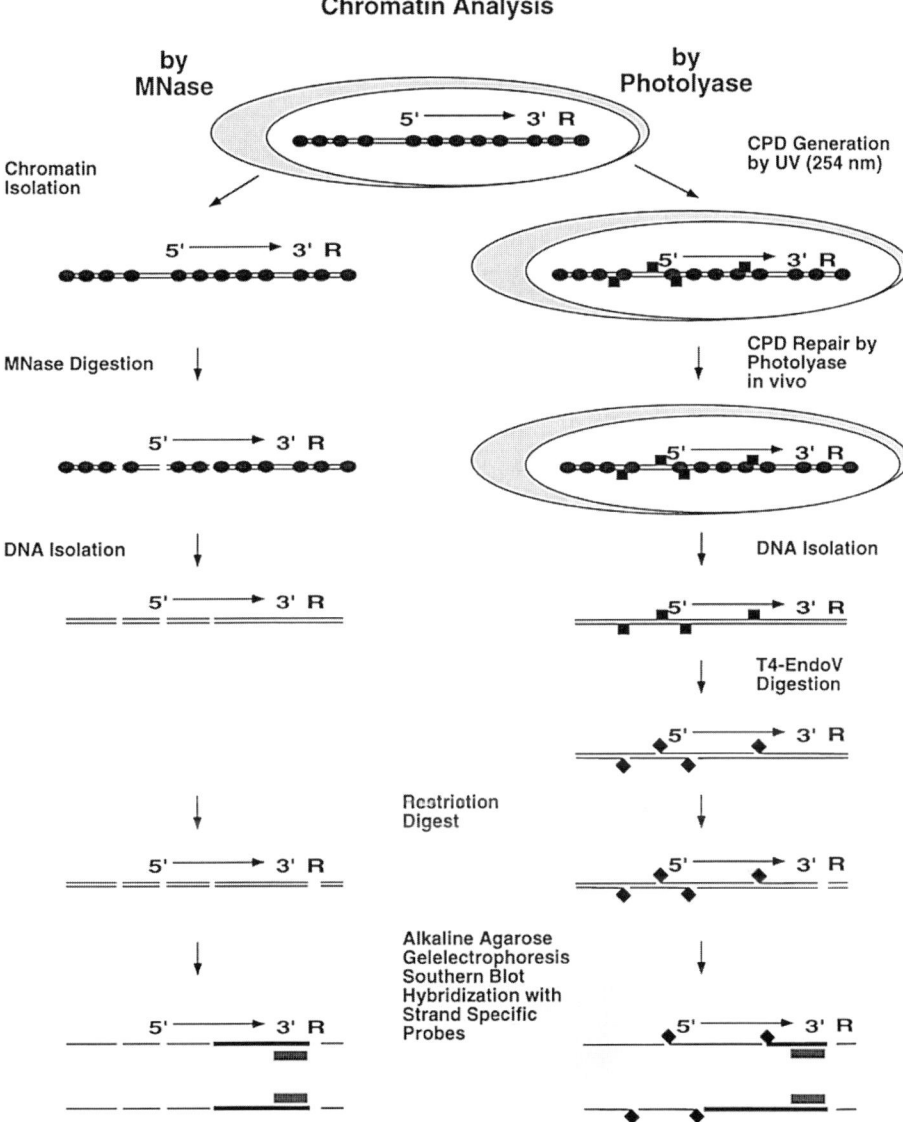

Fig. 1. Comparison of chromatin analysis by MNase footprinting and by CPD repair using photolyase. Schematically illustrated are: a gene covered by nucleosomes (dark dots): 5' and 3' ends and direction of transcription; R, a restriction site for indirect endlabeling; squares, cyclo butane pyrimidine dimers (CPDs) generated by UV light (254 nm); tilted squares, nicks in DNA generated by T4-endoV at CPD sites; big ovals, yeast cells. For practical reasons, the restriction digest is usually done prior to T4-endoV digestion.

5. The cutting sites on each strand of the DNA are displayed by indirect endlabeling from a restriction site using strand-specific probes (14). For that purpose, DNA is cut with a restriction enzyme, which provides the reference point for mapping. The DNA fragments are fractionated by electrophoresis on denaturing alkaline agarose gels and transferred to a membrane using a Southern blot protocol. Strand-specific radioactive probes that abut the restriction site are used for hybridization. The membranes are exposed to X-ray films or PhosphorImager screens.

An example is shown in **Fig. 2**. The bands represent DNA fragments that have the restriction site at one end and the CPD cut at the other end. The length of the fragments is used to calculate the site of CPD (distance from the restriction site). The intensity of the band reflects the CPD yields at that particular site. Decreasing band intensities with increasing photoreactivation time reflects DNA repair.

For chromatin analysis, chromatin is partially purified and cut with MNase. MNase introduces double-strand cuts in linker DNA and in nuclease sensitive regions, whereas nucleosomal DNA is protected against MNase digestion. Since T4-endoV and MNase induce DNA cuts, chromatin analysis by photoreactivation of CPDs in vivo can be compared side by side on the same gel with chromatin analysis by MNase in vitro (**Fig. 2**).

A comparison between chromatin structure and CPD repair by photolyase is shown for a yeast strain FTY117 deficient in NER (*rad1Δ*). FTY117 contains a well-characterized minichromosome (YRpCS1) as a model substrate (**Fig. 2A**). Indirect endlabeling was done from the XbaI site (X) in counterclockwise direction using strand-specific RNA probes. Chromatin analysis is shown in **Fig. 2B,C** (lanes 1–3). It shows positioned nucleosomes (white boxes) separated by linker DNA and nuclease-sensitive regions (5'- and 3' ends of DED1, HIS3, PET56; ARS1, origin of replication). CPD analysis is shown in lanes 4–10. Cells were irradiated with 100 J/m^2 and photoreactivation was done for up to 120 min. Unirradiated DNA (not shown) and mock treated DNA showed a background smear resulting from nicking of DNA during preparation (–T4-endoV; lane 10, **Fig. 2**). In contrast, T4-endoV treated DNA revealed numerous bands of different intensities (+T4-endoV; lanes 4–9). These bands can be assigned to dipyrimidines and polypyrimidine tracts in the DNA sequence. Many strong bands correspond to CPDs in T-tracts in the promoter regions of the DED1–, HIS3–, and PET56– genes, demonstrating that these tracts are "hot spots" of CPD formation.

During incubation in the dark for 120 min (dark control), no repair was observed (lane 9). However, during irradiation with photoreactivating light, more than 90% of CPDs were removed from both strands within 120 min (lanes 5–8), demonstrating CPD repair by photolyase. Inspection of the results at individual sites or clusters of CPDs very strikingly reveals two classes of repair: fast repair, when CPDs are removed within 15–30 min (dots and squares in **Fig. 2**) and slow repair, when CPDs remain detectable for up to 60–120 min. A comparison of CPD repair with the chromatin analysis shows that fast repair strictly correlates with the accessibility of DNA to MNase (bands in chromatin lanes), and slow repair corresponds to inaccessibility to MNase (no bands in chromatin lanes). Hence, chromatin structure regulates the accessibility of photolyase to CPDs. The fact that CPDs in nucleosomes are repaired can be taken as evidence for dynamic properties of nucleosomes that make lesions accessible.

The indirect endlabeling approach permits the display and measurement of whole yeast genes (e.g., HIS3, **Fig. 2**). When CPD repair by photolyase was measured in the transcribed region of the HIS3 gene (excluding the promoter and the 3' end), repair of the transcribed strand was slower than repair of the nontranscribed strand *(7)*. This strand bias of repair is a result of inhibition of photolyase by RNA polymerase, which is stalled at CPDs on the transcribed strand *(8)*, and provides information on the transcriptional state of a chromatin region.

2. Materials

2.1. UV Irradiation and Photoreactivation

1. UV irradiation box (homemade). It contains:
   a. Six germicidal lamps (Sylvania, Type: G15 T8, Mcycrin, Switzerland), arranged in parallel, with peak emission at 254 nm.
   b. An illumination stage of a 55 cm × 55 cm area at a distance of 39 cm from the lamps. The average flux used for the experiments is approx 0.40 mW/cm^2 (using two lamps).
2. Photoreactivation box (homemade). It contains:
   a. Six lamps (Sylvania, Type: F15 T8/BLB), arranged in parallel with peak emission at 375 nm.
   b. An illumination stage (55 cm × 55 cm). The surface of cell suspension is approx 8.5 cm from the lamps.
   c. The illumination stage is connected to a water bath for temperature control.
   d. The average flux used in the experiments is approx 1.4 mW/cm^2 (using four lamps).
3. UVX radiometer with a UVX-25-photocell and a UVX-36 photocell to measure 254 nm and 366 nm, respectively (UVP, San Gabriel, CA).
4. Yeast cultures (*see* **Note 1**) and appropriate media.
5. SD medium: synthetic minimal medium containing dextrose *(16)* supplemented, when required, with amino acids.
6. SG medium: synthetic minimal medium containing galactose (freshly added after autoclaving, sterilized by filtration) supplemented, when required, with amino acids.

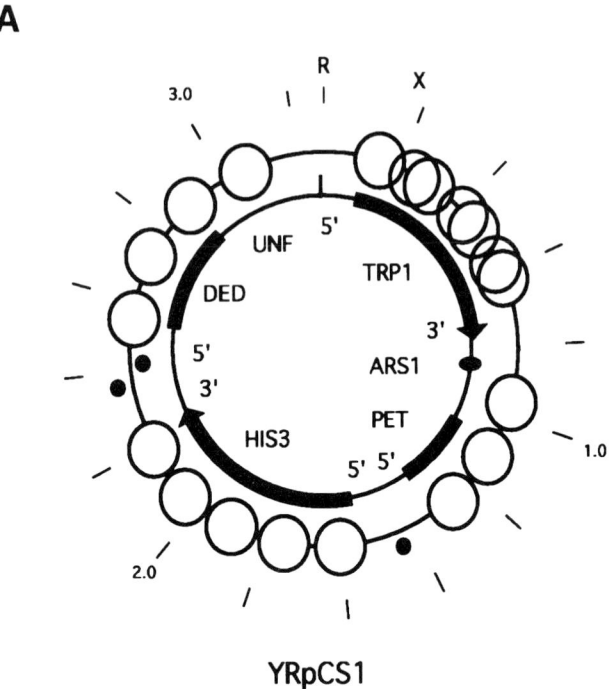

Fig. 2. *(continued on next page)* Chromatin structure modulates CPD-Repair by photolyase. **(A)** The minichromosome YRpCS1 of *S. cerevisiae* strain FTY117 contains the *pet56-HIS3-ded1* sequence with the HIS3 gene and the truncated DED1 and PET56 genes inserted in the UNF region of the TRP1ARS1 circle. UNF denotes the TRP1ARS1 region from the ARS1 consensus element (*oval*) to the *Eco*RI site. Nucleosome positions and nucleosome-free regions are shown as described *(17)*. The TRP1 gene in YRpCS1 shows overlapping nucleosome positions as in the TRP1ARS1 circle *(18)*. *Dots* denote some polypyrimidine regions and polydT-tracts, which are hotspots of CPD formation and which are quickly repaired by photolyase (outside is top strand; inside is bottom strand). Nucleosome positions (circles), the promoter regions (5'), the 3' ends of the genes (3'), and the ARS1 origin of replication (ARS1) are indicated. *Eco*RI (R) and *Xba*I (X) are restriction sites. Map units in base pairs are indicated in 0.2-kb steps. **(B,C)** Chromatin structure and CPD repair by photolyase in the top strand and bottom strand, respectively. The bottom strand is the transcribed strand of the TRP1 and HIS3 genes. FTY117 cells were UV irradiated with 100 J/m^2. Chromatin structure was analyzed by micrococcal nuclease digestion (MNase) of DNA (lane 4) and chromatin (CHR, lanes 2,3) extracted from irradiated cells. Photoreactivation (+ Photoreact) was for 15–120 min (lanes 5–8). CPD distribution and repair were analyzed by T4-endoV cleavage (+ T4-endoV, lanes 4–9). Lane 10 is irradiated DNA (same as lane 1) without T4-endoV cleavage. An aliquot of cells was kept in the dark for 120 min (lane 9). Cleavage sites for MNase and T4-endoV are shown by indirect endlabeling from the *Xba*I site **(A)**. A schematic interpretation of chromatin

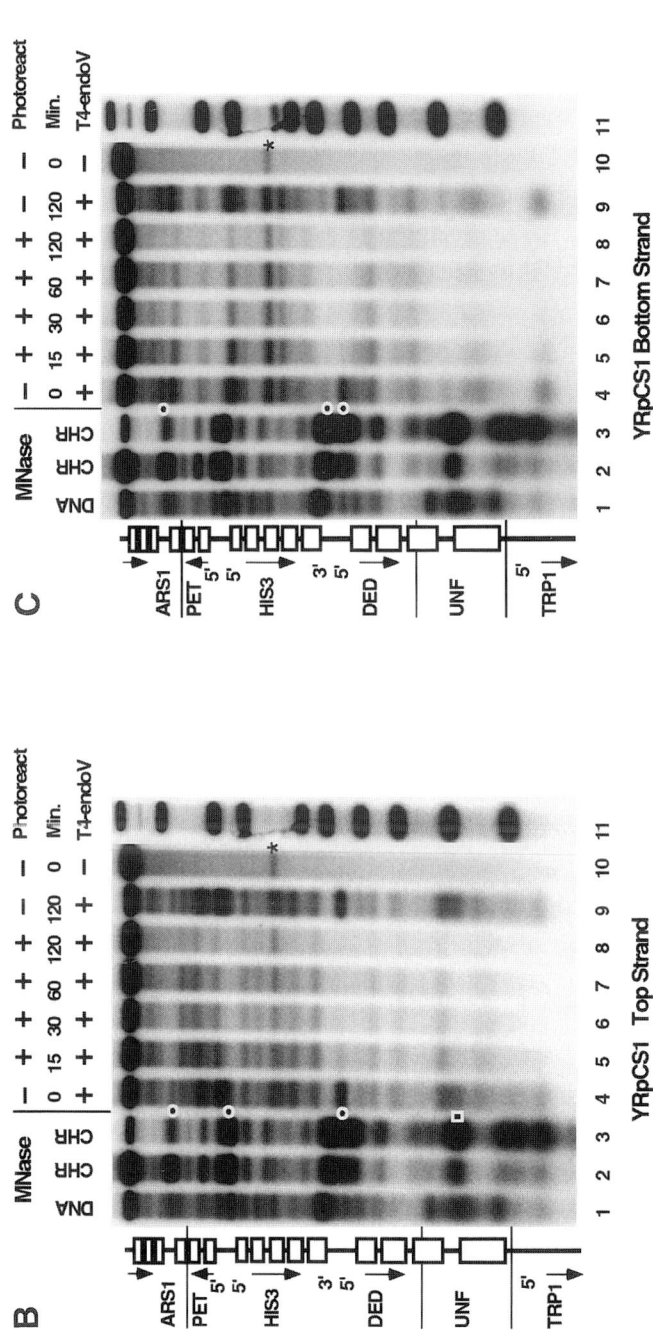

structure is shown (left). Chromatin regions of 140–200 bp that are protected against MNase cleavage represent positioned nucleosomes (*rectangles*), cutting sites between nucleosomes represent linker DNA, long regions with multiple cutting sites represent NSRs (ARS1; 5'PET-5'HIS3; 3'HIS3-5'DED; 5'TRP1). 5' and 3' ends of genes, direction of transcription (*arrows*) are indicated. *Dots* and *squares* indicate fast repair in NSRs and linker DNA, respectively. *Stars* denote crosshybridization with genomic DNA. Size markers (in base pairs, lane 11) are 261, 460, 690, 895, 1122, 1291, 1796, 2093, 2719, 3347. (Adopted from **ref. 7** with permission of Oxford University Press.)

7. Concentrated solutions of appropriate amino acids (*16*).
8. 21.9 cm × 31.2-cm plastic trays.
9. Dark rooms equipped with safety lights (Sylvania GE gold fluorescent light).
10. RC5B centrifuge and GS3 rotor (Sorvall, Digitana, Horgen, Switzerland).

2.2. Purfication of Genomic DNA Using a QIAGEN Protocol

QIAGEN Genomic DNA Handbook, Sept. 1995, (Quiagen, Basel, Switzerland) maxi prep volumes.

1. Qiagen Genomic-Tips: 500/G (cat. no. 10 262).
2. Buffers according to QIAGEN Genomic DNA Handbook (Sept. 1995), equilibrated to RT:
   a. Buffer Y1 (yeast lysis buffer): 1 M Sorbitol, 100 mM EDTA, 14 mM ß-mercaptoethanol (added just before use).
   b. Buffer G2 (general lysis buffer): 800 mM GuHCl, 30 mM EDTA, 30 mM Tris-HCl, 5% Tween-20, 0.5% Triton X-100, pH 8.0.
   c. Buffer QBT (equilibration buffer): 750 mM NaCl, 50 mM MOPS, 15% ethanol, 1.5% Triton X-100, pH 7.0.
   d. Buffer QC (wash buffer): 1.0 M NaCl, 50 mM MOPS, 15% ethanol, pH 7.0.
   e. Buffer QF (elution buffer): 1.25 M NaCl, 50 mM Tris-HCl, 15% ethanol, pH 8.5.
3. Zymolyase 100T (Seikagaku Corporation Bioscience Products AG, Emmenbruecke, Switzerland): 10 mg/mL resuspended in H$_2$O just before use.
4. Proteinase K (Boehringer Mannheim): 10 mg/mL in 50 mM Tris-HCl, pH 8.0, stored at –20°C or 20 mg/mL in H$_2$O, dissolved just before use.
5. RNase A (Boehringer Mannheim): 10 mg/mL in 10 mM Tris-HCl pH 7.5, 15 mM NaCl; stored at –20°C.
6. 10T1E: 10 mM Tris-HCl, 1 mM EDTA, pH 8.0.
7. 1% SDS.
8. SS34 rotor, HB4 rotor (Sorvall), tabletop centrifuge (Jouan CR4.22, Instrumenten Gesellschaft AG, Zuerich, Switzerland).
9. Isopropanol.
10. Cold 70% ethanol.

2.3. Purfication of DNA from Plasmid Containing Cells

1. Proteinase K (Boehringer Mannheim): 10 mg/mL in 50 mM Tris-HCl, pH 8.0, stored at –20°C.
2. RNase A (Boehringer Mannheim): 10 mg/mL in 10 mM Tris-HCl, pH 7.5, 15 mM NaCl; stored at –20°C.
3. Zymolyase 100T (Seikagaku Corporation): 10 mg/mL suspended in 40 mM K-Phosphate pH 7.5, 1M Sorbitol, stored at –20°C.
4. Zymolyase reaction buffer : 250 mM EDTA, pH 8.0, 1 M Sorbitol, 20 mM β-mercaptoethanol, 1 mM PMSF (mixed before use).
5. PMSF (phenylmethane-sulfonyl fluoride; Merck): 0.5 M dissolved in DMSO (methyl sulfoxide; Fluka).

6. ß-mercaptoethanol (Fluka).
7. 20% SDS.
8. 1 M sorbitol, 10 mM EDTA.
9. 50T 20E: 50 mM Tris-HCl, 20 mM EDTA, pH 8.0.
10. 10T1E: 10 mM Tris-HCl, 1 mM EDTA, pH 8.0.
11. 5 M potassium acetate, pH 5.5.
12. P: phenol saturated with 0.1 M Tris-HCl, pH 8.0.
13. D/I: mixture of dichloromethane/isoamyl alcohol (24:1).
14. P/D/I: mixture of phenol/dichloromethane/isoamyl alcohol (25:24:1).
15. 3M sodium acetate, pH 4.8.
16. Isopropanol.
17. Ethanol (99.9%).
18. Elutip-d columns (Schleicher & Schuell).
19. Low-salt solution: 0.2M NaCl, 20 mM Tris-HCl, pH 7.4, 1.0 mM EDTA.
20. High-salt solution: 1.0M NaCl, 20 mM Tris-HCl, pH 7.4, 1.0 mM EDTA.
21. SS34-rotors, HB4-rotors (Sorvall).

2.4. Indirect Endlabeling

1. In vitro transcription kit (Stratagene).
2. Appropriate DNA fragments subcloned in a bluescript vector (Stratagene) (*see* **Note 3**).
3. Restriction enzymes and appropriate 10X buffers.
4. P/D/I: mixture of phenol/dichloromethane/isoamyl alcohol (25:24:1).
5. D/I: mixture of dichloromethane/isoamyl alcohol (24:1).
6. Ethanol (99.9%).
7. 10T1E: 10 mM Tris-HCl, 1 mM EDTA, pH 8.0.
8. T4-endonuclease V (available from Epicentre Technologies (Madison, WI).
9. 10X T4 EndoV buffer (T4RB): 500 mM Tris-HCl, pH 7.5, 50 mM EDTA, pH 8.0.
10. Agarose (Gibco-BRL): ultra pure.
11. 5X Alkaline loading buffer (ALB): 12.5% Ficoll 400, 5 mM EDTA, 0.125% Bromocresol green, 250 mM NaOH (added just before use) *(15)*.
12. Alkaline electrophoresis buffer: 50 mM NaOH, 1 mM EDTA.
13. Zeta-Probe GT Genomic Tested Blotting Membrane (Bio-Rad Laboratories).
14. 3MM Whatman paper.
15. 0.4 N NaOH.
16. 20X SSC: 3 M NaCl, 0.3 M sodium citrate, pH 7.0.
17. 0.5 M sodium phosphate, pH 6.5.
18. 100X Denhardt's solution: 1 g Ficoll, 1 g polyvinylpyrrolidone K30, 1 g BSA in 50 mL H_2O.
19. 20% SDS.
20. 4 mg/mL sonicated herring sperm DNA, stored at –20°C.
21. Prehybridization and hybridization solution: 4X SSC, 50 mM phosphate buffer, pH 6.5, 10X Denhardt's solution, 1% SDS, 0.5 mg/mL sonicated herring sperm DNA (boiled and snap cooled on ice just before use).

22. Size markers prepared by a combination of restriction digests of genomic or plasmid DNA that hybridize to the probe. Alternatively, commercial-size markers might be used, but they need to be hybridized separately.
23. Gel apparatus (for large agarose gels, 20 × 25 cm, Horizon, Gibco-BRL).
24. Magnetic stirrers, peristaltic pump, glass plates.

3. Methods

3.1. UV Irradiation and Photoreactivation of Yeast Cells in Suspension

1. Grow 3–4 × 1 L yeast cultures in the appropriate medium and temperature to an absorbance at 600 nm of about 0.8–1.2 corresponding to approx 10^7 cells/mL (*see* **Note 1**).
2. Harvest cells by centrifugation in a GS3 rotor (Sorvall) at 6000g and 4°C for 8 min.
3. Resuspend cells and pool in minimal medium without amino acids and adjust to approx 2.5–3.5 × 10^7 cells/mL (totaling approx 2.6 L), at room temperature (*see* **Note 2**).
4. Do all steps from irradiation to lysis of spheroplasts in dark rooms equipped with gold fluorescent light to avoid photoreactivation (*see* **Note 4**).
5. Split cell suspension into aliquots of 250 mL and pour into plastic trays (21.9 cm × 31.2 cm) to produce a 4-mm layer.
6. Irradiate at room temperature with 100–150 J/m² of UV light (predominantly 254 nm) generated by germicidal lamps.
7. Immediately after irradiation, pool cells and supplement the medium by addition of the appropriate amino acids or uracil.
8. For photoreactivation, split cell suspension into aliquots. Put 250-mL aliquot on ice (0' repair). Incubate one 250-mL aliquot at room temperature for 120 min in the dark (dark repair, NER). Put two aliquots (500 mL) into two plastic trays. Place the trays on the illumination stage connected to a water bath. Irradiate with photoreactivating light (peak emission at 375 nm) at 1.4 mW/cm². The temperature of the cell suspension during photoreactivation is between 23°C and 26°C.
9. Remove aliquots of 250 mL at 15, 30, 60, 120 min and chill on ice.
10. Harvest cells by centrifugation in a GS3 rotor at 6000g, 4°C for 8 min.

3.2. Isolation of Genomic DNA Using a QIAGEN Protocol (see *Note 5*)

Purification of DNA is described using a Qiagen Kit according to Qiagen Genomic DNA Handbook, Sept. 1995, (maxi prep volumes) in rooms equipped with safety light to avoid photoreactivation.

1. Resuspend yeast pellets in 12 mL 10T1E each, transfer to 50-mL Falcon tubes and collect cells by centrifugation in table top centrifuge at 3900g and 4°C for 2 min.
2. Resuspend cells in 12 mL Buffer Y1, add 500 µL Zymolyase 100T (10 mg/mL) and convert to spheroplasts by incubation at 30°C.
3. To test spheroplasting, mix 50-µL aliquots with 950 µL 1% SDS. Spheroplasting is complete when A_{600} drops from initial approx 0.8–1.0 to below 0.1. Spheroplasting takes 10–45 min (*see* **Note 6**).

4. Collect spheroplasts by centrifugation in tabletop centrifuge at 3900g and 4°C for 2 min.
5. Gently resuspend pellets in 15 mL Buffer G2 supplemented with 200 µg/mL RNase A and incubate for 20 min at 30°C.
6. Add 5 mg Proteinase K and incubate at 50°C for approx 3 h and at room temperature overnight.
7. Centrifuge the lysates in a Sorvall SS34 rotor at 5000g and 4°C for 10 min.
8. Collect the clear supernatants in 50-mL Falcon tubes and discard the pellets.
9. Equilibrate a Qiagen Genomic-tip 500/G with 10 mL Buffer QBT and allow to empty by gravity flow into a 250-mL glass flask.
10. Vortex supernatants for 10 s and load onto the Qiagen tip. They enter the resin by gravity flow.
11. Wash Qiagen Genomic-tips with 2 × 15 mL Buffer QC.
12. Elute genomic DNA with 15 mL Buffer QF (prewarmed to 50°C) into Corex tubes.
13. Precipitate DNA by addition of 0.7 vol (10.5 mL) of isopropanol at RT and centrifugation in a HB4 rotor (Sorvall) at 13,000g and 4°C for 30 min.
14. Wash pellets with 4 mL cold 70% EtOH (after brief vortexing) and recentrifuge in an HB4 rotor (Sorvall) at 13,000g and 4°C for 10 min.
15. Dry pellets in the air in inverted tubes for 10 mins and for approx 5 min at 55°C.
16. Dissolve DNA in 500–600 µL 10T1E at 55°C for 2 h and at room temperature overnight.
17. Check DNA concentration by agarose gel electophoresis and ethidium bromide staining. The yield of a 250-mL culture varies between 30–250 µg genomic DNA.

3.3. Isolation of DNA from Plasmid Containing Yeast (see Note 5)

This procedure should be done in rooms equipped with safety light to avoid photoreactivation.

1. Resuspend pellets of UV irradiated and photoreactivated cells (from **Subheading 3.1.**) in 15 mL Zymolyase reaction buffer, transfer the solutions to 50-mL Falcon tubes.
2. Add 180–300 µL Zymolyase 100T (10 mg/mL) and convert the cells to spheroplasts by incubation at 30°C.
3. Test spheroplasting as described for isolation of genomic DNA (**Subheading 3.2.**).
4. Collect spheroplasts by centrifugation in tabletop centrifuge at 3900g and 4°C for 2 min.
5. Resuspend pellets in 12 mL 1 M Sorbitol 10 mM EDTA, centrifuge as above and resuspend in 7.5 mL 50 T 20E.
6. Add 375 µL 20% SDS and 75 µL Proteinase K (10 mg/mL). Incubate at 65°C for 2 h and at room temperature overnight.
7. Add 3 mL of 5 M potassium acetate, mix well, and incubate on ice for 1 h.
8. Transfer the suspension into 14-mL Falcon tubes and centrifuge in a SS34 rotor (Sorvall) at 12,000g and 4°C for 30 min.
9. Collect the clear supernatant containing DNA in 50-mL Falcon tubes.
10. Purify DNA by sequential extraction once with P, once with P/D/I and once with D/I.

11. Collect the aqueous phase and precipitate DNA with 0.7 vol of isopropanol at room temperature for approx 1 h, followed by centrifugation in HB4 rotor (Sorvall) at 16,000g and 4°C for 30 min.
12. Dry the pellets in vacuo and dissolve DNA in 1.88 mL 10T1E, pH 8.0.
13. Digest RNA with 18.8 µL RNAseA (10 mg/mL) at 37°C overnight.
14. Purify DNA by sequential extractions with P, P/D/I and D/I.
15. Transfer the aqueous phase to Falcon tubes and precipitate the DNA with 0.1 vol sodium acetate (3 M) and 3 volumes of cold ethanol. Incubate at –20°C for 1 h and centrifuge in an HB4 rotor (Sorvall) at 16,000g and 4°C for 30 min.
16. Dry the pellets in vacuo and dissolve the DNA in 300 µL–1 mL low-salt solution at 37°C for 20–30 min.
17. Bind the DNA on Elutip-d (Schleicher and Schuell) and elute with high-salt solution following the the instructions of the supplier.
18. Precipitate DNA and dissolve it in 300 µL 10T1E, pH 8.0.
19. Check DNA-content by agarose gel electophoresis and ethidium bromide staining. The yield of a 250-mL aliquot varies between 6–25 µg total DNA from approx $8.0–10 \times 10^9$ cells.

3.4. Mapping of CPDs by Indirect Endlabeling

1. Digest DNA to completion with the appropriate restriction enzyme (*see* **Note 3**). Purify DNA once with P/D/I and once with D/I.
2. Precipitate the DNA by addition of three volumes of ethanol at –20°C for 1 h followed by centrifugation in an HB4 rotor (Sorvall) at 16,000g and 4°C for 20 min.
3. Dry the pellet in vacuo and redissolve it in 10T1E, pH 8.0 to a DNA concentration of about 50–100 ng/µL (plasmid DNA) and 150–300 ng/µL (genomic DNA).
4. Digest with T4EndoV in a final reaction volume of 40 µL adjusted to 1X T4RB at 37°C for 2.5 h. For mock treated samples replace T4EndoV by 1X T4RB. For T4-endoV digestion, use approx 125 ng of restricted plasmid DNA or about 2.5 µg of restricted genomic DNA.
5. Add 10 µL 5X ALB to stop the reaction and to denature the DNA.
6. Prepare two horizontal agarose gels (A and B, for hybridiziation with the top- and bottom-strand probe, respectively) (20 cm × 25 cm, BRL; 300 mL 1.5% agarose in 50 mM NaCl, 1 mM EDTA, *see* **ref. *15***). Soak the gels in 1X alkaline electrophoresis buffer (for up to 3 h). Prerun the gels at 50–60 V (constant voltage) and 180–270 mA for 3 h at 4°C using peristaltic pump in both directions and magnetic stirrers at both electrodes to circulate the buffer.
7. Load 24 µL DNA samples per slot and allow the DNA to enter the gel at 60–65 V and 210–340 mA for 10–20 min (without buffer circulation). Cover the gel with a glass plate (to stop the gel from floating) and continue electrophoresis with buffer circulation at 50–55 V (constant voltage) and 270–110 mA for 15–17 h at 4°C, until the dye (bromocresol green) migrates about 14–16 cm.
8. For Southern transfer, soak the gel briefly in 0.4 N NaOH. Lay two sheets of 3MM Whatman paper presoaked in 0.4 N NaOH over a plastic plate bridging a tray with 0.4 N NaOH. The paper remains in contact with 0.4 N NaOH on both

sides. Carefully invert the gel and place it on the paper. Cover it with a Zeta-Probe GT membrane (presoaked first in H_2O and then in 0.4 N NaOH). Cover the membrane with four 3MM Whatman papers presoaked in 0.4 N NaOH, two layers of dry 3MM Whatman paper, 10 cm of paper towels, a glass plate, and a weight of approx 250 g. Transfer DNA to the membrane for approx 5–20 h. Replace wet towels after a few hours with fresh ones.

9. After transfer, neutralize the membranes by rinsing them twice for approx 15 min in 2X SSC. Make sure that the pH of the 2X SSC solution falls to seven, if necessary, rinse the membranes again. Bake the membranes at 80°C for 1–2 h.
10. Place the membranes in rotary cylinders with the DNA bound side facing inside and incubate them in 20 mL prehybridization solution per 20 cm × 24-cm membrane at 65–67°C for 4–16 h in a hybridization oven.
11. Prepare strand-specific RNA-probes using an in vitro transcription kit and DNA-templates subcloned in Bluescript vectors (Stratagene) (*see* **Note 3**).
12. Replace prehybridization solution with the same volume of freshly boiled and chilled prehybridization solution. Add the strand-specific RNA probe (approx 3.3–7.5 × 10^7 cpm in 100–400 µL). Hybridize at 65–67°C overnight. Marker lanes may be hybridized in a separate cylinder using the same conditions and DNA probes that were denatured for 5 min in a boiling water bath just before use.
12. Pour off the hybridization solution, wash the membranes six times for 20 min each in 0.6X SSC, 0.1% SDS at 65–67°C.
13. Dry the membranes briefly between two sheets of 3MM Whatman paper and expose them to X-ray films (Fuji) using enhancer screens or to PhosphorImager storage plates (Molecular Dynamics).
14. Measure bands and calculate the distance from the restriction sites (*see* **Note 7**).

4. Notes

1. Yeast cultures are grown in the appropriate medium *(16)*. Selective media are required for maintenance of minichromosomes.
2. The cell density in the liquid culture and UV-absorption properties of the medium need to be considered for DNA-damage induction as well as for photoreactivation. The UV dose should be optimized by quantification of the CPD content in DNA. Optimal conditions are approx 0.5–2 CPDs/1 kb (e.g., **ref. 7**).
3. For indirect endlabeling, a restriction site should be selected which places the region of interest at a distance of 0.3–2.5 kb. To generate strand-specific probes it is recommended to use DNA-fragments (approx 150–200 bp) subcloned in a Blueskript vector (Stratagene). Alternatively, DNA-sequences can be directly amplified from the yeast genome using strand-specific primers. Radioactive probes can be generated using small DNA fragments as substrates and primer extension by Taq polymerase with radioactive dNTPs. Several rounds of primer extension can be used to increase the amount of radioactive DNA. Strand-specificity of indirect endlabel probes will be evident from the results. Since CPD distribution is strand-specific, each strand will produce a characteristic CPD-pattern (*see* **Fig. 2**; **ref. 7**).

4. Using gold fluorescent light as "safety light" is highly recommended, because photoreactivation is a fast process and could be promoted by traces of day light.
5. For preparation of genomic DNA as well as for preparation of DNA from plasmid containing cells, we use either extensive purification by organic extraction (phenol) or more recently protocols by QIAGEN. Both procedures yield DNA which is suitable for T4-endoV cleavage and indirect endlabeling. The yields may vary considerably, but this has not been systematically tested.
6. Spheroplasting may depend on the strain and growth conditions. Spheroplasting needs to be fast in order to reduce the risk of CPD removal by nucleotide excision repair. This risk is eliminated by using *rad1Δ* mutants. Photoreactivation is avoided by working in safety light.
7. A major concern is accuracy in measurements of cutting sites. We prefer 25-cm long 1% alkaline agarose gels. This allows to spread 2.5 kb in the lower 15 cm of the gel and to measure the length of the bands with the required precision. At 2.6 kb (top band of our marker is 2560 bp; **Fig. 1B**), we measure 250 bp/5 mm or 50 bp/mm, whereas in the lower part, we obtain 250 bp/25 mm or 10 bp/mm. Since the band width measured at weak exposures is approx 1 mm, the average mapping results in a precision of approx 20 bp.

Acknowledgment

We thank Dr. M. Smerdon (Washington State University) for initiation of DNA-repair projects in our group and Dr. U. Suter (Institut für Zellbiologie, ETH-Zürich) for continuous support. This work was supported by the Swiss National Science Foundation (Grant Nr 3100-053739.98) and the EUROFAN project (BBW Nr 950191-11).

References

1. Singh, J. and Klar, A. J. S. (1992) Active genes in budding yeast display enhanced in vivo accessibility to foreign DNA methylases: a novel in vivo probe for chromatin structure of yeast. *Genes Develop.* **6,** 186–196.
2. Kladde, M. P. and Simpson, R. T. (1996) Chromatin structure mapping in vivo using methyltransferases. *Meth. Enzymol.* **274,** 214–233.
3. Friedberg, E. C., Walker, G. C., and Siede, W. (1995) DNA repair and mutagenesis. *ASM Press*, Washington, DC.
4. Sancar, A. (1996) No "End of History" for photolyases. *Science* **272,** 48,49.
5. Sancar, A. and Sancar, G. B. (1988) DNA repair enzymes. *Ann. Rev. Biochem.* **57,** 29–67.
6. Yasui, A., Eker, A. P. M., Yasuhira, S., Yajima, H., Kobayashi, T., Takao, M., and Oikawa, A. (1994) A new class of DNA photolyases present in various organisms including aplacental mammals. *EMBO J* **13,** 6143–6151.
7. Suter, B., Livingstone-Zatchej, M., and Thoma, F. (1997) Chromatin structure modulates DNA repair by photolyase in vivo. *EMBO J* **16,** 2150–2160.
8. Livingstone-Zatchej, M., Meier, A., Suter, B., and Thoma, F. (1997) RNA-Polymerase II transcription inhibits DNA repair by photolyase in the transcribed strand of active yeast genes. *Nucleic Acids Res.* **25,** 3795–3800.

9. Gale, J. M., Nissen, K. A., and Smerdon, M. J. (1987) UV-induced formation of pyrimidine dimers in nucleosome core DNA is strongly modulated with a period of 10.3 bases. *Proc. Natl. Acad. Sci. USA* **84,** 6644–6648.
10. Pehrson, J. R. (1989) Thymine dimer formation as a probe of the path of DNA in and between nucleosome in intact chromatin. *Proc. Natl. Acad. Sci. USA* **86,** 9149–9153.
11. Schieferstein, U. and Thoma, F. (1996) Modulation of cyclobutane pyrimidine dimer formation in a positioned nucleosome containing polydA.dT tracts. *Biochemistry* **35,** 7705–7714.
12. Tornaletti, S. and Pfeifer, G. P. (1996) UV damage and repair mechanisms in mammalian cells. *Bioessays* **18,** 221–228.
13. Gordon, L. K. and Haseltine, W. A. (1980) Comparison of the cleavage of pyrimidine dimers by the bacteriophage T4 and Micrococcus luteus UV-specific endonucleases. *J. Biol. Chem.* **255,** 12,047–12,050.
14. Smerdon, M. J. and Thoma, F. (1990) Site-specific DNA repair at the nucleosome level in a yeast minichromosome. *Cell* **61,** 675–684.
15. Maniatis, T., Fritsch, E., and Sambrook, J. (1982) Molecular cloning: A Laboratory Manual. Cold Spring Harbor Laboratory Press. Cold Spring Harbor, NY.
16. Sherman, F., Fink, G. R., and Hicks, J. B. (1986) Laboratory Course Manual for Methods in Yeast Genetics. Cold Spring Harbor Laboratory Press, Cold Spring Harbor, NY.
17. Losa, R., Omari, S., and Thoma, F. (1990) Poly(dA)'poly(dT) rich sequences are not sufficient to exclude nucleosome formation in a constitutive yeast promoter. *Nucleic Acids Res.* **18,** 3495–3502.
18. Thoma, F., Bergman, L. W., and Simpson, R. T. (1984) Nuclease digestion of circular TRP1ARS1 chromatin reveals positioned nucleosomes separated by nuclease sensitive regions. *J. Mol. Biol.* **177,** 715–733.

19

Transcriptional and Structural Analyses of Isolated SV40 Chromatin

Ulla Hansen

1. Introduction

1.1. Simian Virus 40 (SV40) as a Model System for Isolating Transcriptionally Competent Cellular Chromatin

An ideal biochemical source of a defined chromatin template assembled in vivo is the SV40 minichromosomes (*see* **Note 1**). At all stages in the viral lytic cycle, SV40 DNA is complexed with cellular histone and nonhistone proteins to form the episomal chromatin structure called a minichromosome (MC). MCs are therefore the viral template for both the host replication and host transcription machinery. The chromatin properties of MCs reflect transcriptionally competent host chromatin in every respect that has been examined. They contain more highly acetylated histones, high-mobility group proteins, and DNase I hypersensitive sites *(1–4)*. For these reasons, SV40 MCs have long been used as a model system for transcriptionally active chromatin. Because of the transcriptionally competent nature of these templates, they provide chromatin with characteristics distinct from, but complementary to, most in vitro reconstituted chromatin. SV40 MCs are especially useful in approaching questions regarding stages in transcriptional activation from a potentially competent to a fully active state. Furthermore, because of the utility of SV40 as a viral vector for exogenous promoters *(5,6)*, fully functional cellular promoters in a native chromatin context can also be isolated and studied in this manner.

Because SV40 MCs are readily manipulatable, they provide in vivo-assembled chromatin with appropriately positioned nucleosomes, not only for transcriptional studies, but also for structural studies in vitro. The ability to simultaneously monitor both characteristics enables direct experimentation of the relationship between chromatin structure and transcriptional competency *(7)*. From a structural perspective, SV40 chromatin can be analyzed both at a local level, examining the positioning of nucleosomes, and at a global level, examining the condensation state of the MCs. SV40 MCs are amenable to investigating chromatin compaction because of the unique facility with which higher order chromatin structure can be altered (in particular by adding histone H1) without causing nonspecific precipitation of the chromatin templates *(7)*. Unlike most chromatin, MCs are not precipitated by histone H1 at physiological salt concentrations, therefore permitting a functional analysis of the role of linker histones.

1.1.1. Abundance, Isolation, and Purification of SV40 MCs

During a lytic infection in African green monkey kidney cells, SV40 DNA is replicated as an episome up to 100,000 copies per cell *(8)*. Techniques for efficiently isolating MCs from infected cells involve minimal perturbation in their structure *(9–11)*. Thus, biochemical quantities of active chromatin of a defined structure can readily be isolated for in vitro studies (*see* **Note 2**).

The procedure for isolating MCs involves preparation of cell nuclei, incubation of the nuclei in a buffer in which the structural integrity of the nuclei is largely maintained though the MCs can leach out, and purification of the MCs from soluble nuclear proteins by zonal centrifugation through preformed sucrose density gradients. The latter step also separates subpopulations of the MCs. The partially packaged previrions sediment at approx 200 S, compared with 75S–90S for unpackaged MCs. In addition, the actively transcribing and actively replicating genomes are separable, as a shoulder with a slightly higher rate of sedimentation, from the peak of the transcriptionally competent, but not actively transcribing genomes.

In addition to the host core and linker histones, the isolated, native MCs contain a variety of bound, nonhistone proteins. These include, but are not limited to, high-mobility group (HMG) proteins *(3,4)*, transcription machinery *(12,13)*, replication machinery, topoisomerases, the viral large T antigen, and viral capsid proteins (for review, *see* **ref. 14**). When it is desirable to analyze a more-defined template in vitro, all but the core histones can be removed by isolation in elevated salt *(7)*. Subsequently, other chromosomal and nonchromosomal proteins, including histone H1 *(7)*, HMG proteins *(4)*, and specific DNA-binding proteins (G. Sewack and U. Hansen, unpublished observations), can be specifically reconstituted onto the core nucleosomal template. The acetylation state of the core histones can also be manipulated, both in vivo prior to isolation, and in vitro by modification of the purified MCs. Other types of modifications of chromosomal proteins should be equally feasible, though these have not yet been tested.

1.1.2. Biochemical Properties of Isolated SV40 MCs

SV40 (strain 776) MCs contain an average of 24 nucleosomes on the 5243 bp of viral DNA *(15)*. Except for a few strongly positioned nucleosomes, including one covering the major late initiation site *(16)*, the nucleosomes are not translationally positioned at specific locations *(17)*. The enhanced DNase I hypersensitivity of viral promoter/origin sequences in the SV40 MCs reflects phased nucleosomes bordering a relatively nucleosome-free region *(15,18–21)*. The divergent early and late promoters, as well as the viral DNA replication origin, overlap each other within this 400-bp region. All transcriptionally active templates in vivo contain the "nucleosome-free" region *(22–24)*. In fact, contrary to early reports, the bulk of intracellular nonpackaged MCs contain the nucleosome-free promoter region *(9,25)*. As MCs are packaged with viral capsid proteins, nucleosomes become uniformly spread around the template, including over the promoters and origin sequences *(25)*. The previously described isolation of 90S MCs that lack the nucleosome-free promoter region results from unpackaging of previrion structures during preparation of the MCs.

The transcriptional properties of isolated SV40 MCs can be assayed in vitro by addition of either crude transcription extracts or of partially or fully purified transcription factors. In order to analyze only RNA synthesized *de novo* and to exclude measurement of the RNA initiated in cells and associated with the isolated, transcribing MCs as ternary complexes, it is critical to use an assay that detects only labeled RNA synthesized in vitro during the transcription reaction (*see* **Note 3**). As it is essential to maintain the structural integrity of the MCs during the transcription reaction, these circular genomes are added unperturbed to the transcription reaction, resulting in RNA products of varying lengths. To effectively characterize the *de novo* transcripts with such varying 3'-termini, RNAs are detected and quantitated by incorporation of radiolabeled nucleotides during the transcription reaction, hybridization of the RNA with single-stranded complementary DNA, and digestion of the RNA:DNA hybrids with RNase T1, leading to products with a unique 3'-terminus *(26,27)*. As long as the single-stranded DNA probe includes the complete promoter region, the start sites of the transcripts are identified by this protocol as well. In addition, the procedure readily distinguishes artifactual "readthrough" transcripts (which are only minimally produced from minichromosomal templates) from the appropriately initiated transcript(s).

With the use of these techniques, we have demonstrated that transcription in vitro from the SV40 chromatin templates mimics in every respect the in vivo regulation of transcription from the SV40 promoters. In particular, we have established the following major in vitro transcriptional characteristics of these templates:

1. *De novo* initiation of transcription from isolated SV40 MCs in vitro is relatively efficient, being at least as efficient as transcription from naked DNA templates in vitro *(27)*. Thus, these templates do not display the repressive effect of chromatin on transcription, in contrast to chromatin assembled in most in vitro reconstitution systems. Furthermore, the vast majority of transcripts synthesized from SV40 MCs in vitro are initiated *de novo*.
2. Efficient initiation of transcription requires the appropriate chromatin structure at the SV40 MC promoters, the nucleosome-free region that is also required for gene expression in vivo *(13)*. All isolated MCs, whether being actively transcribed when isolated or not, are equally competent to be transcribed in vitro *(13,27)*.
3. The pattern of promoter utilization on the SV40 MC templates mimics transcription observed in vivo during lytic infection, which is distinct from the pattern of transcription supported from naked DNA templates *(27)*.
4. Low levels of basal transcription factors, in particular TFIID, remain on the MC templates, poised to reinitiate transcription *(13)*. However, the vast majority of transcription *in vitro* occurs from MC promoters on which preinitiation complexes are newly assembled.
5. Elongation by RNA polymerase II readily occurs through at least 10 nucleosomes in vitro *(4)*. However, rates of elongation are slower than those observed on naked DNA in vitro *(4,7)*.

Because SV40 MCs reflect transcriptionally competent chromatin in their in vitro characteristics, they provide ideal templates for studies on the induction of chromatin to a transcriptionally active state. Experiments so far have been performed in the absence of replication and of *de novo* chromatin deposition, and they are therefore a model for rapid transcriptional induction in mammalian cells. The potential, although technically challenging and as yet unproven, exists for extending this system to study the effect of DNA replication of SV40 chromatin in vitro *(28,29)* on transcriptional regulation. In summary, SV40 MCs afford a manipulatable chromatin system with which one can not only obtain correlations between structure and activity *(7)* but also can modify the templates (in vitro and in some cases in vivo) with respect to chromosomal proteins *(4)*, transcription factors *(13,27)*, and DNA sequences *(30)* to investigate how various components affect transcriptional potential.

2. Materials
2.1. Preparation and Propagation of Virus
1. DME: Dulbecco's modified Eagle's medium, containing 4.5 g/L glucose and 22 g/L sodium bicarbonate.
2. 10X dissociation buffer *(31)*: 10 mM ethyleneglycol-bis-N,N'-tetra-acetic acid (EGTA), 30 mM dithiothreitol (DTT), 1.5 M NaCl, 0.5 M Tris-HCl, pH 8.5.

3. Hirt extraction buffer *(32)*: 0.6% SDS, 10 mM Tris-HCl, pH 7.9, 10 mM EDTA.
4. Phosphate-buffered saline (PBS): 137 mM NaCl, 2.7 mM KCl, 4.3 mM Na_2HPO_4, 1.4 mM KH_2PO_4. Autoclave before use.
5. Ligation buffer: 50 mM Tris-HCl, pH 7.8, 10 mM $MgCl_2$, 10 mM DTT, 1 mM ATP, 25 µg/mL bovine serum albumin.
6. Simian virus 40: SV40 strain 776 contains the fully sequenced viral genome and is therefore the optimal virus for these experiments.
7. CV-1 cell line: CV-1 cells are an African green monkey kidney cell line that efficiently supports growth of SV40. Our cell line (originally from the laboratory of Phillip A. Sharp, MIT) grows in DME plus 10% calf serum at 37°C in a 5% CO_2 atmosphere. Sometimes, cells grown too long in culture lose their infectivity. The cell line stock should be passaged minimally. Growing cells should be split 2× weekly by 1:5 dilutions and should not be kept in culture for longer than 1–2 mo.
8. CMT cell line: CMT cells *(33)* are derived from CV-1 cells, and they constitutively express SV40 large T antigen. Thus, these cells can complement an SV40 T antigen defect and support growth of a homogeneous viral stock deleted for the SV40 early genes (*see* **Note 4**). The integrated T antigen gene is driven by a metallothionein promoter, which is inducible to higher levels with certain divalent cations (*see* **Note 4**). Propagate these cells as described above for CV-1 cells, except that 10% fetal calf serum should be used in the tissue culture media.

2.2. Isolation of SV40 MCs

1. Rinse buffer: 5.0 mM HEPES-NaOH, pH 7.5, 140 mM NaCl.
2. Lysis buffer *(10)*: 0.25 % Triton X-100, 10 mM EDTA, 10 mM HEPES-NaOH, pH 6.8, 5–500 µM of 5,5'-dithiobis(2-nitrobenzoic acid) (DTNB) (*see* **Note 5**), 0.5 mM phenylmethylsulfonyl fluoride (PMSF), 2 µg/mL antipain, 2 µg/mL aprotinin, 0.5 µg/mL leupeptin. The protease inhibitors (PMSF, antipain, aprotinin, and leupeptin) and DTNB are added immediately before use.
3. 0.5 M PMSF: Prepared in 100% ethanol, and added immediately before use of buffers.
4. 50 mM DTNB: 50 mM 5,5'-bithiobis(2-nitrobenzoic acid), 124 mM Tris base (15 mg Tris base in final vol of 1 mL). The solution, which should be prepared fresh, is buffered by the addition of Tris base, resulting in a pH of between 7.0 and 8.0. If Tris base is not added, the DTNB will not dissolve.
5. Extraction buffer: 0.25% Triton X-100, 0.12 M NaCl, 10 mM EDTA, 10 mM HEPES-NaOH, pH 7.5, 0.5 mM PMSF, 2 µg/mL antipain, 2 µg/mL aprotinin, 0.5 µg/mL leupeptin. The protease inhibitors (PMSF, antipain, aprotinin, and leupeptin) are added immediately before use; 5–500 µM DTNB may be added, as well (*see* **Note 5**).
6. 15% (w/w) sucrose solution: 15% (w/w) sucrose (15.9 g sucrose in final vol of 100 mL; *see* **ref. *34***), 50 mM HEPES-NaOH, pH 7.5.
7. 30% (w/w) sucrose solution: 30% (w/w) sucrose (33.8 g sucrose in final vol of 100 mL; *see* **ref. *34***), 50 mM HEPES-NaOH, pH 7.5.
8. Tris-acetate-EDTA (TAE) buffer: 40 mM Tris-acetate, 2 mM EDTA. The pH should be between 8.2 and 8.5.

2.3. Structural Analyses of SV40 MCs

1. Nucleases: Micrococcal nuclease, deoxyribonuclease I.
2. Dimethyl sulfate.

2.4. Transcriptional Analyses of SV40 MCs

1. Extraction buffer: 7.5 M urea, 0.5% SDS, 10 mM EDTA, 10 mM Tris-HCl, pH 7.8.
2. 5X RNase-free DNase buffer: 100 mM HEPES-NaOH, pH 7.5, 35 mM MgCl$_2$, 250 μM CaCl$_2$.
3. 2X hybridization buffer: 1.5 M NaCl, 100 mM HEPES-NaOH, pH 7.0, 2 mM EDTA.
4. 2X proteinase K solution: 0.4 mg/mL proteinase K, 100 mM Tris-HCl, pH 7.5, 2% SDS, 20 mM EDTA.
5. Phenol:chloroform:isoamyl alcohol: Mix together 50 mL buffered phenol, 50 mL chloroform, and 2 mL isoamyl alcohol. Store in opaque glass container in the refrigerator.
6. T1 RNase buffer: 10 mM HEPES-NaOH, pH 7.5, 0.2 M NaCl, 1 mM EDTA.
7. Sample loading buffer: 80% deionized formamide; 89 mM Tris base, 89 mM boric acid, 2 mM EDTA (1X TBE); 0.05% xylene cyanol, 0.05% bromophenol blue.
8. HeLa whole-cell extract (*see* **Note 6**): Prepare essentially as initially described *(35)*, although 0.1 mL of 1 M NaOH (not 0.01 mL) is added per 10 g ammonium sulfate to the whole-cell extract suspension. As a final step, dialyze into extract buffer: 20 mM HEPES-NaOH, pH 7.9, 100 mM KCl, 12.5 mM MgCl$_2$, 0.1 mM EDTA, 2 mM DTT, 17% glycerol. If it is desired, remove histone H1 from the crude HeLa whole-cell extract by precipitation with 2.26 M ammonium sulfate *(7,36)*, resuspension of the pellet after centrifugation into extract buffer, and dialysis into this same buffer. Protein concentration of the final extract should be in the range of 20 mg/mL. Cellular extracts supporting transcription in vitro are also commercially available.
9. poly [d(I-C)]·[d(I-C)]: Prepare 200 μg/mL solution of poly [d(I-C)]·[d(I-C)] (*see* **Note 7**) in 10 mM Tris-HCl, pH 8.3, 1 mM EDTA for use as carrier DNA in the reactions. If desired, the DNA can be dissolved at a higher concentration into 0.1 M NaCl, 10 mM Tris-HCl, pH 8.3, 1 mM EDTA, heated to 48°C for 5 min and cooled slowly to room temperature in order to facilitate appropriate annealing. However, the additional NaCl must then be taken into account in subsequent reactions.
10. 25X NTPs: 0.5 mM UTP (*see* **Note 7**), 5 mM ATP, 5 mM CTP, 5 mM GTP, 100 mM creatine phosphate (to maintain triphosphate pools in the crude extract). Each NTP stock solution is initially prepared from the di-, tri- or tetrasodium salts in a 10-mM Tris-HCl buffer.
11. tRNA: Dissolve yeast tRNA in 10 mM Tris-HCl, pH 7.9, 1 mM EDTA. Extract with phenol:chloroform:isoamyl alcohol several times, at least 1X more once interface is no longer observed; extract 2X with chloroform. Precipitate with ethanol, and resuspend at 10 mg/mL in 10 mM Tris-HCl, pH 8.3, 1 mM EDTA.
12. Proteinase K stock solution: 5 mg/mL (Boehringer Mannheim) in 10 mM HEPES-NaOH, pH 7.5, 1 mM EDTA. Prepare stock, aliquot into 500-μL samples, quick freeze in dry ice:ethanol bath, and store at –20°C.

13. RNase-free DNase: 2 mg/mL (10,000 U/mL) (Worthington-Cooper) in 50 m*M* HEPES-NaOH, pH 7.5, 10 m*M* MgCl$_2$, 50% glycerol. Store at –20°C.
14. Ribonuclease T1: 5 U/µL (1.4 µg/µL) (Calbiochem) in 30 m*M* Tris-HCl, pH 8.0. Prepare stock, aliquot into 50-µL samples, quick freeze in dry ice:ethanol bath, and store at –20°C.

3. Methods

3.1. Preparation of Wild-Type or Recombinant SV40 Stocks

3.1.1. SV40 as a Vector for Cellular Promoters

1. Clone the cellular promoter with a reporter gene, if desired, (e.g., ß-globin, chloramphenicol acetyltransferase) in place of the SV40 early genes in an SV40 recombinant bacterial plasmid. A convenient plasmid for insertion of exogenous DNA is pSVSX-T (*see* **Note 8**). The parent of this plasmid, pSVSX, is a SV40/pBR322 recombinant plasmid, fused at the respective *Eco*RI restriction sites, containing a single *Xho*I site at position 2666 in the SV40 genome. Subsequent deletion of SV40 sequences 2770 (*Bcl*I) to 5027 (*Eco*NI) removed the early coding sequences, yielding pSVSX-T (**Fig. 1A**). Cloning of exogenous sequences into the unique *Xho*I site results in plasmids such as pSVSpS2 *(30)*, which contains the human pS2 promoter from -1100 to +10 driving rabbit ß-globin coding sequences (**Fig. 1B**). For optimal packaging of recombinant viral DNA into an SV40 virion, the length of the recombinant genome should be no larger than 5–9% greater than the length of a wild-type genome (5243 bp), with more-efficient packaging the closer it is to the wild-type length *(37)*. In the case of pSVSX-T, this translates into a DNA insert of 2260 bp to result in a wild-type length genome and DNA inserts not exceeding 2700 bp (including both the cellular regulatory sequences and any reporter gene) for the highest efficiency of packaging. It is preferable to insert the promoter (and reporter gene) such that the promoter is positioned in an opposite orientation to that of the endogenous t/T genes being replaced in order to distance the inserted transcriptional elements as far from the SV40 enhancer sequences as feasible (1000–2000 bp).
2. Precisely remove the bacterial plasmid sequences from the recombinant viral genome by digestion with an appropriate restriction enzyme (limited digestion with EcoRI, in the case of the construct in **Fig. 1B**). Recircularize the DNA with ligase, using sterile ligation buffer (yielding the recombinant viral genome, as in **Fig. 1B**). The DNA concentration should be low to promote circularization over concatamerization (e.g., 0.75 ng DNA/µL ligation mixture). Heat inactivate the ligase for 10 min at 65°C. Verify by gel electrophoresis that the recircularization was successful.
3. Prepare a stock of the desired recombinant virus using a limiting dilution transfection protocol (*see* **Note 9**), as follows. Prepare tenfold serial dilutions of the ligated DNA in sterile 10 m*M* Tris-HCl, pH 7.8, 10 m*M* EDTA. To 100 µL of each dilution of DNA, add 80 µL of 5 mg/mL DEAE-dextran (sterile and freshly prepared) and 220 µL of 2X DME + 10% fetal calf serum. Use dilutions ranging

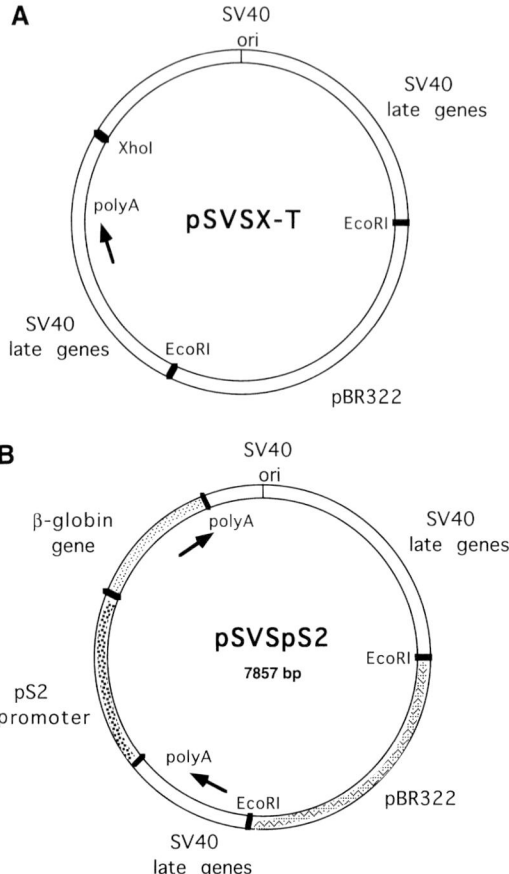

Fig. 1. Diagrams of plasmids instrumental in constructing recombinant SV40 viruses. (**A**) Diagram of pSVSX-T, a fusion of pBR322 and SV40 DNA sequences through their respective *Eco*RI sites. The SV40 late genes are interrupted by pBR322 sequences. The positions of the SV40 late polyadenylation site (polyA), the SV40 origin of DNA replication (ori), and the unique *Xho*I site, for insertion of exogenous, cellular promoters, and reporter genes, are shown. (**B**) Diagram of pSVSpS2, a representative recombinant SV40 genome fused to pBR322 sequences. pSVSpS2 was derived from pSVSX-T by inserting human pS2 promoter sequences (–1100 to +10) and a ß-globin reporter cDNA into the *Xho*I site of pSVSX-T. In addition to landmarks shown in (**A**), the pS2 promoter sequences are shown by dark stippling and the ß-globin sequences by light stippling.

from 7.5 ng DNA to 7.5×10^{-6} ng DNA/100 µL. Trypsinize CMT cells, preparing a single-cell suspension of cells at 5×10^5 cells/mL in DME + 5 % fetal calf serum. Add 400 µL DNA to each 400-µL aliquot of cells, using sterile

microcentrifuge tubes. Invert tubes; incubate in shaking water bath at 37°C for 35 min. Centrifuge 5 min at 14,700g at 4°C in J6 centrifuge, to pellet cells. Wash each tube of cells 2X with 0.5 mL DME, removing supernatant each time with sterile, plugged Pasteur pipet. Resuspend each tube of cells in 10 mL DME, 5% fetal calf serum, 100 μM $ZnCl_2$, 10 μM $CdSO_4$ plus antibiotics. Pipet 1-mL cell suspension into each well of a 24-microwell plate. Use a set of 10 wells/dilution of ligated DNA. Include a mock-transfected, control set of cells for comparison of the morphology of the cells at subsequent times.
4. Monitor the state of the cells daily during the incubation of the microwell plates of cells for the subsequent 2–4 wk. After an initial 4 d of incubation, change the media, and as necessary thereafter, add more medium to the wells to prevent evaporation and pH changes. When cells demonstrate cytopathic effect (CPE) by rounding up and dying, collect the media from each well. The set(s) of cells to analyze further are those transfected by a given dilution of DNA, resulting in the cells in some, but not all, of the wells demonstrating CPE. These cells received the least amount of DNA necessary to achieve viral production and are therefore least likely to have undergone recombination during DNA transfection and most likely to contain unrearranged viral genomes. If the transfection method is capable of delivering one molecule of DNA per cell, one can be 95% confident that CPE in a well resulted from a single DNA molecule entering the cells when only 10% of the cultures receiving this dilution of DNA develop infection.
5. Harvest the virus by collecting the media and cells after pipetting up and down to detach any remaining cells attached to the plate. Sonicate the cellular suspension in a cup sonicator for 1 min to release any attached virus into the supernatant, and sediment the cellular debris from the sonicated suspension by centrifugation at low speed (*see* **Note 10**). To store the viral stock, quick freeze the supernatant in a dry ice:ethanol bath and store in a –70°C freezer. Such SV40 viral stocks are stable for several years with only minimal loss of titer.
6. Analyze the genomic DNA in the virus stocks by incubating 10 μL viral stock, 2 μL 10X dissociation buffer, and 8 μL H_2O at room temperature for 30 min. Add 20 μL of 2X proteinase K and incubate an additional 30 min at 37°C. Extract with phenol: chloroform:isoamyl alcohol, then with chloroform, and precipitate with ethanol.
7. Purify the viral DNA, using anion exchange chromatography (e.g., Qiagen silica gel particles or glass milk), by the manufacturer's or common procedures (*see* **Note 11**).
8. Resuspend the DNA, digest with the appropriate restriction enzymes, and analyze by Southern blotting to determine which virus stock contains the desired, unrearranged, and complete viral genome sequence. To verify length of critical DNA fragments, compare with digestion of the input plasmid DNA. Pick two or more viral stocks to amplify further.
9. Amplify the viral stock, always saving some of the original isolate in case it is needed. To infect, remove medium from 40 15-cm-diameter plates of 70–80% confluent CMT cells and rinse the cells with DME. Pipet dropwise 100 μL of the viral stock uniformly across each plate, rocking briefly back and forth. Add 1 ml DME to prevent evaporation that leads to cell death and adsorb the virus for 1–2 h

at 37°C in the CO_2 incubator. Add 10 mL DME, 2% fetal calf serum, 100 μM $ZnCl_2$, 10 μM $CdSO_4$, and antibiotics to the plates (*see* **Note 12**), and incubate 2–4 wk, until the cells demonstrate uniform CPE. Always maintain mock-infected plates of cells in order to compare cell morphology. Supplement (or remove and replace) the media on the plates every 3–6 d (*see* **Note 13**). Take care to change media on the control plates separately from the infected plates to avoid cross-contamination with virus.

10. Sterilely harvest the virus stock:
    a. Scrape the cells and combine with the media, which contains a large amount of cellular debris, into a large centrifuge bottle.
    b. Sediment the cells and debris by centrifugation at 5300g in a J6 centrifuge for 5 min at 4°C.
    c. Sonicate the cellular pellet, resuspended in a small volume of media, in a cup sonicator for 1 min, to release any attached virus into the supernatant.
    d. Sediment the cellular debris from the sonicated suspension, as above.
    e. Combine the supernatant from the sonicated pellet with the supernatant media from the first centrifugation step.
    f. Aliquot the virus stock, freeze the aliquots in a dry ice:ethanol bath, and store in a –70°C freezer (*see* **Note 10**).

11. Titer the recombinant SV40 stock by the end-point dilution method *(38)* (*see* **Note 14**). Prepare tenfold serial dilutions of the virus stock in DME in a final vol of 200 μL, in sterile microcentrifuge tubes. Make certain to change pipettes for each subsequent dilution. Dilutions should range from 10^{-1}–10^{-8} (*see* **Note 15**). In addition, prepare intermediate dilutions of 5×10^{-6}, 5×10^{-7}, and 5×10^{-8}. For each dilution, ranging from 10^{-3}–10^{-8}, use 100 μL of the virus per well to infect at least four wells of a 24-microwell plate that are 70–80% confluent with CMT cells. Include mock-infected control wells, using 100 μL DME. Following adsorption of the virus at 37°C for 1–2 h, remove the medium and replace with fresh DME plus 2% fetal calf serum, 100 μM $ZnCl_2$, 10 μM $CdSO_4$, and antibiotics. Monitor the state of the cells daily for CPE during the incubation of the microwell plates of cells for the subsequent 2–4 wk. After an initial 4 d of incubation, and as necessary thereafter, add more medium to the wells to prevent evaporation and pH changes. The final percentage of wells that develop CPE at the varying dilutions of virus can be used to determine the 50% tissue culture infectious dose ($TCID_{50}$), which can be converted to plaque-forming units (pfu) by the following equation: $TCID_{50}/mL \times 0.69 = pfu/mL$. An approximate titer of the virus, sufficient for all subsequent procedures in the isolation of MCs, is obtainable by extrapolating from the percentage of wells with CPE at each dilution of virus to what viral concentration would result in a 50% infection rate. A more precise determination of the $TCID_{50}$ (at least 10 wells of cells required per virus dilution), using the methodology of Reed and Muench, is fully described in many laboratory manuals regarding baculovirus *(38)*.

12. To verify that the amplified viral stock is not contaminated with variants, which might ultimately overgrow the population upon propagation (*see* **Note 16**),

recombinant viral DNA from the expanded stock must be prepared and analyzed. Infect 20–30 15-cm-diameter tissue culture plates of CMT cells, at 70–80% confluence, with virus, ideally at a multiplicity of infection (moi) of between 5–10 pfu/cell to ensure infection of 100% of the cells. At lower moi's, fewer cells would be infected resulting in lower yields of DNA (e.g., at a moi of 1 pfu/cell, 63% of the cells are infected, as determined by the Poisson distribution curve). As described above, remove media from the cells and rinse with DME prior to addition of virus and adsorption for 1–2 h in a CO_2 atmosphere at 37°C. Then, add DME plus 2% fetal calf serum, 100 μM $ZnCl_2$, 10 μM $CdSO_4$, and antibiotics to a total volume per plate of 15 mL. Harvest the viral DNA between 48 and 72 h postinfection (*see* **Note 17**):

a. Remove the media.
b. Wash the cells with PBS.
c. Add 2 mL Hirt extraction buffer per plate, rocking to cover the entire plate.
d. Incubate 15 min at room temperature.
e. Add 0.5 mL of 5 M CsCl per plate and mix well.
f. Scrape with rubber policeman into centrifuge tubes (30-mL Corex tubes).
g. Incubate on ice overnight (or for at least 2 h).
h. Centrifuge at 4°C for 60 min at 21,000g in an SS-34 rotor in a high speed centrifuge.
i. Collect the supernatant (the pellets should be firm).
j. Add 0.815 g CsCl/mL of supernatant; then add 0.1 mL of 10 mg/mL ethidium bromide/mL of supernatant.
k. Centrifuge in a NVT90 rotor at 410,000g for 4 h (long enough to pellet contaminating RNA).
l. Remove the lower DNA band, and repeat the CsCl-ethidium bromide centrifugation step if transcription-grade DNA is desired.
m. Extract the DNA solution three times with isopropanol to remove the ethidium bromide.
n. Dialyze the DNA in two changes of 1 L of 10 mM Tris-HCl, pH 8.3, 1 mM EDTA.

The viral DNA can be analyzed by restriction enzyme digestion, Southern blotting, and sequencing, to ensure that the sequence structure is as desired. This DNA also provides an essential reagent for comparing the properties of chromatin and DNA templates when transcribing the corresponding MCs in vitro (*see* **Subheading 3.4.**).

3.1.2. Propagation of SV40

1. Wild type SV40 virus is propagated in CV-1 cells. Our CV-1 cells are propagated in DME + 10% calf serum in a 5% CO_2 atmosphere at 37°C. Infected cells are maintained in DME + 2% calf serum. In order to propagate viral stocks, infect CV-1 cells with SV40 virus at a moi of 0.001 (*see* **Note 16**). Proceed with the infection and harvest as described in **steps 9** and **10** in **Subheading 3.1.1.**, growing the CV-1 cells in the presence of calf serum, and in the absence of zinc and cadmium divalent cations.

2. Wild type SV40 is most directly titered by plaque assay, using CV-1 cells. Prepare tenfold serial dilutions of the virus stock, from 10^{-1}–10^{-8}, into DME. Make certain to change pipettes for each subsequent dilution. Rinse ten subconfluent, 60-mm-diameter plates of CV-1 cells with DME. Add 0.1 mL of each viral dilution dropwise over the entire surface of the plate, in duplicate. Use the 10^{-5}, 10^{-6}, 10^{-7}, and 10^{-8} viral dilutions, as well as duplicate control plates that are mock-infected with just DME. Allow viral adsorption by incubating in the 37°C CO_2 incubator for 1–2 h. During that time, prepare a 1.8% bactoagar solution by heating the agar in water in a microwave oven. Incubate in a 48°C water bath, allowing the agar sufficient time to cool to 48°C. Just prior to overlaying the infected plates of cells, combine equal volumes of 2X DME + 4% calf serum + antibiotics with 1.8% bactoagar. Immediately add 5 mL of the mixture onto each plate of cells, pipetting down the inside wall of the plate so as to avoid any disturbance in the lawn of cells. Permit the agar to set for approx 30 min before moving the plates back into the 37°C incubator. Monitor the plates of cells for the next 1–2 wk for the appearance of plaques. Refeed the plates at least once, by overlaying an additional 2–3 mL bactoagar:DME:2% calf serum:antibiotics over the previous agar. To facilitate final quantitation of the numbers of plaques, prepare the bactoagar:DME:serum solution as before, but including neutral red to a final concentration of approximately 0.01%. Overlay each plate with 2 mL of this mixture. Wrap the set of plates in aluminum foil to exclude light. One to four days later, count the plaques (clear circles on a red background) on plates where plaques are distinct and well separated. This will provide the titer of the original virus stock, as expressed in pfu per milliliter of stock.
3. If the virus stock has not been carefully propagated, and/or analysis of its DNA reveals the presence of multiple viral genomes in the stock, the virus requires purification. This can be accomplished most readily by plaque purification (*see* **Note 18**). Obtain plaques from the viral stock as described (*see* **Subheading 3.1.2., step 2**). On a plate on which plaques are well separated, pick plugs containing plaques by puncturing through the entire agar overlay with a sterile Pasteur pipet. Combine each plug with 1 ml DME plus antibiotics. Harvest by three cycles of freezing in dry ice:ethanol bath and thawing in a 37°C water bath. To store, refreeze in dry ice:ethanol bath and keep in –70°C freezer. To ensure purity of the virus stock, repeat the plaque purification procedure once a plaque isolate has been identified that contains the wild type genomic sequence. Then, amplify the viral stock as described (*see* **Subheading 3.1.2., step 1**).
4. As controls for structural and transcriptional analyses with SV40 MCs, prepare and purify SV40 genomic DNA by the Hirt extraction protocol. This protocol is described (*see* **Subheading 3.1.1., step 12**) for recombinant viruses requiring growth in CMT cells. Perform parallel procedures, except using CV-1 cells and media containing calf serum, without the $ZnCl_2$ or $CdSO_4$. Quantitate the amount of viral DNA by obtaining the UV absorbance of a dilution of the DNA.

3.2. Preparation of SV40 MCs

3.2.1. Isolation of Native, Intracellular SV40 MCs

1. For wild-type SV40, remove media from ten to twenty 15-cm plates of CV-1 cells, at 70–80% confluence. Rinse plates with DME without serum. Infect with virus at a moi of 5–10 pfu/cell, by pipetting the viral stock dropwise around the entire plate, rocking the plate back and forth, and adsorbing for 1–2 h in the CO_2 incubator at 37°C. To each plate, add 15 mL DME, 2% calf serum, and antibiotics, and return the cells to the incubator (*see* **Note 19**). For recombinant SV40 virus, infect 30 15-cm plates CMT cells using the same general protocol, except that 24 h prior to viral adsorption, replace the media with DME, 2% fetal calf serum, 100 μM $ZnCl_2$, 10 μM $CdSO_4$, and antibiotics. Continue to use this media throughout the remainder of the infection process.
2. Twenty-four hours postinfection, remove the media from the cells. Replace with media containing 75 μCi [^{3}H]-thymidine per plate, in order to radiolabel the viral DNA.
3. Forty-eight to seventy-two hours postinfection, at a time when most of the cells remain attached to the plate, remove the media from the plates. Keep the plates on ice for all subsequent steps. Wash each plate with 10 mL rinse buffer, previously stored at 4°C. Remove buffer. It is convenient to process in groups of 10 plates, so that the cells remain cold and are processed rapidly throughout the following procedures.
4. Add 5 mL lysis buffer per plate. Transfer plates to a cold room; scrape with a rubber policeman. Transfer cells and buffer to 50-mL centrifuge tube (cells from five plates per centrifuge tube).
5. Centrifuge at 500g for 10 min at 4°C in a tabletop centrifuge (IEC), to gently pellet the nuclei.
6. Discard the supernatant. Resuspend each pellet in 10 mL lysis buffer + 0.1 M NaCl. Dounce the cells, using 20 strokes, in a 15 mL dounce homogenizer with a "B" pestle (*see* **Note 20**).
7. Centrifuge again at 500g for 10 min at 4°C in a tabletop centrifuge to pellet nuclei.
8. Discard the supernatant. Combine the pellets of nuclei resulting from 10–30 plates. Add 1.0–2.0 mL extraction buffer per pellet (using roughly 1 mL/10–15 plates of cells). Dounce with 10 strokes in a 1 mL Dounce homogenizer with a "B" pestle. Incubate on ice for 3–4 h. Dounce with 1–2 strokes every 15–20 min.
9. Centrifuge at 500g for 10 min at 4°C in the tabletop centrifuge.
10. Load the supernatant onto a 10 mL linear 15–30% (w/w) sucrose gradient. Centrifuge in a Beckman SW41 rotor at 180,000g for 100 min at 4°C.
11. Immediately fractionate the gradients in the cold room, by puncturing the bottom of the tubes with a needle. Collect 0.5 mL fractions.
12. To determine the position of the peaks of SV40 nucleoprotein complexes (both previrions and free MCs), remove 10 μL of each fraction for counting in scintillant with a liquid scintillation counter. **Figure 2** demonstrates a representative profile.
13. To quantitate the yield of MCs (*see* **Note 21**), remove an additional 10 μL of each fraction. Add 2 μL 10% SDS, 1 μL of 1% bromophenol blue dye. Heat at 50°C for 20 min. Load onto a 1.2% agarose gel. Include in the gel samples of increas-

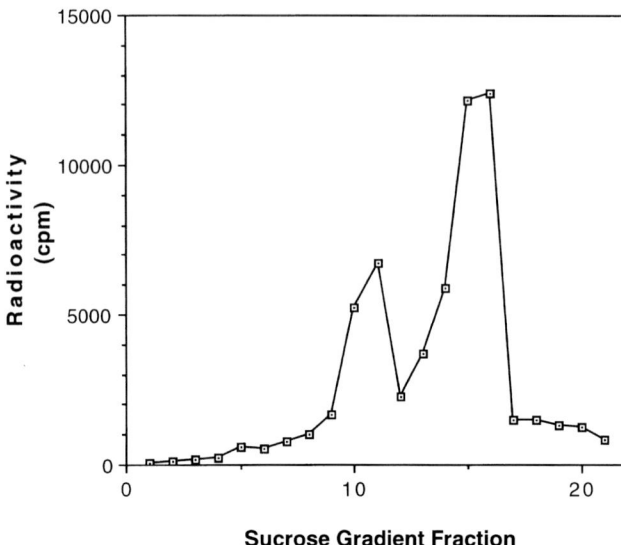

Fig. 2. Sucrose gradient profile of MCs from infection with a defective viral genome. CMT cells were infected with the virus XS10 *(56)*, an SV40 genome that lacks sequences from nucleotide 73 through 272, including SV40 enhancer sequences and three of the six GC-boxes in the SV40 promoter region. XS10 is nonviable in CV-1 cells, as it is unable to support synthesis of the early gene products (SV40 T and t antigens), however it can be propagated in CMT cells. The MCs were isolated and purified by sucrose gradient sedimentation essentially as described in the text. The amount of radioactivity (in cpm, from labeling of the DNA with [^{3}H]thymidine) in each fraction is shown. The bottom of the gradient is represented by fraction 1, and the top of the gradient by fraction 21. The previrions sedimented mainly in fractions 10 and 11, and the free MCs in fractions 14–16. Note the shoulder on the heavy side of the free MCs peak (e.g., fraction 13), representing MCs that are actively replicating and transcribing. This gradient was not sedimented as long as usual; in most cases the peaks sediment 2–3 fractions further down the gradient.

ing dilutions of SV40 DNA or recombinant SV40 viral DNA, as obtained by Hirt extraction (*see* **Subheading 3.1.2.**, **step 4** or **Subheading 3.1.1.**, **step 12**, respectively). Following electrophoresis in TAE buffer, stain with 0.5 µg/mL ethidium bromide, destain with H_2O, and compare the fluorescence of the MCs fractions with that of the known quantities of viral DNA. Typical concentrations of MCs obtained in the peak fractions are: 50–150 µg/mL for wild-type SV40 MCs, and 10–25 µg/mL for recombinant SV40 MCs.

14. Aliquot the sucrose gradient fractions containing MCs into 50-µL samples. Quick freeze in liquid nitrogen. Do not pool the MCs fractions prior to storage, as some fractions contain MCs with differing properties. Store at –70°C. The MCs are structurally and functionally stable in this condition for several months.

3.2.2. Isolation of Core MCs

1. For isolation of MCs associated only with core histones, prepare nuclear extracts containing MCs, as above (*see* **Subheading 3.2.1., steps 1–9**). Add NaCl to a final concentration of 0.5 M, and incubate for 10 min at 4°C. Load the extract onto a 10 mL linear 15–30% (w/w) sucrose gradient, containing 0.5 M NaCl. Centrifuge in a Beckman SW41 rotor at 180,000g for 100 min at 4°C. Collect the fractions and identify the location of the core MCs as described above (*see* **Subheading 3.2.1., steps 11** and **12**). The peak of core MCs sediments substantially less rapidly than native MCs through the gradient (*7*).
2. Pool the fractions of core MCs. Dialyze overnight against 50 mM HEPES-NaOH, pH 7.5, 15% sucrose. Quick freeze aliquots in a dry ice:ethanol bath, and store at –70°C.

3.2.3. Preparation of Core MCs, Reconstituted with Linker Histone

1. Prepare core MCs as described above (*see* **Subheading 3.2.2., step 1**), by sedimentation of the crude MC extract preparation through a linear sucrose gradient in high salt. Quantitate the amount of core MCs by analyzing the DNA content, as described in **step 13** in **Subheading 3.2.1.** Pool the sucrose gradient fractions containing the core MCs.
2. Determine the nucleosome concentration of the core MCs preparation, assuming that there are 24 nucleosomes per SV40 DNA. Add pure histone H1 (Boehringer-Mannheim) to the core MCs solution, at an input ratio of H1/core nucleosome of 1.0 (*see* **Note 22**). Dialyze overnight against 50 mM HEPES-NaOH, pH 7.5, 15% sucrose.
3. Purify the reconstituted histone H1/core MCs by sedimentation through a linear 15–30% (w/w) sucrose gradient, as described in **step 10** in **Subheading 3.2.1.** for native MCs. Identify the peak (*see* Note 22), aliquot the H1/core MCs, freeze, and store as described for native MCs (*see* **Subheading 3.2.1., steps 11–14**).

3.2.4. Preparation of MCs Containing More Highly Acetylated Core Histones

1. Infect cells with wild-type or recombinant virus, and radiolabel with [^{3}H]thymidine (*see* **Subheading 3.2.1., steps 1** and **2**).
2. Four hours prior to harvesting cells for MCs, add 300–400 nM trichostatin A to the media of the cells (*see* **Note 23**).
3. Continue to harvest the cells and isolate the MCs (*see* **Subheading 3.2.1., steps 3–14**).

3.3. Structural Analyses of SV40 MCs

3.3.1. Compaction State of the MCs: Sedimentation Analyses

1. Incubate the isolated MCs in the experimental buffer or with the experimental additional components in 50 µL. Centrifuge the samples through a 2-mL linear sucrose gradient (15–30%, w/w), in 10 mM HEPES-NaOH, pH 7.5, with

additional components as desired, for 65 min at 106,000g in a Beckman TLS-55 rotor at 5°C.

2. Collect 100 µL fractions, and monitor the position of the MCs by measuring the amount of [^{3}H] in each fraction (*see* **Subheading 3.2.1., step 12**).

3.3.2. Nucleosome Positioning: Enzymatic and Chemical Probes

1. Under conditions that lead to an extent of modification of the DNA between 1 bp per 150 bp and 1 bp to 500 bp, incubate the isolated MCs (*see* **Note 24**) with either enzymatic probes (e.g., DNase, micrococcal nuclease) or chemical probes (e.g. dimethyl sulfate, followed by strand scission with piperidine), as described in other chapters in this volume. Extract and isolate the purified DNA.
2. With reagents that lead to a 5'-phosphate terminus on the DNA (e.g., dimethyl sulfate, MPE *[39]*, and DNAse I *[40,41]*), analyze the positions of either single-stranded or double-stranded cleavage sites in the DNA by the standard ligation-mediated polymerase chain reaction (LMPCR) methodology with a nested set of primers (*see* Chapter 31, **refs.** *30,42,43*). For mapping only double-stranded cleavages, useful for analysis of micrococcal nuclease cleavages to map boundaries of nucleosomes *(44)*, ligate double-stranded linkers directly to the cleaved DNA, rather than first extending a specific oligonucleotide to create double-stranded termini before ligation with the linkers. For reagents creating a 5'-hydroxyl terminus (e.g., micrococcal nuclease), phosphorylate the 5'-termini prior to ligation and PCR analysis *(45)*.

3.4. Transcriptional Analyses of SV40 MCs

3.4.1. Transcription in HeLa Whole-Cell Extract

1. Determine the optimal amount of extract, the optimal concentration of purified SV40 viral DNA or recombinant SV40 viral DNA (*see* **Note 25**), and the optimal concentrations of monovalent cation and magnesium for transcription from any particular extract. For HeLa whole cell extracts (*see* **Note 26**), at protein concentrations of approximately 20 mg/mL, begin by titrating the extract amounts using 10 µg/mL DNA per reaction (*see below* for details on each transcription reaction). Subsequently, using the optimal amount of extract, titrate the concentration of viral DNA in the reaction to achieve maximal transcription. Finally, titrate monovalent and divalent cation concentrations using the desired DNA template (*see* **Note 27**). The final salt concentrations, including the salts added in the extract buffer generally fall in the following ranges: 40–60 mM KCl, and 4–6 mM MgCl$_2$. Glycerol (from the extract buffer) should be kept below 8%.
2. As a first step to determine the overall efficiency of transcription from a particular DNA or minichromosomal template, the following protocol produces uniformly labeled RNA transcripts over a one hour time course (*see* **Note 28**). The final total DNA concentration in the transcription reactions should equal the optimal DNA concentration determined in **step 1**. If the MC preparation is not concentrated enough to supply optimal amounts of DNA, add poly [d(I-C)]·[d(I-C)] as carrier DNA (*see* **Note 7**) to bring the total DNA concentration to the appropri-

ate level. Mix together on ice: 0.8 µL of 25X NTPs, 10 µCi [α-^{32}P]UTP, KCl and/or MgCl$_2$ (as required to obtain the final, optimal concentrations), H$_2$O (such that the final reaction volume will ultimately be 20 µL), and poly [d(I-C)]·[d(I-C)]. Add either MCs (up to 7 µL) (*see* **Note 29**) or viral DNA. When a comparison is desired between transcription from the MCs vs. viral DNA, supplement the transcription reaction containing viral DNA with 22.5% sucrose at the same volume as the volume of MCs that are being transcribed. Add HeLa whole cell extract (up to 7 µL), mix, and incubate the reactions for 1 h at 30°C (*see* **Note 30**).

3. Add 20 µL of 2X proteinase K solution to terminate the reaction. Incubate at 37°C for 10 min.
4. Add 200 µL extraction buffer and 2 µL tRNA stock solution. Extract once with an equal volume of phenol:chloroform:isoamyl alcohol; centrifuge. Extract the top layer once with an equal volume of chloroform; centrifuge.
5. Add 200 µL of 2 *M* ammonium acetate to the top layer; vortex. Precipitate the nucleic acids with 1 ml absolute ethanol for 30 min in a dry ice:ethanol bath. Centrifuge at 4°C for 15 min in a microcentrifuge at 15,000*g*. Discard the supernatant and air dry the pellet.
6. Resuspend the pellet in 10 µL 5X DNase buffer and 35 µL H$_2$O. Add 5 µL of 2 mg/mL RNase-free DNase. Incubate at 30°C for 20 min.
7. Add 50 µL of 2X proteinase K solution. Incubate at 37°C for 10 min. Extract once with an equal volume of phenol:chloroform:isoamyl alcohol; centrifuge. Extract the top layer once with an equal volume of chloroform; centrifuge.
8. Add 100 µL of 0.5 *M* ammonium acetate to the top layer; vortex. Precipitate the nucleic acids with 0.5 mL absolute ethanol for 30 min in a dry ice:ethanol bath. Centrifuge at 4°C for 15 min in a microcentrifuge at 15,000*g*. Discard the supernatant and air dry the pellet.

3.4.2. Hybridization and RNA Analysis
(see **Fig. 3** for Diagram of RNA Analysis)

1. For analysis of the *de novo* synthesis of two, separate transcripts from each transcription reaction (often one control and one experimental, or in the case of wild type SV40, the early transcripts and the late transcripts), resuspend the RNA pellet fully in 15 µL of 2X hybridization buffer and 12 µL H$_2$O. Divide sample into two tubes.
2. To each tube, add 1.5 µL of 0.2 mg/mL single-stranded DNA complementary to the transcript of interest (*see* **Note 31**). Heat for 10 min, completely submerged to prevent evaporation and condensation of the liquid, in a 68°C circulating water bath (a lucite submarine is useful for this purpose). Hybridize, completely submerged, for at least 2 h in a 50°C circulating water bath. (It is often useful to perform this incubation overnight.)
3. Rapidly add 200 µL T1 RNase buffer (at 4°C) to quench the hybridization reaction. Add 1 µL of 5 U/µL T1 RNase; mix well. Incubate at 30°C for precisely 30 min (*see* **Note 32**); quench by addition of 5 µL of 5 mg/mL proteinase K and 2 µL of 10 mg/mL tRNA. Incubate for 60 min at 37°C.

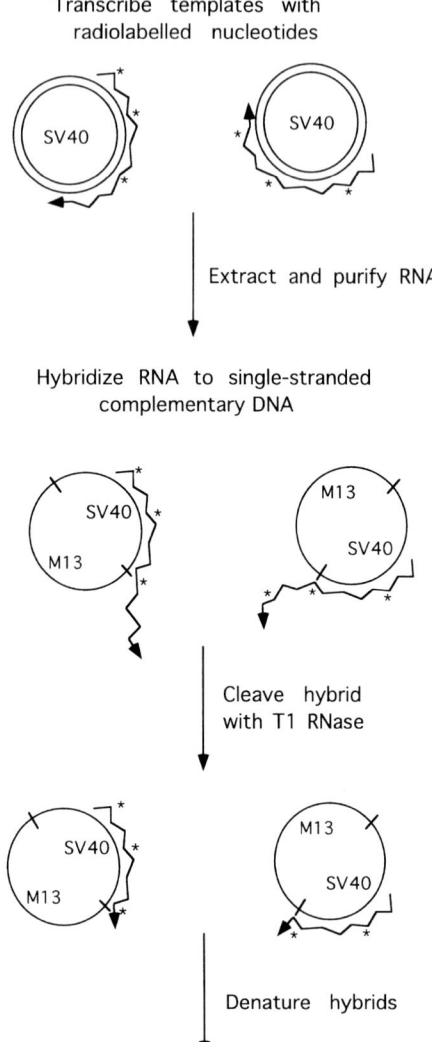

Fig. 3. Diagram of the procedure for analyzing RNA synthesized from MCs. In this case, synthesis of RNA from two separate SV40 genomes is shown. The nucleosomes were omitted in order to simplify the diagram.

4. Add 1 mL absolute ethanol; vortex. Precipitate the nucleic acids by incubation in a dry ice:ethanol bath for 30 min. Centrifuge at 4°C for 15 min in a microcentrifuge at 15,000g. Discard the supernatant and air dry the pellet.

SV40 Minichromosomes: Model for Active Chromatin

5. Resuspend the pellet in 15 µL sample loading buffer. Heat at 80°C for 5–10 min; quick chill in ice water bath.
6. Prepare a 1.5-mm-thick, 25-cm-long, 8.3 M urea-1X TBE-5% polyacrylamide (29:1 acrylamide:bisacrylamide) gel. Pre-electrophorese in 1X TBE running buffer at 400 V for 45 min.
7. Pipet buffer into the sample wells of the gel, to squirt out urea that has settled in the slots. Load samples into the wells; electrophorese in 1X TBE at 400 V. Soak the gel in 5% glycerol for 20 min. Dry and autoradiograph.

4. Notes

1. SV40 is a primate virus that grows lytically in cells of certain species of monkeys, in particular the African green monkey. Human cells are semipermissive for SV40 replication, resulting in a burst size about 10% that from permissive monkey cells. (The apparent defect for full permissivity lies with viral DNA replication and not transcription.) Thus, human cells should also yield adequate quantities of MC for the experiments described below. The virus is nonpathogenic in humans, as determined inadvertently due to its presence as a contaminant in certain lots of the poliomyelitis virus vaccine in the 1950s *(46)*. Subsequent studies of the millions of people inoculated with these vaccines have indicated no disease consequences. Thus, no particular biohazard precautions, above those usually in place for research involving mammalian cells and viruses, need be taken in the laboratory when using either the wild-type virus or most recombinant viruses of the type described here, in which cellular promoters are propagated in the context of a recombinant viral genome.
2. We always isolate MCs from infected, rather than transfected cells, for two reasons. First, viral infection results in all of the DNA being in a native MC structure, whereas this is questionable for transfected DNA. Although under some conditions, transfected SV40 DNA is appropriately packaged into chromatin *(47–50)*, this generally requires that only limited amounts of DNA enter the cells. Obtaining an appropriate chromatin structure on the majority of transfected DNA cannot be assumed without substantial verification. Second, viral infection results in functional DNA entry into 100% of cells, whereas DNA transfection results in functional entry into only 5–30% of cells.

    To counter concerns regarding the high copy number of SV40 MCs in an infected cell and whether this might titrate out important, limiting cellular component(s) that would be desirable to isolate on the SV40 MCs at high stoichiometry, the replication of the virus can be limited (although the yield of isolated MCs would correspondingly decrease). The viral replication inhibitor cytosine arabinoside *(14)* can be added to the medium during viral infection. By varying the time of addition of the inhibitor and/or its concentration, the degree of replication of SV40 MCs in the infected cells can be modulated. For recombinant viruses lacking SV40 large T antigen coding sequences, DNA replication can be prevented by removing the source of T antigen (i.e., infecting CV-1 cells instead of CMT cells).

3. Alternatively, the subpopulation of isolated MCs that are not associated with endogenous RNA can be used for transcription in vitro. SV40 MCs sedimenting to the light side of the sucrose gradient peak do not contain RNA that was initiated in vivo and are not associated with basal transcription factors *(13)*. Thus, transcripts produced from such MC templates can be directly analyzed by any of the usual techniques (primer extension, S1 nuclease analysis, and so forth). However, controls (such as the addition of a-amanitin during the transcription reaction) should always be performed to verify the absence of endogenous RNA on the isolated templates.
4. Although the widely employed COS cells also produce constitutive levels of SV40 large T antigen, it is crucial to use CMT cells, rather than COS cells, for propagation of recombinant SV40 viruses. COS cells contain integrated SV40 regulatory sequences in addition to the integrated T antigen gene. These regulatory sequences are also present in all recombinant viral genomes. In contrast, CMT cells only contain the coding sequences of T antigen, as T antigen expression is driven by the exogenous metallothionein promoter. Therefore, recombination between viral sequences and integrated sequences is possible in COS cells, but not in CMT cells (because the SV40 T antigen coding region is removed in the recombinant viral genomes). Such recombination would generate unwanted viral progeny. CMT cells are also useful because the exogenous metallothionein promoter is inducible by some divalent cations, such as zinc and cadmium, to yield even higher levels of T antigen.
5. The concentration of DTNB required to maintain the structure of the previrions appears to vary, depending on the cells and on the virus used. Thus, a range of concentrations is given here. The use of DTNB in the extraction buffer is not always necessary, but should be tested for any particular source of cells for whether its presence in this buffer affects the resulting MCs sucrose gradient profile.
6. Many different types of transcription extracts can be used in these assays, to study either RNA polymerase II (pol II) or RNA polymerase III (pol III) transcription. Other commonly used transcription systems are: Shapiro nuclear extract *(51)*, Dignam nuclear extract *(52)*, or partially or fully purified basal transcription factors *(13,53,54)*, sometimes supplemented with specific activator and coactivator proteins. For different types of extracts, the optimal conditions for transcription vary.
7. This simple DNA is used as carrier DNA, to bring the amount of total DNA in the crude extract reactions to an optimal concentration for transcription. Generally, transcription from a template is inhibited at low concentrations of DNA, presumably due to the binding to the DNA of inhibitory proteins in the extract. One can obtain a linear response of transcripts to template, however, if carrier DNA is added to optimal levels *(55)*. Note that the alternating heteropolymer, poly [d(I-C)]·[d(I-C)], **not** the complementary homopolymeric strands, poly dI·dC, is used. This carrier DNA is the best when using radiolabeled UTP to label the transcripts, as the transcripts from the carrier DNA will only contain CMP and GMP, and

therefore no background, radiolabeled RNA will be generated from the carrier DNA. [Under these circumstances, the 25X NTPs contain low UTP concentrations, for efficient labeling.] For radiolabeling with either GTP or CTP, poly [d(A-T)]·[d(A-T)] should be used as carrier, instead, and the 25X NTPs mixture should contain low CTP or GTP, respectively. Because of the separate ATP/dATP requirement for initiation of transcription *(53)*, it is not advised to radiolabel transcripts with ATP, unless the reactions are also supplemented with a high concentration of dATP. The average length of the DNA strands in the heteropolymer should be maximized; if strands are too short, this carrier DNA will not stimulate transcription at limiting concentrations of template DNA. Shorter lengths of carrier strands lead to a higher concentration of ends and nicks in the DNA, which apparently limit effective competition for the inhibitory DNA-binding proteins in the extract.

8. pSVSX is a derivative of pSVS *(56)*, a recombinant plasmid, fused at respective EcoRI restriction sites, incorporating the complete 776 SV40 viral genome and a portion of pBR322 containing just the ampicillin resistance gene and the origin of plasmid DNA replication. pSVSX was constructed by fragment exchange of XSLR20 *(57)* sequences into pSVS, thereby inserting a unique XhoI site at a HpaI site at position 2666 in the SV40 genome. As described in the text, pSVSX-T (**Fig. 1A**) was subsequently constructed by removal of T antigen sequences. Although pSVSX-T is functional for construction of novel recombinant viral genomes, it is not ideal. We are currently generating a more widely useful parental plasmid, with the following characteristics:
   a. It will contain a polylinker cloning site at the point of insertion of foreign DNA.
   b. It will fuse SV40 sequences to bacterial plasmid sequences using a restriction enzyme site other than *Eco*RI, which would be less likely to be present in exogenous DNA inserts.
   c. It will contain polyadenylation sequences flanking the inserted DNA in both directions, to maximally insulate it from viral transcriptional controls (pSVSX-T only contains a polyadenylation sequence on one side of the position of insert).

9. The limiting end point dilution method of isolating the virus stock *(38)* ideally generates a viral stock from a single, transfected molecule of DNA. Our experience has shown this protocol to be critical, in order to minimize rearrangements and reiterations in DNA sequences because of recombination that commonly occurs during the transfection process. If such rearranged sequence variants contain a selective advantage during viral replication, they can overgrow the desired genomic sequence. For transfection, the DEAE-dextran protocol is described here. Other transfection protocols can also be used for introducing the ligated recombinant viral DNA into CMT cells, although one must take care to use a protocol that will introduce a minimal number of DNA molecules per cell. Thus, the $CaCl_2$ precipitate transfection protocol is not recommended.

10. Although virus is released into the medium following a lytic cycle, resulting in the death of the cell, virus can readsorb onto cellular debris. Thus, care is taken to

include the cellular debris in preparation of the viral stock. As an alternative to sonication, virus can be released by a freeze-thaw protocol: freeze the combined media and cells in a dry ice:ethanol bath, thaw in a 37°C circulating water bath, and repeat two additional times. Sediment the cellular debris (*see* **Subheading 3.1.1.**, **step 5**), then refreeze and store at –70°C.
11. This DNA purification step may not be necessary for subsequent digestion with certain restriction enzymes.
12. Once cells are infected, it is advisable to lower the percentage of serum in the medium, so as to maintain the viability of the cells without requiring additional cell growth. The lower amount of serum is also complemented by viral infection, as SV40 induces cells both to reenter the cell cycle and to progress into S phase. For CMT cells, the presence of $ZnCl_2$ and $CdSO_4$ induce higher levels of T antigen in the cells, leading to a higher yield of virus in a shorter period of time.
13. Virus that is released after lytic infection of a particular cell is quickly readsorbed onto other uninfected cells, thus replacement of media results in little eventual loss in viral titer. Nonetheless, simple supplementing of the existing medium with additional, fresh medium is preferred.
14. CMT cells cannot be used in the normal fashion for plaque assays, as they do not survive for long periods of time under agar. However, plaque assays are more precise for determining titer. Therefore, if desired, the following modifications can be made to the plaque assay (*see* **Subheading 3.1.2.**, **step 2**) to enable the use of CMT cells in such an assay. Perform all the steps as described (although using fetal calf serum instead of calf serum, and adding $ZnCl_2$ and $CdSO_4$ to the medium) through the addition of the initial bactoagar mixture to the cells. Monitor the state of the cells in subsequent days, and when they begin to look sick (every 4–7 d), carefully remove the agar overlay (without disturbing the cells), replate more CMT cells on top of the existing infected lawn, and overlay once more with bactoagar solution. Stain the cells and count the plaques (*see* **Subheading 3.1.2.**, **step 2**).
15. Normal viral titers of wild type SV40 range from 3×10^7 to 5×10^8 pfu/mL, and of recombinant SV40 viruses from 5×10^5 to 5×10^6 pfu/mL, depending on the growth characteristics of the virus.
16. When SV40 viral stocks are propagated at high multiplicities of infection, the resulting stocks contain fewer and fewer infectious particles, although the number of total particles remains roughly constant. The noninfectious particles contain viral DNA that is deleted, reiterated, inverted, and in some cases mixed with cellular DNA sequences, because of recombination and rearrangements of the DNA during infection. So that they can propagate, the defective viruses must contain viral origin sequences, which are often repeated multiple times per genome. To maintain the viral stock with a high percentage of fully infectious viral genomes, it is therefore essential, when propagating virus for preparation of a stock, to infect at an moi ranging from 0.01–0.001, to ensure that cells are only infected with a single particle, therefore leading to amplification of only fully infectious viruses.

17. The original Hirt extraction method uses NaCl rather than CsCl. In this original protocol, the DNA in the Hirt supernatant is purified by extraction with phenol:chloroform, extraction with chloroform, and is finally precipitated with ethanol. The use of CsCl facilitates the DNA purification, by using CsCl equilibrium density gradient centrifugation. Instead of dialysis of the DNA after purification, one can also precipitate the DNA with ethanol, as long as the DNA-CsCl solution is first diluted to prevent subsequent precipitation of the CsCl.
18. Purification of a virus stock can also be accomplished by the limiting end point dilution method, as described in reference *(38)*. Briefly, infect CV-1 cells with ten-fold serial dilutions of SV40 (or CMT cells with tenfold serial dilutions of recombinant virus; *see* **Subheading 3.1.1., step 11**). Monitor the state of the cells daily for CPE during the incubation of the microwell plates of cells for the subsequent 2–4 wk. After an initial 4 d of incubation, and as necessary thereafter, add more medium to the wells to prevent evaporation and pH changes. One can be 95% confident that CPE in a well resulted from a single infectious virus entering the cells when only 10% of the cultures receiving this dilution of virus develop infection. Analyze the viral supernatants from such wells for the appropriate genomic structure, and then use the desired supernatant to generate large viral stocks, as described above.
19. If the titer of the viral stock is such that one requires five or more milliliters to obtain a high moi, then merely supplement the media to a total of 15 mL, or wait to add media until it is required for cell viability on a later day.
20. Dounce homogenization separates residual "cytoplasmic" contamination, such as endoplasmic reticulum, from the nuclei. The nuclei should be as uncontaminated as possible prior to incubation to release MCs. The extent of cytoplasmic material surrounding the nuclei can be monitored by visualization through a microscope, to ensure that contamination is removed by the homogenization procedure, leaving "clean" nuclei.
21. For reasons that are not clear to us, the efficiency of [^{3}H]thymidine incorporation into minichromosomal DNA varies with different viruses. Because infection by different viruses leads to different specific activities of the resulting viral DNA, the amount of incorporated radioactivity does not necessarily reflect the yield of MCs.
22. When H1 is added at a molar ratio of 1.0 per core nucleosome, the resulting MCs will probably not contain saturating amounts of H1 (in our hands, the incorporated ratio was 0.5 histone H1 per nucleosome). One can incubate the core MCs with more H1, but then it is critical to verify that the associated H1 is appropriately positioned on the nucleosomes *(7)*. In contrast to bulk cellular chromatin, SV40 MCs do not aggregate at physiological salt concentrations in the presence of H1 (e.g., **refs.** *58,59*). In our experiments, the lack of precipitation at normal ionic conditions used in transcriptions was shown by sedimentation analyses and by the absence of inhibition of the level of transcriptional initiation on these templates *(7)*. The H1-core MCs sediment through a preformed sucrose gradient at a rate inbetween those of native MCs and core MCs. The unusual ability to manipulate histone H1 levels on these SV40 MCs without precipitation of the

chromatin is likely to be due to the transcriptional competency of these templates, in agreement with the data of Huang and Cole *(60)*, who identified two populations of H1-containing cellular chromatin, distinct in their solubility properties. The soluble fraction appeared to contain the transcriptionally competent chromatin.
23. As a specific deacetylase inhibitor, trapoxin can also be used *(61)*. Before these highly specific inhibitors were identified, sodium butyrate was often used to produce SV40 MCs with more highly acetylated core histones; however, this inhibitor is known to be less specific, and to effect the state of other chromosomal proteins, as well.
24. The mapping of the chromatin structure and footprinting of protein-DNA interactions can also be performed in vivo on the recombinant viruses, by incubating lysolecithin-treated infected cells with appropriate enzymes or chemical probes *(30)*. A combination of enzymatic and chemical probes will detect periodic cleavages indicative of positioned nucleosomes, in addition to mapping regions on the chromatin protected from cleavage by the specific binding of proteins. LMPCR allows for nucleotide resolution of either single-stranded or double-stranded cleavages *(44)*. With judicious choice of primers complementary only to the recombinant viral genome, the structure of the episomal, viral chromatin, and not that of the chromosomal chromatin would be monitored.
25. For transcription reactions, the viral or plasmid DNA should be highly purified, ideally by equilibrium centrifugation twice through a CsCl density gradient. Residual contaminants (i.e., RNA) in other DNA preparations can significantly inhibit the efficiency of transcription in vitro. In addition, it is inadvisable to treat any DNA preparation to be used for transcription reactions with RNase.
26. *See* **Note 6** for other transcription extracts that can be used in vitro. For ensuring that all transcription in vitro from isolated MCs is initiated *de novo*, supplement partially or fully purified basal transcription factors with a purified pol II resistant to a-amanitin *(27)*. In this system, all transcripts synthesized in the presence of a-amanitin must be newly initiated.
27. Different promoters demonstrate different salt optima *(62)*, therefore titrate separately for each promoter of interest.
28. For other types of analyses, the MCs or DNA can be preincubated with the extract in the absence of NTPs to establish stable preinitiation complexes (a 30 min incubation should be sufficient for maximal complex formation). Then, transcription is initiated rapidly by addition of nucleotides. One can follow the first round of initiation by incubating first, for a short period of time, with a nucleotide mixture containing one radiolabelled NTP at low concentration (pulse period), followed by incubation with high concentrations of all NTPs, to facilitate the elongation process (chase period). By varying the times of preincubation, pulse, and chase, the formation of stable preinitiation complexes, the rate of initiation, and the rate of elongation can be independently measured. To analyze elongation in further detail, the radioactively labelled RNA from each time point can be hybridized to a panel of single-stranded DNA molecules (*see* **ref. 63**), containing increasing lengths of

recombinant viral sequences. In this manner, the periods of time required to generate single, full-length hybrids of increasing lengths can be determined. Note: On DNA templates, sarkosyl is highly useful for studying single rounds of initiation. However, in considering these types of experiments on MC templates the utility of sarkosyl is limited, as it totally disrupts chromatin structure.

29. The sucrose in the MCs preparations will inhibit transcription, although at the sucrose concentration generally used, the reaction will normally remain efficient enough to monitor transcription. If transcription efficiency for any particular template is low, the sucrose can be dialyzed out or separated out by a molecular sieve column. However, we have not tested whether a cycle of freezing and thawing in the absence of sucrose will adversely affect the MCs structure or transcriptional potential. Thus, the MCs should be transcribed immediately after removal of sucrose (unless the investigators carefully monitor the stability to freeze:thaw cycles of MC structure and transcription properties in the absence of sucrose).

30. Do not incubate the transcription reactions performed with HeLa whole cell extract at 37°C. There is an RNase inhibitor in crude extract that is inactivated at 37°C, but remains active at 30°C. The activity of this RNase inhibitor is essential for production of RNA transcripts.

31. It is important to keep the concentration of complementary, single-stranded DNA high, in order to drive the hybridization reaction to completion within a short time frame. A number of avenues can be used for obtaining the desired complementary single-stranded DNA. We have employed two methods, based on phage M13 vectors.

   a. Clone the complementary DNA, mindful of the strandedness desired, into an appropriate phagemid such as pBluescript II-SK(+/-) (Stratagene). Package single-stranded DNA into phage particles by culturing a single colony of bacteria containing the plasmid with helper phage (e.g., VCSM13 interference-resistant helper phage; Stratagene). Pellet the bacteria completely, carefully remove the supernatant, and digest it with DNase-free RNase. Precipitate the phage from the supernatant with polyethylene glycol. Isolate the single-stranded DNA by proteinase K digestion of the phage particles in the presence of SDS, extraction with phenol:chloroform:isoamyl alcohol, extraction with chloroform, and precipitation with ethanol. Expected yield is 1–3 µg single-stranded DNA per milliliter bacterial culture.

   b. Alternatively, and a bit more laboriously, clone the complementary DNA, mindful of the strandedness desired, directly into an M13 vector such as mp18 or mp19 (**64**; New England BioLabs). With these vectors, double-stranded replicative M13 DNA is isolated, restricted, and ligated to the insert DNA. Following transformation into bacteria (JM103), pick plaques, and identify the desired genome. Infect fresh cells and grow a 1 L culture. Spin down the cells, precipitate the phage from the supernatant of the infected cell culture, and isolate the single-stranded phage DNA as above. If higher purity single-stranded DNA is necessary (if the isolated DNA appears to contain contaminants), band the precipitated phage in a CsCl equilibrium density gradient (refractive index of approx 1.363) by adding 0.43 g CsCl/mL of solution prior to centrifugation. Remove the opaque, whitish purified phage from the gradient and dialyze into 10 mM Tris-HCl, pH 8.3, 1 mM EDTA. Prepare the single-stranded DNA from the purified phage as above. Expected yield is 0.5–2 mg single-stranded DNA per liter of bacterial culture.

32. Each time a new stock solution of RNase T1 is prepared, titrate the amount of RNase T1 added to hybrids between the desired RNA and its complementary single-stranded DNA. Choose a concentration of T1 RNase that digests away background radioactive RNA species, yet does not begin to cause a smear of RNA below the expected length of the product. Ideally, the T1 RNase should cleave the nonhybridized RNA after every guanine residue; therefore the product RNA after digestion should be slightly longer than the portion hybridizing to the DNA, as it will extend to the first guanine residue after the end of the hybrid. The optimal concentration of RNase T1 can also vary slightly from one RNA-DNA hybrid to another, depending on the sequence.

References

1. Coca-Prados, M., Vidali, G., and Hsu, M.-T. (1980) Intracellular forms of Simian Virus 40 nucleoprotein complexes. III. Study of histone modifications. *J. Virol.* **36,** 353–360.
2. La Bella, F. and Vesco, C. (1980) Late modifications of Simian Virus 40 chromatin during the lytic cycle occur in an immature form of virion. *J. Virol.* **33,** 1138–1150.
3. La Bella, F., Romani, M., Vesco, C., and Vidali, G. (1981) High mobility group proteins 1 and 2 are present in simian virus 40 provirions, but not in virions. *Nucleic Acids Res.* **9,** 121–131.
4. Ding, H.-F., Rimsky, S., Batson, S. C., Bustin, M., and Hansen, U. (1994) Stimulation of RNA polymerase II elongation by chromosomal protein HMG-14. *Science* **265,** 796–799.
5. Lassar, A. B., Hamer, D. H., and Roeder, R. G. (1985) Stable transcription complex on a class III gene in a minichromosome. *Mol. Cell. Biol.* **5,** 40–45.
6. Hamer, D. H. (1980) DNA cloning in mammalian cells with SV40 vectors, in *Genetic Engineering, Volume 2* (Setlow, J. K. and Hollaender, A., eds.), Plenum Press, New York, pp. 83–101.
7. Ding, H.-F., Bustin, M., and Hansen, U. (1997) Alleviation of histone H1-mediated transcriptional repression and chromatin compaction by the acidic activation region in chromosomal protein HMG-14. *Mol. Cell. Biol.* **17,** 5843–5855.
8. Scott, W. A. (1988) SV40 chromatin structure, in *Molecular Aspects of Papovaviruses* (Aloni, Y., ed.) Martinus Nijhoff Publishing, Boston, pp. 199–217.
9. Ambrose, C., Blasquez, V., and Bina, M. (1986) A block in initiation of simian virus 40 assembly results in the accumulation of minichromosomes containing an exposed regulatory region. *Proc. Natl. Acad. Sci. USA* **83,** 3287–3291.
10. Boyce, F. M., Sundin, O., Barsoum, J., and Varshavsky, A. (1982) New way to isolate Simian Virus 40 nucleoprotein complexes from infected cells: use of a thiol-specific reagent. *J. Virol.* **42,** 292–296.

11. Fernandez-Munoz, R., Coca-Prados, M., and Hsu, M.-T. (1979) Intracellular forms of Simian Virus 40 nucleoprotein complexes. I. Methods of isolation and characterization in CV-1 cells. *J. Virol.* **29,** 612–623.
12. Piette, J., Cereghini, S., Kryszke, M.-H., and Yaniv, M. (1986) Identification of cellular proteins that interact with polyomavirus or Simian Virus 40 enhancers, in *Cancer Cells*, 4th ed. (Botchan, M., Grodzicker, T., and Sharp, P. A., eds.), Cold Spring Harbor Laboratory, Cold Spring Harbor, NY, pp. 103–113.
13. Batson, S. C., Rimsky, S., Sundseth, R., and Hansen, U. (1993) Association of nucleosome-free regions and basal transcription factors with *in vivo*-assembled chromatin templates active *in vitro*. *Nucleic Acids Res.* **21,** 3459–3468.
14. DePamphilis, M. L. and Bradley, M. K. (1986) Replication of SV40 and polyoma virus chromosomes, in *The Papovaviridae, Vol. 1: The Polyomaviruses* (Salzman, N. P., ed.) Plenum, New York, NY, pp. 99–246.
15. Saragosti, S., Moyne, G., and Yaniv, M. (1980) Absence of nucleosomes in a fraction of SV40 chromatin between the origin of replication and the region coding for the late leader RNA. *Cell* **20,** 65–73.
16. Powers, J. H. and Bina, M. (1991) In vitro assembly of a positioned nucleosome near the hypersensitive region in simian virus 40 chromatin. *J. Mol. Biol.* **221,** 795–803.
17. Ambrose, C., Lowman, H., Rajadhyaksha, A., Blasquez, V., and Bina, M. (1990) Location of nucleosomes in Simian Virus 40 chromatin. *J. Mol. Biol.* **214,** 875–884.
18. Jakobovits, E. B., Bratosin, S., and Aloni, Y. (1982) Formation of a nucleosome-free region in SV40 minichromosomes is dependent upon a restricted segment of DNA. *Virology* **120,** 340–348.
19. Scott, W. A. and Wigmore, D. J. (1978) Sites in Simian Virus 40 chromatin which are preferentially cleaved by endonucleases. *Cell* **15,** 1511–1518.
20. Varshavsky, A. J., Sundin, O., and Bohn, M. (1979) A stretch of "late" SV40 viral DNA about 400 bp long which includes the origin of replication is specifically exposed in SV40 minichromosomes. *Cell* **16,** 453–466.
21. Waldeck, W., Föhring, B., Chowdhury, K., Gruss, P., and Sauer, G. (1978) Origin of DNA replication in papovavirus chromatin is recognized by endogeonous endonuclease. *Proc. Natl. Acad. Sci. USA* **75,** 5964–5968.
22. Choder, M., Bratosin, S., and Aloni, Y. (1984) A direct analysis of transcribed minichromosomes: all transcribed SV40 minichromosomes have a nuclease-hypersensitive region within a nucleosome-free domain. *EMBO J.* **3,** 2929–2936.
23. Weiss, E., Ruhlmann, C., and Oudet, P. (1986) Transcriptionally active SV40 minichromosomes are restriction enzyme sensitive and contain a nucleosome-free origin region. *Nucleic Acids Res.* **14,** 2045–2058.
24. Weiss, E., Regnier, E., and Oudet, P. (1987) Restriction enzyme accessibility and RNA polymerase localization on transcriptionally active SV40 minichromosomoes isolated late in infection. *Virology* **159,** 84–93.
25. Blasquez, V., Stein, A., Ambrose, C., and Bina, M. (1986) Simian virus 40 protein VP1 is involved in spacing nucleosomes in minichromosomes. *J. Mol. Biol.* **191,** 97–106.

26. Hansen, U. and Sharp, P. A. (1983) Sequences controlling *in vitro* transcription of SV40 promoters. *EMBO J.* **2**, 2293–2303.
27. Batson, S. C., Sundseth, R., Heath, C. V., Samuels, M., and Hansen, U. (1992) In vitro initiation of transcription by RNA polymerase II on in vivo-assembled chromatin templates. *Mol. Cell. Biol.* **12**, 1639–1651.
28. Gruss, C., Gutierrez, C., Burhans, W. C., DePamphilis, M. L., Koller, T., and Sogo, J. M. (1990) Nucleosome assembly in mammalian cell extracts before and after DNA replication. *EMBO J.* **9**, 2911–2922.
29. Stillman, B. (1986) Chromatin assembly during SV40 DNA replication in vitro. *Cell* **45**, 555–565.
30. Sewack, G. F. and Hansen, U. (1997) Nucleosome positioning and transcription-associated chromatin alterations on the human estrogen-responsive pS2 promoter. *J. Biol. Chem.* **272**, 31,118–31,129.
31. Brady, J. N., Winston, V. D., and Consigli, R. A. (1978) Characterization of a DNA-protein complex and capsomere subunits derived from polyoma virus by treatment with ethyleneglycol-bis-N,N'-tetraacetic acid and dithiothreitol. *J. Virol.* **27**, 193–204.
32. Hirt, B. (1967) Selective extraction of polyoma DNA from infected mouse cell cultures. *J. Mol. Biol.* **26**, 365–369.
33. Gerard, R. D. and Gluzman, Y. (1985) New host cell system for regulated Simian Virus 40 DNA replication. *Mol. Cell. Biol.* **5**, 3231–3240.
34. Griffith, O. M. (1979) *Techniques of Preparative, Zonal, and Continuous Flow Ultracentrifugation.* Beckman Instruments, Palo Alto, CA.
35. Manley, J. L., Fire, A., Cano, A., Sharp, P. A., and Gefter, M. L. (1980) DNA-dependent transcription of adenovirus genes in a soluble whole-cell extract. *Proc. Natl. Acad. Sci. USA* **77**, 3855–3859.
36. Croston, G. E., Kerrigan, L. A., Lira, L. M., Marshak, D. R., and Kadonaga, J. T. (1991) Sequence-specific antirepression of histone H1-mediated inhibition of basal RNA polymerase II transcription. *Science* **251**, 643–649.
37. Chang, X.-B. and Wilson, J. H. (1986) Formation of deletions after initiation of simian virus 40 replication: Influence of packaging limit of the capsid. *J. Virol.* **58**, 393–401.
38. O'Reilly, D. R., Miller, L. K., and Luckow, V. A. (1992) *Baculovirus Expression Vectors: A Laboratory Manual.* W. H. Freeman, New York, NY.
39. Cartwright, I. L. and Elgin, S. C. R. (1989) Nonenzymatic cleavage of chromatin. *Methods Enzymol.* **170**, 359–369.
40. Wu, C. (1989) Analysis of hypersensitive sites in chromatin. *Methods Enzymol.* **170**, 269–289.
41. Bellard, M., Dretzen, G., Giangrande, A., and Ramain, P. (1989) Nuclease digestion of transcriptionally active chromatin. *Methods Enzymol.* **170**, 317–346.
42. Mueller, P. R. and Wold, B. (1989) In vivo footprinting of a muscle specific enhancer by ligation mediated PCR. *Science* **246**, 780–786.
43. Garrity, P. A. and Wold, B. J. (1992) Effects of different DNA polymerases in ligation-mediated PCR: Enhanced genomic sequencing and *in vivo* footprinting. *Proc. Natl. Acad. Sci. USA* **89**, 1021–1025.

44. McPherson, C. E., Shim, E.-Y., Friedman, D. S., and Zaret, K. S. (1993) An active tissue-specific enhancer and bound transcription factors existing in a precisely positioned nucleosomal array. *Cell* **75,** 387–398.
45. Pfeifer, G. P. and Riggs, A. D. (1991) Chromatin differences between active and inactive X chromosomes revealed by genomic footprinting of permeabilized cells using DNase I and ligation-mediated PCR. *Genes Dev.* **5,** 1102–1113.
46. Grodzicker, T. and Hopkins, N. (1980) Origins of contemporary DNA tumor virus research, in *DNA Tumor Viruses, Part 2*, 2nd edition (Tooze, J., ed.) Cold Spring Harbor Laboratory, Cold Spring Harbor, NY, pp. 1–59.
47. Cereghini, S. and Yaniv, M. (1984) Assembly of transfected DNA into chromatin: structural changes in the origin-promoter-enhancer region upon replication. *EMBO J.* **3,** 1243–1253.
48. Innis, J. W. and Scott, W. A. (1983) Chromatin structure of simian virus 40-pBR322 recombinant plasmids in COS-1 cells. *Mol. Cell. Biol.* **3,** 2203–2210.
49. Gilmour, R. S., Gow, J. W., and Spandidos, D. A. (1982) In vivo assembly of regularly spaced nucleosomes on mouse β^{maj}-globin DNA cloned in an SV40 recombinant. *Bioscience Reports* **2,** 1031–1040.
50. Reeves, R., Gorman, C. M., and Howard, B. (1985) Minichromosome assembly of non-integrated plasmid DNA transfected into mammalian cells. *Nucleic Acids Res.* **13,** 3599–3615.
51. Shapiro, D. J., Sharp, P. A., Wahli, W. W., and Keller, M. J. (1988) A high-efficiency HeLa cell nuclear transcription extract. *DNA* **7,** 47–55.
52. Dignam, J. D., Lebovitz, R. M., and Roeder, R. G. (1983) Accurate transcription initiation by RNA polymerase II in a soluble extract from isolated mammalian nuclei. *Nucleic Acids Res.* **11,** 1475–1489.
53. Conaway, R. C. and Conaway, J. W. (1993) General initiation factors for RNA polymerase II. *Annu. Rev. Bioch.* **62,** 161–190.
54. Zawel, L. and Reinberg, D. (1993) Initiation of transcription by RNA polymerase II: A multi-step process. *Prog. Nucleic Acid Res. Mol. Biol.* **44,** 67–108.
55. Hansen, U., Tenen, D. G., Livingston, D. M., and Sharp, P. A. (1981) T antigen repression of SV40 early transcription from two promoters. *Cell* **27,** 603–612.
56. Fromm, M. and Berg, P. (1982) Deletion mapping of DNA regions required for SV40 early region promoter function *in vivo*. *J. Mol. Appl. Gen.* **1,** 457–481.
57. Fromm, M. and Berg, P. (1983) Simian Virus 40 early- and late-region promoter functions are enhanced by the 72-base-pair repeat inserted at distant locations and inverted orientations. *Mol. Cell. Biol.* **3,** 991–999.
58. Oudet, P., Weiss, E., and Regnier, E. (1989) Preparation of Simian Virus 40 minichromosomes. *Methods Enzymol.* **170,** 14–25.
59. Varshavsky, A. J., Bakayev, V. V., Chumackov, P. M., and Georgiev, G. P. (1976) Minichromosome of simian virus 40: presence of histone H1. *Nucleic Acids Res.* **3,** 2101–2113.
60. Huang, H.-C. and Cole, R. D. (1984) The distribution of H1 histone is nonuniform in chromatin and correlates with different degrees of condensation. *J. Biol. Chem.* **259,** 14,237–14,242.

61. Yoshida, M., Horinouchi, S., and Beppu, T. (1995) Trichostatin A and trapoxin: novel chemical probes for the role of histone acetylation in chromatin structure and function. *BioEssays* **17,** 423–430.
62. Sundseth, R. and Hansen, U. M. (1990) A systematic approach to the study of RNA polymerase II mediated transcription in vitro. *DNA Prot. Eng. Tech.* **2,** 57–65.
63. Chodosh, L. A., Fire, A., Samuels, M., and Sharp, P. A. (1989) 5,6-dichloro-1-ß-D-ribofuranosylbenzimidazole inhibits transcription elongation by RNA polymerase II *in vitro*. *J. Biol. Chem.* **264,** 2250–2257.
64. Messing, J. (1983) New M13 vectors for cloning. *Methods Enzymol.* **101,** 20–78.

20

In Vitro Replication of Chromatin Templates

Claudia Gruss

1. Introduction

Eukaryotic genome replication implies that both the DNA and the associated proteins must be duplicated during each S phase. Model systems such as the simian virus 40 (SV40) in vitro replication system have provided important information on the mechanism of DNA replication in mammalian cells (reviewed in **ref. *1***). The SV40 genome is replicated by the host replication machinery in conjunction with a single viral protein, T antigen (T-Ag). Replication of the viral genome begins at the well-defined origin and proceeds in a bidirectional and semidiscontinuous manner. A valuable feature is that SV40 DNA exists in infected cells as minichromosomes which resemble the host cell chromatin. Therefore studies on SV40 DNA replication can address issues such as chromatin replication.

A breakthrough in the establishment of the SV40 in vitro replication system has been the development of conditions for preparing soluble extracts from SV40 infected and uninfected cells that support initiation and complete replication of plasmid DNAs containing the SV40 origin of replication (*2*). These studies were extended by results showing that cell-free extracts from competent human HeLa (*see* **Subheading 3.1.**) or 293 cells support SV40 DNA replication in the presence of the SV40 T-Ag (*3–5*). Large quantities of the SV40 T-Ag are usually obtained through expression in baculovirus vectors (*6*) and rapid immunoaffinity techniques for purification (*see* **Subheading 3.2.**) (*7*).

Besides protein-free SV40 DNA, SV40 minichromosomes or reconstituted chromatin containing the SV40 origin can be used as template in the SV40 in vitro replication system. The crucial point during replication of chromatin templates in vitro is the initiation step. Once initiated the replication machinery proceeds more or less unperturbed through nucleosomally organized DNA. However, the origin of replication has to be free of nucleosomes to allow an

From: *Methods in Molecular Biology, Vol. 119: Chromatin Protocols*
Edited by: P. B. Becker © Humana Press Inc., Totowa, NJ

efficient binding of T-Ag as a requirement for the DNA synthesis in vitro *(8–10)*. Normally 20–25% of the SV40 minichromosomes, extracted from infected cells contain a nucleosome-free region, spanning approx 400 bp around the SV40 origin *(11–13*; *see* Chapter 19*)*. These molecules are active as template in the SV40 in vitro replication system (*see* **Subheading 3.3.1.**).

When chromatin is reconstituted in vitro and used as a template for in vitro replication, it is necessary to prevent the formation of nucleosomes on the SV40 origin during reconstitution. This can be achieved by pre-binding of the SV40 T-Ag to the origin in the presence of cytosolic S100 replication extract and three of the four dNTP's prior to chromatin assembly in *Xenopus* oocyte extracts *(10)* (*see* **Subheading 3.3.2.**).

More recent experiments have shown that so-called chromatin remodelling factors (reviewed in **refs. 14–18**) are able to alter the chromatin structure of a nucleosomal origin in a way that initiation factors gain access to their target sequences and thus can efficiently initiate replication *(19)*. The addition of these factors to reconstituted chromatin allows efficient replication of these templates in the SV40 in vitro replication system (*see* **Subheading 3.3.3.**).

Replication products are normally investigated by neutral gel electrophoresis (**Fig. 1A**) (*see* **Subheading 3.4.1.**). Replication efficiency is determined by the incorporation of radioactive nucleotides by TCA precipitation (**Fig. 1B**) (*see* **Subheading 3.4.2.**).

The SV40 in vitro replication system has served as a useful method for the analysis of the cellular factors required for replication *(1,20,21)*. By using chromatin as template the mechanisms of chromatin replication have been further elucidated (reviewed in **ref. 22**). Thus, the mechanism of the transfer of the parental nucleosomes to the daughter strands has been investigated *(10,23–26)*. The influence of chromatin structure on replication efficiency has been studied with chromatin templates reconstituted with histone H1 *(27)* or the non-histone protein HMG-17 *(28)*. Furthermore the effects of post-translational modifications as the cell-cycle dependent phosphorylation of histone H1 *(29)* and the acetylation of the core histones *(30)* have been investigated. In addition the effects of chromatin structure on the accessibility for topoisomerases and thus the replication efficiency of the templates have been further elucidated by using this system *(31)*.

Fig. 1. *(continued on next page)* In vitro replication of SV40 DNA and SV40 minichromosomes. (**A**) 150 ng of protein-free SV40 DNA (DNA) and 500 ng of native SV40 minichromosomes (Mc) are replicated for 2 h in the SV40 in vitro replication system. Deproteinized replication products are analyzed by agarose gel electrophoresis and autoradiography. HMW, high-molecular-weight DNA; II, relaxed and open circular DNA; I, superhelical form I DNA. (**B**) Replication efficiencies are determined by TCA precipitation and expressed as picomoles dNTP incorporated.

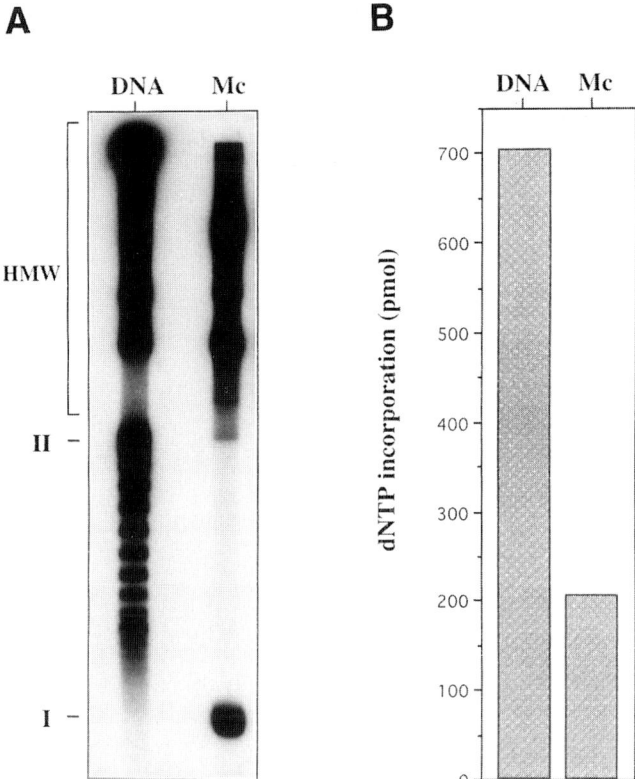

Fig. 1. *(continued)* With both templates a considerable fraction of incorporated radioactivity is found in slowly migrating high-molecular-weight material (HMW), which includes replicative intermediates, DNA dimers, and possibly rolling ring-type structures, which could arise when replication continued after artificial breakage of an replication fork.

Note that, depending on the template completely replicated molecules differ in their structure. Replication of protein-free DNA results in circular, relaxed form II DNA as well as in DNA with up to about 10 superhelical turns *(10,23)*.

In contrast to replicated protein-free DNA, the replicated minichromosomal DNA contains a significantly higher degree of constrained superhelicity (form I DNA), which is a result of the transfer of the parental nucleosomes to the daughter strands and a limited assembly of newly synthesized histones (for detailed discussion *see* **ref. 22**).

Replication efficiency of both templates is different, which is because of an altered accessibility of topoisomerases to chromatin templates (for details *see* **ref. 31**).

2. Materials

1. Cells: The monkey kidney cell line CV-1 (American Type Culture Collection [ATCC], Rockville, MD; cat. no. CCL 70] is used for infection with SV40. The human HeLa cell line (ATCC, cat. no. CCL 5) is used for preparation of cytosolic

S100 replication extract. *Spodoptera frugiperda* (Sf9) cells (INVITROGEN cat. no. B 825-01) are used for infection with recombinant baculovirus (*6*) to express the SV40 T-Ag. CV-1 and HeLa cells are grown on 14.5 cm tissue culture dishes in Dulbecco's modified Eagle's medium (DMEM) supplemented with 5% (v/v) fetal calf serum (FCS) at 37°C in a 5% CO_2 atmosphere. Sf9 cells are grown in 50 mL flasks in Grace insect medium (Gibco-BRL) supplemented with 7.5% (v/v) fetal calf serum (FCS) at 27°C without carbon dioxide supplementation.

2. Phosphate buffered saline (PBS): 20 mM Tris-HCl, pH 7.4, 0.25 M sucrose, 137 mM NaCl, 5 mM KCl, 1 mM $CaCl_2$, 0.5 mM $MgCl_2$. Store at 4°C.
3. Hypotonic buffer/sucrose: 20 mM HEPES-KOH, pH 7.8, 5 mM potassium acetate, 1.5 mM $MgCl_2$, 250 mM sucrose. Autoclave, store at 4°C. Add 0.1 mM DTT freshly to the aliquot used.
4. Hypotonic buffer: 20 mM HEPES-KOH, pH 7.8, 5 mM potassium acetate, 1.5 mM $MgCl_2$. Autoclave, store at 4°C. Add 0.1 mM DTT freshly to the aliquot used.
5. Schwyzer buffer: 100 mM Tris-HCl, pH 9.0, 100 mM NaCl, 5 mM KCl, 1 mM $CaCl_2$, 0.5 mM $MgCl_2$.
6. NET-buffer: 50 mM Tris-HCl, pH 7.5, 150 mM NaCl, 5 mM EDTA.
7. PIPES-buffer: 10 mM PIPES, pH 7.0, 5 mM NaCl, 0.1 mM EDTA. Sterilize by filtration through 0.45 mM filters (Renner). Store in the dark at 4°C.
8. Elution buffer: 20 mM triethylamine, pH 10.8, 10% glycerol, adjust the pH with a few drops 1 N HCl. Sterilize by filtration, prepare freshly just before use.
9. Stock solutions for 10X replication mix: 1 M HEPES-KOH pH 7.8; 0.1 M DTT; 1 M $MgCl_2$; 20 mM each of CTP, UTP, and GTP; 440 mM ATP; 40 mM each of dATP, dGTP, dCTP, and dTTP, 1 M creatine phosphate; creatine kinase 10 µg/µL. Nucleotides (Pharmacia) are dissolved in 100 mM HEPES-KOH, pH 7.8, the pH of the ATP solution might be adjusted with a few drops of 1 M KOH. Nucleotides are stored in small aliquots (20–50 µL) at –70°C. DTT aliquots (–70°C) are thawed only once.
10. Stop solution: 60 mM EDTA, 2% dodecylsulfate (SDS).
11. Oocyte extraction buffer: 20 mM HEPES-KOH, pH 7.5; 5 mM KCl, 1.5 mM $MgCl_2$, 1 mM EGTA, 10% glycerol, 10 mM β-glycerophosphate, 0.5 mM DTT.
12. EX 120: 10 mM HEPES-KOH, pH 7.6, 120 mM KCl, 10% glycerol, 5 mM $MgCl_2$, 0.5 mM EGTA, 10 mM β-glycerophosphate, 1 mM DTT.
13. Agarose loading buffer: 10 mM Tris-HCl, pH 7.8, 1 mM EDTA, 0.04% bromphenolblue, 0.04% xylenecyanol, 2.5% Ficoll 400.
14. TBE-buffer: 45 mM Tris-borate, pH 8.4, 0.5 mM EDTA.

3. Methods

3.1. Preparation of Cytosolic S100 Replication Extract from HeLa Cells

Cytosolic S100 extracts, which contain the cellular replication factors required for in vitro replication of DNA and chromatin templates, are routinely prepared from 20–25 cell culture dishes (14.5 cm) of HeLa cells, which should be 80–90% confluent (*see* **Note 1**). Preparation is done at 4°C as follows:

1. Remove the cell culture medium and wash the cells on the plates twice with cold phosphate-buffered saline (PBS).
2. Scrape off the cells with a piece of stiff rubber, transfer into 15 ml tubes and pellet the cells at 160 g for 5 min at 4 C.
3. Pool the cell pellets, resuspend in 10 mL: hypotonic buffer/sucrose and pellet the cells. Resuspend the cell pellet in 10 mL: hypotonic buffer and pellet the cells again.
4. Resuspend the cell pellet in 500 µL: hypotonic buffer, swell the cells for 10 min and disrupt with 12 strokes in a type S Dounce homogenizer (*see* **Note 2**).
5. Elute the nuclei for 30 min on ice.
6. Remove the nuclei by centrifugation at 16,000g for 10 min at 4°C (e.g., Sorvall HB4 rotor).
7. Centrifuge the supernatant at 100,000g for 1 h at 4°C (e.g., Beckman SW 55 rotor), supernatant = S100.
8. Shock-freeze the S100 extract in small aliquots (150 µL) and store at –70°C (*see* **Note 3**).
9. Protein concentration is determined by the Bio-Rad protein assay according to the manufactures protocol (Bio-Rad) (*see* **Note 4**).

3.2. Preparation of the SV40 T-Antigen

The SV40 T-Ag is purified from insect cells (Sf9), infected with a recombinant baculovirus containing the SV40 T-Ag coding region *(6)* (*see* **Note 5**). Purification is performed by immunoaffinity chromatography *(7)* as follows:

1. Infect 12X 250-mL cell culture flasks of confluent Sf9 cells with 3 mL virus lysate/flask. For this purpose remove the medium, add the virus lysate and incubate for 1 h without shaking at 27°C. Remove the virus lysate, add 10 mL fresh Grace medium/7.5% FCS and incubate for 40 h at 27°C.
2. Remove the cells from the flasks by taping, collect the cells by centrifugation at 250 g for 5 min at 4°C; pool the cell pellets, wash with 50 mL PBS and pellet the cells at 250 g for 5 min. Transfer the cell pellet to a 15-mL tube, wash again with PBS and carefully remove all of the supernatant.
3. Resuspend the cell pellet in 1.5 mL of Schwyzer buffer, containing 0.5% Nonidet P 40 (NP40), 50 µM leupeptin, rotate the tube for 30 min at 4°C.
4. Transfer the cell lysate to an SS34 tube and centrifuge at 38,000g for 15 min at 4°C (SS34 rotor).
5. The supernatant is combined with a 1 mL antibody column (*see* **Note 6**), equilibrated first with 20 vol of Schwyzer buffer, and then with 10 vol of Schwyzer buffer, 0.5% NP40, 50 µM leupeptin. Rotate the column for 4 h at 4°C and store the flow through of the column at –20°C.
6. To remove nonbound proteins extensively wash the column first with 100 mL of NET-buffer, 0.05% NP40 and then with 20 mL of PIPES-buffer.
7. Elute the T-Ag with elution buffer, pH 10.8. For this purpose load 200 µL of elution buffer on the column, collect the fraction and load the next 200 µL buffer.

Collect 15 fractions in eppendorf tubes, containing 20 μL of 0.5 M PIPES, pH 7.0, mix immediately to neutralize the solution (see **Note 7**).
8. Measure the protein concentration of the individual fractions with the Bio-Rad protein assay, pool the T-Ag containing fractions, measure the final protein concentration and add DTT (1 mM) and glycerol (50%).
9. Store the T-Ag in small aliquots (50-μL) at –70°C. The aliquot in use is stored at –20°C. A yield of 100–150 μg T-Ag should be expected per 250-mL flask.
10. Immediately neutralize the antibody column by washing with PBS, store the column in PBS with methiolate.

3.3. In Vitro Replication Assay with Chromatin Templates

3.3.1. In Vitro Replication of SV40 Minichromosomes

Native or salt-treated SV40 minichromosomes can be used as template in the standard SV40 in vitro replication system. Isolation and purification of these minichromosomes is described in detail elsewhere (see Chapter 18, **ref. 32** and **Note 8**).

1. Thaw the S100 extract and the SV40 minichromosomes on ice, and the components of the 10x replication mix.
2. Mix the components of the 10x replication mix (c_{end} 10x mix) in the following order.
   a. 30 μL of 1 M HEPES-KOH, pH 7.8 = 300 mM.
   b. 5 μL of 0.1 M DTT = 5 mM.
   c. 3 μL of 1 M $MgCl_2$ = 30 mM.
   d. 4 μL of 20 mM CTP = 0.8 mM.
   e. 4 μL of 20 mM UTP = 0.8 mM.
   f. 4 μL of 20 mM GTP = 0.8 mM.
   g. 0.95 μL of 40 mM dATP = 380 μM.
   h. 2.5 μL of 40 mM dGTP = 1 mM.
   i. 2.5 μL of 40 mM dCTP = 1 mM.
   j. 2.5 μL of 40 mM dTTP = 1 mM.
   k. 4.5 μL of 440 mM ATP = 20 mM.
   l. 40 μL of 1 M creatine phosphate = 400 mM.
   m. 1.2 μL of creatine kinase (10 μg/μL) = 12 μg.
   Mix carefully, remove the amount needed for the replication assays, add 2 μCi α[^{32}P]dATP (Amersham) per reaction.
3. Mix the individual replication reactions (50 μL) in the following order (see **Note 9**).
   a. H_2O to 50 μL.
   b. S100 extract, 230 μg.
   c. SV40 T-Ag, 1 μg.
   d. 10X replication mix, 5 μL.
   e. Chromatin, 500 ng.
   Mix carefully, spin down and incubate for 2 h at 37°C. Stop the reaction with 30 μL stop solution (see **Note 10**).

3.3.2. In Vitro Replication of Chromatin Reconstituted in Xenopus Oocyte Extracts

DNA, containing the SV40 origin, can be reconstituted into chromatin in *Xenopus* oocyte extracts (*see* Chapter 12; **ref. *33***). To allow initiation of replication, DNA has to be preincubated prior to reconstitution under the following conditions:

1. Incubate 660 ng SV40 T-Ag, 55 µg cytosolic S100 extract proteins and 200 ng of DNA in replication buffer without dGTP and without the ATP regenerating system for 30 min at 37°C.
2. The buffer is then adjusted by spinning the sample through a Sephadex G50 quick spin column (Boehringer), equilibrated in oocyte extraction buffer (*see* **Note 11**).
3. DNA is assembled into chromatin by using *Xenopus* oocyte extracts (30 µL extract/150 ng DNA) for 6 h at 27°C in oocyte extraction buffer, in the presence of 3 mM ATP, 1 mM MgCl$_2$, 40 mM creatine phosphate, creatine kinase (10 µg/µL) and 10 µM aphidicolin (*see* **Note 12**).
4. Aphidicolin is removed by spinning the reconstituted minichromosomes through two subsequent Sephadex G50 quick spin columns equilibrated in 12.5 mM triethanolamine (TEA), pH 7.5 (*see* **Note 13**).
5. These minichromosomes are then used as template in the standard SV40 in vitro replication system (*see* **Subheading 3.3.1.** and **Note 14**).

3.3.3. In Vitro Replication of Chromatin Reconstituted in Drosophila S150 Extracts

DNA, containing the SV40 origin, can be assembled into chromatin in *Drosophila* S150 extracts (*see* Chapter 13). This chromatin contains remodelling factors, which enable the binding of T-Ag to the origin in an ATP-dependent manner and thus an efficient initiation of replication *(19)*. After treatment with Sarkosyl this chromatin is inactive as template and thus, exogenously added remodelling factors can be tested for activity.

1. DNA is reconstituted into chromatin for 6 h at 26°C exactly as described in Chapter 13 *(34)*.
2. The assembly reactions are spun through 2 mL settled Sephacryl S 300 HR spin columns, which are equilibrated in EX120 and prespun at 1100*g* for 2 min.
3. Molecules with a nucleosome-free origin are eliminated by digestion with the restriction endonuclease *Bgl*I, which cuts once within the SV40 origin sequences. To this end, chromatin is incubated with 10 U *Bgl*I/µg chromatin for 1 h at 26°C.
4. Chromatin-associated activities are removed by Sarkosyl-treatment (0.075%) for 10 min at R.T. Sarkosyl is removed by spin column centrifugation as described above *(2)*.
   As this chromatin is inactive as template in the standard SV40 in vitro replication system (*see* **Subheading 3.3.1.**), it can be used as substrate to test the activity of potential chromatin remodelling factors (*see* **Note 15**). Conditions for the purified "Chromatin Accessibility Complex" (CHRAC, **ref. *35***) are as follows:

5. Pre-incubate 200 ng of *Bgl*I digested, Sarkosyl-treated chromatin with 250 ng of highly purified CHRAC *(35)* for 30 min at 26°C.
6. Add ATP (3mM) and 2 µg of the SV40 T-Ag and incubate for further 45 min at 26°C. Use this chromatin as template in the SV40 in vitro replication system, which is modified as follows:
7. Phosphoenolpyruvate (10 mM) and pyruvatekinase (1.8 µg) are used as ATP regenerating system.
8. Replication assay is done in a 190 µL reaction as follows.
   a. H_2O to 190 µL.
   b. S100 extract, 350 µg.
   c. SV40 T-Ag, 2 µg.
   d. 10X replication mix, 19 µL.
   e. Chromatin, 200 ng.

   Mix carefully, spin down and incubate for 2 h at 37°C. Stop the reaction with 110 µL stop solution (*see* **Note 16**).

3.4. Analysis of the Replicated Chromatin

3.4.1. Analysis of the Replication Products by Neutral Gel Electrophoresis

For the analysis of the replication products, they have to be purified as follows:

1. Deproteinize the DNA with 2 µL proteinase K (20 mg/mL), which is added to the stopped replication assay and incubate for 1 h at 55°C.
2. Extract once with phenol and precipitate the DNA with ethanol *(36)*.
3. The dried DNA pellet is resolved in 15 µL agarose loading buffer.
4. Replication products are separated on a 0.8% agarose-gel in 0.5X TBE buffer at 2.5 V/cm for 16 h.
5. The gel is stained for 20 min with ethidium-bromide (2.5 µg/mL) and photographed with a polaroid camera. The gel is then dried for 2 h at 60°C on a vacuum-dryer and depending on the incorporation of nucleotides exposed on X-Ray films for several hours up to some days at –70°C with an intensifying screen in a X-ray cassette.

   The autoradiography of an agarose gel showing the typical distribution of replication products from protein-free DNA and SV40 minichromosomes is shown in **Fig. 1A**.

3.4.2. Measurement of the Incorporated Nucleotides by TCA Precipitation

DNA synthesis is determined by the incorporation of radioactive labeled dATP.

1. Remove 1/10 of the stopped replication assay. Add 10 µL salmon sperm DNA (1 mg/mL) and 1 mL of ice cold 10% trichloracetic acid (TCA), 5% $Na_4P_2O_7$. Incubate for 15 min on ice.
2. By using a filtration apparatus (Millipore), filter the solution through GF/C filters (Whatman), which are prewetted with 5% TCA, 0.5% $Na_4P_2O_7$.
3. Remove nonincorporated nucleotides by washing with 5% TCA, 0.5% $Na_4P_2O_7$. The final wash is with 90% ethanol.

4. Dried filters are counted in 3 mL scintillation liquid (Rotiszint 11eco, Roth) in the ^{32}P-channel of a scintillation counter.
5. To determine the specific activity, 5 µL of the 10X replication mix, diluted 1:30, are pipeted on a GF/C filters and without washing, counted in 3 mL scintillation liquid.

A comparison of the typical incorporation rates for DNA and native SV40 minichromosomes are shown in **Fig. 1B**.

4. Notes

1. Cells should be in logarythmic growth phase, to result in an optimal concentration of replication proteins.
2. Before and after Dounce homogenization, check an aliquot of the cells in the light microscope, to ensure that most of the cells are broken and that the nuclei are still intact.
3. Aliquots are thawed on ice immediately before use and are used only once.
4. Protein concentration of the S100 extract should be in the range of 8–14 mg/mL, to be competent for in vitro replication. Concentration can be varied by adding more or less hypotonic buffer to the cells before Dounce homogenization.
5. For each T-Antigen preparation virus lysate is freshly prepared from one isolated plaques as described *(37)*. The virus lysate is not further propagated.
6. Two milligrams of the T-Ag specific antibody pAB101 *(38)* are coupled to 1 ml cyanogen bromide activated Sepharose 4B (fast-flow) according to the manufactures protocol (Pharmacia).
7. The antibody column with bound T-Ag should not be exposed to pH 10.8 longer than 30 min, because this causes loss of activity of the T-Ag.
8. Protein-free SV40 DNA (150 ng) should be used as template in the SV40 in vitro replication system as positive control for the efficiency of the system. If the system is working properly, incorporation of nucleotides should be in the range of 600–800 pmoles dNTP after 2 h of replication. The efficiency of nucleotide incorporation into chromatin templates will be four- to sixfold lower compared to protein-free DNA (*see* **Fig. 1B**) *(31)*.
9. When different replication assays have to be performed, it is advisable to make a pre-mix containing H$_2$0, S100, T-Ag and replication mix. This pre-mix is prepared just before use.
10. Replication efficiency strongly depends on the salt concentration of the assay. Addition of 25 m*M* salt reduces replication efficiency two- to threefold.
11. Pool at least seven assays (1.4–2 µg DNA) for the quick spin columns to get sufficient recovery.
12. To prevent elongation of pre-initiated molecules during incubation in *Xenopus* oocyte extracts, reconstitution has to be done in the presence of 10 µ*M* aphidicolin.
13. After the second quick spin centrifugation, some samples have to be concentrated in a speed vac to obtain a final concentration of 200 ng chromatin/25 µL.
14. If the pre-incubation step has been successful, around 100 pmoles dNTP will be incorporated into 200 ng of reconstituted chromatin after 2 h of in vitro replica-

tion. Without the pre-incubation step, there will be only background incorporation (around 10 pmoles dNTP) *(10)*.
15. To test for chromatin remodelling activities at the origin, *Bgl*I digested SV40 minichromosomes, which are inactive in the in vitro replication assay, can also be used as template.
16. After remodelling of the nucleosome at the origin, incorporation of nucleotides will be around 200 pmoles dNTP after 2 h of replication.

Acknowledgments

I thank Rolf Knippers, Lothar Halmer, and Vassilios Alexiadis for their comments on the manuscript and Lothar Halmer for preparing the figure.

References

1. Stillman, B. (1989) Initiation of eukaryotic DNA replication in vitro. *Ann. Rev. Cell Biol.* **5,** 197–245.
2. Li, J. J. and Kelly, T. J. (1984) Simian virus 40 DNA replication in vitro. *Proc. Natl. Acad. Sci. USA* **81,** 6973–6977.
3. Li, J. J. and Kelly, T. J. (1985) Simian virus 40 DNA replication in vitro: specificity of initiation and evidence for bidirectional replication. *Mol. Cell. Biol.* **5,** 1238–1246.
4. Stillman, B. W. and Gluzman, Y. (1985) Replication and supercoiling of Simian Virus 40 DNA in cell extracts from human cells. *Mol. Cell. Biol.* **5,** 2051–2060.
5. Wobbe, C. R., Dean, F., Weissbach, L., and Hurwitz, J. (1985) In vitro replication of duplex circular DNA containing the simian virus 40 origin site. *Proc. Natl. Acad. Sci. USA* **82,** 5710–5714.
6. Lanford, R. E. (1988) Expression of simian virus 40 T antigen in insect cells using a baculovirus expression vector. *Virology* **167,** 72–81.
7. Simanis, V. and Lane, D. P. (1985) An immunoaffinity purification procedure for SV40 large T antigen. *Virology* **144,** 88–100.
8. Cheng, L. and Kelly, T. J. (1989) Transcriptional activator nuclear factor I stimulates the replication of SV40 minichromosomes in vivo and in vitro. *Cell* **59,** 541–551.
9. Ishimi, Y. (1992) Preincubation of T antigen with DNA overcomes repression of SV40 DNA replication by nucleosome assembly. *J. Biol. Chem.* **267,** 10,910–10,913.
10. Gruss, C., Wu, J., Koller, T., and Sogo, J. M. (1993) Disruption of the nucleosomes at the replication fork. *EMBO J.* **12,** 4533–4545.
11. Jakobovits, E. B., Bratosin, E., and Aloni, J. (1980) A nucleosome free region in SV40 minichromosomes. *Nature* **285,** 263–265.
12. Saragosti, S., Moyne, G., and Yaniv, M. (1980) Absence of nucleosomes in a fraction of SV40 chromatin between the origin of replication and the region coding for the late leader RNA. *Cell* **20,** 65–73.
13. Sogo, J. M., Stahl, H., Koller, T., and Knippers, R. (1986) Structure of replicating Simian virus 40 minichromosomes. *J. Mol. Biol.* **189,** 189–204.
14. Peterson, C. L. (1996) Multiple switches to turn on chromatin. *Cur. Biol.* **6,** 171–175.
15. Kingston, R. E., Bunker, C. A., and Imbalzano, A. N. (1996) Repression and activation by multiprotein complexes that alter chromatin structure. *Genes Dev.* **10,** 905–920.

16. Felsenfeld, G. (1996) Chromatin unfolds. *Cell* **86,** 16–19.
17. Tsukiyama, T. and Wu, C. (1997) Chromatin remodeling and transcription. *Cur. Op. Gen. Dev.* **7,** 182–191.
18. Pazin, M. J. and Kadonaga, J. T. (1997) SWI2/SNF2 and related proteins:ATP-driven motors that disrupt protein-DNA interactions? *Cell* **88,** 737–740.
19. Alexiadis, V., Varga-Weisz, P. D., Bonte, E., Becker, P. B., and Gruss, C. (1998) Chromatin remodelling by CHRAC enables initiation of DNA replication from a nucleosomal origin. *EMBO J.* **17,** 3428–3438.
20. Fairman, M. P. and Stillman, B. (1988) Cellular factors required for multiple stages of SV40 DNA replication in vitro. *EMBO J.* **7,** 1211–1218.
21. Tsurimoto, T. and Stillman, B. (1991) Replication factors required for SV40 DNA replication in vitro. *J. Biol. Chem.* **266,** 1950–1960.
22. Gruss, C. and Knippers, R. (1996) Structure of replicating chromatin in, *Progressive Nucleic Acid Research and Molecular Biology* (W. E. Cohn and K. Moldave), Academic, San Diego, pp. 337–365.
23. Krude, T. and Knippers, R. (1991) Transfer of nucleosomes from parental to replicated chromatin. *Mol.Cell. Biol.* **11,** 6257–6267.
24. Randall, S. K. and Kelly, T. J. (1992) The fate of parental nucleosomes during SV40 DNA replication. *J. Biol. Chem.* **267,** 14,259–14,265.
25. Sugasawa, K., Ishimi, Y., Eki, T., Hurwitz, J., Kikuchi, A., and Hanaoka, F. (1992) Nonconservative segregation of parental nucleosomes during simian virus 40 chromosome replication *in vitro*. *Proc. Natl. Acad. Sci. USA* **89,** 1055–1059.
26. Quintini, G., Treuner, K., Gruss, C., and Knippers, R. (1996) Role of amino-terminal histone domains in chromatin replication. *Mol. Cell. Biol.* **16,** 2888–2897.
27. Halmer, L. and Gruss, C. (1995) Influence of histone H1 on the *in vitro* replication of DNA and chromatin. *Nucleic Acids Res.* **23,** 773–778.
28. Vestner, B., Bustin, M., and Gruss, C. (1998) Stimulation of replication efficiency of a chromatin template by chromosomal protein HMG-17. *J. Biol. Chem.* **273,** 9409–9414.
29. Halmer, L. and Gruss, C. (1996) Effects of cell-cycle dependent histone H1 phosphorylation on chromatin structure and chromatin replication. *Nucleic Acids Res.* **24,** 1420–1427.
30. Alexiadis, V., Halmer, L., and Gruss, C. (1997) Influence of core histone acetylation on SV40 minichromosomes replication in vitro. *Chromosoma* **105,** 324–331.
31. Halmer, L. and Gruss, C. (1997) Accessibility for topoisomerase I and II regulates the replication efficiency of SV40 minichromosomes in vitro. *Mol. Cell. Biol.* **17,** 2624–2630.
32. Gruss, C. and Knippers, R. (1995) The SV40 minichromosome in, *Methods in Molecular Genetics* (Adolph, K. W., ed.), Academic, Orlando, pp. 101–113.
33. Rodriguez-Campos, A., Shimamura, A., and Worcel, A. (1989) Assembly and properties of chromatin containing histone H1. *J. Mol. Biol.* **209,** 135–150.
34. Becker, P. B., Tsukiyama, T., and Wu, C. (1994) Chromatin assembly extracts from Drosophila embryos. *Methods Cell Biol.* **44,** 207–223.

35. Varga-Weisz, P. D., Wilm, M., Bonte, E., Dumas, K., Mann, M., and Becker, P. B. (1997) Chromatin-remodelling factor CHRAC contains the ATPases ISWI and topoisomerase II. *Nature* **388,** 598–602.
36. Maniatis, T., Fritsch, E. E., and Sambrook, J. (1982) *Molecular Cloning: A Laboratory Manual.* Cold Spring Harbor University Press, Cold Spring Harbor.
37. Gruenwald, S. and Heitz, J. (1993) *Baculovirus Expression Vector System: Procedures and Methods Manual.* PharMingen, San Diego.
38. Stahl, H., Dröge, P., Zentgraf, H., and Knippers, R. (1985) A large-tumor-antigen-specific monoclonal antibody inhibits DNA replication of simian virus 40 minichromosomes in an in vitro elongation system. *J. Virol.* **54,** 473–482.

21

Analysis of HMG-14/-17-Containing Chromatin

Yuri V. Postnikov and Michael Bustin

1. Introduction

In chromatin, each nucleosome contains two specific binding sites for nonhistone chromosomal proteins HMG-14 or HMG-17; however, because the amount of these proteins in the nucleus is limited, only a small fraction of the nucleosomes contain these proteins. A central question on the role of HMG-14/-17 in chromatin structure and function is determination of the organization and distribution of nucleosomes containing these proteins along the chromatin fiber. Here we describe a method suitable to study these questions. It is important to note that the method described has general applicability and can be used to study the organization of any type of unusual nucleosomes in chromatin. For example, it is applicable to studies on the organization of nucleosomes containing minor species, acetylated, or otherwise modified histones, specific nonhistones, or even unusual or modified nucleotides.

The protocol used for studies on the organization of HMG-14/-17 in chromatin *(1)* is described in **Fig. 1**.

The information regarding the organization of HMG-14/-17 proteins in native cellular chromatin illustrates the usefulness of this experimental strategy. With this approach we were able to determine that:

1. In chromatin. nucleosomes contain either zero or two molecules of HMG. Nucleosomes associated with one molecule of HMG were not detected.
2. In chromatin, HMG-14/-17 proteins are bound to nucleosomes as homodimers; that is, the nucleosomes contain two molecules of either HMG-14 or HMG-17 but not a mixture of the two proteins.
3. The presence of HMG-14/-17 does not alter the histone composition of the nucleosomes.

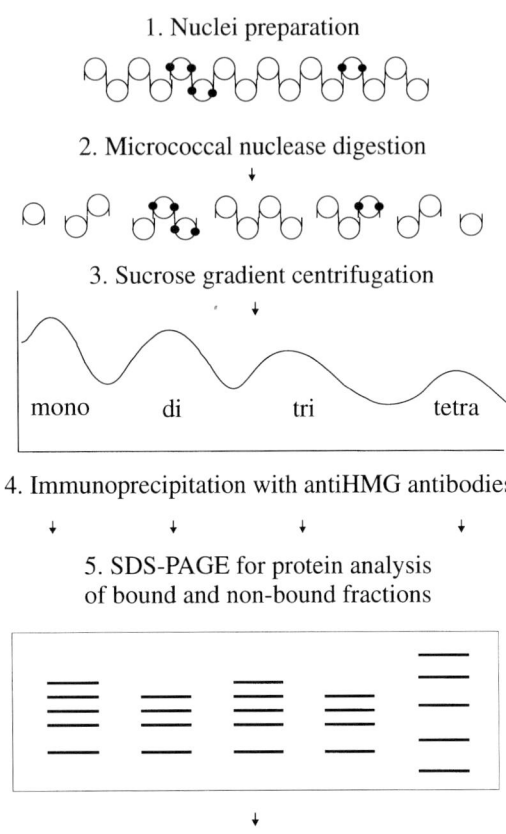

Fig. 1. Experimental strategy for determining the organization of HMG-14/-17 in nucleosomal arrays. (1) Isolation of the nuclei. (2) Nuclei are digested with micrococcal nuclease to generate oligonucleosomes of different sizes. (3) The oligonucleosomes are fractionated on a sucrose gradient. (4) The HMG-17-containing nucleosomes in each oligonucleosome fraction are immunoprecipitated with affinity pure antibody. (5) The protein content in the antibody bound and nonbound oligonucleosome fractions is analyzed by electrophoresis in SDS-containing polyacrylamide gels. (6) The gel is scanned and the data are quantified.

4. Nucleosomes containing HMG-14/-17 are clustered into domains. They are not randomly dispersed along the chromatin fiber.
5. The average size of the domain is six contiguous nucleosomes.

Each domain is homogeneous in its HMG content; that is, in a domain, nucleosomes containing HMG-14 are not intermixed with nucleosomes containing HMG-17. Potentially, this approach will also allow determination of the type of DNA associated with HMG-14/-17 and allow us to determine whether HMG-14/-17 containing nucleosomes are associated with a specific subset of histones or other nonhistone proteins.

2. Materials
2.1. Isolation of Nuclei

1. 1X PBS: 10 mM sodium phosphate, pH 7.5, 140 mM NaCl.
2. 1X HBSS buffer: 340 mM sucrose, 15 mM Tris-HCl, pH 7.5, 15 mM NaCl, 60 mM KCl, 10 mM DTT, 0.5 mM spermine-HCl, 0.15 mM spermidine-HCl, 0.1 mM PMSF.
3. 1X HBSSX buffer: As for 1X HBSS plus 1% Triton X-100 (v/v).
4. 1X TKC buffer: 15 mM Tris-HCl, pH 7.5, 20 mM KCl, 1 mM CaCl$_2$.
5. Glycerol.

2.2. Mild Digestion of Chromatin by Micrococcal Nuclease

1. Micrococcal nuclease (Boehringer Mannheim, Indianapolis, IN) 100 U/µL, stored in 10-mL aliquotes.
2. 1X TKC buffer: 15 mM Tris-HCl, pH 7.5, 20 mM KCl, 1 mM CaCl$_2$.
3. 0.5 M EDTA, pH 8.0.

2.3. Fractionation of Oligonucleosomes on Sucrose Gradients

1. 75% (w/v) sucrose in water, sterile.
2. 50X NTE buffer: 500 mM NaCl, 500 mM Tris-HCl, pH 7.5, 50 mM EDTA, pH 8.
3. Agarose.
4. 1X TAE.
5. 5X Fast DNA Analysis (FDA) solution: 5% (w/v) SDS, 25 mg/L bromophenol blue, 100 mg/mL proteinase K, freshly prepared.
6. Macrosep 10K-cutoff centrifugal concentrators (Filtron, Ann Arbor, MI).

2.4. Isolation of the HMG-14 or HMG-17 Containing Oligonucleosomes Using Specific Antibodies

1. 1X NTE buffer: 10 mM NaCl, 10 mM Tris-HCl, pH 7.5, 1 mM EDTA, pH 8.0.
2. 1X NTEX buffer: As for 1X NTE plus 0.1% (v/v) Triton X-100.
3. Immobilized Protein A on TrisAcryl GF2000 (Pierce, Rockford, IL, cat. no. 20338).
4. Immobilized anti-rabbit IgG sheep antibodies on Dynabeads M-280 (Dynal, Lake Success, NY, cat. no.112.03 or 112.04). Suspension 1:10.
5. 5X SDS loading dye: 125 mM Tris-HCl, pH 6.8, 50 mM DTT, 5% (w/v) SDS, 20% (v/v) glycerol, 25 mg/mL bromophenol blue.
6. Affinity pure rabbit antibodies to human HMG-14 or HMG-17, stored at –20°C in 20-µL aliquots at 1 mg/mL in 1X PBS.
7. Nonimmune rabbit immunoglobulin, 1 mg/mL.
8. 10% (w/v) BSA in 1X PBS in 100-µL aliquots, stored frozen.

2.5. Analysis of the Protein and DNA Content in the Immunofractionated Nucleosomes

1. 40% (w/v) Acrylamide/N,N'-methylene-bis-acrylamide stock solution, 19:1.
2. 1.5 M Tris-HCl, pH 8.8.
3. 10% (w/v) Ammonium persulfate.
4. TEMED.
5. Coomassie staining solution: 1 g/L Coomassie R-250, 50% (v/v) methanol, 10% (v/v) acetic acid.
6. Destaining solution: 20% (v/v) methanol, 5% (v/v) acetic acid.
7. 1X Tris/glycine/SDS buffer: 25 mM Tris-HCl, 190 mM glycine, 0.1% (w/v) SDS, pH 8.3.
8. 5X SDS loading dye: 125 mM Tris-HCl, pH 6.8, 50 mM DTT, 5% (w/v) SDS, 20% (v/v) glycerol, 25 mg/L bromophenol blue.
9. 100% Ethanol.

3. Methods

3.1. Isolation of Nuclei

The following protocol describes the isolation of nuclei from the chicken red blood cells *(2)*.

1. Take 100 mL chicken blood, approx 2×10^{11} nuclei. Wash the chicken red blood cells twice by suspending in 400 mL 1X PBS and spinning 1200g in H-6000A rotor at 4°C for 10 min.
2. Lyse the cells with 200 mL HBSSX buffer on ice. Pellet the nuclei by spinning 4700g in a Sorvall H-6000A rotor at 4°C for 15 min.
3. Wash the nuclei pellet with HBSS buffer 2–3 times as in **step 2**, for 5 min each spin. Expect 10–15 mL of nuclei pellet.
4. Resuspend in 40 mL HBSS buffer plus glycerol (1:1, v/v), store at –20°C (at OD_{260} = 40 or 2 mg/mL) (*see* **Note 1**).

3.2. Mild Digestion of Chromatin by Micrococcal Nuclease (see Note 2)

1. Take 6 mL of stored nuclei, wash in 1X TKC buffer, resuspend in 10 mL 1X TKC.
2. Digest with 10 µL micrococcal nuclease, 37°C, 5 min, do not add EDTA.
3. Chill to 4°C for 2 min, spin 12,000g in Sorvall SS-34 at 4°C for 5 min, save pellet.
4. Resuspend the pellet in 6 mL 0.25 mM EDTA for 12 h or overnight at 4°C, spin as above, discard the pellet. The typical yield is 50–70% of input DNA or 2–3 mg of solubilized chromatin.

3.3. Fractionation of Oligonucleosomes on Sucrose Gradients

1. Prepare sucrose solutions for the linear sucrose gradient:

Components	Top	Bottom
Sucrose, 75%	12% sucrose solution	50% sucrose solution
50X NTE solution	24 mL	100 mL
H_2O	3 mL	3 mL
	up to 150 mL	up to 150 mL

2. Prepare linear 12–50% sucrose gradient in 6 polyallomer tubes for a Beckman SW-28 rotor. Fill the chambers of the gradient mixer with 19 mL each of 12% and 50% sucrose solution and dispense into one tube, repeating the procedure six times, or alternatively with 114 mL each sucrose solution and dispense into six tubes simultaneously using a multichannel peristaltic pump.
3. Layer 1 mL of the micrococcal nuclease digestion mixture over the gradient.
4. Spin at 100,000g for 24 h at 4°C in a SW28 Beckman rotor.
5. Collect 1 mL fractions from the bottom of the tube.
6. Determine the DNA length in each oligonucleosome fraction by analysis in 0.8% agarose gel: take 20 µL from each fraction, add 4 µL of FDA solution and load on a gel.
7. Check OD_{260} in the fractions of interest, desalt and concentrate with a Filtron Macrosep 10 concentrators (according to supplier's recommendation), adjust the final concentration to 0.1 mg/mL DNA in NTE solution (*see* **Note 3**).

3.4. Isolation of the HMG-14 or HMG-17 Containing Oligonucleosomes Using Specific Antibodies (see Note 4)

1. Preincubate 800 µL (80 µg) of purified oligonucleosome preparation with 20 µL immobilized protein A (1:1 suspension, prewashed in 1X NTEX buffer and decanted, not dried) for 1 h at 4°C. Centrifuge 1 min at 80g using Eppendorf 5402 centrifuge and take the supernatant to **step 2**.
2. Mix on ice: 800 µL of preincubated oligonucleosome preparation, 16 µL 5% Triton X-100, 20 µg of immunoaffinity pure antibodies (or nonimmune rabbit IgG in controls). Save 15 µL for control.
3. Incubate 30 min at 4°C.
4. Add the immunoprecipitation mixture to 20 µL immobilized protein A (1:1 suspension, pre-washed in 1X NTEX buffer and decanted, not dried). Incubate overnight or at least 3–4 at 4°C on a vertical rotating wheel or swinging platform (*see* **Note 5**).
5. Spin at 1,000 rpm for 1 min at 4°C using Eppendorf centrifuge. Discard supernatant (save 15 µL for non-bound lane if needed).
6. Add 1 mL of 1X NTEX buffer, incubate at 4°C for 5 min, mix gently but thoroughly. Spin and discard the supernatant. Repeat once.
7. Add 50 µL 1% SDS, heat at 45°C for 5–10 min. Tap lightly the tube to suspend the beads. Spin as above at room temperature. Take the supernatant, mix with 10 µL of FDA solution.

3.5. Analysis of the Protein Content in the Immunofractionated Nucleosomes (see Note 8)

1. Assemble a gel electrophoresis slab unit, seal the edges with 1% agarose or tape.
2. Precool the assembled slab unit and a gradient mixer at 4°C for at least 30 min.
3. Prepare the following solutions without TEMED, keep on ice for at least 15 min.
4. Add 10 µL TEMED to top and bottom acrylamide solutions, mix and dispense equal volumes to the chambers of gradient mixer. For a linear gradient 10–20%

acrylamide, 0.75-mm thick, 15 × 15 cm slab, a total volume of 13 mL is sufficient. For 1-mm thick gel, 15 × 15cm slab, use 18 mL. Using a peristaltic pump pour the gradient from the top, overlay with 0.2–0.4 mL of ethanol.
5. Allow the gel to polymerize and drain off the ethanol.
6. Add 10 μL TEMED to the stacking gel acrylamide solution and pour it onto the top of the gradient gel. Insert the comb and allow the gel to polymerize.
7. Load the gel with the samples and run at 12mA until the bromphenol blue dye reaches the bottom of the gel (4–5 h). The running buffer is 1X Tris/glycine/SDS buffer.
8. Stain the gradient gel with Coomassie staining solution for 30 min, destain it with 20% methanol, 5% acetic acid (*see* **Note 6**).

3.6. Quantitative Analysis for Clustering of HMG-Containing Nucleosomes

1. Scan the Coomassie-stained gel on scanning densitometer (Molecular Dynamics, Sunnyvale, CA) and quantitate the HMG proteins and core histones.
2. Calculate the clustering value C (for definition *see* **Note 7**) of the HMG protein within a nucleosomal array using the formula $C = R(N-1)/(1-R)$ *(1)*. R equals the HMG protein/ core histone ratio. It is calculated by dividing the densitometric value of the HMG by the average of the densitometric value of the core histones (i.e., the total volume of the core histones divided by four). N equals the length of the oligonucleosome fraction taken for analysis. When N is not a integer (the oligonucleosome fraction is not homogeneous), it is possible to calculate N from the relative amounts of the various oligonucleosome species in a sucrose gradient fraction.

4. Notes

1. If the nuclei are to be used immediately and not stored, they should be washed with 1X TKC buffer (*see* **Subheading 3.2.1.**) immediately after washing with HBSS buffer (*see* **Subheading 3.1.3.**).
2. The procedure described leads to the generation of oligonucleosome fragments varying from 1 to 20 nucleosomes in length. Because the distribution of oligonucleosome variants may vary due to the enzyme lot or tissue, we recommend optimization of the experimental conditions in a pilot experiment. The distribution of the oligonucleosome populations in the digest can be evaluated by analyzing the DNA on agarose gels. All the procedures have to be done at low ionic strength so as to:
   a. Minimize the effects of higher order chromatin structure on the digestion kinetics.
   b. Prevent the redistribution of HMG-14/-17 *(3)*.
3. To purify the nucleosome subfraction preparation at N≥4 to homogeneous DNA length, it is necessary to re-run the sucrose gradient using pooled, dialysed and concentrated material obtained after the first gradient. However, as discussed in **Subheading 3.6.**, this may not be always necessary.
4. Critical parameters. The critical step in the procedure is the immunofractionation step. It is imperative that the antibodies react specifically with the antigen and

that the nucleosomes do not bind the support non-specifically *(4)*. We have found that the two immunoadsorbent media described are the best choice, provided that they are freshly prepared and not deteriorated during storage. Appropriate controls with nonimmune IgG have to be included to monitor this step. Immunoprecipitation of the HMG-containing nucleosomes using Protein A-Sepharose or Protein A-Agarose, requires covalent DNA-protein crosslinking prior to the fractionation and harsh buffer conditions *(5,6)*.

A formaldehyde-crosslinking based procedure, described earlier *(5,6)* avoids possible artifacts caused by nonspecific binding to the immunochromatography media. This procedure works better for the analysis of the DNA associated with particular proteins. The possibility to reverse the DNA-protein crosslinks makes it also applicable for the kind of analysis described below. However, on our hands the quantitative evaluation of the proteins, cross-linked to DNA by formaldehyde and immunoprecipitated, was less consistent than in our basic protocol, probably because of the severe chemical modification of the HMG proteins and the core histones.

5. Alternatively, antirabbit IgG sheep antibodies, immobilized on Dynabeads, could be a substitute for immobilized protein A. The protocol should be changed in **step 4** as follows:
   4. Take 1 mL of the Dynabeads suspension and wash the beads twice with 1X NTEX buffer, 0.1% BSA. Decant, do not dry. Add the immunoprecipitation mixture to 100 µL of prewashed decanted Dynabeads. Incubate overnight or at least 3–4 h at 4°C on a vertical rotating wheel or a commercial apparatus that provides bidirectional mixing (like Dynal Sample Mixer).

   Follow the basic protocol described above for immobilized protein A from **steps 5–7**, substituting the centrifugation by magnetic capture, as recommended by manufacturers.
6. Short troubleshooting. Control the recovery of the specific nucleosome fractions by checking the presence of the antigen in nonbound fraction. The amount of the proteins analyzed by SDS-PAGE have to be within the linear range of the densitometric quantitation. Deterioration of the immunoadsorbent may lead to occurrence of nonspecific adsoption of nucleosomes or the impaired yield of the bound product.
7. The method is suitable for the study of the organization of any protein(s) within an oligonucleosomal array, which binds stoichiometrically to core particle. This organization can be expressed as the clustering value C. For HMG proteins, C is the number of contigious nucleosomes containing the HMG-14/-17 protein. The method is based on the fact that chromatin is a repetitive structure, consisting of monoelements—nucleosomes with known stoichiometry of core histones per DNA length.
8. In nuclei HMG-14/-17 proteins comprise only 1–2% of core histones. Therefore the proteins will not be visible in unfractionated chromatin.

References

1. Postnikov, Y. V., Herrera, J. E., Hock, R., Scheer, U., and Bustin, M. (1997) Clusters of nucleosomes containing chromosomal protein HMG-17 in chromatin. *J. Mol. Biol.* **274,** 454–465.

2. Ausio, J., Dong, F., and Van Holde, K. E. (1989) Use of selectively trypsinized nucleosome core particles to analyze the role of the histone tails in the stabilization of the nucleosome. *J. Mol. Biol.* **206,** 451–463.
3. Landsman, D., Mendelson, E., Druckmann, S., and Bustin, M. (1986) Exchange of proteins during immunofractionation of chromatin. *Exp.Cell Res.* **163,** 95–102.
4. Bustin, M. (1989) Preparation and application of immunological probes for nucleosomes. *Methods Enzymol.* **170,** 214–251.
5. Dedon, P. C., Soults, J. A., Allis, C. D., and Gorovsky, M. A. (1991) A simplified formaldehyde fixation and immunoprecipitation technique for studying protein-DNA interactions. *Anal. Biochem.* **197,** 83–90.
6. Solomon, M. J., Larsen, P. L., and Varshavsky, A. (1988) Mapping protein-DNA interactions in vivo with formaldehyde: evidence that histone H4 is retained on a highly transcribed gene. *Cell* **53,** 937–947.

22

Identification and Analysis of Native Nucleosomal Histone Acetyltransferase Complexes

Patrick A. Grant, Shelley L. Berger, and Jerry L. Workman

1. Introduction

Histones are the predominant protein component of chromatin and are subject to a variety of post-translational modifications. Of these, acetylation of the amino-terminal tails of core histones is most intensively studied and is linked to chromatin assembly, the regulation of gene expression, cell cycle progression and cellular transformation *(1)*. Characterization of the enzymes responsible for histone acetylation provides a handle for directly studying such histone modifications in these processes. Two classes of histone acetyltransferases (HATs) have been described; cytoplasmic type B HATs acetylate free histones for subsequent assembly into chromatin *(2)* and nuclear type A HATs, which mediate transcription related acetylation of chromosomal histones *(3)*.

The recent identification and purification of the first nuclear HAT from *Tetrahymena*, a homolog of the yeast transcriptional activator Gcn5, provided the first direct link between histone acetylation and gene activation *(4)*. However it became apparent that recombinant yeast Gcn5 can acetylate only free core histones when presented as a substrate, and not histones contained within nucleosome cores. Only when Gcn5 is associated in high molecular weight complexes can it acetylate nucleosomal histones, the substrate most likely encountered by Gcn5 in the nucleus *(5)*. This observation underscores the importance of purifying native nucleosomal HAT activities to understand the function of the growing number of HAT enzymes.

Here we describe a method to identify and analyze such native HAT complexes from *Saccharomyces cerevisiae*. This procedure relies upon the fortuitous binding of HAT activities to Nickel agarose resin, in the absence of any histidine tags. We describe the preparation of whole cell extracts from yeast,

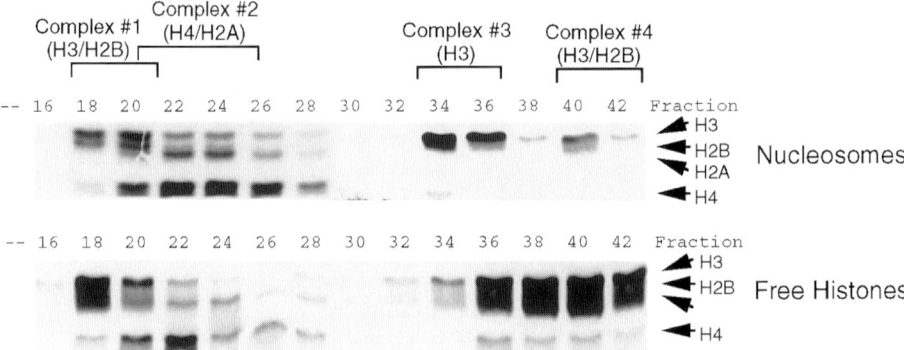

Fig. 1. Separation of distinct HAT complexes which specifically acetylate different nucleosomal or free histones. HAT assays using protein fractions from a Mono Q column run with the Ni-NTA-agarose bound sample, using HeLa nucleosomes (upper panel) or free histones (lower panel) as substrate. Shown are the flourograms of the SDS-PAGE gels from the assays. The arrows indicate the positions of the four core histones. The brackets above the lanes indicate the elution profile of the individual nucleosomal H3/H2B (Complexes #1 and #4), H4/H2A (Complex #2) and H3 HAT (Complex #3) activities *(5)*.

the chromatographic fractionation of these extracts to separate distinct HAT activities and a simple method to assay for nucleosomal HAT activity. This assay has the ability to distinguish between type A and type B HATs, via the use of free histones or nucleosomal histones as substrate (**Fig. 1**). Typically type A HATs are able to acetylate both free and nucleosomal histones, while type B histones acetylate only free histones. The purification and assay procedure described can also be utilized to study the potential effect of specific gene mutations on the function of HAT complexes *(5)*. Also these HAT assays have been used to assay native nuclear HAT activities isolated from mammalian cells.

2. Materials
2.1. Growth of Yeast and Preparation of Whole-Cell Extracts

1. YPD plates: 1% yeast extract, 2% peptone, 2% dextrose, 2% agar, and a pellet of NaOH (around 0.1 g/L). A stirbar should be added to the media before autoclaving. The solution should then be cooled to around 50°C, stirred using a stir plate for 5–10 min and poured into sterile plastic 100 × 15 mm plates.
2. YPD medium: 1% yeast extract, 2% peptone, 2% dextrose (*see* **Note 1**).
3. Extraction buffer: 40mM HEPES-KOH, pH 7.5, 350mM NaCl, 0.1% Tween 20, 10% glycerol, 2 µg/mL pepstatin A, 2 µg/mL leupeptin, 5 µg/mL aprotinin, 1mM PMSF.

4. Bead Beater (Biospec Products).
5. Glass beads (0.5 mm), washed with nitric acid.

2.2. Purification of HAT Activities

1. 10 mL Poly-Prep chromatography column fitted with a 200 mL funnel (Biorad)
2. Ni-NTA agarose resin (Quiagen).
3. Wash buffer: 20 mM Imidazole, pH 7.0, 100 mM NaCl, 0.1% Tween 20, 10% glycerol, 2 µg/mL pepstatin A, 2 µg/mL leupeptin, 5 µg/mL aprotinin, 1 mM PMSF (*see* **Note 2**).
4. Elution buffer: 300 mM Imidazole, pH 7.0, 100 mM NaCl, 0.1% Tween 20, 10% glycerol, 2 µg/mL pepstatin A, 2 µg/mL leupeptin, 5 µg/mL aprotinin, 1 mM PMSF.
5. Mono Q buffers: 50 mM Tris-HCl pH 8.0, 100 mM NaCl, 0.1% Tween-20, 10% glycerol, 2 µg/mL pepstatin A, 2 µg/mL leupeptin, 5 µg/mL aprotinin, 1 mM PMSF, 0.5 mM DTT. A high salt Mono Q buffer should also be prepared using 1M NaCl.
6. Mono Q HR 5/5 anion exchange column (Pharmacia).

2.3. Histone Acetyltransferase (HAT) Assays

1. HAT buffer, 5X: 250 mM Tris-HCl, pH 8.0, 25% glycerol, 0.5 mM EDTA, pH 8.0, 250 mM KCl, 5 mM DTT, 5 mM PMSF, 50 mM sodium butyrate (*see* **Note 2**).
2. ^{3}H Acetyl-coenzyme A at 2–10 Ci/mmol (Amersham).
3. P81 phosphocellulose filter paper circles (Whatman).
4. 50 mM NaHCO$_3$-NaCO$_3$ buffer (pH 9.2).
5. Acetone 100%.
6. Acrylamide (29:1) 40%, electrophoresis grade (Fisher).
7. Tris-HCl/SDS, pH 6.8, 4X: 0.5M Tris, 0.4% SDS. Adjust pH to 6.8 with 1 N HCL.
8. Tris-HCl/SDS, pH 8.8, 4X: 1.5M Tris, 0.4% SDS. Adjust pH to 8.8 with 1 N HCL.
9. Ammonium persulfate 10%.
10. TEMED.
11. Isobutyl alcohol, H$_2$O saturated.
12. Gel unit (Bio-Rad Protean II or Hoeffer SE-600/400 U) with clamp assembly, glass plates, casting stand, electrode assembly, buffer chambers, 10-well comb, 0.75-mm spacers.
13. SDS/sample buffer, 4X: 200 mM Tris-HCl (pH 6.8), 400 mM DTT, 8% SDS, 0.4% bromophenol blue, 40% glycerol.
14. Tris-glycine buffer: 25 mM Tris, 250 mM glycine, 0.1% SDS.
15. Prestained molecular weight standards, broad, or low-range (Bio-Rad or Amersham).
16. Coomassie staining solution: 45% methanol, 0.05% coomassie brilliant blue R-250, 10% acetic acid, 45% distilled water. Dissolve the Coomassie brilliant blue in methanol before adding the other components.
17. Destaining solution: 45% methanol, 10% acetic acid, 45% distilled water.
18. EN^{3}HANCE autoradiography enhancer (NEN Research products).

3. Methods

3.1. Preparation of Whole-Cell Extracts from Yeast

1. Streak a YPD plate with the *Saccharomyces cerevisiae* yeast strain of interest. Incubate the plate at 30°C for 24–48 h.
2. Pick a single yeast colony and inoculate into 10 mL of YPD medium in a sterile 50-mL plastic tube. Grow for approx 8 h at 30°C in a shaking incubator at 300 rpm.
3. Inoculate 6 L of YPD medium with 6 mL of the preculture. Grow yeast in Erlenmyer flasks at 30°C in a shaking incubator at 300 rpm, until the cells reach mid/late log phase (*see* **Note 3**).
4. Pour yeast into 250-mL centrifugation bottles, cool cells on ice, and pellet the cells by spinning at 3000g (4500 rpm in a GSA rotor, Sorvall) for 15 min at 4°C.
5. Resuspend cells in 200 mL cold extraction buffer and repeat **step 4**.
6. Resuspend cell pellet with 25 mL of extraction buffer.
7. Fill a 100-mL bead-beater polycarbonate chamber half-full with 0.5-mm glass beads, washed in extraction buffer. Pour in yeast suspension and fill the container to the mouth with extraction buffer (*see* **Note 4**).
8. Homogenize yeast with a 30-s pulse, following by 1-min pause. Repeat this step an additional nine times.
9. Pour glass bead/yeast suspension into two 50-mL tubes and spin at 1,000 rpm at 4°C in a benchtop centrifuge to remove the glass beads.
10. Spin down the cell debris by centrifugation at 16,000g (11,500 rpm in an SS-34 rotor, Sorvall) for 30 min at 4°C.
11. Pour the supernatant into 70 mL polycarbonate centrifuge bottles (Beckman) and clarify the extract by ultracentrigation at >100,000g (43,000 rpm in a 45 Ti rotor, Beckman) for 1 h at 4°C.

3.2. Purification and Separation of HAT Activities

1. Wash 5 mL (100%) Ni-NTA agarose (Quiagen) with 40 mL extraction buffer and spin at 1,000 rpm in benchtop centrifuge for 5 min. Repeat.
2. Filter yeast extract through cheesecloth and incubate with the Ni-NTA agarose for 3 h at 4°C, on a rotating wheel.
3. Pour mixture into a disposable 10 mL Poly-Prep chromatography column fitted with a 200 mL funnel (Biorad). Collect and save the flow through.
4. Wash the resin with 12.5 mL extraction buffer, at 4°C.
5. Wash the resin with 12.5 mL 20 mM imidazole wash buffer, at 4°C.
6. Elute the bound with 12.5 mL 300 mM imidazole extraction buffer, at 4°C.
7. Load the eluate directly onto a 1 mL Mono Q column (Pharmacia) at a flow rate of 0.5 mL/min (*see* **Note 5**). Collect and save flow through.
8. Wash column with 10 column volumes 100 mM NaCl Mono Q buffer.
9. Elute bound proteins with a 25 mL gradient of 100–500 mM NaCl Mono Q buffer at a flow rate of 0.5 mL/min. Collect 0.5 mL fractions, aliquot, and freeze in liquid nitrogen.

3.3. Assay of Nucleosomal HAT Activities

1. In 30 µL reactions, add 6 µL of 5X HAT buffer, 1 µL of Mono Q fraction, 0.25 µCi of tritiated acetyl CoA (*see* **Note 6**) and 2 µg of nucleosomes or free histones (*see* Chapter 23).
2. Incubate at 30°C for 30 min.
3. Spot 15 µL of each reaction on P81 phosphocellulose filter paper circles (Whatman) and allow to air dry.
4. Wash the filters with 50 mL of 50 mM NaHCO$_3$-NaCO$_3$ buffer, pH 9.2, on a shaking platform for 5 min. Repeat an additional two times.
5. Briefly rinse the filters in 50 mL acetone and allow to air dry. Place filters in scintillation vials, add 4 mL of scintillation fluid and count in a scintillation counter for at least 10 min/sample.
6. Assemble a mini protein gel sandwich, using two clean glass plates separated by 0.75-mm spacers. Pour approx 3.8 mL of separating solution (18% acrylamide, 1X Tris-HCl/SDS, pH 8.8, 0.33% ammonium persulfate, 0.66% TEMED) into gel sandwich. Cover top of gel with H$_2$O saturated isobutyl alcohol and allow gel to polymerize for 20–30 min.
7. Pour off isobutyl alcohol layer and rinse with distilled H$_2$O. Prepare stacking solution (4% acrylamide, 1X Tris-HCl/SDS, pH 6.8, 0.33% ammonium persulfate, 0.66% TEMED) and add approx 1 mL to the gel sandwich. Insert a 10-well, 0.75-mm comb into the stacking gel layer and allow gel to polymerize.
8. Remove comb, attach gel sandwich to upper buffer chamber and fill upper and lower buffer chambers with tris-glycine running buffer. Rinse wells with running buffer.
9. Add 5 µL of 4X SDS/sample buffer to the remaining 15 µL of each HAT reaction.
10. Boil samples in a water bath for 5 min, briefly spin tubes in a microfuge and carefully load samples on the SDS-polycrylamide gel. In a spare well load 5 µL of prestained molecular weight standard.
11. Electrophorese samples at 150 V for 2 h (*see* **Note 7**).
12. Disassemble gel plates and stain gel with approx 50 mL of Coomassie staining solution for 30 min with gentle shaking. Destain gel with destaining solution until histones become visible.
13. Incubate gel in EN^{3}HANCE autoradiography enhancer (NEN Research products, *see* **Note 8**), with gentle agitation for 30–60 min.
14. Rinse gel several times with deionised water and then incubate with deionized water, with gentle shaking for 30 min.
15. Transfer gel to a piece of wet blotting paper, cover with plastic wrap, and dry under vacuum at 60°C for 90 min.
16. Expose dried gel to X-ray film. A readable fluorogram is usually obtained within 24–48 h (*see* **Note 9**).

4. Notes

1. It is preferable to prepare a 20% (10X) solution of dextrose that has been autoclaved separately, and then added to the other ingredients after autoclaving to prevent darkening of the medium and to promote optimal growth.

2. The extraction, wash, elution and Mono Q buffers should be prepared freshly and filtered through a 0.45 µ filter and stored at 4°C or on ice prior to use. A 100-mM stock solution of PMSF can be prepared in isoproponol and stored in a light-proof container. A 1 M stock solution of imidazole can be prepared with distilled water and stored at 4°C for several months. Stock solutions of the proteinase inhibitors aprotinin, leupeptin, and pepstatin A and the deactylase inhibitor sodium butyrate can be prepared as follows, stored at –20°C for several months: aprotinin 10 mg/mL in 10 mM HEPES-KOH, pH 7.5, pepstatin A at 2 mg/mL in methanol, leupeptin at 2 mg/mL in distilled water, sodium butyrate at 1 M in distilled water. Appropriate care should be taken when handling the above reagents as they are toxic or harmful by inhalation, ingestion or contact with skin.
3. The density of the cells is determined using a spectrophotometer, measuring the optical density (OD) at 600 nm. For reliable measurements cultures should be diluted such that OD at A_{600} is less than 1.0. An OD of 1.5–2.0 represents mid/late log phase for most yeast strains (approx 5×10^7 cells/mL).
4. Exclusion of air is important to prevent frothing and for proper operation of the bead-beater. Place the chamber into the ice jacket containing an ice/water suspension. It is important to keep the suspension cool and pause between each homogenization pulse, as a temperature increase of about 10°C occurs per minute of active use.
5. Prior to use, the Mono Q column should be equilibrated by washing with 10 column volumes of 100 mM NaCl Mono Q buffer, followed by 10 vol of 1 M NaCl buffer and finally with 10 vol of 100 mM NaCl Mono Q buffer. It is highly recommended that a fast protein, peptide and polynucleotide liquid chromatography (FPLC) system be used with the Mono Q column.
6. Caution should be taken when handling this radioisotope, wear gloves and monitor for contamination of work area with swipe tests. Dispose of radioactive materials in appropriate liquid and solid waste containers.
7. Histones migrate in the 10–20,000 Da range and since electrophoresis times may vary, optimal separation of histones is best judged by the migration of the prestained molecular weight markers.
8. EN^{3}HANCE is harmful if swallowed, by inhalation, causes severe burns and is combustible. This product should be used with extreme caution, preferably in a sealed container and in a air-flow hood.
9. Better results are obtained using Fuji RX X-ray film and by developing the flourogram manually for several minutes until signals are visible.

References

1. Wade, P. A., Pruss, D., and Wolffe, A. (1997) Histone acetylation: chromatin in action. *Trends Biochem.* **22,** 128–132.
2. Roth, S. Y. and Allis, C. D. (1996) Histone acetylation and chromatin assembly: A single escort, multiple dances? *Cell* **87,** 5–8.
3. Brownell, J. E. and Allis, C. D. (1996) Special HATs for special occasions: Linking histone acetylation to chromatin assembly and gene activation. *Curr. Opin. Genet. Develop.* **6,** 176–184.

4. Brownell, J. E., Zhou, J., Ranalli, T., Kobayashi, R., Edmondson, D. G., Roth, S. Y., and Allis, C. D. (1996) Tetrahymena histone acetyltransferase A: a homolog to yeast Gcn5p linking histone acetylation to gene activation. *Cell* **84,** 843–851.
5. Grant, P. A., Duggan, L., Côté, J., Roberts, S. M., Brownell, J., Candau, R., Ohba, R., Owen-Hughes, T., Allis, C. D., Winston, F., Berger, S. L., and Workman, J. L. (1997) Yeast Gcn5 functions in two multisubunit complexes to acetylate nucleosomal histones: Characterization of an ADA and the SAGA (Spt/Ada) complex. *Genes Dev.* **11,** 1640–1650.

23

Analysis of Nucleosome Disruption by ATP-Driven Chromatin Remodeling Complexes

Tom Owen-Hughes, Rhea T. Utley, David J. Steger, Joshua M. West, Sam John, Jacques Côté, Kristina M. Havas, and Jerry L. Workman

1. Introduction

In vivo DNA is associated with the proteins that constitute chromatin. This means that, any process that requires access to the genetic material must do so within the context of chromatin. It is now clear that there is a complex cellular machinery dedicated to regulating chromatin structure and that the function of this machinery represents an important step in gene regulation *(1–3)*.

Over the last few years a number of activities that function in the creation of receptive chromatin configurations have been described. A subset of these are known to alter chromatin in an ATP dependent reaction. To date all these ATP dependent chromatin remodeling activities have been found to be multiprotein complexes containing the yeast SNF2 protein or a polypeptide with homology to it *(2)*.

A diverse range of assays for the function of these complexes have been described. These include the creation of GAGA factor dependent DNase hypersensitivity at the HSP70 promoter *(4)*, increases in global restriction enzyme accessibility *(5)*, and nucleosome spacing *(6)*. All these approaches have proved to be valuable assays of chromatin remodeling. This chapter describes the use of purified mono and multi-nucleosome templates for studying the action of the ATP dependent chromatin remodeling activities. This system has the advantage that it uses well defined chromatin templates assembled using purified components.

In this chapter protocols are described to enable the purification of large quantities of chromatin or histones. These protocols have been modified from previously published protocols *(7–9)* to enable HeLa nuclear pellets produced as a by product of the preparation of HeLa nuclear extract to be used as source material. These reagents are then used to reconstitute chromatin templates. Finally, a series of assays for the action of ATP dependent chromatin remodeling activities are described.

2. Materials
2.1. Purification of HeLa Oligonucleosomes

1. Pellet of HeLa Nucleosomes (*see* **Note 1**).
2. 0.4 M NaCl HB: 20 mM HEPES, pH 7.5, 1 mM EDTA, 1 mM 2-mercaptoethanol, 10% glycerol, 0.5 mM PMSF, 5 µg/mL aprotinin, 2 µg/mL leupeptin, 2 µg/mL pepstatin.
3. 0.4 M NaCl HB with 0.2% NP-40, 0.6 M NaCl HB, 0.1 M NaCl HB.
4. Dounce homogenizer with type B pestle (Wheaton, Millville, NJ).
5. Branson sonifier 450.
6. Spectra/por 6–8 kDa cut off dialysis membrane.
7. Micrococcal nuclease (Sigma, St. Louis, MO).
8. 100 mM CaCl$_2$.
9. 0.5 M EGTA pH 8.0.
10. 5 M NaCl.
11. 100 mL Sepharose CL-6B (Amersham Pharmacia, Piscataway, NJ) column (1.6 × 70 cm) equilibrated with 0.6 M NaCl HB.
12. 15% SDS PAGE mini-gel (Bio-Rad, Hercules, CA) and electrophoresis equipment.
13. 5X SDS loading buffer (25 mM Tris-Cl [pH 6.8], 0.1 M 2-mercaptoethanol, 10% SDS, 50% glycerol).
14. SDS running buffer (25 mM Tris, 250 mM glycine, pH 8.3, 0.1% SDS).
15. TAE (0.04 M Tris-acetate, 0.001 M EDTA), agarose, agarose gel (11 × 11 × 0.5 cm) electrophoresis equipment.
16. UV spectrophotometer and quartz cuvette.

2.2. Purification of Core Histones

1. Pellet of HeLa Nucleosomes (*see* **Note 1**).
2. 0.4 M NaCl HB: 20 mM HEPES, pH 7.5, 1 mM EDTA, 1 mM 2-mercaptoethanol, 10% glycerol, 0.5 mM PMSF, 5 µg/mL aprotinin, 2 µg/mL leupeptin, 2 µg/mL pepstatin.
3. 0.4 M NaCl HB with 0.2% NP-40.
4. 0.4 M NaCl HAP (50 mM sodium phosphate [pH 6.8], 1 mM 2-mercaptoethanol, 0.5 mM PMSF, 5 µg/mL aprotinin, 2 µg/mL leupeptin, 2 µg/mL pepstatin). 0.6 M NaCl HAP.
5. Dounce homogenizer with type B pestle (Wheaton).
6. Branson sonifier 450.
7. 5 M NaCl.

8. Bio-Gel HTP Hydroxylapatite DNA grade (Bio-Rad).
9. 15% SDS PAGE mini-gel (Bio-Rad) and electrophoresis equipment.
10. 5X SDS loading buffer: 25 mM Tris-HCl, pH 6.8, 0.1 M 2-mercaptoethanol, 10% SDS, 50% glycerol.
11. SDS running buffer: 25 mM Tris, 250 mM glycine, pH 8.3, 0.1% SDS.
12. UV spectrophotometer and quartz cuvet.

2.3. Reconstitution by Dilution Transfer from HeLa Oligonucleosomes

1. 2.5 µg HeLa nucleosomes (*see* **Subheading 3.1.**).
2. Initial dilution buffer: 10 mM HEPES, pH 7.9, 1 mM EDTA, 0.5 mM PMSF.
3. Probe DNA.
4. 5 M NaCl.
5. Final dilution buffer: 10 mM Tris-HCl, pH 7.5, 1 mM EDTA, 0.1% NP-40, 5 mM EDTA, 5 mM dithiothreitol, 0.5 mM PMSF, 20% glycerol, 100 µg/mL bovine serum albumin fraction V (Sigma).

2.4. Reconstitution by Dilution Transfer from Histone Octamers

1. Histone octamers (*see* **Subheading 3.2.**).
2. Reagents of **Subheading 2.3., steps 2–5**.

2.5. Analysis of Nucleosome Disruption by DNase I Digestion

1. 10X binding buffer: 200 mM HEPES, pH 7.5, 5 mM DTT, 1 mM PMSF, 1 mg/mL BSA, 50% glycerol.
2. 1 M NaCl.
3. 100 mM MgCl$_2$.
4. Fractions containing chromatin remodeling activity (*see* **Note 10**).
5. DNase I, RNase free (Boehringer Mannheim, Indianapolis, IN).
6. DNase I dilution buffer: 20 mM Tris-HCl, pH 7.5, 50 mM NaCl, 1 mM DTT, 100 µg/mL BSA (Sigma), 50% Glycerol.
7. DNase I stop buffer: 20 mM Tris-Cl, pH 7.5, 50 mM EDTA, 2% SDS, 200 µg/mL proteinase K (Sigma), 0.25 mg/mL tRNA (Sigma).
8. 0.2 M NaCl.
9. 100% ethanol.
10. 80% ethanol.
11. Sequencing gel loading buffer: 95% formamide, 0.1% xylene cyanol, 0.1% bromophenol blue.
12. TBE: 0.09 M Tris-borate, 0.002 M EDTA.
13. 8% acrylamide (acrylamide: bis = 19:1), 8 M Urea, 1X TBE sequencing gel.
14. Autoradiography equipment.
15. DSB: 3% SDS 50 mM Tris-HCl, pH 8, 0.1 M EDTA, 25% glycerol, 0.05% bromophenol blue and xylene cyanol, 200 µg/mL proteinase K (Sigma).
16. 1.25% agarose gel (25 × 12 × 0.5 cm).
17. Electrophoresis equipment.
18. Fix solution: 10% acetic acid, 10% methanol.

2.6. Stimulation of Factor Binding by DNase I Digestion

1. Same as for **Subheading 2.5.** with the addition of sequence specific transcription factor and probe DNA containing the appropriate binding sites (*see* **Note 11**).

2.7. Stimulation of Factor Binding by Gel Shift Assay

1. 10X binding buffer: 200 mM HEPES, pH 7.5, 5 mM DTT, 1 mM PMSF, 1 mg/mL BSA, 50% glycerol.
2. 1 M NaCl.
3. 100 mM MgCl$_2$.
4. Fractions containing chromatin remodeling activity (*see* **Note 10**).
5. Competitor (*see* **Note 12**).
6. TBE: 0.09 M Tris-borate, 0.002 M EDTA.
7. 4% acrylamide (acrylamide: bis = 29:1), 0.5X TBE gel (18 × 18 × 0.1cm) and electrophoresis equipment.
8. Autoradiography equipment.

3. Methods
3.1. Preparation of Chromatin from HeLa Cell Nuclei

1. Resuspend nuclei (*see* **Note 1**) (10 mL equivalent to 12 L cells) in 50 mL HB + 0.4 M NaCl. Use 1 or 2 strokes of a dounce homogenizer to assist with resuspension.
2. Pellet nuclear material by centrifugation at 20,000g (15,000 rpm SS34 rotor) for 5 min.
3. Repeat **steps 1** and **2** four times.
4. Resuspend nuclear material in 50 mL 0.4 M NaCl, 0.2% NP-40 HB.
5. Pellet nuclear material by centrifugation at 20,000g (15,000 rpm SS34 rotor) for 5 min.
6. Repeat **steps 4** and **5**.
7. Wash pellet twice more with 0.4 M NaCl HB without NP-40.
8. Thoroughly resuspend the Pellet in a small volume approx 20 mL 0.6 M NaCl HB 50 strokes of a dounce homogenizer may be used.
9. Sonicate on ice in pulses of 15 s (50% Duty cycle) for 1 min using setting 6 (Branson sonifier 450). The resulting chromatin suspension should have an average length of approx 4 kb as determined by agarose gel electrophoresis after protease digestion (*see* **Note 2**) and have a greatly reduced viscosity.
10. Dialyze the suspension into HB + 0.1 M NaCl (*see* **Note 3**).
11. Add 0.03 vol of 100 mM CaCl$_2$.
12. Titrate 50-µL aliquots with a range of MNase diltuions, typically 1U, 0.1U, 0.01U. Remove 5 µL samples at various time points and stop with DSB. Digest with proteinase K for 60 min at 50°C as described in **Note 3** to determine conditions under which chromatin with an average length of about 400 bp can be produced.
13. Digest the bulk of the chromatin. Typically 20 U miccrococcal nuclease (Sigma) were used to digest 240 mg of the suspension at 25°C for 5 min.
14. Stop the digestion by adding 0.1 vol of 0.5 M EGTA, pH 8.0, and returned to 4°C.

15. Add 2 *M* NaCl dropwise while mixing to obtain a final concentration of 0.6 *M*.
16. Pellet insoluble material by centrifugation at 20,000*g* (15,000 rpm SS34 rotor) for 15 min.
17. Apply no more than 10 mL to a 100 mL Sepharose CL-6B column (1.6 × 70 cm) that had been equilibrated with HB + 0.6 *M* NaCl.
18. Run the column at 0.2 mL/min while collecting 4-mL fractions. The absorbance of fractions is monitored using a chart recorder and the peak fractions subject to further analysis.
19. The average length of the chromatin fragments is determined by 1% agarose gel electrophoresis following digestion with DSB (*see* **Note 2**).
20. The stoichiometry of the histones is determined by 15% SDS PAGE. All four histones should be present in equal ratios. Fractions containing histone H1 are not suitable for use in reconstitution of H1 depleted chromatin templates.
21. The concentration of the chromatin is determined by measuring the DNA concentration from the absorbance at 260 n*M* of samples diluted in 2 *M* NaCl. If necessary the chromatin may be concentrated (*see* **Note 4**). This procedure is expected to yield approx 50 mg of chromatin from 12 L cells.

3.2. Purification of HeLa Core Histones

1. Core histones can be prepared from extensively washed and sonicated nuclear pellets prepared as described above (**Subheading 3.1., steps 1–6**).
2. Pellets are washed twice with 50 mL 0.4 *M* NaCl HAP (50 m*M* sodium phosphate, pH 6.8, 1 m*M* 2-mercaptoethanol, 0.5 m*M* PMSF).
3. Pellet nuclei by centrifugation at 20,000*g* (15,000 rpm SS34 rotor) for 5 min.
4. Resuspend nuclei in 50 mL HAP + 0.6 *M* NaCl.
5. Sonicate on ice in pulses of 15 s (50% Duty cycle) for 1 min using setting 6 (Branson sonifier 450). The resulting chromatin suspension should have an average length of approx 2 kb as determined by agarose gel electrophoresis after protease digestion (*see* **Note 2**) andhave a greatly reduced viscosity.
6. Mix gently with approx 10 g (dry weight) Bio-Gel HTP Hydroxylapatite DNA grade (Bio-Rad) equilibrated with 0.6 *M* NaCl HAP in batch for 30 min at 4°C.
7. The Bio-Gel is allowed to settle (approx 20 min 4°C), and then a firm pellet is formed by centrifugation at 3000*g* (3750 rpm GH 3.7 rotor) for 5 min. The supernatant is removed, but may be saved for use in the purification of histone H1 *(8)*.
8. Wash twice with 250 mL of 0.6 *M* NaCl HAP.
9. Resuspend the pellet with 2 vol HAP + 2.5 M NaCl. Add additional 5 *M* NaCl dropwise with constant gentle mixing to bring the bring the final NaCl concentration to 2.5 *M* (after compensating for the volume of the Bio-Gel).
10. Pellet the HAP and save the supernatant which contains histones.
11. Extract the Bio-Gel 2 more times with HAP + 2.5 M NaCl.
12. Determine the concentration of histones by measuring the absorbance at 230 nm. ($A230 = 3.3$ is equivalent to 1 mg/mL).
13. Check the integrity, purity, and stoichiometry of the histones by SDS-PAGE electrophoresis.

14. If necessary dialyze the histones to lower salt (but note that histone octamers are not stable below approx 1.2 M NaCl) or concentrate histones using centriprep concentrators (Amicon, Beverly, MA). Aliquots may be frozen and stored at −80°C indefinitely. This procedure yields approx 50 mg core histones when starting with 12 L cells.

3.3. Reconstitution of Radiolabeled DNAs by Dilution Transfer from Purified HeLa Oligonucleosomes

1. 2.5 µg nucleosomes are mixed with probe DNA (*see* **Notes 5** and **6**) in 10 µL of initial dilution buffer +1 M NaCl, 10 mM HEPES, pH 7.9, 1 mM EDTA, 0.5 mM PMSF and incubated at 37°C for 15 min.
2. The reaction is transferred to 30°C and serially diluted with 1.8 µL, 3.5 µL, 4.7 µL, 13 µL, and 17 µL initial dilution buffer with a 150-min incubation at 30°C with each dilution.
3. The reaction is diluted with 50 µL final dilution buffer and incubated at 30°C for 15 min.
4. The reconstituted DNA may be stored at 4°C or frozen an stored at -80°C (*see* **Note 7**). The integrity of the reconstituted DNA should be tested (*see* **Note 8**).

3.4. Reconstitution from Histone Octamers

1. A 10 µL reaction containing histones and DNA (*see* **Note 6**) are mixed at a 1:1 ratio (*see* **Note 9**) of nucleosome binding sites to histone octamers in initial buffer brought to 2 M NaCl and incubated at 37°C for 15 mins
2. The reaction is transferred to 30°C and serially diluted with 3.3 µL, 6.7 µL. 5 µL, 3.6 µL, 4.7 µL, 6.7 µL, 10 µL, 30 µL, and 20 µL initial dilution buffer with a 15 min incubation at 30°C with each dilution.
3. The reaction is diluted with 100 µL final dilution buffer and incubated at 30°C for 15 min.
4. The reconstituted DNA may be stored at 4°C or frozen an stored at −80°C (*see* **Note 7**). The integrity of the reconstituted DNA should be tested (*see* **Note 8**). An example of this is shown in **Fig. 1**.

3.5. Analysis of Nucleosome Disruption by DNase I Digestion

1. Binding reactions containing 2 µL 10X binding buffer, 2 µL 10 mM Mg-ATP, approximately 25 ng chromatin (C10,000 cpm), 3 mM MgCl$_2$, chromatin remodeling activity (*see* **Note 10**), KCl to 50 mM, and distilled water to a final volume of 20 µL are incubated at 30°C for 30 min (the chromatin and remodeling activity are added last).
2. The reactions are brought to room temperature and 1 µL DNase I (0.1U) diluted in DNase I dilution buffer added. The optimum DNase I concentration should be determined by titration. Typically 0.1U DNase are required to digest 25 ng chromatin.
3. Digest for 1 min at room temperature.
4. Stop the digestion by addition of 25 µL DNase I stop buffer.
5. Incubate at 50°C for 1 h.
6. Add 150 µL of 0.2 M NaCl and 600 µL 100% ethanol. The reactions are chilled on ice or at −20°C for 15 mis or longer.

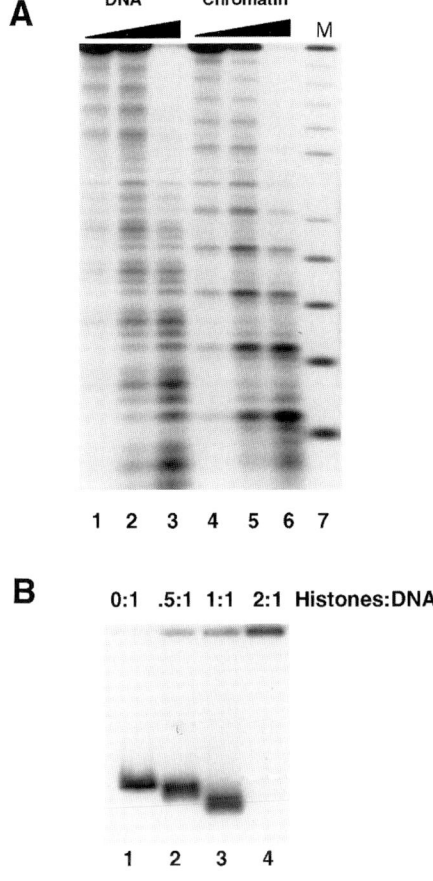

Fig. 1. Reconstitution of a nucleosomal array using core histones. A DNA fragment containing binding sites for 12 nucleosomes was reconstituted as described in **Subheading 3.4.** (**A**) 10 ng probe DNA assembled using a 1:1 ratio of histones: nucleosome binding sites (lanes 4–6) or DNA not assembled with nucleosomes (lanes 1–3). Lanes 1–3 were digested with 0.01 mU MNase, lanes 4–6 contained 0.02 mU. Digestion times were 0.33 min lanes 1,4; 1 min lanes 2,5; 3 min lanes 3,6. The chromatin shows a repeated nuclease digestion pattern distinct from that on DNA and consistent with the presence of positioned nucleosomes. (**B**) DNA or chromatin assembled at the histone: DNA ratios indicated was subject to 1.2% agarose gel electrophoresis. The compaction of the chromatin results in a greater mobility than for free DNA until a 1:1 ratio. When DNA is substoichiometric it is more likely to become insoluble *(15)* hence DNA is retained in the wells in lane 4.

7. Precipitate DNA by centrifugation at 14,000 rpm in a benchtop microcentrifuge for 30 min.
8. Take off the supernatant and wash the pellet with 70% ethanol.

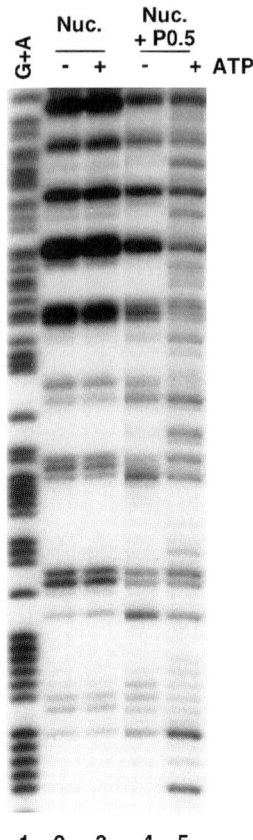

Fig. 2. Detection of nucleosome disruption by DNase I digestion. Lanes 2 and 3 show the characteristic nucleosomal 10 bp repeated DNase I digestion pattern observed when an ScaI to AvaI restriction fragment of the sea urchin 5S rRNA gene *(16)* is assembled into nucleosomes as described in **Subheading 3.3.** This 10 bp repeated pattern is disrupted in the presence of both ATP and appro 2 µg of the material eluting between 0.3 and 0.5 M KCl from a phosphocellulose column loaded with HeLa nuclear extract.

9. Remove all the 70% ethanol (use speedvac if necessary), and resuspend the sample in 5 µL loading dye.
10. Denature the sample by heating to 95°C for 1 min.
11. Load the sample on a prerun 8% sequencing gel. Run at 60 W for sufficient time to resolve sequences of interest.
12. The gel can be transferred to an X-ray cassette on a sheet of used X-rayfilm and exposed at –80°C with an intensifying screen. **Figure 2** shows an example of the results obtained in experiments performed in the presence of crude fractions containing a chromatin remodeling activity.

13. This protocol can be modified for use in the study of muti-nucleosome templates by adding 5 µL DSB after **step 3**. Samples are incubated at 50°C for 1 h and then loaded onto a 1.25% TAE agarose gel. Following electrophoresis the gels are incubated in fix solution for 20 mins and dried down onto blotting paper using a slow temperature ramp. The digestion products can then be resolved by autoradiography.

3.6. Measurement of the Stimulation of Factor Binding by DNase I Digestion

1. Binding reactions are as in **Subheading 3.5.1.**, but also include dilution's of a sequence specific transcription factor (*see* **Note 11**). The reactions are incubated at 30°C for 30 min.
2. Binding reactions are processed as described in **Subheading 3.5.**, **steps 2–13**. An example of the stimulation of factor binding that can be detected using this approach is shown in **Fig. 3**.

3.7. Measurement of the Stimulation of Factor Binding by Gel Shift

1. Binding reactions as for **Subheading 3.5.** are incubated at 30°C for 30 min.
2. Competitor material is added to reactions to release chromatin remodeling activities and/or other DNA-binding activities present in the fractions being tested from nucleosomes. The most appropriate competition substrate and conditions should be determined for each application (*see* **Note 12**).
3. The competition reactions can be loaded directly on a 4% polyacrylamide 0.5X TBE gel and run at 150 V for 2.5 h.
4. The gel is dried onto blotting paper and viewed by autoradiography. This approach has sucessfully been used to detect stimulation of factor binding mediated by the yeast SWI/SNF complex *(13)*.

4. Notes

1. Typically cells for this purpose are grown to a density of 6.4×10^5 in 50–100 L of Jocklik's minimal essential medium plus 5% calf serum in spinner flasks. The cells are collected by centrifugation for 10 min at 2500g and resuspended in PBS (4.3 mM Na$_2$HPO$_4$, 1.4 mM KH$_2$PO$_4$, 137 mM NaCl, 2.7 mM KCl, pH 7.5). Nuclear extracts were made as described by *(10)*. The insoluble nuclear pellets obtained as a result of this procedure could be stored indefinitely at –80°C. Unless indicated otherwise all steps in the preparation of histones and nucleosomes were performed at 4°C or on ice and all buffers contained protease inhibitors (0.5 mM PMSF, 5 µg/mL aprotinin, 2 µg/mL leupeptin, 2 µg/mL Pepstatin).
2. To determine the length of DNA fragments following sonication or small scale MNase digestion, 10-µL aliquots containing approximately 2 µg DNA (determined from absorbance at 260 nm in 2 M NaCl) were digested with 5 µL DSB (3% SDS 50 mM Tris, pH 8.0, .1 M EDTA, 25% glycerol, 0.05% bromophenol blue and xylene cyanol), 200 µg/mL proteinase K (Sigma), incubated at 50°C for 60 min, and analyzed by 1% agarose TAE gel electrophoresis.

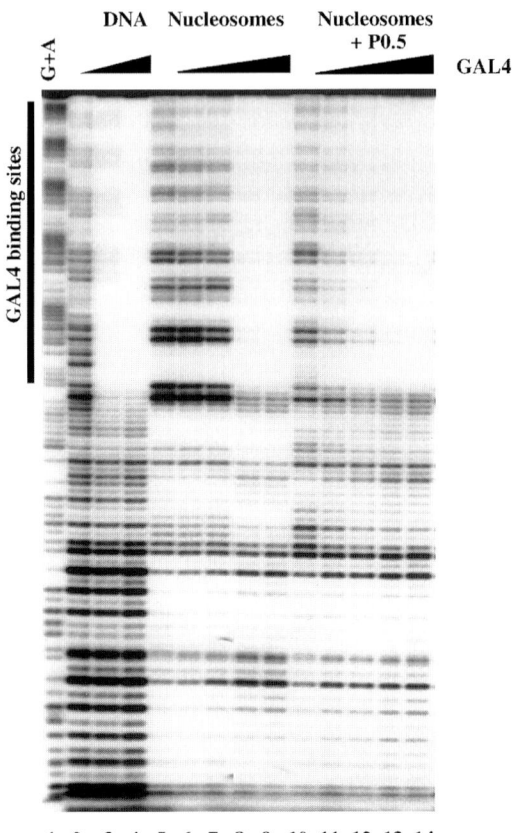

Fig. 3. Stimulation of factor binding assayed by DNase I digestion. Binding reactions were performed as described in **Subheading 3.4.** The nucleosome length probe DNA was derived from pG5H *(13)*. It includes 5 GLA4 binding sites the position of which are indicated. It was reconstituted as described in **Subheading 3.3.** GAL4-VP16 dimers were present at the following concentrations: Lane 2, 5, and 10 none; lanes 3, 6, and 11 2.1 n*M*; lanes 7 and 12, 7n*M*; lane 4 21 n*M*; lanes 8, 13 70 n*M*; lanes 9, 14, 210 n*M*. Lanes 10–14 containing proteins eluting between 0.3–0.5 *M* KCl from a phosphocellulose column loaded with HeLa nuclear extract show increased protection when compared to nucleosomal DNA bound with GAL4 alone (lanes 5–9).

3. Some chromatin will precipitate upon dialysis to low salt. Care should be taken to remove as much of the suspension as possible from the dialysis tubing.
4. A chromatin concentration of approx 1 mg/mL is required for reconstitution reactions (*see* **Subheading 3.3.** and **3.4.**). Where necessary chromatin samples can be concentrated using Centriprep concentrators (Amicon). Dialysis to low salt is not recommended as it may result in the loss of chromatin by precipitation.

5. This method requires that the donor chromatin be in excess of the radiolabelled DNA. Normally this excess should be approx 100-fold to ensure efficient reconstitution. However, where the probe DNA contains multiple nucleosome positioning sequences we have observed quantitative reconstitution when using much lower ratios.
6. None of the assays described here require large quantities of reconstituted material to be prepared. Of more importance is the incorporation of sufficient radioactivity to enable the structure of the template to be studied in subsequent chromatin remodeling assays. Standard procedures can be used for this. End labeling (using polynucleotide kinase or klenow) is a requirement for DNase I digestion assays. Labeling by PCR in the presence of hot nucleotides can be used to produce very hot DNA probes for gel shift analysis. When probe DNA's of high-specific activity are assembled with nucleosomes, low concentrations of chromatin can be used which may enhance the sensitivity of assays for chromatin remodeling. In this case attention must be paid to reports that nucleosomes assembled onto some DNA sequences have been reported to become unstable at low dilution's *(11)*. However, we have found that nucleosomes assembled onto other DNA sequences retain their integrity during binding reactions at concentrations as low as 6.25 ng/mL (note that nucleosomes were stored at a concentration of 250 µg/mL at 0.1 M NaCl in the presence of BSA prior to use in binding reactions). When templates of capable of accommodating >1 nucleosome are used assembly by these methods requires the use of nucleosome positioning sequences to ensure that nucleosomes are correctly spaced. We have used nucleosome positioning sequences derived from the Sea urchin 5S rRNA gene for this purpose *(12–14)* Typical quantities of DNA used in reconstitution's are as follows: mononucleosomes for DNase I footprinting 5 ng DNA, 1×10^6 cpm; Mononucleosomes for gel shift 5 ng DNA, 1×10^6 cpm; end labeled arrays 500 ng, 1×10^6 cpm.
7. Whereas reconstituted mononucleosomes can be stored at 4°C or –80°C, we have found that multi nucleosome arrays keep better at 4°C.
8. To test the level of reconstitution assembled DNA's should be subject to analysis by gel shift in 1% agarose (multinucleosome fragments) or 4% acrylamide (mononucleosomes). In addition nuceosome structure should be probed by nuclease digestion (*see* **Subheading 3.3.**; *[14]*) An example is shown in **Fig. 1**.
9. When reconstituting DNA with core histones the ratio of core histones to DNA is critical. Too few histones may result in incomplete reconstitution and too many may result in the precipitation of the assembled DNA fragment (*see* **Fig. 1**). This is not a problem when DNA's are assembled by transfer from nucleosomes as there is always an excess of DNA present.
10. DNase I digestion is suitable as an assay to detect the disruption of nucleosomes mediated by crude fractions including HeLa nuclear extract.
11. It should be anticipated that the factor will have an affinity for nucleosomal DNA in the range 10- to 100-fold lower than that for naked DNA. The position of factor binding sites within a nucleosome may affect the stimulation observed.

A site centered approx 20 bp from one end of a nucleosome may be a good starting point.
12. We have used a mixture of 0.5 µg HeLa nucleosomes and 0.5 µg calf thymus DNA to release the ySWI/SNF complex from DNA *(13)*. In this case reactions were performed as described in **Subheading 3.7.**, the competitor added and incubation continued for an additional 30 min at 30°C. For other chromatin remodeling activites the optimum source of competitor material and competition conditions should be determined. For example competition at 37°C in the presence of higher salt concentrations may be appropriate where fractions contain proteins that bind DNA tightly. In some cases it may help to remove ATP during competition by treatment with apyrase.

References

1. Owen-Hughes, T. and Workman, J. L. (1994) Experimental analysis of chromatin function in transcriptional control. *Crit. Rev. Euck. Gene Exp.* **4,** 403–441.
2. Tsukiyama, T. and Wu, C. (1997) Chromatin remodeling and transcription. *Curr. Opin. Genet. Dev.* **7,** 182–191.
3. Kingston, R. E., Bunker, C. A., and Imbalzano A. N. (1996) Repression and activation by multiprotein complexes that alter chromatin structure. *Genes Dev.* **10,** 905–920.
4. Tsukiyama, T. and Wu, C. (1995) Purification and properties of an ATP-dependent nucleosome remodeling factor. *Cell* **83,** 1011–1020.
5. Varga-Weisz, P. D., Wilm, M., Bonte, E., Dumas, K., Mann, M., and Becker, P. B. (1997) Chromatin-remodelling factor CHRAC contains the ATPases ISWI and topoisomerase II. *Nature* **388,** 598–602.
6. Ito, T., Bulger, M., Pazin, M. J., Kobayashi, R., and Kadonaga, J. T. (1997) ACF, an ISWI-containing and ATP-utilizing chromatin assembly and remodeling factor. *Cell* **90,** 145–155.
7. Simon, R. H. and Felsenfeld, G. (1979) A new procedure for purifying histone pairs H2A + H2B and H3 + H4 from chromatin using hydroxylapatite. *Nucleic Acids Res.* **6,** 689–696.
8. Côté J., Utley, R. T., and Workman J. L. (1995) Basic Analysis of transcription factor binding to nucleosomes. *Methods Molec. Genet.* **6,** 108–127.
9. Utley, R. T., Owen-Hughes, T. A., Juan, L. J., Côté J., Adams, C. C., and Workman, J. L. (1996) In vitro analysis of transcription factor binding to nucleosomes and nucleosome disruption/displacement. *Methods Enzymol.* **274,** 276–291.
10. Abmayr, S. M. and Workman, J. L. (1993). Preparation of nuclear and cytoplasmic extracts from mammalian cells, in *Current Protocols in Molecular Biology,* vol. 2, Wiley, New York, pp. 12.1.1–12.1.7.
11. Godde J. S. and Wolffe, A. P. (1995) Disruption of reconstituted nucleosomes. The effect of particle concentration, $MgCl_2$ and KCl concentration, the histone tails, and temperature. *J Biol. Chem.* **270,** 27,399–27,402.
12. Owen-Hughes, T. and Workman, J. L. (1996) Remodeling the chromatin structure of a nucleosome array by transcription factor-targeted trans-displacement of histones. *EMBO J.* **15,** 4702–4712.

13. Owen-Hughes, T., Utley, R. T., Côté, J., Peterson, C. L., Workman, J. L. (1996) Persistent site-specific remodeling of a nucleosome array by transient action of the SWI/SNF complex. *Science* **273,** 513–516.
14. Steger, D. J., Owen-Hughes, T., John, S., and Workman, J. L. (1997) Analysis of transcription factor-mediated remodeling of nucleosomal arrays in a purified system. *Methods* **12,** 276–285.
15. Schwarz, P. M., Felthauser, A., Fletcher, T. M., and Hansen, J. C. (1996) Reversible oligonucleosome self-association: dependence on divalent cations and core histone tail domains. *Biochemistry* **35,** 4009–4015
16. Simpson, R. T., Thoma, F., and Brubaker, J. M. (1985) Chromatin reconstituted from tandemly repeated cloned DNA fragments and core histones: a model system for study of higher order structure. *Cell* **42,** 799–808.

24

Nucleosome Remodeling Factor NURF and In Vitro Transcription of Chromatin

Gaku Mizuguchi and Carl Wu

1. Introduction

A central problem in the control of eukaryotic gene expression is how the compaction of DNA in chromatin is overcome to allow the initiation and elongation of transcription *(1–6)*. Current studies reveal that multiple mechanisms are involved in counteracting chromatin-mediated repression, including DNA structure, histone modification, and the action of nonhistone regulators of nucleosome structure *(7–14)*. Recently, novel chromatin remodeling factors: the SWI/SNF complex, RSC, NURF, CHRAC, and ACF have been isolated, whose action is dependent on the energy of ATP hydrolysis *(15–20)*. Here we present an integrated chromatin assembly-transcription system *(21–23)* whereby the functional consequences of such nucleosome remodeling activities may be analyzed by in vitro transcription of reconstituted chromatin templates devoid of endogenous ATP-dependent nucleosome remodeling activities. We also describe procedures that reveal transcriptional activation of chromatin mediated by a chimeric DNA-binding activator GAL4-HSF and the *Drosophila* Nucleosome Remodeling Factor NURF *(24)*.

To evaluate the requirements for activation of a preassembled chromatin template, we employ the following experimental protocol. A plasmid containing five tandemly repeated GAL4 binding sites immediately upstream of the TATA box and the adenovirus E4 minimal core promoter is first reconstituted in a transcriptionally inert nucleosome array using a crude, chromatin assembly extract derived from early *Drosophila* embryos and an ATP-regenerating system. Two related versions of the chromatin assembly extract are effective in reconstituting chromatin for transcriptional activation (S-150: *see* Chapter 13 and **ref. *21*** and S-190: **ref. *23***). After chromatin assembly for 6 h, ATP and proteins not associated with the reconstituted chromatin are removed by Sepharose CL4B gel filtration chromatography (exclusion limit 20 MDa). Notably, ATP-dependent nucleosome remodeling activities are found to cofractionate with the reconstituted chromatin. The partially purified chromatin is incubated with a saturating amount of GAL 4 derivatives in the presence or absence of fresh ATP for the energy-dependent chromatin remodeling step (30 min). The remodeled chromatin is then directly assayed for transcription by incubation (30 min) with a Drosophila soluble nuclear fraction *(22)* to form preinitiation complexes at 26°C. This soluble nuclear fraction contains a highly active RNA polymerase II transcriptional apparatus, is deficient in histones and other inhibitory proteins, and has weak ATP-dependent chromatin remodeling activity, allowing unambiguous analysis of the earlier remodeling step *(23)*. Transcription is finally enabled by the addition of all four NTPs, and the RNA products are analyzed by primer extension after 10 min of transcription. For experiments where ATP-dependent chromatin remodeling activities are to be inactivated, the assembled chromatin is treated with the detergent Sarkosyl before Sepharose CL-4B chromatography.

The integrated chromatin assembly-transcription assay should be of utility in investigating the functional properties of known nucleosome remodeling activities, and also provide a foundation for the discovery of new chromatin modifying activities responsible for gene activation in a chromatin context.

2. Materials

1. *Escherichia coli*: strain BL21 (DE3) pLysE (Novagen, Madison, WI).
2. LB medium.
3. 100 µg/mL Ampicillin.
4. Bacterial expression plasmid: pGM1 (for GAL4 [1-147]) and pGM7 (for GAL4-HSF) *(24)*.
5. 0.1 M IPTG.
6. Phosphate-buffered saline (PBS).
7. 0.22-µm syringe filter (Millex-GV, Millipore, MA).
8. Heparin Sepharose CL6B (Pharmacia, Piscataway, NJ).
9. HEMGZ buffer: 25 mM HEPES-KOH, pH 7.6, 0.1 mM EDTA, 12.5 mM MgCl$_2$, 10% (v/v) glycerol, 20 µM ZnSO$_4$, and freshly added 1 mM DTT, 0.25 mM PMSF, 1 mM sodium bisulfite.
10. 10% (v/v) NP40.
11. HEMGNZ buffer: HEMGZ buffer plus 0.1% (v/v) NP40.
12. Streptavidin-Dynabeads M280 (Dynal, Norway) bound to a 5' terminal biotinylated GAL4 recognition site: for a detailed protocol see Zhong et al. *(25)*.
13. Template plasmid for chromatin assembly and transcription reaction: pGIE-0 *(26)*.
14. *Drosophila* core histones: core histones from *Drosophila* embryos are purified according to Simon and Felsenfeld *(27)*.

15. 10X McNAP: 30 mM ATP (pH to 8.0 with Tris base), 26 mM MgCl$_2$, 300 mM creatin phosphate (Boehringer Mannheim), 10 µg/mL creatin phosphokinase (Boehringer Mannheim); prepared freshly.
16. Extraction buffer R: 10 mM HEPES-KOH, pH 7.6, 10 mM KCl, 1.5 mM MgCl$_2$, 10% (v/v) glycerol, and freshly added 1 mM DTT, 0.2 mM PMSF.
17. *Drosophila* S190 extract for chromatin assembly: a detailed protocol for preparation of the chromatin assembly extract from *Drosophila* embryos is according to Kamakaka et al. *(23)*.
18. Extraction buffer 5/50 (ExB 5/50): 10 mM HEPES-KOH, pH 7.6, 50 mM KCl, 5 mM MgCl$_2$, 10% (v/v) glycerol, 1 mM DTT.
19. Monoject 3cc Leur Lock Syringe (No.71305) (PGC No. 79-4205-09).
20. Sepharose CL4B, washed and stored in ExB 5/50 at 4°C (*see* **Note 1**).
21. 20 mg/mL BSA (Molecular Biology Grade; Boehringer Mannheim).
22. 5% (w/v) N-lauroylsarcosine, sodium salt (Sarkosyl) (Sigma, St. Louis, MO).
23. Micrococcal nuclease (MNase; Boehringer Mannheim): 50 U/µL in extraction buffer R, freshly diluted from a concentrated stock stored at –80°C.
24. 5X MNase stop solution: 2.5% (w/v) Sarkosyl (N-lauroylsarcosine), 100 mM EDTA.
25. 10 mg/mL RNase A (DNase free).
26. 2X Proteinase K buffer: 1% (w/v) SDS, 20 mM EDTA, 20 mM Tris-HCl, pH 8.0, 100 mM NaCl.
27. 10 mg/mL Proteinase K.
28. 20 mg/mL Glycogen (Molecular Biology Grade; Boehringer Mannheim).
29. TE buffer: 10 mM Tris-HCl, pH 7.4, 1 mM EDTA.
30. Orange G loading solution: 50% (v/v) glycerol, 5 mM EDTA, 0.3% (w/v) Orange G.
31. Denaturation solution: 0.5 M NaOH, 1.5 M NaCl.
32. Neutralizing solution: 0.5 M Tris-HCl, pH 7.5, 3 M NaCl.
33. Gene Screen nylon membrane (DuPont/NEN, Boston, MA).
34. Hybridization probe: 5 pmol of a 20- to 30-base oligonucleotide 5' terminally labeled with [γ-^{32}P] ATP (7,000 Ci/mM; ICN, Costa Mesa, CA) and T4 polynucleotide kinase. The sequences of the oligonucleotide probes used are as follows: promoter probe (-57/-25) : (5'-3') GACTCTAGAGGATCCCCAGTC CTATATATA ; distal probe (-900/-874) : (5'-3') TAGGCGTATCACGAGGCC CTTTCGTCT.
35. Hybridization solution: 6X SSC, 2% SDS, 0.1 mg/mL denatured salmon sperm DNA.
36. Wash solution: 6X SSC, 0.5% SDS.
37. Nucleosome remodeling factor (NURF) (P-11 or glycerol gradient fraction): purified from a *Drosophila* embryo nuclear extract according to Tsukiyama and Wu *(18)*.
38. PRIME RNase inhibitor (5 prime 3 prime Inc., Boulder, CO).
39. 20% (w/v) polyvinyl alcohol (Sigma, P-8136).
40. Soluble nuclear fraction (SNF): a detailed protocol for preparation of the soluble nuclear fraction from *Drosophila* embryos is given by Kamakaka and Kadonaga *(28)*.
41. 25 mM each of ATP, GTP, CTP, and UTP mixture (Pharmacia).
42. Transcription reaction stop solution: 20 mM EDTA , pH 8.0, 0.2 M NaCl, 1% (w/v) SDS, 0.25 mg/mL yeast t-RNA.

3. Methods

3.1. Induction and Purification of Recombinant GAL4 Activator Proteins in E. coli

We use GAL4-HSF, a GAL4 DNA binding domain fused to the constitutive activating region from the *Drosophila* heat-shock transcription factor, HSF as a strong transcriptional activator. GAL4-HSF expressed in bacteria shows strong activation of a hybrid UASGAL4-Adenovirus E4 promoter in vitro *(24)*.

3.1.1. Expression of Recombinant GAL4 Activator Proteins

1. Expression is induced in *E. coli* BL21 (DE3) pLysE using cells freshly transformed with pGM1 or pGM7 expression plasmid. Typically, cells are grown at 37°C in 500 mL of LB medium with 100 μg/mL ampicillin to an OD_{600} of 0.5, induced with 0.4 mM IPTG, and grown for an additional 2 h at 37°C.
2. Centrifuge cells for 10 min at 5000 rpm at 4°C in a Beckman JA10 rotor (3300g; Beckman, Palo Alto, CA).
3. Resuspend the pellet in 50 mL of ice-cold PBS. Centrifuge cells again. Freeze cells in liquid nitrogen, and store at –80°C.

3.1.2. Purification of Recombinant GAL4 Activator Proteins

All procedures are performed at 4°C.

1. Resuspend the frozen bacterial pellet in 25 mL of 0.2 M KCl-HEMGZ and disrupt by ultrasonication (approx 42W, 30 s, 8 times) on salt -ice water.
2. Remove insoluble material by centrifugation for 20 min at 12,000 rpm in a Beckman JA20 rotor (11,000g).
3. Filter the supernatant through 0.22-mm syringe filter and load onto a 8-mL Heparin-Sepharose CL6B column pre-equilibrated with 0.2 M KCl-HEMGZ. Wash with more than four bed volumes of the same buffer until OD_{280} returns to the baseline. Elute the GAL4 derivatives with 0.6 M KCl-HEMGZ. The yield of the GAL4 derivatives is approx 0.3–0.4 mg from a 500-mL culture, and the purity of the GAL4 protein should be approx 50% as judged by SDS-PAGE and Coomassie blue staining.
4. Add NP40 (from 10% stock solution) to the 0.6 M KCl-HEMGZ fraction to a final concentration of 0.1%. Dialyze the fraction against 2-L of 0.1 M KCl-HEMGNZ for 4 h at 4°C. Check the conductivity of the sample until it is equal to the dialysis buffer.
5. Apply approx 150 μg (approximately half of total) of GAL4 derivatives to Streptavidin-Dynabeads M280 (5 mg equivalent, Dynal, Norway) bound to a 5' terminal biotinylated GAL4 recognition site:
   5' biotin-GATCCAGATCGGAGTACTGTCCTCCGGTACA-3'
   3'-CTAGGTCTAGCCTCATGACAGGAGGCCATGT-5'
   Incubate the beads and protein fraction for 30 min at 4°C.
6. The beads are washed with 6 mL each of 0.1 M KCl-HEMGNZ and 0.2 M KCl-HEMGNZ and GAL4 protein is eluted with 600 μL of 1.0 M KCl-HEMGNZ.

The 1.0 *M* KCl-HEMGNZ fraction is dialyzed against 1 L of 0.3 *M* KCl-HEMGNZ for 3 h. Check the conductivity of sample until equal to the dialysis buffer. The purity of eluted GAL4 derivatives is more than 95%, as judged by SDS-PAGE and Coomassie Blue staining. The binding capacity of each GAL4 derivative should be determined by DNaseI footprint analysis and by the gel mobility shift assay *(24)*. DNA binding units are defined according to the minimal amount of protein required to saturate the DNA. One DNA binding unit of GAL4 derivative is equivalent to 0.5 pmole protein.

3.2. Nucleosome Assembly Using Drosophila Embryo Extracts

For nucleosome assembly on a template plasmid, the S-190 extract prepared from 0–6 h *Drosophila* embryos is used as described *(23)* (*see* **Note 2**).

Add in order:

1. *Drosophila* S190 nucleosome assembly extract: 30 µL (approx 1.0 mg protein equivalent, exact amount that gives proper nucleosome deposition and spacing is empirically determined for extract).
2. *Drosophila* core histones: 1 µL (0.8–1.0 µg). Incubate on ice for 30 min. Then, add the following components:
3. Extraction buffer R: 57 µL.
4. 10X McNAP: 10 µL.
5. pGIE-0 template DNA: 2 µL (1 µg).

Mix gently at each step. Incubate reaction at 26°C for 6 h.

3.3. Partial Purification of Assembled Chromatin by Spin Column

Assembled plasmid chromatin is partially purified by gel filtration spin column to remove the majority of proteins and ATP from the reaction mixture. This technique allows the demonstration of the existence of an ATP-dependent nucleosome remodeling process *(24,26,29)*.

3.3.1. Spin Column Setup

1. To make the spin column, a disposable Monoject 3-cc syringe is used. To make the bottom disk of the column, a cotton plug is taken from the top piece of a 5 mL disposable plastic pipet using clean forceps. Take 1/3 of the cotton and put it into an empty syringe using clean forceps. Push the cotton to bottom of syringe with a long neck Pasteur pipet and make it even at the bottom.
2. Add approx 1 mL of water and push the wet cotton very hard using the syringe plunger. Remove plunger carefully; otherwise, the flat cotton disk will dislodge.

3.3.2. Spin Column Method *(see* **Note 3***)*

1. Apply Sepharose CL4B resin on the bottom disk carefully using a 5-mL plastic pipet filling from the bottom up. Do not pour the resin from the top of the column directly. Drain extraction buffer from the column. The height of the packed resin should be 2 cm from the bottom. Check the cm ruler on the syringe.

2. Put the column on a 15 mL Blue cap Falcon 2095 polystylene tube (Becton Dickinson, Lincoln Park, NJ) and spin at 1250 rpm ($400g$) for 2 min in a Beckman J6B rotor.
3. Apply 100 µL of chromatin assembly reaction mixture on the resin carefully. Place the column in a 14 mL Falcon 2059 polypropylene tube (Becton Dickinson, Lincoln Park, NJ). Spin the column at 1250 rpm ($400g$) for 2 min. Recover the eluate (100–120 µL) and add BSA to approx 0.6 mg/mL. This chromatin fraction can be stored for at least 17 h at 4°C. The amount of DNA in the chromatin fraction is estimated by agarose gel electrophoresis and ethidium bromide staining with a standard of known concentration. The yield after purification is approx 60–70%.

3.4. Inactivation of ATP-Dependent Nucleosome Remodeling Activities by Sarkosyl Treatment

An ATP-dependent and Sarkosyl-sensitive nucleosome remodeling process is required for nucleosome disruption on the UAS GAL4-Adenovirus E4 hybrid promoter with GAL4 activators as well as on the *hsp70* promoter with the GAGA factor *(24,26,29)*. Sarkosyl sensitivity of the ATP-dependent nucleosome remodeling reaction can be shown by treatment of preassembled chromatin with the detergent before addition of the GAL4 activators and fresh ATP. The remaining Sarkosyl can be removed by means of a gel filtration spin column from the reaction mixture. This technique is performed as described previously *(18)* with minor modifications.

1. For Sarkosyl treatment, 1 µL of 5% (v/v) stock solution of Sarkosyl is added to 100 µL of the assembly reaction mixture to a final concentration of 0.05% and incubated at room temperature for 5 min.
2. 100 µL of Sarkosyl-treated chromatin is immediately applied to a prepacked spin column, and processed as described above. The amount of DNA in the eluted chromatin fraction (approx 120 µL) is estimated by agarose gel electrophoresis and ethidium bromide staining with DNA concentration standards, and by measuring the absorbance at 260 nm. The yield of the plasmid DNA is 60–70%. Approximately 20 µL aliquots (100 ng of DNA equivalent) of the chromatin fractions are each used for a set of micrococcal nuclease digestion or in vitro transcription assays.

3.5. MNase Digestion Assay for NURF Activity

The requirement of NURF for nucleosome remodeling can be shown by adding purified NURF back to the Sarkosyl treated chromatin. Partial digestion with MNase followed by sequential Southern blot hybridization are used to assess the NURF-dependent chromatin remodeling of specific region on preassembled chromatin template. The blot is hybridized with oligonucleotide probes containing the sequence corresponding to the promoter region (promoter probe) or approx 900 bp upstream of the promoter (distal probe).

1. Combine 20 µL of Sarkosyl-treated chromatin (100 ng of DNA equivalent) with 10 µL of ExB 5/50 containing 0.5 mg/mL of BSA.
2. Add in order: 0.3 µL of 50 mM ATP (final 0.5 mM), 0.5 µL of NURF (100 ng/mL P-11 fraction or 50 ng/mL glycerol gradient fraction), and 0.5 pmole (0.3–0.5 µL) of GAL4 activator. Incubate for 30 min at 26°C to facilitate the promoter-specific nucleosome remodeling on chromatin template.
3. Add 0.6 µL of 0.1 M CaCl$_2$ to 30 µL of assembled chromatin at room temperature. Mix gently. Add 0.75 µL of 0.25 U/µL of MNase. Mix quickly.
4. Incubate at room temperature. Take a 16-µL of aliquot at 1, 3, 15 min of digestion and add to 4 µL of 5X MNase stop solution containing 0.2 mg/mL of RNase A. Incubate at 37°C for 30 min.
5. Add 50 µL of 2Xx proteinase K buffer, 28 µL of H$_2$O, and 2.0 µL of 10 mg/mL proteinase K. Allow protease digestion to occur overnight in a 37°C incubator.
6. Precipitate DNA by adding 100 µL of 5 M ammonium acetate, 1 µL of Glycogen (20 mg/mL), and 600 µL of ethanol, spin for 15 min in a microfuge, and wash pellet with 80% ethanol. Dissolve pellet in 6 µL of TE. Load DNA with 1.5 µL of Orange G loading solution on a 1.3% agarose gel in 0.5% TBE. Electrophorese at approx 7 V/cm until the Orange G dye migrates 10 cm.
7. Soak gel in denaturation solution for 45 min. Rinse the gel twice briefly with water. Treat in neutralizing solution for 45 min.
8. Transfer DNA onto a Gene Screen nylon membrane overnight in 10X SSC. Crosslink DNA by UV irradiation. Wet the membrane in water and prehybridize for >90 min in hybridization solution at approx 5°C above the Tm of the DNA probe.
9. Introduce the ^{32}P-labeled oligonucleotide probe (promoter probe or distal probe) in the prehybridization mixture. Incubate the blot at the same temperature for >3 h. Wash blot at the hybridization temperature in the wash solution for 2X 15 min. The blot is exposed to Kodak X-OMAT film at –80°C. The hybridization solution can be stripped by treating the blot in 0.5 M KOH at 42°C for 1–2 h.

3.6. In Vitro Transcription of Chromatin Remodeled by NURF

This transcription reaction is performed essentialy as described elsewhere *(30)*, with minor modifications. The resulting RNA products are analyzed by the primer extention assay as described *(31)*.

1. In the transcription reaction tube, mix in order: 20 µL (100 ng DNA equivalent) of Sarkosyl-treated chromatin template, 0.4 µL of 25 mM ATP (final 0.5 mM), 0.5 µL of NURF (100 ng/mL P-11 fraction or 50 ng/mL glycerol gradient fraction), and 0.5 pmole (0.3–0.5 µL) of GAL4 activator. Incubate for 30 min at 26°C to facilitate the promoter-specific nucleosome remodeling on chromatin template.
2. In a separate microfuge tube, prepare a 67-µL cocktail per transcription reaction containing 26.7 mM HEPES, pH 7.6, 6 mM MgCl$_2$, 44.8 mM KCl, 0.6 mM DTT, 3% (v/v) glycerol, 3.73% (w/v) polyvinyl alcohol, 1 U of PRIME RNase Inhibitor.
3. Add 67 µL of the cocktail solution into the transcription reaction tube and mix by gentle tapping.

4. Add 10 μL of *Drosophila* soluble nuclear fraction (approx 60–70 μg) to the reaction tube and mix by gentle tapping. Incubate at 26°C for 30 min for the formation of preinitiation complex on the promoter.
5. Add 2 μL of 25 mM each of ATP, GTP, CTP, and UTP (final 0.5 mM each) to the reaction mixture. Incubate at 26°C for 10 min for the initiation of transcription.
6. The reaction is terminated by the addition of 100 μL of transcription reaction stop solution, followed by the incubation with 0.1 mg/mL of proteinase K at 37°C for 30 min.
7. After the addition of 200 μL of 5 M ammonium acetate, an equal volume of phenol-chloroform (1:1,v/v) is added to the reaction mix, and vortexed for 1 min to extract the RNA products.
8. Samples are centrifuged at 16,000g (14,000 rpm) at room temperature in an Eppendorf microcentrifuge for 25 min to ensure complete separation of phenol and aqueous phases. The upper (aqueous) phase is transferred to a fresh tube. Extra care should be taken to avoid the interface.
9. An equal volume of chloroform is added, mixed by vortexing for 1 min, and centrifuged at 16,000g (14,000 rpm) at room temperature in an Eppendorf microcentrifuge for 5 min.
10. The upper (aqueous) phase is transferred to a fresh tube. 2.5 vol of ethanol are added, mixed well, and incubated on dry ice for 15 min to precipitate RNA. Samples are centrifuged at 16,000g (14,000 rpm) at room temperature in an Eppendorf microcentrifuge for 15 min. RNA pellet are washed with 80% (v/v) ethanol, and dried. Prolonged drying may make the RNA pellet difficult to dissolve.
11. Transcripts are detected using a ^{32}P-labeled AdE4 primer (+72/+99) in a primer extension assay and cDNA products are analyzed on a 6% denaturing polyacrylamide gel as described previously *(24,31,32)* (*see* **Note 4**). Quantitation of the radioactivity is performed on Fuji BioImage Analyzer (Fuji, Japan).

4. Notes

1. For equilibration of Sepharose CL4B resin with extraction buffer (ExB 5/50), a Corning Filter System (200 mL or 500 ml; Corning Inc, Corning, NY) is used. Pour the resin to the top reservoir of the filter unit and wash the resin with 8–10 packed resin volume of ExB5/50 under low vacuum. Do not let the resin dry to a cake.
2. Instead of the S-190 nucleosome assembly extract from 0–6 h *Drosophila* embryos, the S-150 extract from 0–2 h *Drosophila* embryo (*see* Chapter 13; *21*) (with the final centrifugation modified to 190,000g), is also feasible for this entire assay including Sarkosyl treatment and in vitro transcription of chromatin templates *(21)*.
3. The dimensions of a Monoject 3-cc syringe are 8.5 mm × 6.0 cm (diameter x height). This column dimension is almost the same as the SizeSep 400 spin column (Pharmacia, Piscataway, NJ), which is also feasible for the spin column purification of chromatin.
4. The sequence of AdE4 primer (+72/+99) for primer extension assay is as follows: (5'-3') CTTCACAGCGGCAGCCTAACAGTCAGCC.

References

1. Grunstein, M. (1990) Histone function in transcription. *Annu. Rev. Cell Biol.* **6,** 643–678.
2. Kornberg, R. D., and Lorch, Y. (1992) Chromatin structure and transcription. *Annu. Rev. Cell. Biol.* **8,** 563–587.
3. Van Holde, K., Zlatanova, J., Arents, G., and Moudrianakis, E. (1995) Elements of chromatin structure: histones, nucleosome, and fibres, in *Chromatin Structure and Gene Expression* (Elgin, S. C. R., ed.), Oxford University Press, Oxford, UK, pp. 1–26.
4. Fletcher, T. M., and Hansen, J. C. (1996) The nucleosomal array: structure/function relationships. *Rev. Eukar. Gene Exp.* **6(2&3),** 149–188.
5. Koshland, D., and Strunnikov, A. (1996) Mitotic chromasome condensation. *Annu. Rev. Cell Biol.* **12,** 305–333.
6. Ramakrishnan, V. (1997) Histone structure and the organization of the nucleosome. *Annu. Rev. Biophys. Biomol. Struct.* **26,** 83–112.
7. Becker, P. B. (1994) The establishment of active promoters in chromatin. *Bioessays* **16,** 541–547.
8. Owen-Hughes, T., and Workman, J. L. (1994) Experimental analysis of chromatin function in transcription control. *Crit. Rev. Eukaryot. Gene. Expr.* **4,** 403–441.
9. Paranjape, S. M., Kamakaka, R. T., and Kadonaga, J. T. (1994) Role of chromatin structure in the regulation of transcription by RNA polymerase II. *Annu. Rev. Biochem.* **63,** 265–297.
10. Wolffe, A. P. (1994) Transcription: in tune with the histones. *Cell* **77,** 13–16.
11. Kornberg, R. D., and Lorch, Y. (1995) Interplay between chromatin structure and transcription. *Curr. Opin. Cell. Biol.* **7,** 371–375.
12. Felsenfeld, G. (1996) Chromatin unfolds. *Cell* **86,** 13–19.
13. Brownell, J. E., and Allis, C. D. (1996) Special HATs for special occasions: linking histone acetylation to chromatin assembly and gene activation. *Curr. Opin. Genet. Dev.* **6,** 176–184.
14. Tsukiyama, T., and Wu, C. (1997) Chromatin remodeling and transcription. *Curr. Opin. Genet. Dev.* **7,** 182–191.
15. Carlson, M., and Laurent, B. C. (1994) The SNF/SWI family of global transcriptional activators. *Curr. Opin. Cell. Biol.* **6,** 396–402.
16. Peterson, C. L. Multiple switches to turn on chromatin? (1996) *Curr. Opin. Genet. Dev.* **6,** 171–175.
17. Cairns, B. R., Lorch, Y., Li, Y., Zhang, M., Lacomis, L., Erdjument-Bromage, H., Tempst, P., Du, J., Laurent, B., and Kornberg, R. D. (1996) RSC, an essential, abundant chromatin-remodeling complex. *Cell* **87,** 1249–1260.
18. Tsukiyama, T., and Wu, C. (1995). Purification and properties of an ATP-dependent nucleosome remodeling factor. *Cell* **83,** 1011–1020.
19. Varga-Weisz, P. D., Wilm M., Bonte, E., Dumas, K., Mann, M., and Becker, P. (1997) Chromatin remodeling factor CHRAC contains the ATPases ISWI and topoisomerase II. *Nature* **388,** 598–602.
20. Ito, T., Bulger, M., Pazin, M. J., Kobayashi, R., and Kadonaga, J. T. (1997) ACF, an ISWI-containing and ATP-utilizing chromatin assembly and remodeling fac-

tor. *Cell* **90,** 145–155.
21. Becker, P. B. and Wu, C. (1992) Cell-free system for assembly of transcriptionally repressed chromatin from *Drosophila* embryos. *Mol. Cell. Biol.* **12,** 2241–2249.
22. Kamakaka, R. T., Tyree, C. M., and Kadonaga, J. T. (1991) Accurate and efficient RNA polymerase II transcription with a soluble nuclear fraction derived from Drosophila embryos. *Proc. Natl. Acad. Sci. USA* **88,** 1024–1028.
23. Kamakaka, R. T., Bulger, M., and Kadonaga, J. T. (1993. Potentiation of RNA polymerase II transcription by Gal4-VP16 during but not after DNA replication and chromatin assembly. *Genes. Dev.* **7,** 1779–1795.
24. Mizuguchi, G., Tsukiyama, T., Wisniewski, J., and Wu, C. (1997) Role of nucleosome remodeling factor NURF in transcriptional activation on chromatin. *Mol. Cell.* **1,** 141–150.
25. Zhong, M., Wisniewski, J., Fritsch, M., Mizuguchi, G., Orosz, A., Jedlicka, P., and Wu, C. (1996) Purification of the heat shock transcription factor of *Drosophila*. *Methods Enzymol.* **274,** 113–119.
26. Pazin, M. J., Kamakaka, R. T., and Kadonaga, J. T. (1994) ATP-dependent nucleosome reconfiguration and transcriptional activation from preassembled chromatin templates. *Science* **266,** 2007–2011.
27. Simon, R. H., and Felsenfeld, G. (1979. A new procedure for purifying histone pairs H2A + H2B and H3 + H4 from chromatin using hydroxylapatite. *Nucleic Acids Res.* **6,** 689–696.
28. Kamakaka, R. T., and Kadonaga, J. T. (1994) The soluble nuclear fraction, a highly efficient transcription extract from *Drosophila* embryos. *Methods Cell. Biol.* **44,** 225–235.
29. Tsukiyama, T., Becker, P. B., and Wu, C. (1994) ATP-dependent nucleosome disruption at a heat-shock promoter mediated by binding of GAGA transcription factor. *Nature* **367,** 525–532.
30. Pazin, M. J., Sheridan, P. L., Cannon, K., Cao, Z., Keck, J. G., Kadonaga, J. T., and Jones, K. A. (1996) NF-kappa B-mediated chromatin reconfiguration and transcriptional activation of the HIV-1 enhancer in vitro. *Genes. Dev.* **10,** 37–49.
31. Becker, P. B., Rabindran, S. K., and Wu, C. (1991) Heat shock-regulated transcription in vitro from a reconstituted chromatin template. *Proc. Natl. Acad. Sci. USA* **88,** 4109–4113.
32. Kerrigan, L. A., Croston, G. E., Lira, L. M., and Kadonaga, J. T. (1991) Sequence-specific transcriptional antirepression of the *Drosophila* Kruppel gene by the GAGA factor. *J. Biol. Chem.* **266,** 574–582.

25

An SDS-PAGE-Based Enzyme Activity Assay for the Detection and Identification of Histone Acetyltransferases

James E. Brownell, Craig A. Mizzen, and C. David Allis

1. Introduction

Posttranslational acetylation of the core histone amino-terminal tails correlates with both chromatin assembly and gene expression. This energy-intensive and reversible process is mediated by the opposing activities of histone acetyltransferase (HAT) and deacetylase (HD) enzyme systems, both of which have only recently begun to be characterized at the molecular level *(1,2)*.

HATs catalyze the transfer of acetyl groups from acetylCoA onto the ε-amino groups of specific lysine residues within the amino-termini of each of the core histones. Acetylation neutralizes the charge of ε-amino groups, resulting in a reduction of the positive charge density within these histone domains. HDs, in contrast, remove these acetyl moieties, restoring the ε-amine and increasing the positive charge density within the histone tails. These effects suggest that acetylation modulates the electrostatic interaction of the termini with DNA and possibly other nonhistone chromatin proteins, leading ultimately to alterations in chromatin structure and transcriptional activity *(3)*.

Abundant experimental evidence suggests a fundamental role for chromatin structure in the regulation of gene expression *(4)* and strong correlations exist between overall levels of histone acetylation and the transcription of specific genetic loci. Significantly, the expression of specific genes has been linked to changes in chromatin acetylation at those loci *(5)*. Understanding the HATs and HDs that affect at least some of these changes and the regulatory mechanisms that direct these enzymes has therefore recently become the focus of intense research interest.

In this chapter, an enzyme assay is described that has proven useful in the initial identification and characterization of polypeptides possessing HAT activity. This SDS-PAGE-based acetyltransferase activity assay was originally developed during attempts in our laboratory to isolate HAT-related polypeptides from macronuclear extracts of the ciliated protozoan, *Tetrahymena thermophila*. A major problem encountered in identifying and isolating HATs from *Tetrahymena*, and other natural sources, is that these enzymes are present in cells in extremely low quantities. Thus, conventional purification schemes, utilizing standard enzyme activity assays, have not identified any HAT-related polypeptides to date.

We therefore developed an acetyltransferase activity gel assay to directly identify catalytically active HAT polypeptides by virtue of their ability to transfer [^{3}H]-acetate from [^{3}H]-acetylCoA to core histone substrates incorporated in polyacrylamide gels. Similar methodology utilizing [^{32}P]-ATP rather than [^{3}H]-acetylCoA has been employed previously for the detection of protein kinases *(6)*. As depicted in **Fig. 1**, protein samples containing HAT activity are resolved by SDS-PAGE using gels in which histones (or control protein substrates) have been added prior to polymerization. Following electrophoresis and processing, the gels are incubated in the presence of [^{3}H]-acetylCoA to allow the acetylation reaction to proceed within the gel matrix itself. Following removal of unincorporated [^{3}H]-acetylCoA, the position of catalytically active HATs is then indicated by the fluorographic detection of the [^{3}H]-acetate labeled histone reaction products within the gel. Excision of "acetylated" bands from parallel gels prior to the fluorographic processing, followed by re-electrophoresis of the "products" into a second SDS gel, revealed that histones are indeed being acetylated in this procedure *(7)*. Thus, this approach utilizes enzymatic activity to distinguish catalytically active, HAT-related polypeptides from other proteins present in complex protein mixtures. Moreover, since sample proteins are resolved in this technique just as in conventional SDS-PAGE, an immediate estimate of the apparent molecular mass of HAT active polypeptides can be determined.

This "in-gel" HAT assay enabled us to identify and isolate a 55-kDa polypeptide (p55) from *Tetrahymena* macronuclei that possessed acetyltransferase activity specific for histones and as such, represents the first transcription-associated HAT (i.e. type A or nuclear origin) ever identified *(7)*. Activity gel assays were crucial to the isolation of amounts of p55 sufficient to obtain peptide microsequence data that was employed in cloning the p55 gene *(8)*. DNA sequence analysis revealed that p55 was highly similar to yeast GCN5, and recombinant Gcn5p was subsequently shown to possess HAT activity in the activity gel assay and other HAT assays *(8,9)*. Other workers have employed the activity gel assay to identify B-type HATs implicated in cytoplasmic acetylation of newly synthesized histones *(10,11)*. Collectively, these findings demonstrate that the acetyltransferase activity gel assay can rep-

SDS-PAGE-Based Enzyme Activity Assay

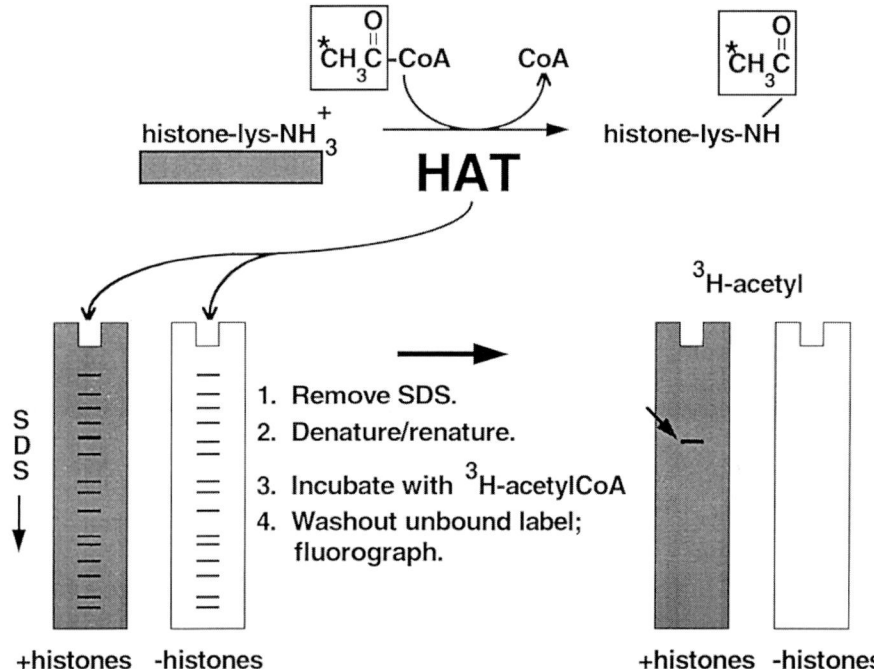

Fig. 1. Schematic diagram of the histone acetyltransferase activity gel assay procedure. Samples containing HAT activity (i.e., nuclear extracts or recombinant proteins) are resolved by SDS-PAGE using gels in which histones (shaded), or negative control protein substrates (i.e., bovine serum albumin; clear) have been added prior to polymerization. Following electrophoresis the gels are processed to renature sample proteins and then incubated with radiolabeled acetylCoA. Sample polypeptides that possess intrinsic histone acetyltransferase activity which is renatured under the assay conditions, catalyze the transfer [^{3}H]-acetyl groups into histone substrates but do not acetylate non-substrate proteins. Following fluorography, HATs can be visualized as discrete bands which denote the position of [^{3}H]-acetate labeled histones in the gel. Since this technique is derived from SDS-PAGE, the apparent molecular mass of HAT polypeptides can be determined by direct comparison to standard proteins electrophoresed under the same conditions.

resent a powerful component of a reverse-genetics approach to identify and isolate HAT-related polypeptides and ultimately their genes.

2. Materials
2.1. Activity Gels and SDS-PAGE

1. Activity gels employ SDS-PAGE as described by Laemmli (**12**) except for the inclusion of substrate proteins in the resolving gel solution. Reagents for casting and running activity gels should be of analytical-grade or similar quality. In our

laboratory, we have used Tris, glycine, glycerol, and dithiothreitol purchased from Sigma, and SDS and acrylamide purchased from BDH. Reagents of comparable grade from other suppliers should provide equivalent results.
2. Enzyme substrate proteins. Stock solutions of calf thymus histones (Type II-A; Sigma #H-9250) and bovine serum albumin (BSA, Sigma #A-4503) at 10 mg/mL in water are stored frozen at –20°C (stable for months).
3. SDS-PAGE apparatus and power supply. A mini-gel format is recommended for optimal assay reproducibility and to minimize the amount of costly isotopic acetylCoA consumed per assay. We have used the Hoefer SE 200 apparatus which provides a resolving gel of approx 6×8 cm.
4. SDS-PAGE sample loading buffer: adjust samples to final concentrations of 50 mM Tris-HCl, pH 6.8, 2% SDS, 25% glycerol, 1% 2-mercaptoethanol, and 0.02% bromphenol blue.

2.2. Denaturation/Renaturation Procedure

1. Buffer 1: 50 mM Tris-HCl, pH 8.0, 20% (v/v) isopropanol (analytical grade or similar quality), 0.1 mM EDTA, 1 mM dithiothreitol. Prepare immediately prior to use.
2. Buffer 2: 50 mM Tris-HCl, pH 8.0, 8 M urea (ICN Ultra Pure Urea no. 821527), 0.1 mM EDTA, 1 mM dithiothreitol. Prepare immediately prior to use.
3. Buffer 3: 50 mM Tris-HCl, pH 8.0, 0.005% Tween-40 (Sigma), 0.1 mM EDTA, 1 mM dithiothreitol. Prepare immediately prior to use.
4. Orbital shaker.
5. Plastic food storage containers. We typically use rectangular containers with approximate dimensions of $14.0 \times 18.0 \times 3.0$ cm with tightly sealing lids.

2.3. Acetylation Reaction

1. Buffer 4: 50 mM Tris-HCl, pH 8.0, 10% (v/v) glycerol, 0.1 mM EDTA, 1 mM dithiothreitol.
2. [^{3}H]-acetylCoA or [^{14}C]-acetylCoA (labeled exclusively on the acetyl moiety, available from ICN, Amersham, and NEN-DuPont). Highest available specific acitivity products are preferred (*see* **Note 4**).
3. Heat-sealable plastic bags and heat sealing apparatus as commonly employed in nucleic acid hybridization analyses.

2.4. Postreaction Processing and Detection

1. Coomassie Blue-R250 staining and destaining solutions or 5% (w/v) trichloroacetic acid.
2. Gel permeable enhancement reagent for fluorography of [^{3}H]-labeled proteins (Dupont-NEN Entensify #NEF992).
3. Kodak X-OMAT film and cassettes.

3. Methods
3.1. Preparation of Enzyme Samples

Crude or partially purified enzyme samples recovered from natural sources (i.e., nuclei), as well as recombinant proteins, are readily assessed for HAT

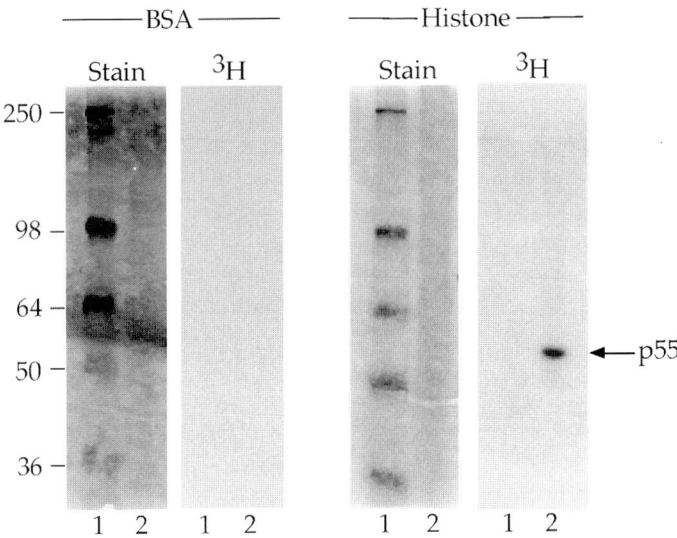

Fig. 2. Typical HAT activity gel results. Prestained molecular weight markers (lane 1) and crude *Tetrahymena* macronuclear extract (approx 10 µg total protein, lane 2) were electrophoresed on 8% acrylamide SDS-PAGE mini-gels containing BSA (left panels) or histones (right panels) and processed to detect HAT activity as described in the text. Following incubation with [^{3}H]-acetylCoA, the gels were stained with Coomassie Blue R250 and destained prior to fluorography. The *Tetrahymena* macronuclear HAT p55 was readily detected on the fluorogram of the histone containing gel (indicated by an arrow) after 6 d of exposure at –70°C. Note that the p55 band was not detected by Coomassie staining in the corresponding stained gel. No autoradiographic signals were apparent on the fluorogram of the BSA gel. The dense background staining evident in the upper portion of the BSA gel is due to the inclusion of 0.1 mg/mL of the substrate proteins in the appropriate upper reservoir buffers during electrophoresis. The relative molecular masses of the standard proteins in thousands of daltons are indicated on the left.

activity using the acetyltransferase activity gel assay. As with most enzymes, sample handling and preparation time should be kept to a minimum. Any sample preparation method may be employed provided it is compatible with SDS-PAGE and detection of acetyltransferase activity (*see* **Notes 1** and **2**). The amount of sample required to detect activity will vary among different HAT preparations and must be determined empirically. In the case of *Tetrahymena* p55, HAT activity can be readily detected from as little as 1 µg of crude macronuclear extract in which p55 represents much less than 1% of the total protein in the sample (*see* **Fig. 2**). A convenient positive control is recombinant yeast GCN5p (available commercially from Upstate Biotechnology Inc.).

As little as 100 ng of the 6xHis-fusion protein expressed in bacteria and purified on Ni-agarose gives a positive signal after six days of autoradiography.

3.2. Histone Acetyltransferase Activity Gel Assay

1. The resolving portion of the activity gel containing 1 mg/mL (final) calf thymus histone is prepared and allowed to polymerize at room temperature for at least one hour. A control activity gel containing 1 mg/mL (final) BSA (a protein not known to be acetylated *in vivo*) is prepared in parallel (*see* **Note 3**).
2. Prior to electrophoresis, conventional stacking gels that do not contain substrate proteins are polymerized with plastic or Teflon combs to form sample wells onto both the histone and BSA-containing gels.
3. To help replenish substrate proteins in the gel matrix as they are depleted during the course of electrophoresis, 0.1 mg/mL histones (or BSA for control assays) can be included in the upper reservoir buffer. This addition is crucial for reliable detection of high molecular weight HATs *(13)*.
4. Electrophoresis is conducted at room temperature using standard SDS-PAGE conditions.

3.3. Gel Processing (Denaturation/Renaturation)

After electrophoresis is completed, activity gels are treated sequentially to remove SDS and then to fully denature and gradually renature sample proteins resolved in the gel prior to the acetylation reaction. These steps are required for at least partial recovery of HAT activity that is detected in subsequent assay steps.

The buffer volumes and wash times listed below have been determined for mini-gels approx 6 cm × 8 cm and 1 mm thick. Recovery of activity in different gel formats may require adjustment of these parameters. Following electrophoresis, carefully remove the gels from the electrophoresis apparatus, transfer each gel to a separate plastic container and wash as outlined below.

1. Remove SDS by washing in buffer 1 (containing 20% isopropanol) at room temperature. Four 15-min washes, each employing 100 mL buffer 1 are performed with gentle agitation on an orbital shaker.
2. Denature sample proteins by washing gels in buffer 2 at room temperature (as above, four 15-min washes with 100 mL buffer 2 using gentle agitation).
3. Renature sample proteins slowly by incubating gels with buffer 3 at 4°C without agitation. Rinse gels once in 100 mL buffer 3 for 15 min at 4°C without agitation, followed by a 12-h (overnight) incubation in 100 mL buffer 3 at 4°C without agitation. The following morning, gels are washed a third time in 100 mL buffer 3 for 30 min at 4°C without agitation. A fourth and final 30-min wash in 100 mL buffer 3 is performed at room temperature without agitation to allow the gel and buffer to equilibrate at room temperature.

3.4. Acetylation Reaction

1. Equilibrate each gel in 100 mL buffer 4 at room temperature for 15 min with gentle agitation on an orbital shaker.

2. Place each gel in a heat sealable plastic bag that is just slightly larger than the gel itself and add 3 mL of fresh buffer 4 containing 5 µCi of [^{3}H]-acetylCoA (*see* **Note 4**). Remove as much air as possible from the bag and seal it. Thoroughly distribute the [^{3}H]-acetylCoA containing reaction buffer about the gel.
3. Incubate at 30°C for 1 h to allow the acetylation reaction to proceed (optimal reaction times will vary depending on the amount of enzyme activity recovered during the renaturation step).

3.5. Postreaction Processing for Fluorography

Following the acetylation reaction step, the gel can be stained using Coomassie Blue-R250 if desired. Staining allows sample proteins bands to be visualized over the background stain contributed by substrate proteins present throughout the gel, and stops the acetylation reaction. The gel should be destained exhaustively (overnight) with several changes of destain solution to reduce levels of background staining and remove unbound radiolabel. In lieu of staining, gels may also be washed using 5% trichloroacetic acid, again with several changes. In either case, thorough washing prior to fluorography is recommended to reduce background signal.

Process gels for fluorography and expose to film. The exposure time required is a function of the amount of activity loaded onto the gel and the degree to which activity has been renatured, as well as the quality of the acetylCoA used (*see* **Note 4**), and can range from as little as 18 hours to as long as a month.

3.6. Interpretation of Results

It should be noted that while the HAT activity gel method is useful for identifying catalytically active HAT polypeptides, in some instances polypeptides that are not HATs may also incorporate low levels of [^{3}H]-acetyl groups under the assay conditions. Sample polypeptides must therefore also be tested in negative control gels such as those that contain BSA (or some other nonsubstrate protein), or in which no protein substrate has been included in the gel. Incorporation of [^{3}H]-acetyl groups by sample polypeptides under these circumstances can be judged to be artifactual due to either nonenzymatic protein acetylation or autoacetylation of sample polypeptides. Moreover, minor radiolabeled bands detected in histone gels should also be viewed with caution. However, histone dependent, major radiolabeled bands are likely to be HATs, given that known HATs demonstrate high specificity for histone substrates. Particularly promising potential HATs are those polypeptides whose autoradiographic band intensity on a histone-containing activity gel is disproportionately greater than the band intensity in Coomassie staining (e.g., p55 in **Fig. 2**).

3.7. Other Applications and Possible Modifications of the Assay Strategy

Currently, little information is available regarding whether acetyltransferases specific for nonhistone substrates in vivo are detected in the assay as described here employing free histone substrate. The acetyltransferase activities of *Tetrahymena* p55, yGCN5p and $TAF_{II}250$ were originally detected by virtue of the ability of these enzymes to catalyze acetate transfer to histones in the activity gel assay *(7,8,13)*, whereas those of PCAF and p300/CBP were detected in conventional enzyme assays *(14,15)*. However, recent reports have demonstrated that $TAF_{II}250$, PCAF and p300 also acetylate specific nonhistone proteins in vitro *(16,17)*, suggesting the intriguing possibility that substrates other than histones may be physiological targets for these activities. These findings are reminiscent of the well-known ability of protein kinases that are presumably specific for non-histone substrates in vivo to utilize histone substrates in vitro. Thus, the in-gel assay described here using histone substrate may have utility in detecting other types of acetyltransferases provided that relevant "test" substrates are available *(see below)*. This notion is supported by the fact that discrete sequence motifs of yGCN5p are conserved in acetyltransferases that are known to possess different substrate specificities *(18)*.

While the ease of preparation and commercial availability of histones makes them a preferred choice to test initially for substrate suitability, it seems likely that there will be instances in which acetyltransferases (or putative acetyltransferases) are discovered that do not utilize histone substrates. In such cases, it may be more economical to screen putative substrates in conventional assays prior to their use in activity gels. If the cost or availability of a candidate substrate precludes incorporating it into activity gels, it may be possible to use a synthetic mimic (e.g., peptide) of the substrate, coupled to an appropriate carrier protein (e.g., BSA) as the substrate for activity gels.

In our experience, activity gels have proven superior to methods in which proteins resolved by SDS-PAGE are electroblotted to PVDF or NC membranes and then renatured to detect HAT activity. However, transblotting/membrane immobilization techniques may be of particular utility in screening expression libraries for novel acetyltransferase activities. Membranes, on which histones or other substrates have been immobilized, could be employed in conjunction with autoradiography to identify clones expressing acetyltransferases. Since we have demonstrated that at least in one case, sufficient HAT activity is retained to permit autoradiographic detection following nitrocellulose immobilization *(19)*, the reverse technique may be of use in identifying physiological substrates for acetyltransferase activities.

4. Notes

1. Bacterially expressed, recombinant HAT proteins dissolved in SDS sample buffer can be run directly in the assay without further purification (several positive control HATs are being commercialized by Upstate Biotechnology; Lake Placid, NY). Thus recombinant proteins which may not be readily solubilized upon cell

lysis and extraction can be added directly in SDS-PAGE sample buffer, electrophoresed, and assessed for HAT activity. Uninduced bacteria and vector only controls should be included in the same assay gels to detect non-specific acetyltransferases present in some bacteria.
2. Several commonly used protease inhibitors are also potent inhibitors of histone acetyltransferase activity *(7)*. In particular, samples to be tested for HAT activity should not be exposed to iodoacetamide, N-ethylmaleimide, or Hg-containing compounds (all known to affect sulfhydryl residues). Other inhibitors including PMSF, leupeptin, aprotinin, bestatin, pepstatin and benzamidine show no apparent inhibitory effect on HAT activity of *Tetrahymena* p55 or yGcn5p and can be added to sample buffers to protect against proteolysis.
3. Prepare histone or BSA substrate proteins as 10-fold concentrated solutions in distilled water, and then dilute the thoroughly dissolved proteins into the resolving gel recipe. At the final concentration employed here (1 mg/mL), histones form a very fine precipitate in the presence of SDS, but this does not appear to adversely affect the assay. Although nucleosomes can be reasonably expected to represent the bone fide in vivo substrates for nuclear HATs, one limitation of the activity gel assay is that the denaturing properties of SDS-PAGE make it incompatible with the use of nucleosomal histone substrates. Thus, HATs which acetylate nucleosomal, but not free histones, will not be detected under assay conditions in which free histones are provided as enzyme substrates. Moreover, it is becoming clear that several HATs, including Gcn5p in yeast, exist in multisubunit complexes with other cellular factors (e.g., Ada and Spt proteins) *(20)*. Because it is likely these interactions are disrupted upon solubilization and electrophoresis in SDS-PAGE, then HAT activities in which multisubunit interactions are required for catalytic activity with free histone substrates will not be detected by this method.
4. In our laboratory, we have used [^{3}H]-acetylCoA exclusively, however [^{14}C]-acetylCoA can also be used. The specific activity of radiolabeled acetylCoA can vary dramatically from production lot to production lot, and typically ranges as follows: [^{14}C]-acetylCoA (40-62 mCi/mmole) or [^{3}H]-acetylCoA (2-15 Ci/mmole). It is recommended that material of the highest specific activity available be used in acetyltransferase assays. AcetylCoA is a high-energy compound (the thioester is subject to spontaneous hydrolysis) and should be immediately aliquotted upon receipt and stored per the manufacturers instructions.

References

1. Wade, P. A. and Wolffe, A. P. (1997) Chromatin: Histone acetyltransferases in control. *Curr. Biol.* **7,** R82–R84.
2. Pazin, M. J. and Kadonaga, J. T. (1997) What's up and down with histone deacetylation and transcription? *Cell* **89,** 325–328.
3. Allfrey, V. G., Faulkner, R., and Mirsky, A. E. (1964) Acetylation and methylation of histones and their possible role in the regulation of RNA synthesis. *Proc. Natl. Acad. Sci. USA* **51,** 786–794.

4. Felsenfeld, G. (1996) Chromatin unfolds. *Cell* **86**, 13–19.
5. Hebbes, T. R., Thorne, A. W., and Crane-Robinson, C. (1988) A direct link between core histone acetylation and transcriptionally active chromatin. *EMBO J.* **7**, 1395–1402.
6. Hutchcroft, J. E., Anostario, M., Harrison, M. L., and Geahlen, R. L. (1991) Renaturation and assay of protein kinases after electrophoresis in sodium dodecyl sulfate-polyacrylamide gels. *Methods Enzymol.* **200**, 417–423.
7. Brownell, J. E. and Allis, C. D. (1995) An activity gel assay detects a single, catalytically active histone acetyltransferase subunit in Tetrahymena macronuclei. *Proc. Natl. Acad. Sci. USA.* **92**, 6364–6368.
8. Brownell, J. E., Zhou, J., Ranalli, T. A., Kobayashi, R., Roth, S. Y., and Allis, C. D. (1996) Tetrahymena histone acetyltransferase A: a homolog to yeast GCN5 linking histone acetylation to gene activation. *Cell* **84**, 843–851.
9. Kuo, M.-H., Brownell, J. E., Sobel, R. E., Ranalli, T. A., Cook, R. G., Edmondson, D. G., Roth, S. Y., and Allis, C. D. (1996) Transcription-linked acetylation by Gcn5p of histones H3 and H4 at specific lysines. *Nature* **383**, 269–271.
10. Parthun, M. R., Widom, J., Gottschling, D. E. (1996) The major cytoplasmic histone acetyltransferase in yeast: links to chromatin replication and histone metabolism. *Cell* **87**, 85–94.
11. Chang, L., Loranger, S. S., Mizzen, C., Ernst, S. G., Allis, C. D., and Annunziato, A. T. (1997) Histones in transit: cytosolic histone complexes and diacetylation of H4 during nucleosome assembly in human cells. *Biochemistry* **36**, 469–480.
12. Laemmli, U. K. (1970) Cleavage of structural proteins during the assembly of the head of bacteriophage T4. *Nature* **227**, 680–685.
13. Mizzen, C. A., Yang, X.-J., Kokubo, T., Brownell, J. E., Bannister, A. J., Owen-Hughes, T., Workman, J., Wang, L., Berger, S. L., Kouzarides, T., Nakatani, Y., and Allis, C. D. (1996) The TAF$_{II}$250 subunit of TFIID has histone acetyltransferase activity. *Cell* **87**, 1261–1270.
14. Yang, X.-J., Ogryzko, V. V., Nishikawa, J., Howard, B. H., and Nakatani, Y. (1996) A p300/CBP-associated factor that competes with the adenoviral oncoprotein E1A. *Nature* **382**, 319–324.
15. Ogryzko, V. V., Schiltz, R. L., Russanova, V., Howard, B. H., and Nakatani, Y. (1996) The transcriptional coactivators p300 and CBP are histone acetyltransferases. *Cell* **87**, 953–959.
16. Gu, W. and Roeder, R. G. (1997) Activation of p53 sequence-specific DNA binding by acetylation of the p53 C-terminal domain. *Cell* **90**, 595–606.
17. Imhof, A., Yang, X.-J., Ogryzko, V. V., Nakatani, Y., Wolffe, A. P., and Ge, H. (1997) Acetylation of general transcription factors by histone acetyltransferases. *Curr. Biol.* **7**, 689-692.
18. Neuwald, A. F. and Landsman, D. (1997) GCN5-related histone N-acetyltransferases belong to a diverse superfamily that includes the yeast SPT10 protein. *Trends Biochem. Sci.* **22**, 154–155.
19. Allis, C. D., Chicoine, L. C., Glover, C. V. C., White, E., and Gorovsky, M. A. (1986) Enzyme activity dot blots: a rapid and convenient assay for acetyltransferase

or protein kinase activity immobilized on nitrocellulose. *Anal. Biochem.* **159,** 58–66.

20. Grant, P. A., Duggan, L., Cote, J., Roberts, S. M., Brownell, J. E., Candau, R., Ohba, R., Owen-Hughes, T., Allis, C. D., Winston, F., Berger, S. L., and Workman, J. L. (1997) Yeast GCN5 functions in two multisubunit complexes to acetylate nucleosomal histones: characterization of an Ada complex and the SAGA (Spt/Ada) complex. *Genes Dev.* **11,** 1640–1650.

26

Analysis of DNaseI Hypersensitive Sites in Chromatin by Cleavage in Permeabilized Cells

Rein Aasland and A. Francis Stewart

1. Introduction

Treatment of nuclei with limited amounts of DNaseI can be used to reveal sites in chromatin that are hypersensitive (HS) to the nuclease *(1,2)*. DNaseI HS sites are thought to correspond to sites where the regular nucleosome structure is perturbed, e.g., by binding of proteins to chromatin such that DNA becomes more accessible to the nuclease. Some HS sites have indeed been reconstituted in vitro during assembly of nucleosomes in the presence of transcription factors. Mapping of DNaseI HS sites and monitoring their time course of appearance (or disappearance) has been successfully used to identify many regulatory elements. Analysis of HS sites is an attractive method when searching for regulatory regions because one can scan genomic regions without detailed prior assumptions of where to look.

Usually, nuclease digestion is performed with isolated nuclei. This classical protocol has been described in detail elsewhere *(1,3)*. Here we describe a simplified protocol *(4)* where the DNaseI treatment takes place as the cells are permeabilized with the nonionic detergent Nonidet P-40 (NP40). We describe two variations of the protocol. In the first variation, cells grown on tissue plates are overlaid with a sucrose-containing buffer which also includes NP40 and DNaseI. After a brief incubation (4 min) at room temperature, the combined permeabilization and nuclease treatment is quenched by addition of a SDS-containing solution. In the other variation, cells are first trypsinised and then resuspended in the sucrose-, NP40-, and DNaseI-containing buffer. After permeabilisation and cleavage, the reaction is quenched as above. The latter variant is preferable if the cells in culture do not grow as a flat monolayer attached to the plate. The cell lysate is subsequently treated overnight with proteinase K in order to inactivate the nuclease. Crude DNA is isolated by phenol:chloroform extraction and ethanol precipitation.

From: *Methods in Molecular Biology, Vol. 119: Chromatin Protocols*
Edited by: P. B. Becker © Humana Press Inc., Totowa, NJ

The DNA is cleaved with an appropriate restriction enzyme, separated by electrophoresis in an agarose gel, and blotted onto a positively charged nylon membrane. The hypersensitive sites are revealed by indirect end-labeling *(1,2)*, i.e., hybridization is performed with a radiolabeled probe derived from one end of the genomic restriction fragment being studied (*see* **Fig. 1**).

DNaseI HS sites can be detected at distances from the end of the restriction fragment ranging from about 1 kb to 10–15 kb limited only by the resolution of the agarose gel electrophoresis system. As illustrated in **Fig. 1**, the ability to map multiple HS sites requires that the DNaseI treatment is *partial* such that a *population of fragments* is generated where cleavage products from all HS sites are represented. To achieve this, parallel reactions are carried out using different amounts of DNaseI.

The experiments are usually designed such that one compares the chromatin structure in cells in two (or more) different states (e.g., before or after induction of transcription of a gene). Because there is always a certain level of variability inherent in the experiments, it is essential to compare the cleavage patterns within one experiment and preferably on the same gel. For each experiment, we use at least three concentrations of DNaseI chosen so that the hypersensitivity is revealed (*see* **Fig. 2**; see how HS sites III form and disappear with increasing amounts of DNaseI). In cases where multiple HS sites form on one restriction fragment, the sites closest to the probe will be revealed at the expense of the sites further away (*see* **Fig 2**, compare sites I-II to sites III-V). In such cases, alternative probing strategies should be devised to investigate the sites far from the probe. Using a probe from the other end of the restriction fragment is often useful. If restriction sites are chosen close to a HS site, they may resolve into multiple weaker HS-sites. For high-resolution mapping of HS sites, one would use genomic DNaseI footprinting or related techniques.

The protocol as described here has been established with rat FTO-2B hepatoma cells *(4)* and murine embryonal carcinoma cells. The protocol has also been used with several other cell types. In each case, it is important to find the optimal concentration of NP40. The method should in principle be applicable to tissues provided that single cell suspensions can be obtained. The detergents lysolecithin and saponin have also been used to achieve cell permeabilisation with this method *(4)*. Other nucleases than DNaseI, such as micrococcal nuclease and restriction enzymes have also been used.

2. Materials

2.1. Treatment of Cells with DNaseI and Proteinase K Treatment

1. PBS: 8 g/L NaCl, 0.2 g/L KCl, 1.44 g/L Na_2HPO_4, 0.24 g/L KH_2PO_4, pH 7.4.
2. NP40-Buffer: 15 mM Tris-HCl, pH 7.5, 60 mM KCl, 15 mM NaCl, 5 mM $MgCl_2$, 0.5 mM EGTA, 300 mM sucrose. Before use, make it 0.5 mM β-mercaptoethanol, 0.2% NP40 (Fluka). NP40 is added from a 25% stock.

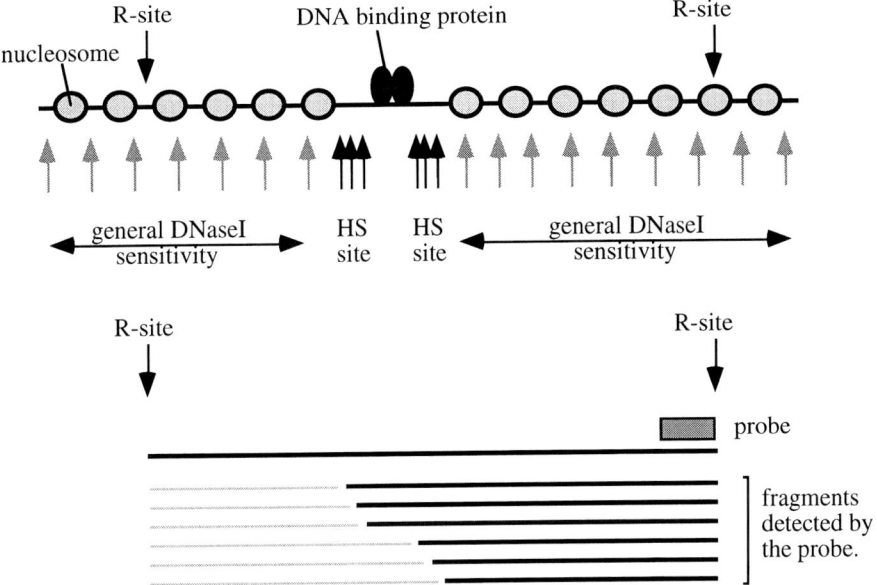

Fig. 1. Mapping of DNaseI HS sites. *(Top)* A nucleosomal array that is disrupted by a DNA binding protein is indicated. Hypersensitive sites may form in the vicinity of the DNA-bound protein. "R-site" indicates the position of two recognition sequences for the restriction enzyme used in the indirect endlabeling. *(Bottom)* The black thick lines represents DNA fragments detected by the probe. The thin gray lines represent DNA fragments not recognized by the probe. The position of the probe on the restriction fragment is indicated with the shaded box. *See* text for further details.

3. DNaseI, 20 units (U)/µL (Worthington) prepared by dissolving the freeze dried enzyme in PBS and adding glycerol to 50%. Store at −20°C.
4. Polypropylene centrifugation tubes, 14 mL "snap-cap" polypropylene tubes (Falcon #2059).
5. Trypsin-EDTA: 0.5 g/L trypsin, 0.2 g/L EDTA (Gibco)
6. 6X Lysis buffer: 6 % SDS, 300 mM Tris-HCl, pH 8.0, 120 mM EDTA.
7. Proteinase K (Boehringer Mannheim) 20 mg/mL in water. Store at −20°C.

2.2. Purification and Quantitation of DNA

1. Phenol/chloroform/isoamyl alcohol (25:24:1).
2. 10 M NH$_4$-acetate.
3. Ethanol, 96%.
4. Ethanol, 70%, ice cold.
5. TE: 10 mM Tris-HCl, pH 7.5, 0.1 mM EDTA.
6. 0.5X TBE: 0.045 M Tris-borate, 0.1 mM EDTA.
7. Agarose, electrophoresis grade.
8. Ethidium bromide, 10 mg/mL.

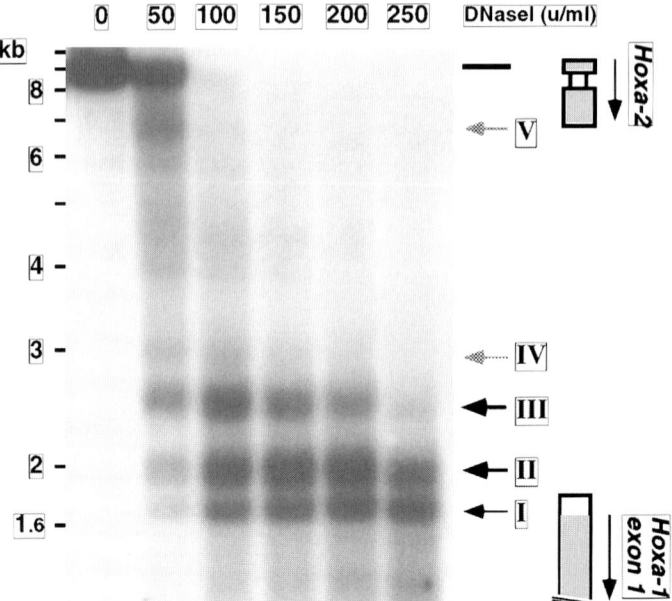

Fig. 2. Mapping of DNaseI hypersensitive sites in the promoter of the murine *Hoxa-1* gene. Murine embryonal carcinoma cells (c1003) were subjected to DNaseI cleavage using method variation B (*see* **Subheading 3.1.2.**, cleavage of trypsinized cells). The amount of DNaseI used is indicated. After cleavage, the DNA was isolated and cut with *Eco*RI and probed with a riboprobe containing 346 nt from one end of an *Eco*RI fragment spanning the *Hoxa-2* and *Hoxa-1* genes. Hypersensitive sites are indicated in roman numerals. Size marker is 1-kb ladder (Gibco-BRL).

2.3. Restriction Enzyme Digest

1. Restriction enzyme with suitable 10X reaction buffer.
2. 3 *M* NaAcetate
3. Ethanol, 96%, ice cold.
4. Ethanol, 70%, ice cold.
5. 1X Genomic DNA sample buffer: 0.05% Bromophenol blue, 0.05% Xylene xyanol, 6% glycerol, 10 m*M* EDTA.

2.4. Gel Electrophoresis, Southern Blotting, and Hybridization

1. Agarose, electrophoresis grade.
2. 1X TAE: 20 m*M* Tris-acetate, pH 8.0, 1 m*M* EDTA.
3. ^{32}P-labeled size marker, 1 Kb ladder (Gibco), labeled with [α^{32}P]dCTP using DNA polymerase Klenow fragment (*4*). Approx 5000 cpm per lane.
4. Charged blotting membrane (PALL Biodyne B or Qiagen QIABRANE Nylon Plus).

5. Radiolabeled riboprobes (6) and hybridization solutions according to Church and Gilbert (7).

3. Methods
3.1. Treatment of Cells with DNaseI and Proteinase K Treatment
3.1.1. Method Variant A: Cleavage in Permeabilized Cells In Situ

1. Prepare NP40 Buffer/DNaseI solution by adding the appropriate volume of DNaseI stock (20 U/µL) to a tube containing 1.5 mL NP40-Buffer. Mix by pipetting, cap the tube, invert twice, and immediately proceed to **step 2**.
2. For each 6 cm tissue culture plate (approx 3–5 × 10^6 cells, *see* **Note 1**), remove the culture medium by aspiration and gently add 1 mL of the NP40-Buffer/DNaseI solution to the plate. Incubate exactly 4 min at room temperature (*see* **Note 2**).
3. Quench the reaction by adding 0.2 mL 6X lysis buffer. Swirl the plate to mix, and leave at room temperature.
4. After all DNaseI incubations are completed and quenched, add to each lysate 12 µL of 20 µg/µL Proteinase-K. Mix well; pipet up and down 2 to 3 times using a 1-mL pipet if necessary. Transfer the lysate to a centrifugation tube and incubate at 37°C overnight.

3.1.2. Method Variant B: Cleavage in Permeabilized Trypsinized Cells

1. For each 10-cm tissue culture plate (approx 6–10 × 10^6 cells, *see* **Note 1**), remove and store culture medium, rinse once with PBS. Add 2 mL Trypsin-EDTA and incubate at 37°C for 3–5 min. Dislodge the cells with a "gentle kick from the side" (if necessary, aid monolayer disruption by pipetting), then immediately add back the culture medium to quench trypsinization.
2. Transfer cells to a centrifugation tube and collect cells by centrifugation at 1000 rpm (174*g*) at 4°C for 5 min.
3. Discard the medium and leave the tubes draining upside down. Add the appropriate volume of DNaseI stock (20 U/µL) to a tube containing 3 mL NP40-Buffer. Mix by pipetting up and down twice, cap the tube and invert twice. Add 2 mL of the NP40-Buffer/DNaseI solution to the cell pellet. Incubate exactly 4 minutes at room temperature. As the incubation starts, use the pipet to gently disperse the pellet. Two or three "strokes" with the pipet should be enough (*see* **Note 2**).
4. Quench the reaction by adding 0.4 mL 6X Lysis Buffer. Mix and leave at room temperature.
5. After all DNaseI incubations are completed and quenched, add to each lysate 24 µL 20 µg/µL Proteinase-K. Mix well, pipet up and down 2 to 3 times using a 1-mL pipet if necessary. Incubate at 37°C overnight.

3.2. Purification and Quantitation of DNA

1. Extract each sample once with 1 vol of phenol/chloroform/isoamylalcohol. Spin at 3500 rpm (2200*g*) at room temperature for 15–20 min.

2. Transfer aqueous phases to new tubes and add 1/3 vol 10 M NH$_4$Acetate and precipitate the DNA by adding 2 vol of ethanol. Mix by inverting the tubes a couple of times. Leave at –20°C for 1–2 h or overnight.
3. Collect DNA by centrifugation at 3500 rpm (2200g) at 4°C for 15–20 min.
4. Pour off supernatants. *Keep an eye on the pellets, they sometimes slide out of the tubes.* Dissolve in 2 mL TE. Dissolution can be aided by incubation at 45°C.
5. Add 1/3 volume 10 M NH$_4$Acetate and precipitate the DNA by adding 2 vol of ethanol. Mix by inverting the tubes a couple of times. Store at –20°C for 1–2 h or overnight.
6. Collect the DNA by centrifugation as in **step 3**.
7. Pour off supernatants. *Keep an eye on the pellets.* Add 3 mL of 70% ice cold ethanol, and wash the pellet with a brief swirl on the vortexer.
8. Collect the DNA by centrifugation as in **step 3**.
9. Pour off supernatants, this time leaving the tubes upside-down. *Keep an eye on the pellets.* Let drain for approx 10 min. Using a pair of forceps wrapped with a piece of tissue paper, wipe off the tube walls (*do not touch the pellet!*) before reverting the tubes to the upright position. Leave them to dry for a few more minutes (*see* **Note 3**) before dissolving DNA in 0.5–0.8 mL TE (*see* **Note 1**). Aid dissolution of DNA by incubation at 45°C and/or on a rotary shaker. Leave the samples in the cold until further processing.
10. Quantitate DNA by making 1:20 dilutions into 1 mL TE and measuring absorbance at 260 and 280 nm. Calculate "DNA-equivalents" assuming 1 A$_{260}$ = 50 µg/µL (*see* **Note 3**).
11. Optional step: check cleavage by electrophoresis of 2 µg DNA on 0.75% agarose mini-gels in 0.5X TBE containing 0.5 µg/µL ethidium bromide.

3.3. Restriction Enzyme Digest

1. Assemble restriction enzyme digestion reactions as follows: Add 40 µL of the appropriate 10X restriction enzyme buffer, 120 µg DNA and water to a final volume of 400 µL. Add 80–100 U of restriction enzyme, mix well and spin the samples briefly before incubating at 37°C (preferably in an incubator, *not* a water bath *nor* in a dry block). After a couple of hours of digestion, mix the samples again, spin and return to the incubator and incubate overnight (*see* **Note 4**).
2. Add 1/10 volume of 3 M sodium acetate and 2 vol of ethanol. Mix well and store at –20°C for at least 15 min.
3. Collect DNA by centrifugation at 13,000 rpm (12,000g) at 4°C for 10–20 min. Pour off supernatants and wash once with 70% ice cold ethanol. Spin again if necessary.
4. Leave the samples upside-down to drain for some 10–15 min *(but never let them dry out)*. Wipe off the tube walls, and dissolve the DNA in 60 µL 1X Genomic DNA Sample-buffer (*see* **Note 5**). Aid dissolution by heating to 45°C and mixing, Samples can be vortexed briefly, but only once. Verify that the DNA is dissolved by pipetting. It is essential that the DNA is properly dissolved before electrophoresis.

3.4. Gel Electrophoresis, Southern Blotting, and Hybridization

1. Pour a thick 0.75–1.25% agarose gel (approx $0.8 \times 20 \times 20$ cm) in 1X TAE without ethidium bromide (*see* **Note 5**). Thick well formers (10×2 mm) are required to hold the 60–100 µL samples and also give better resolution. Make sure that the gel sets with an even thickness, from side-to-side and front-to-back.
2. Cover the gel with 1X TAE and load samples. Include a ^{32}P-labeled size marker. Run electrophoresis at 1.2–1.5 V/cm (24–30 V) overnight with buffer recycling, until the bromophenol blue tracking dye has migrated some 80% across the gel. If the buffer volume is large (>1.5 L), it is not neccessary to recirculate.
3. Apply standard methods *(5)* for Southern blotting and hybridization *(7)* with riboprobes *(6)* to visualize the HS-sites (*see* **Note 6**).

4. Notes

1. The original protocol was established with rat hepatoma cells *(4)*, which were treated with the NP40 buffer/DNaseI while still on the plate. This corresponds to method variant A (*see* **Subheading 3.1.1.**). With embryonal carcinoma (EC) cells, however, we noticed that this protocol always left a large fraction of the cells untouched by DNaseI, probably because of the heterogeneous nature of EC cell culture. Preparation of a single cell suspension after trypsinization (method variant B; *see* **Subheading 3.1.2.**) avoids this problem. The number of cells used per DNaseI-digest must be optimized for each type of cells. Method variant A as described here is designed for 6-cm tissue culture plates with approx $3–5 \times 10^6$ cells per plate. Method variant B is designed for 10-cm plates with approx $6–10 \times 10^6$ cells per plate. The volume used to dissolve the DNA after the second ethanol precipitation (*see* **Subheading 3.2.**, **step 7**) should be adjusted according to the number of cells used.
2. In the experiments described here, we use 0.2% NP40. The optimal concentration of NP40 must be determined for each cell type. Other detergents such as lysolecithin or saponin can also be used *(4)*. With a new cell type, it is recommended to test various permeabilization methods simply by applying the permeabilization buffer to the cells on a plate and watching the action under the microscope. Toward the end of the 4-min incubation period, one should see the nuclei "stand out" of the dissolving cell.

    DNaseI treatment is carried out at *room temperature*, that is $21°C \pm 4°C$. If the temperature is outside this range, adjust the temperature of the NP40-buffer and perform the nuclease cleavage in a water bath with controlled temperature.

    Once the logistical aspects of the method have been established, it may be convenient to do, e.g., three pairs of DNaseI treatments in an interleaved protocol. Six samples can then be processed in approx 30 min.
3. If the DNA pellets are left to dry too much, they will dissolve very slowly. Dissolution of high molecular weight DNA is faster at higher temperature, hence our suggestion to incubate the dissolving DNA at 45°C and/or on a rotary shaker.
    Note that the DNA as prepared here is not pure and that it may contain equal amounts of RNA. The quantitation by UV spectroscopy is therefore not reflecting the true DNA concentration.

4. Certain enzymes quite frequently give only partial digestions. In these cases, add more enzyme after the overnight digestion and incubate for a further 3 h before proceeding.
5. The agarose electrophoresis described here has been optimized to allow separation of large amounts of DNA as is required to see the bands corresponding to HS sites. The use of a loading buffer with 10 mM EDTA (*see* **Subheading 2.3.5.**) is important. It is our experience that TAE is an adequate electrophoresis buffer when used with recycling and electrophoresis at low voltage.
6. It is our experience that charged blotting membranes are best for this purpose. Hybridization with the Church and Gilbert buffers *(7)* has proven superior to the "complex cocktail" methods. We routinely use ^{32}P-labeled riboprobes *(6)* for these experiments. Random primed or nick-translated DNA probes can also be used.

References

1. Wu, C. (1980) The 5' ends of *Drosophila* heat shock genes in chromatin are hypersensitive to DNase I. *Nature* **286,** 854–860.
2. Nedospasov, S. A. and Georgiev, G. P. (1980) Non-random cleavage of SV40 DNA in the compact minichromosome and free in solution by micrococcal nuclease. *Biochem. Biophys. Res. Commun.* **92,** 532–539
3. Wu, C. (1989) Analysis of hypersensitive sites in chromatin. *Methods Enzymol.* **170,** 269–289.
4. Stewart, A. F. and Schütz, G. (1991) A simpler and better method to cleave chromatin with DNase1 for hypersensitive site analysis. *Nucleic Acids Res.* **19,** 3157.
5. Sambrook, J., Fritsch, E. F., and Maniatis, T. (1989) *Molecular Cloning: A Laboratory Manual.* Cold Spring Harbor Laboratory Press, Cold Spring Harbor, NY.
6. Gilman, M. (1993) Ribonuclease protection assay, in *Current Protocols in Molecular Biology* (Ausubel, F. M., et al., eds.), Wiley, NY, pp. 4.7.1–4.7.8.
7. Church, G. M. and Gilbert, W. (1984) Genomic sequencing. *Proc. Natl. Acad. Sci. USA* **81,** 1991–1995.

27

Mapping of Nucleosome Positions in Yeast

Magdalena Livingstone-Zatchej and Fritz Thoma

1. Introduction
1.1. Nucleosomes

The structural and functional subunits of chromatin are nucleosome cores. In a nucleosome core 145 bp of DNA are coiled around the outer surface of an octamer of histone proteins which consists of a tetramer of 2(H3·H4) and two H2A·H2B dimers *(1)*. DNA extending from the nucleosome core to the next nucleosome is called linker DNA. It varies in length from about 20 to 90 bp in different organisms or tissues or between individual nucleosomes. Histone H1 may be associated with linker DNA at the site where the DNA leaves the nucleosome. While core histones are well conserved and present in all eukaryotic organisms, H1 is most variable and may even be missing in some organisms such as yeast *Saccharomyces cerevisiae*. Nucleosomes are built from many different DNA sequences and may contain histone variants (subtypes) and modified histones (e.g., acetylated) which can affect their structural and dynamic properties (reviewed in **ref. 2**).

1.2. Nucleosome Positions

A nucleosome position (also referred to as "translational position") is defined as the specific location of a histone octamer on the DNA-sequence. This implies that the centre, the ends as well as the "rotational setting" are known with nucleotide precision. The rotational setting describes the orientation of the DNA on the histone octamer. The inner surface of DNA faces the histones and is not accessible to proteins, while the outer surface is exposed to the proteins in solution. Octamers may occupy either unique positions or they may move which results in their multiple positions. If nucleosome positions differ by a translational shift of multiples of 10 bp, the rotational setting is preserved for that part of the DNA which remains in the nucleosome. Any other translational shift will also change the rotational setting. "Nucleosome positioning" refers to mechanisms or properties of a system that positions nucleosomes. Nucleosome positions are determined by sequence dependent mechanical properties of DNA, boundary effects, and chromatin folding (reviewed in **refs. *3,4***).

1.3. Mapping of Nucleosome Positions on the DNA-Sequence

Since DNA is partially protected in nucleosomes, but more readily accessible in linker regions, nucleases and chemical reagents are used to cleave DNA where it is accessible. In this chapter, chromatin isolation and footprinting procedures are described to characterize nucleosome positions in yeast chromatin using micrococcal nuclease (MNase) and DNaseI as footprinting enzymes. To locate positioned nucleosome, the cleavage sites obtained by chromatin digestion are mapped on the DNA-sequence using a low resolution technique (indirect endlabeling). A nucleotide resolution genomic footprinting technique is described elsewhere *[5]*). The cutting sites in chromatin are compared with those in deproteinized DNA and the local chromatin structure is deduced (**Fig. 1**).

1.4. Cleavage Reagents

Most commonly used is micrococcal nuclease (MNase, staphylococcal nuclease). It requires Ca^{2+}, has endo- and exonuclease activity, digests RNA, cuts single stranded DNA faster than double stranded DNA, and it shows sequence preference by cutting pA and pT faster than pC or pG *(6)*. In chromatin, MNase preferentially and efficiently introduces double stranded cuts in the linker DNA *(7)*. This property makes it the first choice for mapping nucleosome positions. The sequence preference of the enzyme allows us to determine positions of the cutting sites on naked DNA and to analyse which of those sites

---

Fig. 1. *(continued on next page)* Mapping nucleosome positions by indirect endlabeling. (**A**) Schematic outline of the approach. Arrowheads in DNA point to potential cleavage sites by the nuclease (due to sequence specificity). Arrowheads in Chromatin point to potential accessible sites in chromatin, while other sites are inaccessible due to folding of DNA into nucleosomes by histones (ovals). (**B**) Nucleosome positions in a yeast minichromosome mapped by indirect endlabeling. Minichromosomes called YRpTRURAP *(16)* were partially purified. Chromatin (Chrom) and deproteinized control DNA (DNA) were digested with MNase or with DNase I. The cutting sites were mapped from an *Eco*RI site, using a 186 bp *Eco*RI-*Xba*I probe. The URA3 gene, the 5'part of the TRP1 gene, the ARS1 origin of replication are indicated. Positioned nucleosomes are indicated as white boxes. Dots indicate major cleavage sites in chromatin, squares are sites wich are protected in

Nucleosome Mapping in Yeast

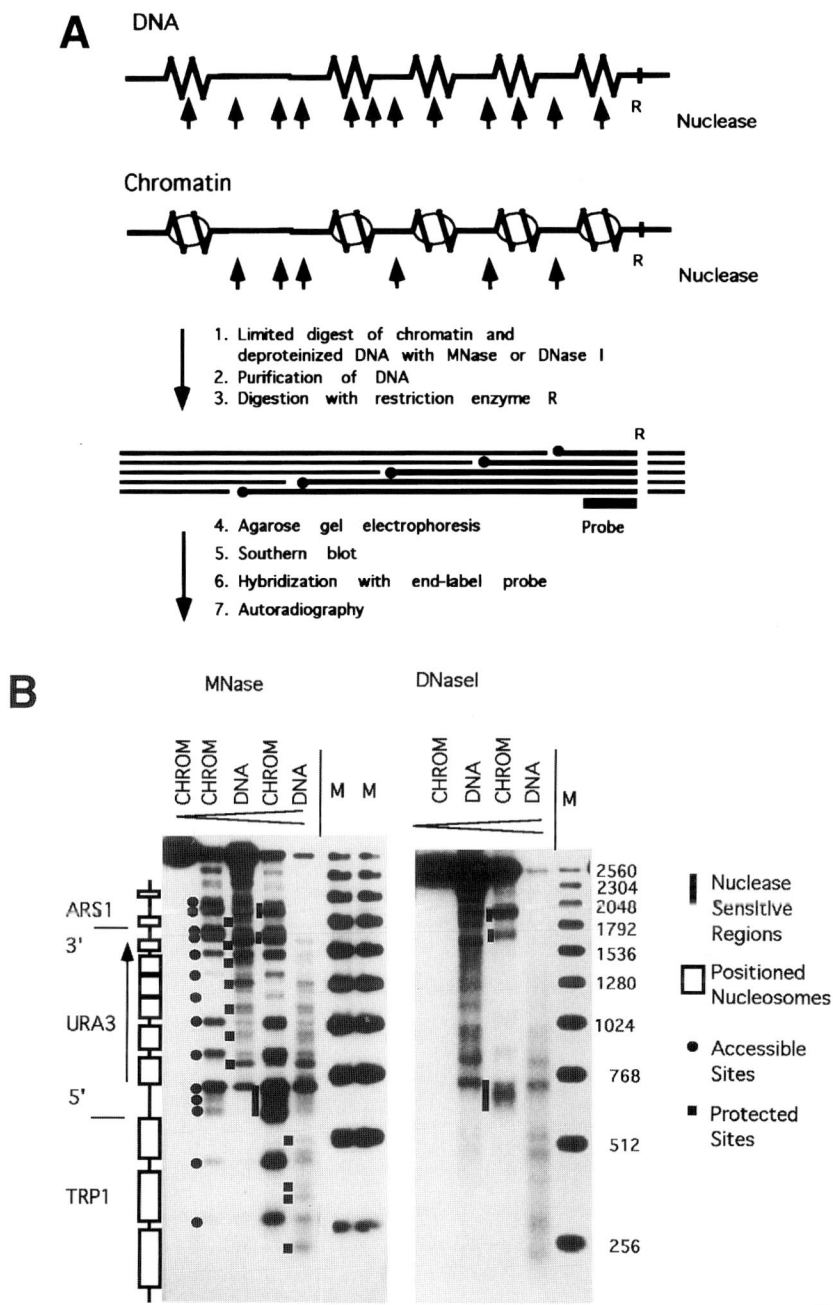

chromatin, bars indicate nuclease sensitive regions. A DNA ladder consisting of multiples of 256 bp is used for calibration (M) *(8)*. (Adapted from **ref. *16*** with permission of Academic Press, London).

are accessible in chromatin and which sites are protected *(8)*. DNase I (bovine pancreas) is a double strand specific endonuclease with sequence specificity for AT-rich regions. It requires divalent cations (Mg^{2+}, Ca^{2+}, Mn^{2+}), attacks DNA in the minor groove and cuts one of the two strands. A double strand break is observed if a second cut occurs on the other strand in close proximity of the first cut *(6)*. In nucleosomes, DNase I leads to a cutting every 10 bp whenever one of the DNA-strands faces the surface *(9)*. Hence, DNase I is the appropriate enzyme to analyse the rotational setting. Apart from the enzymes, various chemical reagents which induce DNA strand breaks were tested. Most frequently used is methidiumpropyl-EDTA (MPE·Fe(II)) for which detailed protocols have been published *(10)*. In general, at sites of ambiguous and unclear interpretation with one reagent, a combination of MNase, DNaseI and MPE·Fe(II) is recommended.

1.5. Mapping of Cleavage Sites by Indirect Endlabeling

The cleavage sites in deproteinized DNA and in chromatin may be displayed, measured and compared at low resolution using an indirect endlabeling technique (**Fig. 1**; *11–14*). Deproteinized DNA and chromatin are digested with MNase or DNaseI to a limited extent, the DNA is purified and cut to completion with a restriction enzyme that cleaves near the region of interest (0.2 to 3 kb) and provides a reference end. The double stranded DNA fragments are separated by electrophoresis in agarose gels, denatured, transferred to a membrane and hybridized to a probe which abuts the restriction site. Autoradiography reveals a series of bands in chromatin lanes and in DNA lanes. Each band corresponds to a DNA fragment with the reference restriction site at one end and the MNase cut at the other end. The length of the band allows to calculate the distance of the MNase cut from the reference site.

1.6. Chromatin Preparation Procedures

A method for preparation of yeast genomic chromatin and a method for partial purification of yeast plasmid chromatin are described. In both methods, yeast *S. cerevisiae* cells are converted to spheroplasts and spheroplasts are lysed in buffers which should maintain chromatin structure. For partial purification of plasmid chromatin, most of the large genomic chromatin is removed by centrifugation and soluble proteins like putative nucleases or proteases are removed by fractionation on a Sephacryl S-300 column *(8,15)*. The yield varies depending on the plasmid copy number and possibly on its size (no systematic study was made on that). This protocol has been successfully used for plasmids of 1453 bp (TRP1ARS1 circle) up to about 4 kb *(4,8,15–20)*. Clean plasmid chromatin may be obtained by a more extensive isolation protocol *(21)*. For preparation of genomic chromatin, the spheroplasts are lysed in a

Nucleosome Mapping in Yeast

Ficoll containing buffer, and the genomic chromatin is pelleted by centrifugation. Alternative procedures and buffer systems have been described *(22–24)*.

A major concern is that chromatin preparation procedures, selection of buffers, or interactions of nucleases or chemicals with chromatin might lead to a rearrangement of nucleosomes or to partial disruption of chromatin structures. However, both methods described here gave similar results with respect to nucleosome positions when the same gene was analysed *(25)*. Moreover, a comparison of chromatin structure with DNA-repair by photolyase in vivo showed that photolyase rapidly repairs DNA-lesions in "open" nuclease sensitive regions and linker DNA, while DNA-damage in nucleosomes is slowly repaired. Hence photolyase in the living cell sees the same structural features in chromatin as MNase after partial purification of chromatin (*26*; see also chapter 18). This substantiates that the in vitro analysis of chromatin structure by MNase digestion indeed reflects a chromatin structure as it exists in living cells.

1.7. Interpretation of Chromatin Structures Using Nuclease Digestion and Indirect Endlabeling

A few important points and rules are listed below which may help to understand complex mapping patterns.

1. The bands in the "DNA" lanes show the sequence preference of MNase and DNase I (**Fig. 1B**; *2*).
2. Since nucleosome core DNA is resistant to MNase digestion, a precisely positioned nucleosome is expected to protect about 140 bp of DNA against double strand cutting by MNase. Clear examples are given in the beginning of the inactive GAL-URARIB gene (**Fig. 2**, marked as boxes), or in a yeast minichromosome YRpTRURAP which contains the TRP1ARS1 sequences interrupted by the whole URA3 gene (**Fig. 1B**, boxes). The distances between cutting sites in the linkers flanking a positioned nucleosomes can vary from approx 140–200 bp, since linker length is variable between individual nucleosomes and since MNase shows sequence specificity. Note for example, that three central nucleosomes of the URA3 gene are tightly packed covering a region of only 460 bp and cleavage by MNase is weak between those nucleosomes (**Fig. 1B**).
3. Longer regions with frequent cutting sites (strong bands, several bands) and no obvious protection are called Nuclease Sensitive Regions (NSRs) or hypersensitive sites *(11)*. **Figure 1B** shows NSRs at the 5'- and 3'-ends of the URA3 gene and at the ARS1 origin of replication (bars). These regions may bind regulatory proteins such as the origin of replication complex *(27)* and exclude nucleosome formation.
4. Due to sequence specificity, MNase may not efficiently cleave at some places although the DNA is accessible. In such a case, alternative enzymes or chemical reagents which have different sequence specificities may be applied. A prominent example is the regulatory region of the GAL1-10 promoter. Neither protein

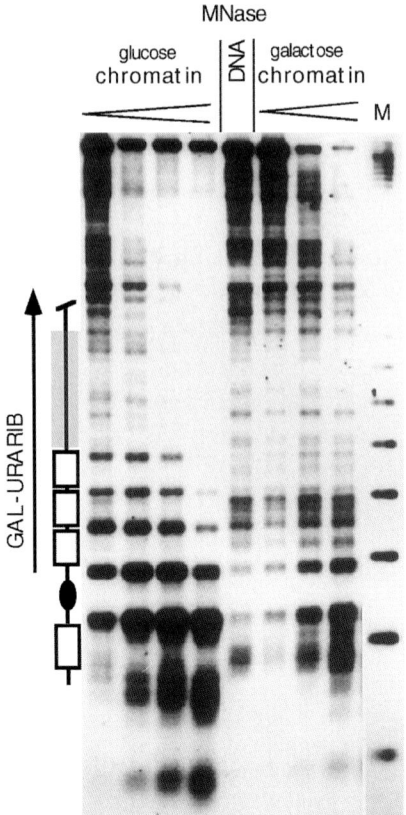

Fig. 2. Nucleosome positioning in a chromosomal gene. GAL-URARIB is an artificial gene inserted in the yeast genome which is transcriptionally active in galactose and repressed in glucose *(29,30)*. Genomic chromatin and DNA were digested with MNase and the cutting sites are displayed by indirect endlabeling. Nucleosomes are positioned in the 5' region (white boxes), no obvious protection is observed in the 3' region suggesting a random arrangement of nucleosomes (shaded area). Nucleosome positioning is lost when the gene is transcribed in galactose. In the GAL1-10 promoter (black oval) sequences are not cut by MNase under those conditions (DNA-lane), hence, information on presence or absence of a nucleosome must be obtained by alternative cleavage reagents. A DNA ladder consisting of multiples of 256 bp is used for calibration (M) *(8)*. (Adapted from **ref. 29** with permission of Oxford University Press).

free DNA nor chromatin are efficiently cut by MNase (**Fig. 2**, black oval), but MPE·Fe(II) and DNaseI cleavage showed that the whole region is not packaged in a nucleosome *(22,28)*.

5. Assuming random arrangement of nucleosomes on the DNA with no preference for positioning, the cutting pattern in chromatin is expected to be similar to that

of deproteinized DNA showing a modulation by the sequence specificity of the enzyme only. Such a situation was found at the end of the artificial yeast gene GAL-URARIB (**Fig. 2**, indicated by the shaded box) or in the 5' region of the GAL-URARIB gene, when the gene is transcribed in galactose media. (The presence of nucleosomes in that region was inferred from psoralen crosslinking experiments *[29–30]*).

6. It needs to be emphasized that the footprinting pattern reflects an average stucture of the whole population. The chromatin structure of individual regions in individual cells might be different.
7. Nucleosomes are dynamic structures which may dissociate and reassemble, or move along the DNA sequence. Hence, cutting sites might be protected in a fraction of the population only. As a consequence of nucleosome mobility or instability, the cutting pattern obtained at low levels of digestion may change at higher levels of digestion as observed in the TRP1 gene of the TRP1ARS1 circle *(8)*. The chromatin population may be heterogenous with respect to the cell cycle, replication, or transcriptional activity and, hence, produce a complex pattern which might be interpreted as not positioned nucleosomes.

2. Materials

2.1. Preparation of Plasmid Chromatin

1. Suspension of yeast cells (*see* **Note 1**).
2. Yeast wash buffer: 40 m*M* K-phosphate, pH 7.5, 1 *M* sorbitol, 0.5 m*M* PMSF (added just before use).
3. PMSF (phenylmethane-sulfonyl fluoride; Merck): 0.5 *M* dissolved in DMSO (methyl sulfoxide, Fluka).
4. Water (deionized and autoclaved).
5. ß-mercaptoethanol (Fluka).
6. Zymolyase 100T (Seikagaku Kogyo Corporation Bioscience Products AG, Emmebruecke, Switzerland): 10 mg/mL suspended in yeast wash buffer without PMSF, stored at –20°C.
7. Buffer A: 20 m*M* Tris-HCl, pH 8.0, 150 m*M* NaCl, 5 m*M* KCl, 1 m*M* EDTA, 1 m*M* PMSF (added just before use).
8. Triton X-100 (Fluka): 20% in water.
9. 1% SDS.
10. Sephacryl S-300 column (Pharmacia): 70–80 mL (2.5 cm × 13.5–15.5 cm) equilibrated with buffer A; connected with a UV-monitor and fraction collector (ISCO, Instrumenten Gesellschaft AG, Zuerich, Switzerland).
11. GS3-rotor, SS34 rotor, RC5B-Centrifuge (Sorvall, Digitana, Horgen, Switzerland).

2.2. DNA Extractions

1. 10% SDS.
2. Proteinase K (Boehringer Mannheim): 10 mg/mL in 50 m*M* Tris-HCl, pH 8.0, stored at –20°C.
3. P: phenol saturated with 0.1 *M* Tris-HCl, pH 8.0.

4. D/I: dichloromethane/isoamyl alcohol (24:1).
5. P/D/I: phenol/dichloromethane/isoamyl alcohol (25:24:1).
6. RNase A (Boehringer Mannheim): 10 mg/mL in 10 mM Tris-HCl, pH 7.5, 15 mM NaCl; stored at –20°C.
7. tRNA from *E. coli* (Sigma): 10 mg/mL in 10T1E; stored at –20°C.
8. 10T1E: 10 mM Tris-HCl, 1 mM EDTA, pH 8.0.
9. Ethanol (99.9%).
10. Sodium acetate (Fluka): 3 M, pH 4.8.
11. HB4-, SS34 rotors (Sorvall).

2.3. MNase and DNaseI Digestions

1. MNase (Nuclease S7 from Staphylococcus aureus, Boehringer Mannheim): 15 U/µL in 1 mg/mL BSA (a dilution in water of 20 mg/mL BSA (Boehringer Mannheim) in 50 mM Tris-HCl, 100 mM NaCl, 0.25 mM EDTA, 1 mM ß-mercaptoethanol, 50% glycerol, pH 7.5) stored in small aliquots at –20°C
2. DNaseI (grade I) (Boehringer Mannheim): 4 U/µL in 150 mM NaCl, 50% glycerol, stored at –20°C.
3. 0.1 M CaCl$_2$.
4. 0.2 M MgCl$_2$.
5. 10% SDS.
6. 0.5 M EDTA pH 8.0.
7. Reagents for DNA-extractions (*see* **Subheading 2.2.**).
8. Buffer A (*see* **Subheading 2.1.**).

2.4. Preparation of Genomic Chromatin and DNA

1. Some reagents are as for preparation of plasmid chromatin and DNA (*see* **Subheading 2.2.**).
2. Prespheroplasting solution: 2.8 mM EDTA(Na$_2$), pH 8.0, 0.7 M ß mercaptoethanol.
3. 1 M sorbitol.
4. ß-mercaptoethanol (Fluka).
5. Ficoll solution: 18% Ficoll 400 (Sigma), 20 mM KH$_2$PO$_4$, pH 6.8, 1 mM MgCl$_2$, 0.25 mM EGTA (K$_2$), 0.25 mM EDTA (K$_2$), 1 mM PMSF (freshly added).
6. Dounce hand-homogenizer (30 mL, type S; B. Braun, Melsungen AG., D-3508 Melsungen, Germany).
7. RCB5 centrifuge and HB4-, SS34-, GS3 rotors (Sorvall); table-top centrifuge (Jouan CR4.22, Instrumenten Gesellschaft AG).

2.5. Indirect Endlabeling

1. Restriction enzymes with appropriate reaction buffers.
2. Radioactive probes generated by random priming using oligolabeling or nick translation kits (Pharmacia).
3. Agarose (GIBCO BRL): ultra pure.
4. TBE: 89 mM Trizma Base (Sigma), 89 mM Boric acid, 2 mM EDTA.
5. Electrophoresis buffer: TBE and 0.5 µg/mL ethidium bromide.

6. Loading buffer: TBE, 8% glycerol, 0.01% bromophenol blue, 0.01% xylene cyanole.
7. 0.4 N NaOH.
8. 2X SSC: 0.3 M NaCl, 0.03 M sodium citrate, pH 7.0.
9. Prehybridization solution: 0.25 M phosphate buffer, pH 7.2, 7% SDS, 1 mM EDTA, pH 8.0, 50 µg/mL tRNA.
10. tRNA from *E. coli* (Sigma): 10 mg/mL in 10T1E; stored at –20°C
11. Hybridization solution: 0.25 M phosphate buffer, pH 7.2, 7% SDS, 1 mM EDTA, pH 8.0.
12. Washing solution I: 40 mM phosphate buffer, pH 7.2, 5% SDS, 1 mM EDTA, pH 8.0 (prewarmed overnight in the 70°C oven).
13. Washing solution II: 40 mM phosphate buffer, pH 7.2, 1% SDS, 1 mM EDTA, pH 8.0 (prewarmed overnight in the 70°C oven).
14. 3MM Whatman paper; paper towels.
15. DNA-size marker: DNA-ladder consisting of multiples of 256 bp Sea urchin 5S-DNA *(8)*.

3. Methods
3.1. Partial Purification of Yeast Plasmid Chromatin

1. Grow 3X 1 L yeast cultures in the appropriate medium and temperature to an absorbance at 600 nm of about 0.8–1.2 corresponding to approx 10^7 cells/mL (*see* **Note 1**). Harvest cells by centrifugation in a GS3 rotor at 6000g and 4°C for 8 min.
2. Resuspend the cell pellet in 200–300 mL water and collect by centrifugation as in **step 1**.
3. Resuspend the cell pellet in 80 mL yeast wash buffer. Convert to spheroplasts by addition of 144 µL ß-mercaptoethanol, 1 mL Zymolyase 100T (10 mg/mL) and incubation in a 30°C water bath with occasional shaking.
4. To test spheroplasting, mix 50-µL aliquots with 950 µL 1% SDS and measure absorbance at 600 nm. Spheroplasting is complete when A_{600} drops from initial approx 1.4 to below 0.2. Spheroplasting takes 10–45 min (*see* **Note 2**).
5. Carry out subsequent steps at 4°C or on ice.
6. Collect spheroplasts by centrifugation in a GS3 rotor at 6000g for 8 min. Gently resuspend spheroplasts in 100 mL yeast wash buffer using a 10 mL disposable pipet with a cut tip, and recollect them by centrifugation as above.
7. Lyse spheroplasts by resuspending in 20 mL buffer A containing additional 0.2% Triton X-100. Transfer lysed spheroplasts to a SS34 tube and keep on ice for 5–15 min.
8. Pellet genomic chromatin and insoluble debris by centrifugation in a SS34 rotor at 27,000g for 30 min. Carefully remove the white top layer. Collect the yellowish supernatant (S1; about 20 mL) which contains the plasmid.
9. Load 10 mL of the S1 supernatant on a 70–80 mL Sephacryl S-300 column and elute with buffer A. Plasmid chromatin and some genomic chromatin fragments elute with the turbid fractions of the void volume (*see* **Note 3**).
10. Collect about 15–17 mL starting with the first turbid fraction *(chromatin pool I)* and use immediately for digestion with MNase and DNaseI.

11. After the first run, wash the column with about 150 mL buffer A, load the rest of S1 (approx 10 mL) and fractionate as above to yield *chromatin pool II*. Use *chromatin pool II* to extract DNA for the "naked DNA-controls" ("plasmid DNA").

3.2. Extraction of Plasmid DNA

1. Add 150 µL SDS (10%) and 1.5 mg Proteinase K to 15 mL *chromatin pool II* and incubate at 50°C for 2 h and overnight at room temperature.
2. Extract DNA once with P, once with P/D/I and once with D/I.
3. Add 150 µL RNase A and incubate at 37°C for 30 min.
4. Extract DNA once more with P/D/I and once with D/I.
5. Add 30µL tRNA, precipitate with 3 vol of cold ethanol at –20°C for at least one hour and centrifuge in an HB4 rotor at 16,000g and 4°C for 40 min.
6. Dry the DNA pellets in vacuo, dissolve DNA in 10T1E, reprecipitate with 1/10 of the volume of NaAcetate (3 M) and 2.5–3 vol of ethanol. Finally dissolve DNA in 300 µL 10T1E (20 µL correspond to 1 mL *chromatin pool*).
7. Use 10 µL to check DNA on a 1% agarose gel. Plasmid DNA (a few ng) and some copurified genomic DNA should be visible by ethidium bromide staining (*see* **Note 4**).

3.3. MNase Digestion of Plasmid Chromatin (see Note 5)

1. Make 5 aliquots (2 mL) of the *chromatin pool I* on ice.
2. Add 100 µL CaCl$_2$ (0.1 M) and incubate at 37°C for 5 min.
3. Add different amounts of MNase (0, 2, 10, 50, 250 U/mL) and incubate at 37°C for 5 min.
4. Stop digestion by addition of 200 µL SDS (10%), 50 µL EDTA (0.5 M), 25 µg Proteinase K. Incubate at 50°C for 2 h and at room temperature overnight.
5. Extract DNA once with P, once with P/D/I and once with D/I.
6. Add 20 µL RNaseA and incubate at 37°C for 30 min.
7. Extract DNA once more with P/D/I and D/I.
8. Add 2 µL tRNA, precipitate DNA by addition of 3 vol of ethanol at –20°C for at least 1 h, and centrifuge in an HB4 rotor at 16,000g and 4°C for 40 min.
9. Dry DNA pellets and dissolve in 360 µL 10T1E. Reprecipitate with 1/10 of the volume of NaAcetate (3 M) and 2.5 vol of ethanol as described above. Dissolve the final DNA pellets in 200 µL 10T1E. For indirect endlabeling, 10 µL per gel lane are sufficient.

3.4. MNase Digestion of Plasmid DNA

1. Mix 40 µL of plasmid DNA with 300 µL buffer A.
2. Add 15 µL CaCl$_2$ (0.1 M) and incubate for 5 min at 37°C
3. Add different amounts of MNase (0, 0.4, 2, 10 U/mL chromatin pool) and incubate for 5 min at 37°C.
4. Stop reaction by addition of 30 µL SDS (10%), 7 µL EDTA (0.5 M).
5. Extract DNA with P, P/D/I, D/I. Add 2 µg tRNA as a carrier. Precipitate DNA as described above and dissolve it in 40 µL 10T1E. About 2 µL per gel lane are sufficient for indirect endlabeling.

3.5. DNaseI Digestion of Plasmid Chromatin

1. Make 5 aliquots (1 mL) of *chromatin pool I* on ice.
2. Add 25 μL MgCl$_2$ (0.2 M) to each tube and incubate at 37°C for 5 min.
3. Add DNase I (0, 0.5, 5, 25, 125 U/mL) and incubate at 37°C for 5 min.
4. Stop reaction by addition of 100 μL SDS (10%), 25 μL EDTA (0.5 M), 12 μg Proteinase K.
5. Extract, digest with RNase and precipitate DNA as described for MNase digestion of plasmid chromatin (*see* **Subheading 3.3.**) and dissolve it in 100 μL 10T1E. About 10 μL per gel lane are sufficient for indirect endlabeling.

3.6. DNaseI Digestion of Plasmid DNA

1. Mix 40 μL of plasmid DNA with 300 μL buffer A.
2. Add 7.5 μL MgCl$_2$ (0.2 M) and incubate at 37°C for 5 min.
3. Add DNase I (0, 0.1, 0.5, 2, 10 U/mL chromatin pool) and incubate for another 5 min at 37°C.
4. Stop reaction by addition of 30 μL SDS (10%), 7 μL EDTA (0.5 M).
5. Extract and precipitate DNA as described for MNase digestion of plasmid DNA (*see* **Subheading 3.4.**) and dissolve it in 40 μL 10T1E. About 2 μL per gel lane are sufficient for indirect endlabeling.

3.7. Preparation of Yeast Genomic Chromatin (see Note 6)

1. Grow 3X 1 L yeast cultures to an A$_{600}$ of about 0.6–1.2. Collect cells by centrifugation in a GS3 rotor at 6000g at 4°C for 8 min.
2. Suspend pellets in 100 mL water, transfer suspension into two preweighed 50-mL Falcon tubes, and collect by centrifugation in a table-top centrifuge at 1700g at room temperature for 5 min.
3. Estimate the volume of the pellet (approx 5 mL). Resuspend the pellet in two volumes of prespheroplasting solution and incubate gently shaking at 30°C for 30 min.
4. Pellet cells by centrifugation at 1700g at room temperature for 5 min. Resuspend cells in 20 mL 1 M sorbitol per tube and pellet as above.
5. Weigh pellets and resuspend in freshly prepared 1 M sorbitol, 5 mM ß-mercaptoethanol solution (5 mL/g of cells).
6. Convert cells to spheroplasts by addition of 1 mL Zymolyase 100T (10 mg/mL) and incubation in a 30°C water bath with occasional shaking. Check spheroplasting as described (*see* **Subheading 3.1.**).
7. Carry out subsequent steps at 4°C or on ice.
8. Collect spheroplasts by centrifugation at 3000g for 5 min in a table-top centrifuge, and resuspend them in 50 mL 1 M sorbitol using a 10-mL pipet with a cut tip. Pellet and weigh spheroplasts.
9. Resuspend spheroplasts in Ficoll solution (in 7 mL/g spheroplasts). Lyse spheroplasts by one stroke with a tight Dounce hand homogenizer and transfer the suspension into two SS34 tubes.
10. Pellet the genomic chromatin (and insoluble debris) by centrifugation at 27,000g in a SS34 rotor at 4°C for 30 min. Remove the white top layer and discard the supernatant.

11. Resuspend the pellets in buffer A (7 mL/g spheroplasts) to give a *Crude Chromatin Pool*. 12 mL are immediately used for MNase digestions; the rest is used to extract genomic DNA.

3.8. Extraction of Genomic DNA

1. Add 1.6 mL SDS (10%), 0.4 mL EDTA (0.5 M), 100 µg Proteinase K to 20 mL of the *chromatin pool*, incubate at 50°C for 2 h and at room temperature overnight.
2. Pellet insoluble debris by centrifugation in a SS34 rotor at 12,000g at room temperature for 30 min.
3. Extract, digest with RNase A and precipitate DNA as described for extraction of plasmid DNA (*see* **Subheading 3.2.**). Dissolve DNA in 400 µl 10T1E (roughly 0.5 µg/µL; 20 µL correspond to 1 mL *chromatin pool*). Remove insoluble material by centrifugation in a Microfuge at 15,000g at room temperature for 5 min.

3.9. MNase Digestion of Genomic Chromatin

1. Prepare 2-mL aliquots of the *chromatin pool* on ice.
2. Add 100 µL CaCl$_2$ (0.1M) and incubate at 37°C for 5 min.
3. Add different amounts of MNase and incubate at 37°C for 5 min. The amount of enzyme required depends on strain and growth conditions. 0–105 U/mL and 0–38 U/mL are used for chromatin from cells grown in selective media with glucose and galactose, respectively.
4. Stop digestion by addition of 200 µL SDS (10%), 50 µL EDTA (0.5 M), 12 µg Proteinase K and incubate at 50°C for 2 h and at room temperature overnight.
5. Remove insoluble debris by centrifugation in a SS34 rotor at 12,000g at room temperature for 30 min.
6. Extract DNA from the supernatant, digest with RNase A and precipitate as described for MNase digestion of plasmid chromatin (*see* **Subheading 3.3.**). Finally, dissolve DNA in 200 µL 10T1E. 10 µL are sufficient for one gel lane of indirect endlabeling.

3.10. MNase Digestion of Genomic DNA

1. Mix 40 µL of genomic DNA (corresponding to approx 2 mL *chromatin pool*) with 300 µL buffer A.
2. Add 15 µL CaCl$_2$ (0.1 M) and incubate for 5 min at 37°C.
3. Add different amounts of MNase (1.5–15 U/mL *chromatin pool*) and incubate for 5 min at 37°C.
4. Stop reaction by addition of 30 µL SDS (10%), 7 µL EDTA (0.5 M).
5. Extract DNA once with P, once with P/D/I and once with D/I.
6. Precipitate DNA with 1/10 of the volume of NaAcetate (3 M) and 2.5 vol of ethanol, incubate at –20°C for at least 1 h, centrifuge in an HB4 rotor at 16,000g and 4°C for 40 min. Dry DNA pellets in vacuo and redissolve in 200 µL 10T1E. 10 µL are sufficient for one gel lane of indirect endlabeling.

3.11. Mapping Cutting Sites by Indirect Endlabeling (see Note 7)

1. Choose restriction sites and probes for indirect endlabeling (*see* **Note 8**).
2. Generate radioactive probes by random priming using oligolabeling or nick translation kits.
3. Digest DNA samples of MNase digested chromatin and MNase digested control DNA with the appropriate restriction enzyme. Stop the reactions with EDTA. Precipitate DNA with ethanol. Dry DNA in vacuo and dissolve it in 10 μL Loading Buffer.
4. Prepare horizontal gels (300 mL, 20 cm × 25 cm, BRL) which contain 1% agarose in TBE. Add 0.5 μg/mL ethidium bromide just before agarose solidifies. Submerge gels in electrophoresis buffer. Load samples and run electrophoresis at 50 V (constant voltage) and 20–25 mA for 15 h at room temperature.
5. For Southern transfer using Pall B membranes, soak the gels in 0.4 N NaOH for about 15 min.
6. Lay two sheets of 3 MM Whatman paper presoaked in 0.4 N NaOH over a plastic plate bridging a tray with 0.4 N NaOH. The paper remains in contact with 0.4 N NaOH on both sides.
7. Carefully invert the gel and place it on the paper. Cover it with Pall B membrane presoaked briefly in 0.4 N NaOH, four 3MM Whatman papers presoaked in 0.4 N NaOH, two layers of dry 3MM Whatman paper, 10 cm of paper towels, a glass plate and a weight of approx 250 g. The gel should not be compressed and flattened during DNA-transfer. Transfer to the Pall B membrane is for approx 16–20 h. Replace wet towels after few hours with fresh ones.
8. After transfer, mark the slots and marker lanes with a pen. Neutralize the membranes by rinsing them twice for about 15 min in 2X SSC. Make sure that the pH of the 2X SSC solution falls to seven, if necessary, rinse the membranes again. Air dry the membranes between Whatman paper for at least 30 min. Cut the membranes to allow separate hybridizations with different probes.
9. Place the membranes in rotary cylinders with the DNA bound side facing inside of the glass and incubate them in 20 mL prehybridization solution per 20 cm × 24 cm membrane at 65°C for 1 h in a hybridization oven (Bachofer, Laboratoriums Geräte, D-7410 Reutlingen, Germany).
10. Replace prehybridization solution with the same volume of prewarmed hybridization solution.
11. Denature labeled DNA for 5 min in a boiling water bath and snap cool it on ice. Add it to the hybridization solution and hybridize at 65°C overnight.
12. Pour off the hybridization solution and rinse the membranes with prewarmed washing solution I. Wash the membranes twice for 30 min at 65°C in washing solution I, and twice for 30 min at 65°C in washing solution II.
13. Air dry the membranes between two sheets of 3 MM Whatman paper and expose them to x-ray films (Fuji) using enhancer screens or to PhosphorImager Storage Plates (Molecular Dynamics).
14. Identify the bands and measure the lengths of the bands using the DIGIGEL program (DNASTAR, Madison). The cutting site on the DNA-sequence is calculated as the distance from the restriction site (*see* **Note 9**).

4. Notes

1. Yeast cultures are grown in the appropriate medium *(33)*. Selective media are required for maintenance of minichromosomes.
2. Spheroplasting may depend on the strain and growth conditions. If spheroplasting needs to be fast in order to reduce the risk of alterations of gene expression and chromatin structure, more Zymolyase may be added.
3. Centrifugation is used to remove a fraction of the cellular debris and the large fraction of the genome. The Sephacryl column is used to rapidly separate plasmid chromatin from small contaminants such as proteases. The degradation of chromatin has never been monitored. However the consistency of chromatin footprints as well as in vivo results with photolyase *(26)* argue against a degradation of chromatin during this preparation protocol.
4. Yields of minichromosomes may depend on the strain, the copy number and size of the minichromosomes. This has never been tested systematically.
5. Chromatin is digested to a limited degree for a short time (5 min) with variable amounts of enzyme. Incubation without enzyme is a control for endonuclease activity. No such activity was found in our preparations. Nuclease digestion may depend on the strain and growth conditions. We noticed that chromatin of cells grown in galactose is more rapidly digested by MNase than chromatin of cells grown in glucose.
6. This protocol is based on a procedure described by Almer and Hörz *(31)* with minor modifications *(32)*.
7. Other gel electrophoresis conditions, membranes (Nitrocellulose; Schleicher and Schuell (Therwil, Switzerland) or Zeta-Probe GT ; Bio-rad Labratories), transfer and hybridization conditions may be used.
8. Choice of restriction sites and endlabel probes is critical. The restriction site as well as the probe should not be located in a nuclease sensitive site. To generate probes it is recommended to use subcloned DNA-fragments (approx 150–200 bp). Alternatively, DNA-sequences can be directly amplified from the yeast genome using strand specific primers. Radioactive probes can be generated using small DNA fragments as substrates and primer extension by Taq polymerase with radioactive dNTPs. Several rounds of primer extension can be used to increase the amount of radioactive DNA.
9. A major concern is accuracy in measurements of cutting sites. We prefer 25 cm long 1% agarose gels run in TBE with ethidium bromide. This allows to spread 2.5 kb in the lower 15 cm of the gel and to measure the length of the bands with the required precision. At 2.6 kb (top band of our marker is 2560 bp; **Fig. 1B**), we measure 250 bp/5 mm or 50 bp/mm, whereas in the lower part, we obtain 250 bp/ 25 mm or 10 bp/mm. Because the band width measured at weak exposures is approx 1 mm, the average mapping results in a precision of approx 20 bp.

Acknowledgment

We are grateful to Dr. U. Suter (Institut für Zellbiologie, ETH-Zürich) for continuous support. This work was supported by the Swiss National Science Foundation.

References

1. Luger, K., Mäder, A. W., Richmond, R. K., Sargent, D. F., and Richmond, T. J. (1997) Crystal structure of the nucleosome core particle at 2.8 Å resolution. *Nature* **389**, 251–260.
2. Wolffe, A. (1995) *Chromatin*. Academic Press, San Diego, CA.
3. Simpson, R. T. (1991) Nucleosome positioning: occurrence, mechanisms, and functional consequences. *Prog. Nucleic Acid Res. Mol. Biol.* **40**, 143–184.
4. Thoma, F. (1992) Nucleosome positioning. *Biochimica et Biophysica Acta* **1130**, 1–19.
5. Thoma, F. (1996) Mapping of nucleosome positions. *Methods Enzymol.* **274**, 197–214.
6. Bellard, M., Dretzen, G., Giangrande, A., and Ramain, P. (1989) Nuclease digestion of transcriptionally active chromatin. *Methods Enzymol.* **170**, 317–346.
7. Noll, M. and Kornberg, R. D. (1977) Action of micrococcal nuclease on chromatin and the location of histone H1. *J. Mol. Biol.* **109**, 393–404.
8. Thoma, F., Bergman, L. W., and Simpson, R. T. (1984) Nuclease digestion of circular TRP1ARS1 chromatin reveals positioned nucleosomes separated by nuclease sensitive regions. *J. Mol. Biol.* **177**, 715–733.
9. Lutter, L. C. (1979) Precise location of DNaseI cutting sites in the nucleosome core determined by high resolution gel electrophoresis. *Nucleic Acids. Res.* **6**, 41–55.
10. Cartwright, I. L. and Elgin, S. C. R. (1989) Nonenzymatic cleavage of chromatin. *Methods Enzymol.* **170**, 359–369.
11. Wu, C. (1980) The 5'ends of *Drosophila* heat shock genes in chromatin are hypersensitive to DNaseI. *Nature* **286**, 854–860.
12. Nedospasov, S. A. and Georgiev, G. P. (1980) Non-random cleavage of SV-40 DNA in the compact minichromosome and free in solution by micrococcal nuclease. *Biochem. Biophys. Res. Commun.* **92**, 532–539.
13. Wu, C. (1989) Analysis of hypersensitive sites in chromatin. *Methods Enzymol.* **170**, 269–289.
14. Nedospasov, S. A., Shakhov, A. N., and Georgiev, G. P. (1989) Analysis of nucleosome positioning by indirect end-labeling and molecular cloning. *Methods Enzymol.* **170**, 408–420.
15. Thoma, F. and Simpson, R. T. (1985) Local protein-DNA interactions may determine nucleosome positions on yeast plasmids. *Nature* **315**, 250–252.
16. Thoma, F. (1986) Protein-DNA interactions and nuclease sensitive regions determine nucleosome positions on yeast plasmid chromatin. *J. Mol. Biol.* **190**, 177–190.
17. Losa, R., Omari, S., and Thoma, F. (1990) Poly(dA)'poly(dT) rich sequences are not sufficient to exclude nucleosome formation in a constitutive yeast promoter. *Nucleic Acids Res.* **18**, 3495–3502.
18. Bernardi, F., Zatchej, M., and Thoma, F. (1992) Species specific protein–DNA interactions may determine the chromatin units of genes in *S. cerevisiae* and in *S. pombe*. *EMBO J.* **11**, 1177–1185.
19. Thoma, F. and Zatchej, M. (1988) Chromatin folding modulates nucleosome positioning in yeast minichromosomes. *Cell* **55**, 945–953.
20. Tanaka, S., Halter, D., Livingstone-Zatchej, M., Reszel, B., and Thoma, F. (1994) Transcription through the yeast origin of replication ARS1 ends at the ABFI bind-

ing site and affects extrachromosomal maintenance of minichromosomes. *Nucleic Acids Res.* **22,** 3904–3910.
21. Pederson, D. S., Venkatesan, M., Thoma, F., and Simpson, R. T. (1986) Isolation of an episomal yeast gene and replication origin as chromatin. *Proc. Natl. Acad. Sci. USA* **83,** 7206–7210.
22. Lohr, D. (1984) Organization of the GAL1-GAL10 intergenic control region chromatin. *Nucleic Acids Res.* **12,** 8457–8474.
23. Almer, A., Rudolph, H., Hinnen, A., and Hörz, W. (1986) Removal of positioned nucleosomes from the yeast PHO5 promoter upon PHO5 induction releases additional upstream activating DNA elements. *EMBO J.* **5,** 2689–2696.
24. Buttinelli, M., DiMauro, E. D., and Negri, R. (1993) Multiple nucleosome positioning with unique rotational setting for the Saccharomyces cerevisiae 5S rRNA gene in vitro and in vivo. *Proc. Natl. Acad. Sci. USA* **90,** 9315–9319.
25. Tanaka, S., Livingstone-Zatchej, M., and Thoma, F. (1996) Chromatin structure of the yeast URA3 gene at high resolution provides insight into structure and positioning of nucleosomes in the chromosomal context. *J. Mol. Biol.* **257,** 919–934.
26. Suter, B., Livingstone-Zatchej, M., and Thoma, F. (1997) Chromatin structure modulates DNA repair by photolyase in vivo. *EMBO J.* **16,** 2150–2160.
27. Diffley, J. F. X. and Cocker, J. H. (1992) Protein DNA interactions at a yeast replication origin. *Nature* **357,** 169–172.
28. Fedor, M. J., Lue, N. F., and Kornberg, R. D. (1988) Statistical positioning of nucleosomes by specific protein-binding to an upstream activating sequence in yeast. *J. Mol. Biol.* **204,** 109–127.
29. Cavalli, G. and Thoma, F. (1993) Chromatin transitions during activation and repression of galactose-regulated genes in yeast. *EMBO J.* **12,** 4603–4613.
30. Cavalli, G., Bachmann, D., and Thoma, F. (1996) Inactivation of topoisomerases affect transcription dependent chromatin transitions in rDNA but not in a gene transcribed by RNA-polymerase II. *EMBO J.* **15,** 590–597.
31. Almer, A. and Hörz, W. (1986) Nuclease hypersensitive regions with adjacent positioned nucleosomes mark the gene boundaries of the PHO5/PHO3 locus. *EMBO J.* **5,** 2681–2687.
32. Bernardi, F., Koller, T., and Thoma, F. (1991) The ade6-gene of the fission yeast *Schizosaccharomyces pombe* has the same chromatin structure in the chromosome and in plasmids. *Yeast* **7,** 547–558.
33. Sherman, F., Fink, G. R., and Hicks, J. B. (1986) *Laboratory Course Manual for Methods in Yeast Genetics.* Cold Spring Harbor Laboratory Press, Cold Spring Harbor, NY.

28

Analysis of DNA Topology in Yeast Chromatin

Randall H. Morse

1. Introduction

Topological measurements have been used to investigate chromatin structure since the discovery that closed circular DNA molecules differing only in the number of supercoils that they contain can be resolved by gel electrophoresis *(1)* and the recognition that nucleosomes, the fundamental units of chromatin, confer defined changes in topology on closed circular DNA templates *(1,2)*. The number of supercoils introduced into a closed circular DNA molecule per nucleosome can be calculated by comparing unassembled and nucleosome-assembled DNA. Nucleosomes can be counted by electron microscopy, and the total number of supercoils analyzed by gel electrophoretic comparison of naked and nucleosome-assembled DNA relaxed by topoisomerase (the latter after stripping off the histones). This kind of analysis has revealed that one nucleosome confers almost exactly one negative supercoil into DNA both in vitro *(2–4;* see Chapter 6) and in vivo, in yeast as well as in higher eukaryotes *(5–7)*. This change is somewhat smaller than expected, since DNA wraps around the histone octamer approx 1.75 turns, leading to a calculated change of -1.65 in linking number *(8)*. Some of the excess change is accounted for by a compensating alteration in the twist of nucleosomal DNA, but a part of it remains unaccounted for *(9)*. Nevertheless, it is an empirical fact and allows the number of nucleosomes on a closed circular DNA molecule to be inferred by comparing the linking number of relaxed plasmid chromatin to that of relaxed naked DNA.

Since each nucleosome confers one negative supercoil to DNA, loss of nucleosomes or alteration of nucleosome structure is expected to change DNA topology. This property has been exploited in numerous studies of chromatin structure and function (for a thorough discussion of studies on DNA topology in chromatin structure and gene expression done prior to 1992, *see* **ref.** *10*).

From: *Methods in Molecular Biology, Vol. 119: Chromatin Protocols*
Edited by: P. B. Becker © Humana Press Inc., Totowa, NJ

For example, a variety of mutations in the histone proteins have been shown to alter topology of plasmid chromatin in the yeast *Saccharomyces cerevisiae* *(11–13)*, as has histone depletion *(14)*. Acetylation of the histone amino-termini has been reported to alter nucleosome topology in vitro *(4,15)*. Removal of the same amino-termini by trypsinization was reported not to alter linking number *(16)*; whether these two observations are compatible or not has not been determined. Conflicting in vivo experiments on the effect of histone acetylation on topology have been reported *(17,18)*. More recently, chromatin topology has been studied in plasmids containing binding sites for transcriptional activators in frog oocytes and in yeast; in both systems, loss of negative supercoiling consistent with nucleosome loss or alteration was observed in the presence of the transcriptional activator, and in both cases required an activation domain *(19,20)*.

Topological studies have also been used in structural investigations of chromatin. Naked DNA undergoes thermal untwisting; this small helical unwinding that occurs with increasing temperature (approx 0.011 radial degrees/°C/base pair) results in increased negative supercoiling for molecules relaxed at increasing temperatures *(21,22)*. This effect is suppressed by incorporation of DNA into nucleosomes, such that each nucleosome prevents about 160-180 bp of DNA from thermally untwisting *(16,23–25)*. In the yeast *Saccharomyces cerevisiae*, however, this suppression is reduced, such that the equivalent of only approx 60 bp of DNA per nucleosome is prevented from thermally untwisting *(7,26)*. These effects of nucleosomal packaging on thermal untwisting of DNA have also been interpreted as arising from changes in nucleosome shape *(27–29)*. This explanation, however, requires opposite shape changes for yeast and higher eukaryotic nucleosomes in their responses to temperature, and also requires the extent of deformation of yeast nucleosomes to be proportional to the change in temperature, which seems not entirely plausible. Regardless of the explanation, it is important to be aware that altering the temperature at which plasmid chromatin is prepared from yeast cells will affect DNA topology (*see* **Note 1**).

One considerable advantage that topological analysis has over other methods of analyzing chromatin structure is that small samples can be used and isolation of material is done rapidly. These features have allowed kinetic analysis of minichromosome topology which showed that although topoisomerase relaxes torsional stress in yeast minichromosomes extremely rapidly at room temperature and above (1–5 min), no change in topology is observed due to transcription of the same templates, leading to the inference that nucleosomes recover their normal structure quickly following polymerase passage *(30)*. Kinetic analysis was also used to show that the change in topology caused by activator binding to a yeast episome mentioned above is complete within

45 min, indicating that DNA replication is not required for the alteration of chromatin structure *(19)*.

1.1. Theory

Experiments predating the separation of individual DNA topoisomers by gel electrophoresis demonstrated that the free energy associated with DNA supercoiling is proportional to the square of the extent of supercoiling (Eq. 1) *(31,32)*.

$$\Delta G_{sc} = K \cdot (\Delta Lk)^2 \qquad (1)$$

This behavior is in accordance with Hooke's Law, suggesting that the energy associated with supercoiling arises from elastic deformation of structure. Together with classical thermodynamic considerations, equation (1) implies that the concentration of individual topoisomers at equilibrium (i.e., after being relaxed by DNA topoisomerase) will conform to a Gaussian distribution *(21,22*; also **ref. 33**, pp. 38ff). Specifically,

$$-(RT/\Delta Lk_i) \cdot \ln (I_i/I_{max}) = K (\Delta Lk_i + 2 w_T) \qquad (2)$$

where $R = 1.99$ cal deg^{-1} mol^{-1} (the gas constant), T is the temperature, ΔLk_i is the difference in linking number between topoisomer *i* having relative concentration I_i and the maximally abundant topoisomer having relative concentration I_{max}, and ω_T is the number of nonintegral turns by which the precise center of the topoisomer distribution differs from Lk_{max}, the linking number of the maximally abundant topoisomer. Plotting $(1/\Delta Lk_i) \cdot \ln (I_i/I_{max})$ vs ΔLk_i therefore yields a straight line having slope m and intercept $2m \cdot \omega_T$; dividing the intercept by two times the slope yields ω_T, which then allows determination of the precise center of the topoisomer distribution. Since analysis of plasmid chromatin from living cells has revealed that minichromosome topoisomers also conform to a Gaussian distribution, the same analysis can be applied to closed circular DNA packaged as chromatin *(24,25,34)*.

2. Materials

2.1. DNA Isolation from Yeast

1. Media for growing yeast.
2. Acid-washed glass beads (425–600 micron).
3. 5 mg/mL proteinase K, 5% SDS (this may be stored indefinitely in 1 mL aliquots at –20°C and thawed by warming in a 37°C water bath).
4. TE solution: 10 m*M* Tris-HCl, pH 8.0, 0.5 m*M* EDTA.

2.2. DNA Purification

1. Phenol.
2. Chloroform.
3. 4 *M* ammonium acetate.

4. Ethanol.
5. TE solution (see **Subheading 2.1.**).
6. 10 mg/mL RNase A.

2.3. Agarose Gel Electrophoresis

1. Agarose (high-quality for blotting, such as GTG agarose from FKB).
2. 60 mg/mL solution of chloroquine diphosphate (chloroquine is toxic and should be handled with caution) in deionized water (this can be stored indefinitely in the dark at 4°C; we make this up in a 50-mL conical tube and wrap the tube in aluminum foil).
3. 50X TAE buffer: 2 M Tris-acetate, 50 mM EDTA (242 g Tris base, 57.1 glacial acetic acid, 100 mL 0.5 M EDTA, pH 8.0 per liter of solution *(35)*.
4. Loading dye: 18% Ficoll 400 plus 0.3% xylene cyanol (see **Note 2**).

2.4. Southern Blotting

1. Nylon membrane for blotting (e.g., Genescreen or Duralon UV).
2. 0.5 M HCl (concentrated HCl is 12 M).
3. 0.5 M NaOH (dilute from 10 M NaOH solution).
4. 3 mm paper.
5. 20 X SSC *(35)*: 3 M NaCl, 0.3 M sodium citrate, pH 7.0 (175 g NaCl, 88.2 g sodium citrate dihydrate dissolved in 800 mL H_2O; pH adjusted with HCl and volume adjusted to 1 L).
6. 6 X SSC.
7. 0.9 M Tris, 1.5 M HCl (add 12.5 mL concentrated HCl to 87.5 mL 1 M Tris-HCl, pH 8.0).
8. Paper towels.
9. Plastic wrap.
10. UV crosslinker.
11. Sealable plastic bag and heat sealer.

2.5. Hybridization

1. Radioactively labeled probe specific to the DNA you wish to analyze (use any of the commercially available kits for random prime labeling of DNA).
2. 1 M NaP$_i$: 134 g Na$_2$HPO$_4$·7H$_2$O (71 g anhydrous) plus 4 mL phosphoric acid per liter (pH will be 7.2) (add sodium phosphate to water; if you add water to the sodium phosphate it becomes very difficult to dissolve).
3. Hybridization buffer: 52.5 mL 1 M NaP$_i$ pH 7.2, 12 mL H_2O, 35 mL 20% SDS, 0.2 mL 0.5 M EDTA, pH 8.0, 1 g bovine serum albumin (dissolve by 5–10 min incubation in 65°C water bath; this may form some precipitate on standing but can be redissolved at 65°C).
4. Church Gilbert wash #1: 350 ml H_2O, 20 mL 1 M NaP$_i$, 125 mL 20% SDS, 1 mL 0.5 M EDTA, pH 8.0.
5. Church Gilbert wash #2: 910 mL H_2O, 40 mL 1 M NaP$_i$, 50 mL 20% SDS, 2 mL 0.5 M EDTA, pH 8.0.

6. 0.1% SDS/0.2 X SSC at 65°C.
7. X-ray film or phosphorimager screen.
8. Scanning densitometer or phosphorimager.
9. Spreadsheet computer program.

3. Methods
3.1. Comparison of Topology Among Species Having Relatively Small Differences in Linking Number

This protocol is suitable for many kinds of topological analysis, for example examining effects of mutant histones, activator binding, and thermal untwisting on topology (*see* **Subheading 1.**). Modifications which allow comparison of species having larger differences in linking number, for example to count nucleosomes by comparing linking number of plasmid chromatin and relaxed naked DNA, are given in a brief section following this one.

3.1.1. Preparation of DNA

1. Grow yeast cells in appropriate media (10 mL per sample to be analyzed) to $A_{600} = 0.5$–1.5. Before taking samples out of shaking incubator, prepare 1.5-mL microfuge tubes: label tubes; add acid-washed glass beads (425–600 micron) to 0.3–0.5 mL level of tube; add 100 µL 5 mg/mL proteinase K, 5% SDS. When this is done, spin down yeast cells in desktop centrifuge 2000 rpm (800g) 2 min at room temperature. Pour off supernatant and resuspend pellets in TE solution. Transfer resuspended pellets to labeled microfuge tube containing glass beads and proteinase K/SDS. Vortex 1 min at top speed (*see* **Note 3**); place on ice 15 s; vortex 1 min.
2. Hold the microfuge tube upside down and make a small hole in the bottom with a hot dissecting needle (heated over a flame) (*see* **Note 4**). Collect the lysed cell suspension by placing the tube onto a clean labeled 1.5-mL microfuge tube and spinning 8 K rpm in a microfuge 5–10 s. Discard the tube with glass beads (the suspension should now be in the lower microfuge tube) and spin the lower microfuge tube 8–10 min at full speed in microfuge. Since there is proteinase K and SDS in the sample it does not matter whether these spins are done at 4°C or room temperature. Transfer 320 µL of the supernatant to a new labeled 1.5-mL microfuge tube and incubate 37°C (up to 50°C is fine) at least 2 h.
3. Extract sample with phenol and chloroform. Add 80 µL 4 *M* ammonium acetate to sample and 1 mL 100% ethanol; precipitate. Spin sample 10 min at top speed in microfuge at 4°C, discard supernatant (it can be poured off), and take up the sample in 30 µL TE solution plus 1 µL 10 mg/mL RNase A (*see* **Note 5**).

3.1.2. Gel Electrophoresis and Blotting (35,36)

1. Electrophorese DNA on a topoisomer-resolving gel: Pour a large (21 cm or more from wells to end of gel) agarose gel containing 40 µg/mL chloroquine diphosphate (*see* **Note 6**). The agarose concentration to be used depends on the size of

the closed circular DNA being analyzed: for a 1.4–2 kb plasmid, use 1.5% agarose; for 2–3 kb use 1.2%; for 3–4.5 kb use 1%; for 4–6 kb use 0.8% and for 6–8 kb use 0.7% (e.g., for the 2 micron plasmid).
2. Use TAE buffer in the gel and running buffer, and add chloroquine (133 µL of 60 mg/mL chloroquine diphosphate solution for a 200 mL gel) after letting the melted agarose cool to 55–65°C. Use the same chloroquine concentration in the gel and running buffer; it is not necessary to recirculate the gel buffer solution during electrophoresis.
3. Add 1 µL of loading dye to 15 µL of each sample (it is not necessary to add chloroquine to the samples), load samples and run at 2.5 V/cm (measure cm from electrode to electrode) for 18–20 h (*see* **Note 7**). The gel can be run at room temperature and need not be shielded from light. Do not add ethidium bromide to the gel or running buffer (*see* **Note 8**)!
4. Cut out part of gel to be blotted (discard empty lanes; it is also desirable to slice away the extreme edges of the gel, as the menisci make it not flat and more difficult to achieve uniform contact with the nylon membrane) and measure its size.
5. Soak gel in 0.5 *N* HCl 2 × 10 min in a glass or plastic tray. No shaking is necessary.
6. Soak in 0.5 *N* NaOH 2 × 10 min.
7. Meanwhile, with a razor blade and gloves, cut out a piece of nylon membrane such as Genescreen (Dupont) or Duralon UV (Stratagene) the same size as the gel. Also cut three pieces of 3 mm paper to the same size, and three pieces slightly larger. Soak the membrane in 6 x SSC.
8. Soak gel in 20 X SSC plus a few mL of 0.9 *M* Tris/1.5 *M* HCl to neutralize (neutralization does not have to be complete for nylon; somewhere in the pH 6.0–8.0 range is fine). Soaking can be from 5–10 min to a half hour.
9. Lay gel, bottom up, on a flat plastic gel tray. Lay the nylon membrane over the gel (be sure you've marked your gel in some way; by loading lanes asymmetrically, or cutting a corner of the gel and the membrane, so you know the orientation when it comes time to look at the autoradiogram) and roll out gently but firmly with a plastic pipet. It is not necessary to apply much pressure here, and too much pressure can crack the gel; the rolling is done only to remove bubbles. Soak the three 3 mm paper sheets that are larger than the gel in 6 X SSC and lay over gel plus membrane and roll out bubbles gently but firmly. Turn gel plus membrane plus 3 mm paper onto a stack of paper towels; for larger gels these need to be unfolded (*see* **Note 9**). You're blotting! Soak 3 mm paper pieces the same size as gel in 20 X SSC and lay over the top of the gel and roll out bubbles. Cover with a film wrap such as Saran Wrap, then a paper towel, then a flat plastic gel tray. Put a light weight (an empty 1-L bottle is good) on top.
10. After 15–20 min, remove weight, gel tray, paper towel and Saran wrap. Pick up gel plus 3 mm paper on both sides and remove wet paper towels from underneath. Replace blot, wet the top with 20X SSC and replace topping. You can use a heavier weight, such as two half-full 1-L bottles.
11. When blotting is complete (2 h to overnight), briefly immerse membrane in 6X SSC (to clean of agarose fragments, stray fruit flies, and so forth), keeping track of which

side the nucleic acid is on, then lay the membrane face up on a piece of 3 mm paper. Irradiate in a UV crosslinker (e.g., a Stratalinker from Stratagene or the equivalent). (You're irradiating the side the material is on to crosslink it to the membrane).
12. Put the blot in a sealable plastic pouch. The easiest way to do this is to slide it in on a piece of 3 mm paper without touching the membrane to the pouch, then touch the moist membrane against the bag and allow surface tension to keep the membrane in the bag while you remove the 3 mm paper. You are ready to hybridize.

3.1.3. Hybridization

Hybridize blot with a probe specific to the closed circular DNA you wish to analyze:

1. Prepare a double-stranded, radioactively labeled probe using DNA which is specific to the closed circular DNA you wish to analyze, using any of the many commercially available kits for random-primed labeling.
2. Add hybridization mix to the bag containing your blot (approx 4 mL for one pint bag, 10 ml for one quart, 15 mL for 2 quarts; enough to ensure no bubbles). Push bubbles to top and heat-seal the bag. Place in bath or hybridization oven at desired temperature, usually 65°C, for 5 min to 2 h.
3. Denature labeled double-stranded probe by heating in a screwcap microfuge tube at 95°C for 3 min. Put probe on ice. Add probe to a few milliliters, 2–5 mL, depending on blot size, hybridization buffer in a plastic tube at room temperature.
4. While the probe is denaturing at 95°C, take the blot out of the bath or hybridization oven, slice a corner of the pouch (if you can slice through only one side of the pouch this makes adding the probe easier), and insert a pipet tip to hold the opening open after squeezing or pipeting out hybridization buffer. Use a 5- or 10-mL pipet to add hybridization buffer containing labeled probe when ready and reseal. Replace in bath or hybridization oven and allow to hybridize from 8–24 h.
5. Prepare Church Gilbert wash buffers #1 (500 mL) and #2 (1 L) and 0.2 X SSC/ 0.1% SDS (500 mL) and place at desired temperature (usually 65°C) to prepare for washing blot.
6. Wash blot: slice bag open over a plastic tray, carefully remove membrane from bag, dispose of radioactive bag and buffer, and wash blot 2 X 5 min at room temperature using hot (65°C) Church Gilbert wash #1. Wash again at room temperature 2X 5 min with hot (65°C) Church Gilbert wash #2. Then add fresh wash #2 and place blot in 65°C oven (or bath) 10–25 min (shaking is not necessary). Pour off #2 wash, and wash at 65°C for 10–30 min with 0.2 X SSC/0.1% SDS.
7. Remove blot from tray, lay on 3 mm paper. Wrap in Saran wrap and expose to film or to phosphorimager screen.

3.1.4. Analysis

1. Obtain the intensities of the individual topoisomer bands (*see* **Fig. 1**) either by using a phosphorimager, or by using a densitometer to scan an autoradiogram (*see* **Note 10**).

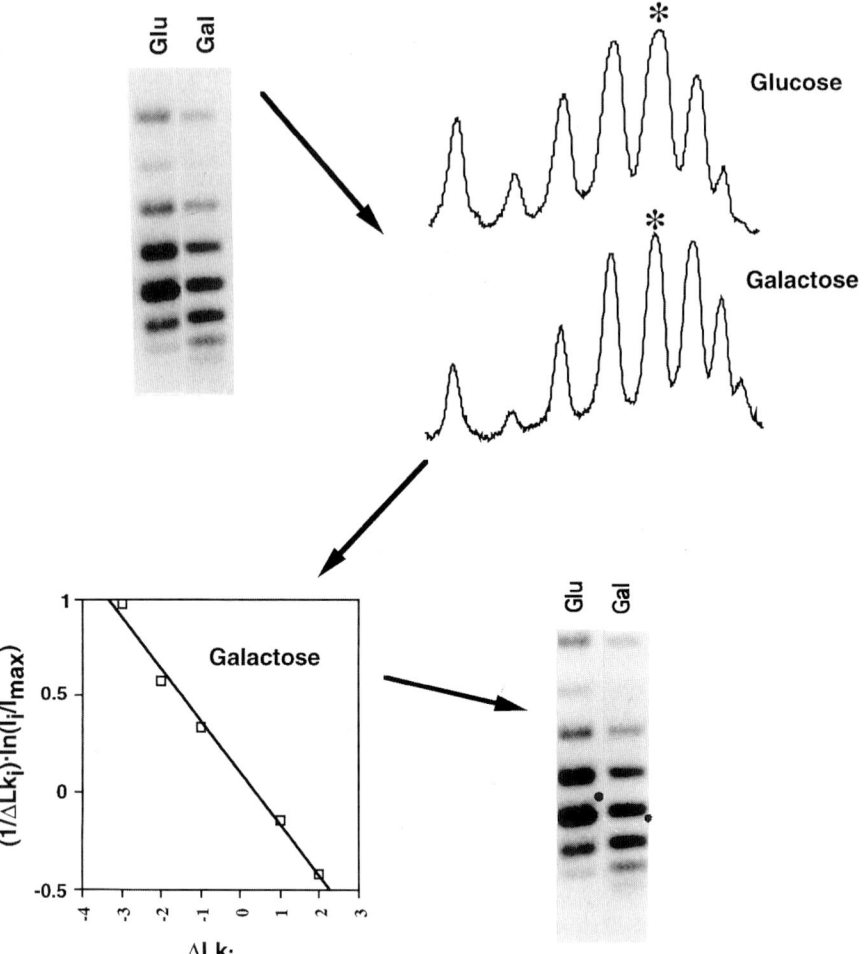

Fig. 1. Finding the centers of topoisomer distributions. Yeast harboring the TALS episome were grown in glucose and galactose medium, as indicated, and TALS topoisomers were resolved on a 1.5% agarose gel containing 40 µg/mL chloroquine diphosphate (*19*; TALS is 1.8 kb). The linear scans which show the relative intensities of the topoisomers were obtained by scanning the autoradiogram on a flatbed scanner and using the public domain program NIH Image (available at http://rsb.info.nih.gov/nih-image/) to trace individual lanes. The bands marked by asterisks are the most intense. (The traces are for illustrative purpose; intensities of individual bands were obtained using a phosphorimager.) Plotting $(1/\Delta Lk_i) \cdot \ln(I_i/I_{max})$ vs. ΔLk_i (for simplicity, only the galactose lane is shown) yields a straight line having slope –0.267 and intercept 0.106 (r^2 = 0.992); thus $b/2m$ = –0.20, and the center is 0.20 downward (more positively supercoiled) from the most abundant topoisomer, as shown by the dot. A similar plot for the glucose lane yields $b/2m$ = 0.447, so the net difference in linking number is 0.6.

2. Use a spreadsheet program such as Cricketgraph to calculate the Gaussian center of the topoisomer distribution in each lane as follows:
    a. Determine the band having the greatest intensity; assign it a relative linking number (Lk) of zero. Assign linking numbers to the bands having more negative linking number (more slowly migrating in a 40 µg/mL chloroquine gel) –1, –2, –3, etc, and the bands corresponding to more positive linking number +1, +2, etc.
    b. Plot $(1/\Delta Lk_i) \cdot \ln(I_i/I_{max})$ versus ΔLk_i, where ΔLk_i corresponds to the assigned relative linking number of the band *i* having relative intensity I_i, and I_{max} is the intensity of the maximally intense band. Note that I_{max} corresponds to $\Delta Lk_i = 0$ and so cannot be used in the plot (**Fig. 1**).
    c. Find the slope (m) and intercept (b) of the resulting straight line (*see* **Notes 11** and **12**). Calculate b/2m; this will be a number between –0.5 and 0.5 if your assignment of the most intense band was correct (that is, if this is the band closest to the center of the distribution). Regardless of whether you correctly assigned the band or not, the exact center of the distribution can be found a distance b/2m from the band you assigned as the most intense, with the direction being towards more negative linking number if b/2m is positive, and towards more positive linking number if b/2m is negative.
    d. You can now calculate differences in topology from one sample to another by examining the relative shifts in linking number from one lane to another, taking care that you have assigned the shift measured by b/2m relative to the correct topoisomer band in each lane (*see* **Note 13**).

3.2. Comparison of Topology Among Species Having Relatively Large Differences in Linking Number

In some cases it may be desirable to compare topology between molecules having larger differences in linking number, for example to determine the average number of nucleosomes present on a minichromosome by comparing the number of supercoils present in plasmid chromatin relative to relaxed, naked DNA *(37)*. In such cases it may be advantageous to use 2-dimensional electrophoresis (**Fig. 2**; **ref. *38***). Normally, molecules which are highly positively or negatively supercoiled under the conditions of electrophoresis are compressed as rapidly migrating species, while those having a nearly relaxed configuration under electrophoresis conditions may be compressed together with the slowly migrating nicked circular molecules. By electrophoresing the same molecules at different chloroquine concentrations, these compressed species can be resolved as individual topoisomers on a single gel. Thus, although this technique does not increase sensitivity, it does allow resolution of more topoisomer species on a single gel which can aid in assessing large linking number differences. This protocol differs from that already given by the following modifications.

1. Preparation of relaxed naked DNA and of "connector" DNA. These can be prepared from purified plasmid DNA (the same species being analyzed), if avail-

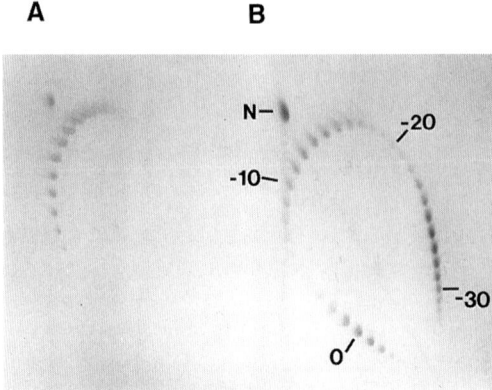

Fig. 2. Using 2-D gel electrophoresis to determine linking number of plasmid chromatin relative to naked DNA. Relaxed, naked DNA was run together with plasmid chromatin SV40 DNA and a "connector" DNA (**B**), and the connector DNA was run alone (**A**). The first dimension was run downward at 1.3 µg/mL chloroquine diphosphate, and the second dimension was run left to right at 0.3 µg/mL chloroquine diphosphate, using 0.7% agarose. Linking numbers, relative to the center of the relaxed, naked DNA sample (assigned a value of zero) are indicated. From **ref. *37***, with permission.

able, or from the material purified as in parts 1–3 above if not (for example for 2 micron plasmid or other plasmids having only yeast sequences). For a truly accurate count of nucleosome number, the same buffer conditions and temperature should be used to relax naked DNA and isolated plasmid chromatin; the preparation of the latter is beyond the scope of this chapter. At the least, the naked DNA should be relaxed at the same temperature as the yeast cells from which plasmid chromatin DNA was prepared were grown (usually 30°C). DNA and minichromosomes can be relaxed using commercially available topoisomerase I under manufacturer's suggested conditions (*see* **Note 14**).

For small plasmids (< 2 kb) it may be possible to compare relaxed DNA and DNA from plasmid chromatin on a single gel; for larger plasmids it is likely that a connector DNA having intermediate average linking number will be needed to "fill the gap" between the topoisomers coming from naked DNA and plasmid chromatin. This may be prepared by relaxing naked DNA in the presence of chloroquine; use 60–80 µg/mL initially, and alter this concentration as needed if the "connector" still leaves a gap.

2. Instead of using 40 µg/mL chloroquine diphosphate in the gel and running buffer, use 0 or 0.3 µg/mL (*see* **Note 15**). Combine relaxed naked DNA and DNA from plasmid chromatin in a single tube along with connector DNA, if needed. If connector DNA is used load a sample in a separate lane (allow two or three lanes separation between samples to allow easier cutting out of the lanes). You may also want to run relaxed naked DNA alone to facilitate identification on the final

2-D blot. Following electrophoresis cut out strips corresponding to the lanes to be analyzed. Soak these in TAE buffer including 1.3 µg/mL chloroquine diphosphate for 1–2 h. Set the agarose strip on a gel plate at the origin (where the comb would normally be) and pour a new gel containing TAE buffer and 1.3 µg/mL chloroquine diphosphate; take care not to pour the gel too hot. Two lanes can be run in this dimension; if you use two allow approx 5 cm in between the agarose strips. Run at 2.5 V/cm for about 15 h. Proceed with blotting, hybridization and analysis as above.

4. Notes

1. Because temperature can affect DNA topology, especially of yeast chromatin (compared to chromatin from higher eukaryotes *[7,26]*), it is important to process samples relatively quickly. Similar caution is recommended for topological analysis from higher eukaryotic organisms *(39)*. Thus, samples should not be removed from the shaker incubator prior to processing, and should not be kept on ice but rather at room temperature while processing. We generally process no more than six to eight samples at a time to allow rapid preparation of DNA. If extremely rapid DNA isolation is needed, for example for kinetic analysis, yeast can first be spheroplasted and allowed to recover in buffer containing 1 *M* sorbitol to preserve osmolarity; DNA can then be isolated extremely rapidly by lysing spheroplasts with hot buffer containing guanidinium hydrochloride *(30)*.
2. We have observed bromphenol blue to interfere with binding of DNA to nylon membranes. We therefore strongly recommend omitting it from the loading dye.
3. To achieve maximum cell breakage, vortex by hand and be sure that glass beads are vigorously distributed throughout the tube during vortexing. (If the beads are mostly staying in the bottom of the tube the vortexing is too mild.)
4. Although a flamed syringe works well with thin wall tubes such as are used for PCR, we have found it much easier to make a hole in a 1.5-mL microfuge tube using a heated dissecting needle.
5. The majority of yeast nucleic acid is RNA. Hence, you should have a visible white pellet after the first precipitation (prior to RNase treatment). If a second precipitation is done a much smaller pellet will be seen. However it is not necessary to reprecipitate following RNase treatment before electrophoresis.
6. We recommend using 40 µg/mL chloroquine diphosphate for plasmid chromatin, as this usually gives the best resolution in our hands. At this chloroquine concentration, DNA topoisomers from plasmid chromatin run as positively supercoiled molecules, and more rapidly migrating species are more positively (= less negatively) supercoiled. This also means that relaxed naked DNA will be compressed as a rapidly migrating (much less negatively supercoiled) species, and that is why different chloroquine concentrations are recommended for comparing topology of relaxed naked DNA to plasmid chromatin DNA (*see* **Subheading 3.2.**). A given plasmid chromatin may be best resolved with a different chloroquine concentration; lowering the chloroquine concentration will first cause a given topoisomer to move upward in the gel (slower mobility) until it has similar

mobility to nicked circular DNA; lowering the chloroquine concentration further will cause topoisomers to migrate with increased mobility and to run as negatively supercoiled molecules. We nearly always can achieve good resolution with either 30 μg or 40 μg/mL chloroquine diphosphate. An advantage of using relatively high concentrations such as this (and the reason that chloroquine is easier to use than ethidium bromide, in addition to its lower toxic/carcinogenic potential) is that no gradient builds up during prolonged electrophoresis.

7. When loading samples, it is a good idea to leave an empty lane between two samples near the right or left hand side of the gel to allow identification of lanes in the resulting autoradiogram. Be sure to note which lane is left empty! Note also that it is crucial to run the gel slowly or else topoisomers will not be well resolved.

8. Ethidium bromide alters DNA linking number by intercalating into DNA and, at the concentrations used for staining DNA in gels, will prevent topoisomers from being resolved. If you have occasion to visualize topoisomers by ethidium staining, the gel must be soaked in an ethidium bromide solution after electrophoresis is complete.

9. To separate the wet gel from the plastic tray, you can slide the tray horizontally if the gel is not too large. If this is difficult (usually), use a spatula to separate an edge of the gel from the tray, then carefully lift the tray away while sliding the spatula under to separate the rest of the gel. If this does not go smoothly you may want to roll out the gel and membrane again to remove bubbles.

10. An advantage the phosphorimager offers here is an increased dynamic range; thus, measured intensities from the phosphorimager can usually safely be assumed to be proportional to topoisomer concentration. If intensities are measured by scanning film with a densitometer, it is important that the film exposure be in the linear response range. Preflashed film or its equivalent (e.g., Kodak XAR film) can be used with optical densities at 600 nm of individual bands in the 0.2–2.0 range.

11. This analysis should yield an excellent fit to a straight line, with $r^2 > 0.97$. It may happen that a topoisomer distribution does not yield a good fit to a straight line because one or two points deviate from what would otherwise be a good fit. This can occur if there is a spot on the autoradiogram or if linearized plasmid, which generally runs a bit faster than nicked circular molecules, is accidentally included in the analysis. It can also be due to using bands which are very faint and which therefore include too much background noise. These defects can be corrected by simply dropping the offending point from the analysis. If more than one or two (at most) points do not fit the straight line, or if it is not obvious which points do not fit, this may be an indication of other problems. For example, the sample may not have been at equilibrium or background may be too high. Repeating the experiment several times will reveal if a sample not fitting a Gaussian distribution is inherent (for example, in a kinetic analysis during changing conditions). If this is so the center can be estimated from a less than optimal straight line fit, or by determining the point in the distribution corresponding to the "center of gravity," the point at which there is equal intensity corresponding to topoisomers of higher and lower linking number.

12. The slope times the length of the closed circular DNA molecule in bp corresponds to the parameter NK. This parameter reflects the breadth of the topoisomer distribution, corrected for plasmid size, and is usually about 400–700 (in bp RT) for plasmid chromatin. Interestingly, this is smaller than the number found for relaxed naked DNA, which is about 1150 for plasmids larger than about 1.5 kb *(21,22)*. This reflects a broader distribution for plasmid chromatin than for naked DNA, which is likely to be due to heterogeneity in nucleosome number or structure and may reflect a dynamic structure for plasmid chromatin in vivo *(34)*.
13. To obtain quantifiable results with standard deviations, it is necessary to repeat topoisomer analysis in independent experiments. It should go without saying that results from analysis of a single isolation and analysis of DNA topology should never be presented.
14. If relaxation of naked DNA goes to completion, an excellent fit to a Gaussian distribution should be obtained (*see* **Note 11**), with NK about 1150 bp · RT (*see* **Note 12**). If a trailing edge of topoisomers is observed either with naked DNA or plasmid chromatin relaxation by topoisomerase was probably incomplete.
15. At these chloroquine concentrations, naked DNA molecules will migrate as positively supercoiled species, and DNA from plasmid chromatin migrates as negatively supercoiled species. For a given plasmid, it may be necessary to experiment with different chloroquine concentrations in one or both dimensions.

Acknowledgment

I thank Dr. Len Lutter for providing **Fig. 2**. This work was supported by NIH grant GM51993.

References

1. Keller, W. and Wendel, I. (1974) Stepwise relaxation of supercoiled SV40 DNA. *Cold Spring Harbor Symp. Quant. Biol.* **39,** 199–208.
2. Germond, J. E., Hirt, B., Oudet, P., Gross-Bellard, M., and Chambon, P. (1975) Folding of the DNA double helix in chromatin-like structures from simian virus 40. *Proc. Natl. Acad. Sci. USA* **72,** 1843–1847.
3. Simpson, R. T., Thoma, F., and Brubaker, J. M. (1985) Chromatin reconstituted from tandemly repeated cloned DNA fragments and core histones: a model system for study of higher order structure. *Cell* **42,** 799–808.
4. Norton, V. G., Imai, B. S., Yau, P., and Bradbury, E. M. (1989) Histone acetylation reduces nucleosome core particle linking number change. *Cell* **57,** 449–457.
5. Keller, W., Muller, U., Eicken, I., Wendel, I., and Zentgraf, H. (1978) Biochemical and ultrastructural analysis of SV40 chromatin. *Cold Spring Harbor Symp. Quant. Biol.* **42,** 227–243.
6. Pederson, D. S., Venkatesan, M., Thoma, F., and Simpson, R. T. (1986) Isolation of an episomal yeast gene and replication origin as chromatin. *Proc. Natl. Acad. Sci. USA* **83,** 7206–7210.
7. Morse, R. H., Pederson, D. S., Dean, A., and Simpson, R. T. (1987) Yeast nucleosomes allow thermal untwisting of DNA. *Nucleic Acids. Res.* **15,** 10,311–10,330.

8. White, J. H., Cozzarelli, N. R., and Bauer, W. R. (1988) Helical repeat and linking number of surface wrapped DNA. *Science* **241,** 323–327.
9. Hayes, J. J., Tullius, T. D., and Wolffe, A. P. (1990) The structure of DNA in a nucleosome. *Proc. Natl. Acad. Sci. USA* **87,** 7405–7409.
10. Freeman, L. T. and Garrard, W. T. (1992) DNA supercoiling in chromatin structure and gene expression. *Crit. Rev. Euk. Exp.* **2,** 165–209.
11. Lenfant, F., Mann, R. K., Thomsen, B., Ling, X., and Grunstein, M. (1996) All four core histone N-termini contain sequences required for the repression of basal transcription in yeast. *EMBO J.* **15,** 3974–3985.
12. Smith, M. M., Yang, P., Santisteban, M. S., Boone, P. W., Goldstein, A. T., and Megee, P. C. (1996) A novel histone H4 mutant defective in nuclear division and mitotic chromosome transmission. *Mol. Cell. Biol.* **16,** 1017–1026.
13. Wechsler, M. A., Kladde, M. P., Alfieri, J. A., and Peterson, C. L. (1997) Effects of Sin- versions of histone H4 on yeast chromatin structure and function. *EMBO J.* **16,** 2086–2095.
14. Kim, U. J., Han, M., Kayne, P., and Grunstein, M. (1988) Effects of histone H4 depletion on the cell cycle and transcription of Saccharomyces cerevisiae. *EMBO J.* **7,** 2211–2219.
15. Norton, V. G., Marvin, K. W., Yau, P., and Bradbury, E. M. (1990) Nucleosome linking number change controlled by acetylation of histones H3 and H4. *J. Biol. Chem.* **265,** 19,848–19,852.
16. Morse, R. H. and Cantor, C. R. (1986) Effect of trypsinization and histone H5 addition on DNA twist and topology in reconstituted minichromosomes. *Nucleic Acids Res.* **14,** 3293–3310.
17. Thomsen, B., Bendixen, C., and Westegaard, O. (1991) Histone hyperacetylation is accompanied by changes in DNA topology in vivo. *Eur. J. Biochem.* **201,** 107–111.
18. Lutter, L. C., Judis, L., and Paretti, R. F. (1992) The effects of histone acetylation on chromatin topology in vivo. *Mol. Cell. Biol.* **12,** 5004–5014.
19. Stafford, G. A. and Morse, R. H. (1997) Chromatin remodeling by transcriptional activation domains in a yeast episome. *J. Biol. Chem.* **272,** 11,526–11,534.
20. Wong, J., Shi, Y.-B., and Wolffe, A. P. (1997) Determinants of chromatin disruption and transcriptional regulation instigated by the thyroid hormone receptor: hormone-regulated chromatin disruption is not sufficient for transcriptional activation. *EMBO J.* **16,** 3158–3171.
21. Depew, R. E. and Wang, J. C. (1975) Conformational fluctuations of DNA helix. *Proc. Natl. Acad. Sci. USA* **72,** 4275–4280.
22. Pulleyblank, D. E., Shure, M., Tang, D., Vinograd, J., and Vosberg, H.-S. (1975) Action of nicking-closing enzyme on supercoiled and nonsupercoiled closed circular DNA: formation of a Boltzmann distribution of topological isomers. *Proc. Natl. Acad. Sci. USA* **72,** 4280–4284.
23. Morse, R. H. and Cantor, C. R. (1985) Nucleosome core particles suppress the thermal untwisting of core DNA and adjacent linker DNA. *Proc. Natl. Acad. Sci. USA* **82,** 4653–4657.

24. Ambrose, C., McLaughlin, R., and Bina, M. (1987) The flexibility and topology of simian virus 40 DNA in minichromosomes. *Nucleic Acids Res.* **15**, 3703–3721.
25. Lutter, L. C. (1989) Thermal unwinding of simian virus 40 transcription complex DNA. *Proc. Natl. Acad. Sci. USA* **86**, 8712–8716.
26. Saavedra, R. A. and Huberman, J. A. (1986) Both DNA topoisomerases I and II relax 2µ plasmid DNA in living yeast cells. *Cell* **45**, 65–70.
27. White, J. H., Gallo, R., and Bauer, W. R. (1989) Dependence of the linking deficiency of supercoiled minichromosomes upon nucleosome distortion. *Nucleic Acids Res.* **17**, 5827–5835.
28. White, J. H., Gallo, R. and Bauer, W. R. (1989) Effect of nucleosome distortion on the linking deficiency in relaxed minichromosomes. *J. Mol. Biol.* **207**, 193–199.
29. Bauer, W. R., Hayes, J. J., White, J. H., and Wolffe, A. P. (1994) Nucleosome structural changes due to acetylation. *J. Mol. Biol.* **236**, 685–690.
30. Pederson, D. S. and Morse, R. H. (1990) Effect of transcription of yeast chromatin on DNA topology in vivo. *EMBO J.* **9**, 1873–1881.
31. Bauer, W. and Vinograd, J. (1970) Interaction of closed circular DNA with intercalative dyes. II. The free energy of superhelix formation in SV40 DNA. *J. Mol. Biol.* **47**, 419–435.
32. Hsieh, T.-S. and Wang, J. C. (1975) Thermodynamic properties of superhelical DNAs. *Biochemistry* **14**, 527–535.
33. Bates, A. D. and Maxwell, A. (1993) *DNA Topology*. IRL, Oxford University Press, Oxford, UK.
34. Morse, R. H. (1991) Topoisomer heterogeneity of plasmid chromatin in living cells. *J. Mol. Biol.* **222**, 133–137.
35. Sambrook, J., Fritsch, E. F., and Maniatis, T. (1989) *Molecular Cloning: A Laboratory Manual* Cold Spring Harbor Laboratory Press, Cold Spring Harbor, NY.
36. Church, G. and Gilbert, W. (1984) Genomic sequencing. *Proc. Natl. Acad. Sci. USA* **81**, 1991–1995.
37. Drabik, C. E., Nicita, C. A., and Lutter, L. C. (1997) Measurement of the linking number change in transcribing chromatin. *J. Mol. Biol.* **267**, 794–806.
38. Lee, C.-H., Mizusawa, H., and Kakefuda, T. (1981) Unwinding of double-stranded DNA helix by dehydration. *Proc. Natl. Acad. Sci. USA* **78**, 2838–2842.
39. Givens, R. M., Saavedra, R. A., and Huberman, J. A. (1996) Topological complexity of SV40 minichromosomes. *J. Mol. Biol.* **257**, 53–65.

29

DNA Methyltransferases as Probes for Chromatin Structure in Yeast

Michael P. Kladde, Mai Xu, and Robert T. Simpson

1. Introduction

Expression of DNA methyltransferases (MTases) in yeast, which has a naturally unmethylated genome, enables the study of chromatin structure in intact cells. Initial studies, employing in vivo expression of *dam* MTase, which catalyzes the production of N^6-methyladenine (^{6me}A) at GATC sites, were limited by the occurrence of a target site only once every ~300 bp in native yeast sequences *(1,2)*. This limited resolution of *dam* MTase, although possible to circumvent through the introduction of additional target sites *(3)*, poses an obvious barrier to its usefulness. In addition, with no existing genomic sequencing methodology for detection of ^{6me}A, the methylation status of potential target sites must be ascertained with methylation-sensitive restriction endonucleases (e.g., *Dpn*I and *Dpn*II).

A significant advance was made possible with the cloning of the gene encoding M.*Sss*I, which catalyzes *de novo* modification of CG sites with 5-methylcytosine (^{5me}C) *(4)*. Expression of M.*Sss*I in yeast has provided an eightfold increase in resolution over *dam*, with one native site occurring about every 35 bp *(5)*. The operational resolution is actually greater as steric occlusion of MTases occurs in the vicinity of DNA-bound factors, as well as within, their specific binding sites. Detection of methylation, and hence the degree of accessibility of each CG site to M.*Sss*I in chromatin relative to protein-free DNA, is determined by the elegant strategy of Frommer et al. which leads to a positive display of ^{5me}C following bisulfite treatment of isolated DNA (**Fig. 1**) *(6,7)*.

Using M.*Sss*I to probe chromatin structure offers a number of advantages over conventional methodologies:

From: *Methods in Molecular Biology, Vol. 119: Chromatin Protocols*
Edited by: P. B. Becker © Humana Press Inc., Totowa, NJ

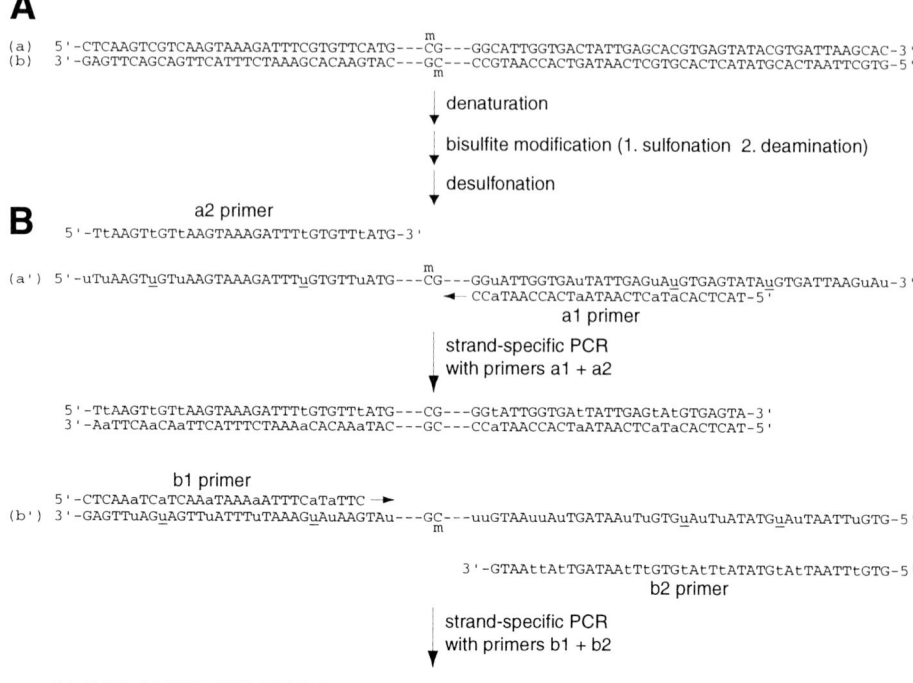

Fig. 1. Detection of 5-methylcytosine by bisulfite deamination. (**A**) DNA containing 5-methylcytosine (^{m}C) is denatured to upper (a) and lower (b) single-strands and treated with sodium metabisulfite. During the bisulfite treatment, cytosines are first sulfonated to produce cytosine sulfonate and then deaminated to produce a sulfonyl adduct of uracil. The DNA is treated with alkali to effect desulfonation, removal of the sulfonyl moiety. (**B**) In the resulting modified upper (a') and lower (b') DNA strands, each cytosine, but not 5-methylcytosine, is converted to uracil (indicated in lower case). Underlined uracil residues denote a mixture of predominantly uracil and some 5-methyl-cytosine at positions of CG target sites. In naming conventions of Frommer et al. (*6*), the region of interest from each of the deaminated DNA strands, which are no longer complementary, is amplified in separate PCR reactions by primer pairs 'a1' and 'a2,' and that of 'b1' and 'b2,' respectively. In the design of the a2 and b2 primers, all cytosines are changed to thymines (C to t). Conversely, the a1 and b1 primers are synthesized with transitions of each guanine to adenine (G to a). Cytosines, remaining in the population of amplified PCR products, are detected by direct thermal cycle sequencing using ^{32}P end-labeled primers a1 or b1 in the presence of ddGTP. All the indicated primers yield high quality footprints and representative data generated with the b1 and b2 primers is shown in **Fig. 2**.

1. Expression of M.*Sss*I does not alter the growth characteristics of yeast, allowing experiments to be performed in living cells.
2. Nuclear DNA is not damaged (e.g., alkylated) or degraded.

Footprinting with DNA Methyltransferases

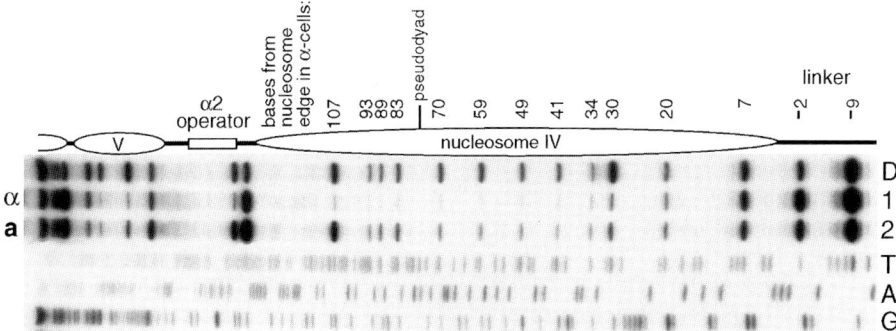

Fig. 2. In vivo detection of positioned nucleosomes and DNA-bound α2 repressor by M.SssI. A minichromosome containing the α2 operator was transformed into isogenic **a**- and α-cells. Previously, in nuclei, micrococcal nuclease demonstrated the presence of positioned nucleosomes (indicated by ellipses) adjacent to the operator in α-cells, but not in **a**-cells *(18)*. The top of the gel is on the left. Note the considerable blockage of methylation in α-cells (lane 1) relative to protein-free DNA (D) that begins at 30 bp from the nucleosome edge and becomes more apparent at the pseudodyad, indicating that M.SssI can be used *in vivo* to detect positioned nucleosomes. The protection against methylation in **a**-cell chromatin (lane 2) relative to naked DNA is not as dramatic, indicating the presence of nucleosomes which are not as organized as in α-cells. Within the operator, note protection of the CG site in α-cells, which express the α2 repressor, relative to control DNA and **a**-cells where the repressor is absent.

3. Preparation of nuclei is avoided thereby averting loss of DNA-bound proteins or complexes.
4. Nucleosomes and DNA-nonhistone interactions can be detected simultaneously.
5. The use of PCR enables a reduction in yeast culture size of 100-fold, making it feasible to handle many samples in a single experiment.

We have focused on chromatin analysis with M.SssI in this chapter, the same protocols can be applied when employing additional cytosine-5-MTases that may be discovered in the future. The potential availability of enzymes that recognize short target sequences would also lead to further increases in resolution and the general utility of the footprinting strategy.

2. Materials

All procedures require a supply of:

1. Distilled deionized water, sterile and nonsterile.
2. Sterile 1.5- and 0.5-mL microcentrifuge tubes and micropipet tips.
3. Sterile, disposable 15- and 50-mL polypropylene, conical centrifuge tubes.

Store all enzymes at −20°C in a nondefrosting freezer and all other reagents at room temperature, unless indicated otherwise.

2.1. Construction of Yeast Strains Expressing DNA Methyltransferases

2.1.1. Isolation and Preparation of MTase Expression Plasmid DNA

1. 2X YT medium: 10 g yeast extract, 16 g Bacto tryptone, 5 g NaCl/L. Dissolve components in distilled water, measure to volume, pH to 7.5, and autoclave for 20–30 min. For plates, add 15 g agar/L just prior to autoclaving. After autoclaving, cool until comfortable to the touch, add 2 mL of 50 mg/mL ampicillin stock (*see* **item 2**) per L, stir thoroughly, and pour plates aseptically. After cooling, invert the plates and leave overnight at room temperature. Plates can be wrapped and stored for 4–6 wk at 4°C. Liquid medium can be stored at room temperature. For growing *Escherichia coli* harboring plasmids, ampicillin should be added to liquid medium just prior to use.
2. Ampicillin stock solution (50 mg/mL): Titrate with concentrated NaOH to dissolve completely and then sterilize by filtration. The antibiotic solution is stable for approximately 6 mo when stored at –20°C. *Warning:* Wear gloves and a particle mask to avoid unnecessary exposure which can lead to hypersensitivity.
3. Appropriate strain of *E. coli* (*see* **Note 1**) made competent for DNA transformation.
4. Plasmid pMPK1-*Sss*I-19 which, under control of the yeast *GAL1* promoter, expresses M.*Sss*I with the SV40 nuclear localization sequence fused to its N-terminus. The plasmid is designed for integration as a single-copy at a *lys2* locus.
5. Plasmid pMPK1. The expression vector lacking M.*Sss*I coding sequences. The plasmids and sequences of pMPK1 and pMPK1-*Sss*I-19 are available upon request.
6. Sterile 50% (v/v) glycerol. Sterilize by filtration and store at 4°C.
7. Suitable tubes for cryogenic storage at –70°C.
8. Promega Wizard™ Minipreps DNA purification kit. *Caution:* The resin contains 7 M guanidine-HCl which is an irritant. Wear gloves and eye protection when handling.
9. 3-mL Luer-lok syringes.
10. Restriction endonuclease *Xho*I.
11. NEB2 10X restriction enzyme buffer: 500 mM NaCl, 100 mM Tris-HCl, pH 7.9, 100 mM MgCl$_2$, 10 mM dithiothreitol. Supplied with *Xho*I by New England Biolabs. Store at –20°C or 4°C.
12. 10.0 M NH$_4$OAc. The pH should be approx 7.8 and does not need to be adjusted.
13. Absolute ethanol.
14. 10X gel loading buffer: 15% (w/v) Ficoll 400, 0.25% bromophenol blue and/or 0.25% xylene cyanole FF. Make one with both dyes and one containing only the xylene.
15. DNA size-markers: A mixture containing 0.1 mg/mL of each λ and ΦX174 RF bacteriophage DNA digested with *Hin*dIII and *Hae*III, respectively, and 1X gel loading buffer containing both dyes. Store at 4°C.
16. 0.1X TE: 1.0 mM Tris-HCl, pH 8.0, 0.1 mM EDTA. Sterilize by autoclaving or filtration.
17. 70% ethanol:30% 0.1X TE: Add 13 mL 0.1X TE to a sterile 50-mL conical polypropylene tube and bring to 50 mL with absolute ethanol.

Footprinting with DNA Methyltransferases

18. Agarose with a low coefficient of electroendosmosis.
19. 20,000X ethidium bromide: 10 mg/mL, store in foil-wrapped bottle. *Caution:* Ethidium bromide is mutagenic/possibly carcinogenic. Handle with gloves and consult your institution about safe disposal.
20. 50X TAE gel running buffer: 242 g Trizma base, 57 mL glacial acetic acid, 16.8 g EDTA (anhydrous)/L.
21. 0.7% (w/v) agarose in 1X TAE. Add ethidium bromide to a final concentration of 0.5 µg/mL.
22. Apparatus for horizontal, submerged agarose gel electrophoresis and power supply.
23. Transilluminator for viewing DNA gels (i.e., emitting UV light near 300 nm).

2.1.2. Creating a lys2 Selectable Marker in a Yeast Strain of Interest

1. Appropriate strain(s) of *Saccharomyces cerevisiae* (*see* **Note 2**).
2. YPD: 10 g yeast extract, 20 g peptone, 20 g glucose/L. Dissolve in distilled water, measure to volume, and autoclave for 20 min. For plates, add 20 g agar/L prior to autoclaving. Pour plates aseptically, invert after cooling, and dry at room temperature for 2–3 d. Plates can be wrapped and stored for at least several months at 4°C. Liquid medium can be stored at room temperature.
3. YPG (4% galactose): 10 g yeast extract, 20 g peptone, 40 galactose/L. The galactose should be of high quality (≥ 98%) to avoid contamination with glucose. Autoclave for 20 min.
4. Complete synthetic medium (CSM): 6.7 g Bacto-yeast nitrogen base with ammonium sulfate and without amino acids, 20 g glucose or galactose, and appropriate CSM supplement mix (Bio101, La Jolla, CA; the amount to be added is indicated on the label of the container). Dissolve all components to volume and autoclave for 20 min. Prepare and store plates as indicated in item 2. Media lacking any defined nutrient (i.e., dropout media) can be prepared this way; hence, CSM-URA is complete synthetic medium minus uracil. Plates to be used in selection of prototrophs following yeast electroporation should also contain 1.0 *M* sorbitol which is added prior to autoclaving. To construct the *lys2*-Δ1 marker, plates 1.0 *M* in sorbitol of both CSM-URA and CSM-LEU will be needed.
5. Plasmid p*lys2*::R-*URA3*-R *(5)*. The plasmid and its sequence is available on request.
6. Restriction endonucleases, *Not*I and *Xho*I.
7. Plasmid pHM53 *(8)*.
8. 1.0 *M* sorbitol: Autoclave for 20 min and store at 4°C.
9. Sterile, disposable 0.45 µm-pore filter unit, 100-mL capacity.
10. Electroporation apparatus and cuvets (*see* **Note 3**).
11. Means of replica plating. We usually use sterile wood applicators.
12. Sterile 50% (v/v) glycerol. See **Subheading 2.1.1.**, **item 6**.
13. Suitable tubes for cryogenic storage at –70°C.

2.1.3. Integration of MTase Expression Plasmid in Yeast

1. Appropriate strains of *S. cerevisiae* (*see* **Note 4**).
2. Prepared pMPK1-*Sss*I-19 and/or pMPK1. See **Subheading 3.1.1.**

3. YPD, plates and liquid medium. *See* **Subheading 2.1.2.**, **item 2**.
4. CSM-LYS, 1.0 M sorbitol plates. *See* **Subheading 2.1.2.**, **item 4**.
5. Sterile 1.0 M sorbitol. Store at 4°C.
6. Electroporation apparatus and cuvettes (*see* **Note 3**).

2.2. Screening for Yeast Strains Expressing Functional Cytosine-5-DNA MTases

2.2.1. Isolation of Total DNA from Yeast

1. Liquid YPG (4% galactose) or suitable CSM dropout (4% galactose) medium.
2. Sterile 1.5-mL microcentrifuge tubes with screw caps.
3. TE: 10 mM Tris-HCl, pH 8.0, 1 mM EDTA. Sterilize by autoclaving or filtration.
4. Acid-washed glass beads (425–600 µm in diameter).
5. Smash buffer: 10 mM Tris-HCl, pH 8.0, 2% Triton X-100, 1 mM EDTA, 1% SDS, 100 mM NaCl.
6. Phenol saturated with 0.1 M Tris-HCl, pH 8.0. Store at 4°C. Caution: Highly corrosive and toxic. Wear gloves and safety glasses. Dispose of properly.
7. Chloroform:isoamyl alcohol ($CHCl_3$:IAA). 24:1 (v/v). Chloroform is a suspected carcinogen. Wear gloves and safety glasses. Dispose of appropriately.
8. Vortexer with platform and insert capable of holding several 1.5-mL microcentrifuge tubes.

2.2.2. Deamination of DNA by Treatment with Bisulfite

1. Degassed, distilled water. Fill a bottle to full capacity with boiling distilled water, seal air-tight, and cool to room temperature. The water can be degassed a few days in advance.
2. Glass scintillation vials, 20-mL capacity.
3. 3 N NaOH. Make fresh just before use. Weigh a few NaOH pellets and dissolve in appropriate volume of the degassed, distilled water (0.4 g NaOH makes 3.3 mL). Avoid excessive vortexing which will cause aeration of this and the following two solutions. *Caution:* Concentrated solutions of NaOH are caustic. Wear gloves and eye protection.
4. 3 mg/mL sheared, calf thymus DNA. Dissolve in sterile TE (*see* **Subheading 2.2.1.**, **item 3**) and shear 10X with a 25 gage needle. Store at 4°C.
5. 0.5 M EDTA, pH 8.0. Add solid EDTA to distilled water, titrate to pH 8.0 with NaOH, and sterilize by filtration or autoclaving.
6. 3X sample denaturation buffer. Just before use, make an appropriate volume (depending on the number of samples) of reagent in the ratio of: 3 µL freshly prepared 3 N NaOH:0.7 µL 3 mg/mL sheared, calf thymus DNA:0.5 µL 0.5 M EDTA, pH 8.0:5.8 µL distilled water (i.e., 10 µL denaturation buffer will be added to each sample).
7. 100 mM hydroquinone: Make fresh just before use. Weigh approximately 0.05 g hydroquinone and dissolve in appropriate volume of the degassed, distilled water (0.05 g hydroquinone makes 4.5 mL). *Caution:* Hydroquinone is toxic and a suspected carcinogen, wear gloves and eye protection. Do not breath dust. Dispose of safely.

8. Saturated sodium metabisulfite solution (pH 5.0). Make fresh just before use. Put a small stir bar into the scintillation vial, add 100 µL of the freshly made 100 mM hydroquinone, and then dump in approx 5 g of sodium metabisulfite from a previously unopened vial (*see* **Note 5**). Add 7 mL of the degassed, distilled water, stir immediately, and quickly add 1.2 mL of the freshly made 3 N NaOH solution. At this point, the pH at room temperature should be very close to 5.0 and should be adjusted to pH 5.0 with the 3 N NaOH. Prewarm the solution to 50°C before adding to samples. A variable (depending on the weight of sodium metabisulfite in a particular vial) of amount of undissolved solid will usually be present in the final solution. *Caution:* Bisulfite-containing solutions are mutagenic, wear gloves and safety glasses. Dispose of solutions properly.
9. Mineral oil.
10. 6 M guanidine thiocyanate. *Caution:* Irritant, wear gloves and eye protection.
11. Promega Wizard PCR Preps resin (catalog #A7181) (*see* **Note 6**). *Caution:* The resin contains 6M guanidine thiocyanate which is an irritant.
12. Promega Wizard Minicolumns (catalog #A7211).
13. Promega Vac-Man Laboratory Vacuum Manifold (catalog #A7231).
14. 3-mL Luer-Lok syringes.
15. 80% isopropanol.
16. Sterile 0.1X TE, pH 8.0.
17. Desulfonation solution: Just before use, make up an appropriate volume in the ratio of: 7 µL 3 N NaOH:1 µL 3 mg/mL sheared, calf thymus DNA (i.e., 8 µL of this solution will be added to each sample).
18. 10 0 M NH$_4$OAc (~7.8). *See* **Subheading 2.1.1., item 12**.
19. Absolute ethanol.
20. 70% ethanol:30% 0.1X TE, pH 8.0. *See* **Subheading 2.1.1., item 17**.

2.2.3. PCR Amplification from Deaminated DNA to Identify Yeast Strains Expressing M.SssI (see **Note 7**)

1. Oligonucleotides MKO2b1 (5'-CTCGTCTTCACCAATCGCG-3') and MKO2b2 (5'-TGTATTTATATATTTGTTAATAGATTAAAAATATCGTTTCGT-3') to be used as PCR primers (*see* **Note 8**).
2. dNTP mix: 2.5 mM in dATP, dCTP, dGTP, and dTTP diluted in sterile, distilled water (*see* **Note 9**). Store at –20°C in aliquots of 50-100 µL.
3. 10X Taq DNA polymerase buffer: 100 mM Tris-HCl, pH 8.3, 500 mM KCl, 30 mM MgCl$_2$, 0.5% NP-40, 0.5% Tween 20.
4. Taq DNA polymerase (*see* **Note 10**).
5. Ampliwax PCR Gem 50 wax beads (Perkin-Elmer catalog #N808-0150).
6. PCR thermocycler.
7. 50X TAE gel running buffer. *See* **Subheading 2.1.1., item 20**.
8. 1.5% agarose buffered with 1X TAE and containing 0.5 µg/mL ethidium bromide (*see* **Subheading 2.1.1., item 21**).
9. DNA size-markers (**Subheading 2.1.1., item 15**).

10. 10X gel loading buffer containing xylene cyanole FF dye only (**Subheading 2.1.1., item 14**).
11. Transilluminator for viewing DNA gels (i.e., emitting UV light near 300 nm).
12. Apparatus for horizontal, submerged agarose gel electrophoresis and power supply.
13. Sterile 50% (v/v) glycerol. *See* **Subheading 2.1.1., item 6**.
14. Suitable tubes for cryogenic storage at –70°C.

2.3. In Vivo Footprinting in Yeast using M.SssI

2.3.1. Yeast Cell Culturing and Isolation of Total DNA

1. Appropriate yeast strain(s) expressing M.*Sss*I or other suitable cytosine-5-DNA MTase (*see* **Note 11**).
2. YPD plates. *See* **Subheading 2.1.2., item 2**.
3. All items listed in **Subheading 2.2.1.** for isolating total yeast DNA.

2.3.2. In Vitro Methylation of DNA with M.SssI

1. *E. coli* GM2163 (*dam⁻ dcm⁻*) for isolation of methylation-free plasmid DNA (*see* **Note 12**).
2. Plasmid or PCR fragment amplified from wild-type DNA that contains the region of DNA under study.
3. NEB2 10X restriction enzyme buffer. *See* **Subheading 2.1.1., item 11**.
4. M.*Sss*I DNA methyltransferase (New England Biolabs).
5. 32 mM S-adenosyl methionine (SAM). Supplied with the M.*Sss*I. Store at –20°C.
6. 0.1X TE. *See* **Subheading 2.1.1., item 16**.

2.3.3. Deamination of DNA

See **Subheading 2.2.2.**

2.3.4. PCR Amplification from Deaminated DNA for Footprinting a Region of Interest

The same materials indicated in **Subheading 2.2.3.**, excluding the oligonucleotides MKO2b1 and MKO2b2, will be required. Instead of a 1.5% agarose gel, one of 0.7–1.0% should be used.

You will also need:

1. Appropriate oligonucleotides (*see* **Fig. 1** and **Notes 9** and **14**) for amplifying the region under study from total yeast and control DNA which have been previously been treated with bisulfite.
2. Promega Wizard PCR Preps resin (*see* **Note 6**). *Caution:* The resin contains 6 M guanidine thiocyanate which is an irritant. Wear gloves and safety glasses.
3. Promega Wizard PCR Preps minicolumns.
4. Promega Wizard PCR Preps direct PCR product purification buffer: 50 mM KCl, 10 mM Tris-HCl, pH 8.8, 1.5 mM MgCl$_2$, 0.1% Triton X-100.
5. 3-mL Luer-lok syringes.
6. 80% isopropanol.
7. 0.1X TE, pH 8.0. *See* **Subheading 2.1.1., item 16**.

2.3.5. Direct Sequencing of Purified PCR Products by Thermocycling

1. Appropriate oligonucleotide to be used as a sequencing primer (*see* **Note 13** and **Fig. 1**).
2. [γ-^{32}P]ATP (3000 Ci/mmole). *Caution:* Use procedures for the safe handling of radioisotopes.
3. 10X T4 polynucleotide kinase buffer: 700 mM Tris-HCl, pH 7.6, 100 mM MgCl$_2$, 50 mM dithiothreitol.
4. T4 polynucleotide kinase.
5. Sephadex G-50 spin columns: Rehydrate Sephadex G-50 in sterile 0.1X TE, pH 8.0, producing a thick slurry. Tightly pack a 1-mm thick plug of sterile, siliconized glass wool in a 1-mL syringe barrel and, using a Pasteur pipet, fill the syringe completely with slurry. The columns (and unused slurry) can be stored for months at 4°C after wrapping with parafilm. Prior to use, warm to room temperature, remove the stopper and parafilm, and suspend inside a disposable, 15-mL conical tube. Centrifuge at 1000g, discard the eluate, add 100 µL 0.1X TE, and repeat the centrifugation. Discard the eluate and place a sterile 1.5-mL screw cap microcentrifuge tube inside the 15-mL conical tube and insert the column. *Note:* If columns are purchased commercially, it is important to re-equilibrate them with three washes of sterile 0.1X TE, pH 8.0.
6. Thermocycler (preferably with a heated lid).
7. Sequitherm DNA polymerase (Epicentre Technologies, Madison, WI). The remaining items should be stored at –20°C and, after thawing, kept on ice.
8. 10X Sequitherm DNA polymerase buffer: 0.5 M Tris-HCl, pH 9.3, 25 mM MgCl$_2$.
9. 50X footprinting dNTP mix: 250 mM each dATP, dCTP, and dTTP (dGTP is omitted) in sterile, distilled water (*see* **Note 9**).
10. 100X ddGTP. 5 mM in sterile, distilled water (*see* **Note 9**).
11. 3X sequencing termination mixes: Each mix contains 15 mM each dATP, dCTP, dTTP, and 7-deaza-dGTP. In addition, respective termination mixes contain: A mix, 0.23 mM ddATP; C mix, 0.3 mM ddCTP; and T mix, 0.9 mM ddTTP. (The G mix is not useful in these studies because most of the cytosines are removed from the DNA by the bisulfite treatment.)
12. 3X sequencing reaction stop/loading buffer: 95% (v/v) deionized formamide, 10 mM EDTA, pH 7.6, 0.025% xylene cyanole FF and bromophenol blue.
13. Reagents and equipment for sequencing gels.

3. Methods

All procedures are to be performed at room temperature unless specified otherwise.

3.1. Constructing MTase-Expressing Strains of S. cerevisiae

3.1.1. Isolation and Preparation of M.SssI Expression Plasmid DNA

1. Introduce the M.*Sss*I expression plasmid, pMPK1-*Sss*I-19, into a suitable strain of *E. coli* (*see* **Note 1**) made competent for transformation.

2. Spread the transformed cells on a 2X YT plate containing 100 μg/mL ampicillin and incubate overnight at 37°C.
3. Select a single colony and inoculate 7 mL of 2X YT containing 100 μg/mL ampicillin and incubate with shaking (300 rpm) overnight at 37°C.
4. The next day, make a stock of the cells for cryogenic storage by pipeting 0.7 mL of the culture into a suitable cryovial, add 0.35 mL 50% glycerol, vortex well, and store at –70°C.
5. Isolate the plasmid DNA from the cells using Promega Minipreps according to the manufacturer's suggestions except, for each column to be used, process 3 mL of cells and elute the plasmid DNA from the column with 50 μL 0.1X TE, prewarmed to 70°C. Store the eluted DNA at 4°C.
6. Electrophorese 3 μL miniprep DNA and 5 μL DNA size-markers on a 0.7% agarose gel. View on a transilluminator to estimate the DNA concentration.
7. Digest the DNA with *Xho*I as specified by the manufacturer.
8. Ascertain that the plasmid has been completely linearized by checking a small aliquot of the digestion on a gel in parallel to the input, uncut DNA.
9. If necessary (*see* **Note 14**), ethanol precipitate the DNA.

3.1.2. Creating a lys2 Selectable Marker in a Yeast Strain of Interest (see **Note 2**)

This section can be skipped if your experimental strain already contains a *lys2* marker. Asepsis should be maintained throughout the procedure.

1. Isolate plasmids p*lys2*::R-*URA3*-R *(5)* and pHM53 *(8)* as indicated in **Subheading 3.1.1.**
2. Digest plasmid p*lys2*::R-*URA3*-R with *Not*I and *Xho*I (*see* **Note 15**).
3. Check a small aliquot of the digestion on a 0.7% TAE agarose gel to ascertain complete cutting and store at –20°C.
4. Inoculate 10 mL YPD in a sterile, 50-mL conical tube with a Lys⁺ Gal⁺ *ura3 leu2* strain of *S. cerevisiae* (*see* **Note 2**). Leave the lid loose and secure it with tape to allow aeration.
5. The next morning, the culture should have an optical density at 600 nm (OD_{600}) of approx 1–2. Add 10 mL YPD to the cells and incubate 2 h at 30°C with shaking.
6. Prepare the cells for electroporation by washing 3X with 25 mL of ice-cold 1.0 *M* sorbitol (*see* **Note 16**). Keep at 4°C.
7. On ice, pipet 40 μL of the cell suspension into a sterile 1.5-mL microcentrifuge tube that contains 5 μL of the *Not*I/*Xho*I-digested p*lys2*::R-*URA3*-R.
8. Carefully avoiding bubbles, gently mix the DNA and cells and transfer the sample between the electrodes of a sterile electroporation cuvette.
9. Deliver a pulse of 1 kV at 330 μF capacitance with the electroporator.
10. Quickly wash the cells off the electrodes into the cuvette chamber with 0.4 mL ice-cold 1.0 *M* sorbitol and shake gently.
11. Spread the entire sample on CSM-URA (glucose) plate containing 1.0 *M* sorbitol. After the liquid soaks into the agar, invert the plate, and incubate at 30°C for 3–4 d.

12. Select about four resulting colonies and streak for single colonies on CSM-URA (glucose). Incubate at 30°C for 2–3 d.
13. Test a single, clonal colony from each re-streaked transformant for inability to grow on CSM-LYS, to demonstrate correct replacement of *LYS2* sequences with R-*URA3*-R (*see* **Note 17**).
14. To reclaim the Ura⁻ phenotype, we utilize the strategy of Roca et al. *(8)* by electroporating *lys2*::R-*URA3*-R cells with pHM53, containing the *Z. rouxii* recombinase under control of the *GAL1* promoter, and plating them on CSM-LEU (glucose) containing 1.0 M sorbitol.
15. After streaking resulting colonies on CSM-LEU (glucose), cells from a single colony are inoculated into 10 mL CSM-LEU (2 % galactose) and incubated for approximately 16 h at 30°C with shaking.
16. To promote loss of plasmid pHM53, the cells are outgrown for 2 d by diluting an aliquot (50 µL) into 10 mL YPD which has been supplemented with 200 µL of 2 mg/mL L-leucine. Dilute the culture in the same medium whenever the cells reach an OD_{600} ~ 3.
17. The outgrown cells are streaked for single colonies on YPD and resulting colonies are replica plated for a Ura⁻ Leu⁻ phenotype. The final *lys2-Δ1 leu2 ura3* strain can be grown in YPD to an OD_{600} approx 1 and an aliquot preserved cryogenically (*see* **Subheading 3.1.1.**, **step 4**).

3.1.3. Integration of MTase Expression Plasmid in Yeast

1. An appropriate strain of yeast (*see* **Note 4**) should be prepared for electroporation as indicated in **Subheading 3.1.2.**, **steps 4–6**.
2. On ice, pipet 40 µL of the cell suspension into a sterile 1.5-mL microcentrifuge tube that contains 5 µL (approx 1 µg) of *Xho*I-digested pMPK1-*Sss*I-19, mix, and electroporate as indicated in **Subheading 3.1.2.**, **items 8–10**. Plate on CSM-LYS (glucose) containing 1.0 M sorbitol.
3. Streak resulting transformants for single colonies on CSM-LYS (glucose) (*see* **Note 18**).

3.2. Screening for Yeast Strains with Functional Cytosine-5-DNA MTases

3.2.1. Isolation of Total DNA from Yeast

The procedure is a modified version of that given in Rose et al. *(9)*. All steps are carried out at room temperature unless stated otherwise.

1. Inoculate 5 mL of YPG (4% galactose) or appropriate CSM dropout (4% galactose) medium and incubate with shaking overnight at 30°C (*see* **Note 19**).
2. The next day, when the culture has reached an $OD_{600} \geq 2$, pellet the cells by centrifugation for 3 min at 1000*g* (*see* **Note 20**).
3. Decant the supernatant, resuspend the cell pellet in 0.7 mL TE, pH 8.0, and transfer the cells to a 1.5-mL screw-cap microcentrifuge tube containing 0.3 g of acid-washed glass beads (*see* **Note 21**).

4. Pellet the cells again by centrifugation for 5 s at 10,000 rpm in a microcentrifuge. Remove and discard the supernatant. (The glass beads will be interspersed with the cells to allow rapid resuspension in the next step.)
5. Add 0.2 mL each of smash buffer and buffer-saturated phenol. Vortex vigorously for 8 min.
6. Add 0.2 mL each of TE and $CHCl_3$:IAA. Vortex vigorously for 4 min.
7. Centrifuge at 14,000 rpm in a microcentrifuge for 5 min. Transfer the aqueous, upper phase, carefully avoiding the white, interphase material, to a new 1.5-mL microcentrifuge tube that contains 0.2 mL 10.0 M NH_4OAc. Vortex and incubate on ice for 2 h to overnight. A white, insoluble precipitate will form that is *not* DNA nor RNA, but which appears to inhibit a number of enzyme reactions *(10)*.
8. To pellet the precipitate, centrifuge the samples at 16,000g in a microcentrifuge for 5 min. Transfer the supernatant to a new 1.5-mL microcentrifuge tube that contains 0.6 mL 2-propanol and vortex. The nucleic acid (DNA and a lot of RNA) should precipitate immediately making the solution slightly cloudy.
9. Pellet the nucleic acid by centrifugation at 16,000g for 5 min. Completely remove and discard the supernatant.
10. Add 0.4 mL 70% ethanol:30% 0.1X TE, vortex briefly, and centrifuge at 16,000g for 2 min.
11. Remove and discard the supernatant, briefly dry the pellet (*see* **Note 14**), and resuspend the pellet in 50 μL sterile 0.1X TE. Store the samples at –20°C. It is not necessary to get rid of the RNA.

3.2.2. Bisulfite Deamination of DNA

1. Degas the distilled water sufficiently in advance so that it has cooled to room temperature and prepare the 3 N NaOH and 3X sample denaturation buffer.
2. Add 10 μL 3X sample denaturation buffer to a 0.5-mL microcentrifuge tube that contains 20 μL of each sample of total genomic DNA or control DNA (*see* **Subheading 3.3.2.**) and vortex. The samples can be safely left at room temperature while preparing the remaining solutions. Save the remainder of 3 N NaOH, it will be needed to make the desulfonation solution.
3. Prepare the 100 mM hydroquinone and the saturated sodium metabisulfite solution and pre-warm it to 50°C.
4. Denature the samples by incubation at 98°C for 5 min. Maintain the samples at 98°C (i.e., do not cool them) during addition of the saturated sodium metabisulfite solution (*see* **Note 22**).
5. Stir the pre-warmed saturated sodium metabisulfite solution, add 0.2 mL directly into the sample and vortex immediately. After bisulfite has been added to each tube, overlay each reaction with 100 μL mineral oil (add 2 drops from a dropper), and incubate at 50°C for 6 h in the dark (*see* **Note 23**).
6. Preheat a heating block to 95°C.
7. To desalt the DNA, transfer the metabisulfite solution to a 1.5-mL microcentrifuge tube (*see* **Note 24**).

8. Attach one Promega minicolumn onto a 3-mL Luer-Lok syringe barrel for each sample and attach to the Promega vacuum manifold.
9. Add 1 mL Promega Wizard PCR Preps resin to each sample (*see* **Note 6**). Mix and transfer to a syringe fitted with a minicolumn.
10. Apply vacuum to draw the resin suspension into the minicolumn. Wash the resin 3X with 1 mL 80% isopropanol making sure to evacuate each wash *completely* before adding the subsequent wash. *Note:* Dispose of the solution in the manifold properly.
11. Detach the minicolumn from the syringe barrel and press it into a 1.5-mL microcentrifuge tube. Remove residual isopropanol by centrifugation at 16,000g for 2 min in a microcentrifuge.
12. Transfer the minicolumn to a new 1.5-mL microcentrifuge tube and apply 52 µL 0.1X TE, preheated to 95°C, to the minicolumn and leave for 5 min.
13. Centrifuge the minicolumn for 20 s to elute the DNA.
14. Transfer the eluted DNA, avoiding any pelleted resin that passed through the minicolumn, to a 1.5-mL microcentrifuge tube containing 8 µL desulfonation solution. Vortex and incubate at 37°C for 15 min.
15. Neutralize each sample by adding 18 µL 10.0 M NH$_4$OAc and vortexing.
16. Add 0.2 mL absolute ethanol, vortex, and incubate at least 5 min to overnight at –70°C.
17. Pellet the DNA by centrifugation at 14,000 rpm (16,000g) for 20 min at room temperature in a microcentrifuge.
18. Remove and discard the supernatant, being careful not to disturb the pellet.
19. Wash the pellet by adding 0.3 mL 70% ethanol:30% 0.1X TE, vortex, centrifuge at 14,000 rpm for 2 min, and remove the supernatant.
20. Dry the pellet slightly and resuspend the DNA in 50 µL sterile 0.1X TE. Store the samples at –20°C.

3.2.3. "Hot Start" PCR Amplification from Deaminated DNA to Identify SssI⁺ Yeast Strains (see **Note 7**)

1. Make an appropriate volume of the following two reaction mixtures:

Lower mixture (µL)		Upper mixture (µL)
4.25	sterile, distilled water	30.25
1.25	10X Taq buffer	5
4	dNTP mix	—
1.5	20 µM primer MKO2b1	—
1.5	20 µM primer MKO2b2	—
—	Taq DNA polymerase (5U/µL)	0.25

2. Pipet 12.5 µL of the lower mixture into a 0.5-mL microcentrifuge tube and add a Perkin-Elmer Ampliwax PCR Gem 50 wax bead. Incubate in the thermocycler for 5 min at 80°C followed by 1 min at 25°C.
3. Layer 35.5 µL upper mixture onto the solidified wax. Add 2 µL of deaminated DNA and vortex gently.

4. Place the reactions in a thermocycler (pre-heated to 94°C) and subject to: 1 cycle of 94°C for 3 min; followed by 30 cycles of 94°C for 45 s, 60°C for 45 s, and 72°C for 1 min; and a final extension cycle of 72°C for 4 min.
5. Analyze 10 µL of each reaction on a 1.5% TAE agarose gel containing 0.5 µg/mL ethidium bromide. Samples positive for methylation by M.SssI will produce a prominent amplification product of 258 bp. Negative samples will usually lack any product or, at most, contain a faint band.

3.3. In Vivo Footprinting with M.SssI

3.3.1. Yeast Culturing Conditions and Isolation of Total DNA

1. From a YPD plate (*see* **Note 18**), inoculate cells into 10 mL of YPG (4% galactose) or appropriate CSM dropout (4% galactose) medium and incubate with shaking (300 rpm) overnight at 30°C (*see* **Note 25**).
2. The next day, reseed the cells into new 4% galactose medium at $OD_{600} = 1$ and incubate at 30°C with shaking for 16 h (*see* **Note 26**).
3. Isolate total yeast DNA as described in **Subheading 3.2.1.**

3.3.2. In Vitro Methylation of DNA with M.SssI and Deamination

1. Transform a plasmid containing the region of interest into *E. coli* GM2163 and isolate pure miniprep DNA (*see* **Subheading 3.1.1.**). Alternatively, the relevant region can be present on a purified PCR product amplified from wild-type DNA.
2. If using a plasmid, the DNA must first be digested with an appropriate restriction endonuclease (i.e., one that does not cleave within or between the sites of primer binding). Digest 1.1 µg DNA in a volume of 10 µL (*see* **Note 27**).
3. In parallel to uncut plasmid, check a small aliquot (0.5 µL) of the digestion on a 0.7% TAE agarose gel to ensure complete linearization of the plasmid. It is not necessary to remove the restriction enzyme prior to in vitro methylation.
4. Set up a methylation reaction (50 µL total volume) that contains the 1 µg of digested plasmid DNA (or uncut PCR product) in 1X restriction enzyme buffer, 2.5 µL 3.2 m*M* SAM (diluted from the 32 m*M* stock), 4 µL NEB2 10X restriction enzyme buffer, and sterile, distilled water to bring the volume to 49.5 µL. Preincubate the reaction at 37°C for 5 min (*see* **Note 28**).
5. Add 0.5 µL M.SssI (1 U), vortex, and incubate at 37°C. Remove 16.7 µL at 6, 13, and 30 min, transfer to a 1.5-mL microcentrifuge tube, and heat inactivate the M.SssI at 70°C for 15 min.
6. Add 0.3 mL TE. The methylated DNA can be stored at –20°C and, if thawed, kept on ice.
7. For deamination, pipet 10 µL (~10 ng DNA) in vitro methylated DNA and 10 µL sterile, distilled water into a 0.5-mL microcentrifuge tube and proceed as indicated in **Subheading 3.2.2.**

3.3.3. Deamination of Yeast Total DNA

1. If the region under study is present on a plasmid, the DNA must first be digested with an appropriate restriction endonuclease (i.e., one that does not cleave within

or between the sites of primer binding) (*see* **Note 29**). Set up digestions containing 16.5 µL total yeast DNA (approx one-third of that isolated in **Subheading 3.2.2.**), 2 µL 10X restriction buffer, and 1.5 µL (30 U) enzyme and incubate them at 37°C for approx 3 h. Digestion of the DNA can be omitted if the region of interest is chromosomal or located on a linear minichromosome (e.g., YAC).
2. Prepare deaminated yeast DNA by bisulfite treatment as indicated in **Subheading 3.2.2.** If the DNA was digested by a restriction enzyme, the 3X sample denaturation buffer can be added directly to the digestions. Deamination of appropriate control DNA samples (e.g., in vitro methylated DNA; **Subheading 3.3.2.**) should also be considered (*see* **Note 30**).

3.3.4. PCR Amplification from Deaminated DNA for Footprinting a Region of Interest

1. Set up 'hot start' PCR amplification as in **Subheading 3.2.3.**, except include 2 µL each of appropriate primers (*see* **Fig. 1** and **Note 13**) in the 'lower mixture,' and consequently only 3.25 µL of sterile, distilled water (*see* **Note 31**).
2. Place the reactions in a thermocycler (pre-heated to 94°C) and subject to: 1 cycle of 94°C for 3 min; followed by 30 cycles of 94°C for 45 s, 5°C below the calculated T_m (*see* **Note 13**) for 45 s, and 72°C for 1 min; and a final extension cycle of 72°C for 4 min. Keep the reactions at 4°C when they are completed.
3. Analyze 3 µL of each PCR reaction on an agarose gel to verify that the product is homogeneous and of the expected size (*see* **Note 32**).
4. Transfer the PCR reaction from beneath the wax to a 1.5-mL microcentrifuge tube that contains 0.1 mL of Wizard PCR Preps direct PCR product purification buffer, and vortex.
5. Process further as indicated in **Subheading 3.2.2.**, **items 12–15**. Elute the DNA from the resin by applying 45 µL of 0.1X TE pre-heated to 70°C, wait 1 min, and centrifuge at 14,000 rpm for 20 s in a microcentrifuge.
6. Transfer the samples to new microcentrifuge tubes avoiding any pelleted resin.
7. Analyze 5 µL of each PCR reaction on an agarose gel to estimate the DNA concentration (*see* **Note 33**) which should be approx 5–10 ng/µL.

3.3.5. Thermocycle Sequencing of Purified PCR Products (see **Note 34**)

1. Set up a reaction (20 µL final volume) to end-label a 'b1' or 'a1' primer (**Fig. 1**): Include 2 µL 10X T4 polynucleotide kinase buffer, 15-20 pmol primer, 70-100 µCi [γ-^{32}P]ATP (3000 Ci/mmole), 1 µL T4 polynucleotide kinase (10 U), and sterile, distilled water to volume. Incubate at 37°C for 45–60 min.
2. Remove the unincorporated radionucleotide by passage through a Sephadex G-50 spin-column. The incorporation of ^{32}P into the primer should be at least 50%. The primer should be kept on ice or can be stored at –20°C for at least a week.
3. Pipet 50-100 fmol of each purified PCR product prepared in **Subheading 3.3.4.** into a 0.5-mL microcentrifuge tube and place on ice. For a 600 bp PCR product, 100 fmol is approx 40 ng DNA.

4. Each thermal cycle sequencing reaction (final volume of 8–10 µL) should contain: Sterile, distilled water to volume, 1X Sequitherm buffer, 5 µM each dATP, dCTP, and dTTP, 50 µM ddGTP, 1.0-1.2 pmol ^{32}P-end-labeled 'b1' or 'a1' primer, and 1.25 U Sequitherm DNA polymerase. Set up a mixture on ice, adding the components in the order listed (*see* **Note 35**).
5. Reactions (8–10 µL final volume) for sequencing ladders contain: 1X Sequitherm buffer , 0.1 pmol DNA, 1.0-1.2 pmol ^{32}P-end-labeled 'b1' or 'a1' primer, 1X sequencing termination mix, and 1.5 U Sequitherm DNA polymerase.
6. The reactions are placed in a thermocycler (preheated to 95°C) and the cycling parameters are: one cycle of 3 min at 95°C followed by 10 cycles of 0.5 min at 95°C, 0.5 min at 5°C below the calculated T_m of the primer, and 1 min at 72°C. Activate the heated lid function of the thermocycler or, if not available, overlay each reaction with 10 µL mineral oil.
7. When the thermocycling is completed, keep the reactions at 4°C, add 0.5 vol of 3X sequencing reaction stop/loading buffer, and vortex.
8. Denature the samples at 70°C for 5 min, quick chill them on ice, and electrophorese 3 µL on a denaturing 6% polyacrylamide (19:1 acrylamide:bisacrylamide), 50% urea gel. The remainder of each sample can be stored at –20°C.
9. After electrophoresis, dry the gel and visualize by phosphorimager or autoradiography.

4. Notes

1. Any commonly used laboratory strain (e.g., DH5α, XL1-Blue, SURE, and so forth) will suffice.
2. The procedure in this section minimally requires a *ura3* marker in a Lys$^+$ strain. A *leu2* marker is also necessary if the *ura3* marker is to be reclaimed. As M.*Sss*I is under control of *GAL1*, is it essential that prospective strains are *GAL2* (many are *gal2*) and *GAL4*. This can usually be determined simply by screening for growth on medium containing galactose as a sole carbon source (e.g., YPG).
3. If equipment for electroporation is not available, use the DMSO-enhanced lithium acetate transformation protocol *(11)*.
4. Strains need to contain a *lys2* marker for targeted integration of the M.*Sss*I expression plasmid. It is preferable to have a nonreverting allele available such as *lys2-Δ1* which we have outlined (**Subheading 3.1.2.**), but with the screen for functional MTase integrants described in **Subheading 3.2.**, reverting alleles such as *lys2-801*^{amber} do not present difficulty.
5. Do not expose solid sodium metabisulfite to the atmosphere until just prior to use. We purchase 0.5 kg (Aldrich) and aliquot it into 5-g capacity glass vials in an oxygen- and water-free chemical safety hood. It is not necessary to weigh each aliquot precisely in the safety hood, pour the reagent through a funnel, filling each vial near capacity. Secure the vial lids tightly and store the vials in the dark in a vessel that contains Drierite. It is not necessary to flood the chamber with an inert gas. The reagent is stable for at least 2 yr. Single-use, 5-g quantities of sodium metabisulfite can also be purchased.

6. The resin can be diluted twofold with 6 M guanidine thiocyanate. Promega Wizard DNA cleanup kit has also been used in the desalting *(7)*.
7. The procedure is an adaptation of methylation-specific PCR (MSP) *(12)*. The primers are designed to screen positively only for strains producing ^{5me}C in CG sites at *GAL1*.
8. Primers are synthesized with a Beckman Oligo 1000 DNA synthesizer. After synthesis, oligonucleotides are deprotected, lyophilized, and resuspended in 0.1 mL 0.1X TE. The oligonucleotides are then precipitated by adding 25 µL 10.0 M NH$_4$OAc, vortexing, and addition of 0.3 mL absolute ethanol. After mixing, the precipitate is pelleted by centrifugation for 10 min at 14,000 rpm in a microcentrifuge at room temperature. The supernatant is removed completely and the pellet is washed with 0.3 mL absolute ethanol, dried, resuspended in 0.15 mL 0.1X TE, and stored at –20°C. As each primer contains a skewed base composition, the concentration (in µM) of each oligonucleotide is obtained from: $A_{260} \times DF \times 10^6/\Sigma_e$, where DF is the dilution factor and Σ_e is the total sum of the individual extinction coefficients for each base within the oligonucleotide. The individual extinction coefficients are: A, 15400; G, 13700; C, 9100; and T, 9700 M^{-1} cm^{-1} *(13)*.
9. High purity nucleotides are essential. We use the ultrapure grade of Pharmacia; deoxynucleotide kit (catalog no. 27-2035-01) and ddGTP (catalog no. 27-2075-01). We have tested one other source of ddGTP, but it yielded an inferior termination efficiency.
10. A high quality grade of *Taq* DNA polymerase is required. The enzyme currently distributed by Fisher Scientific yields excellent, reproducible results.
11. The following M.*Sss*I expressing strains are available on request: YPH500ΔL.19-2, *MAT*a *ade2*-101 *ura3*-52 *his3*-Δ200 *leu2*-Δ1 *trp1*-Δ63 *lys2*-Δ1:*LYS2-GAL1*promoter-*Sss*I (derived from YPH500)*(14)*; MKY49 (S288C background), *MAT*a *ade2*-101 *his3*-Δ200 *ura3*-52 *lys2*-801:*LYS2-GAL1*promoter-*Sss*I; MKY52 (W303 background), *MAT*a *ade2*-1 *can1*-100 *his3*-11,15 *leu2*-3,112 *trp1*-1 *ura3*-1 *lys2*-Δ1:*LYS2-GAL1*promoter-*Sss*I.
12. GM2163 can be obtained without charge from New England Biolabs (catalog no. 401-P) with any standard order.
13. Design the primers with appropriate transitions (G to a, or, C to t) as depicted in **Fig. 1**. In our experience, in order of most to least importance, the considerations in designing the primers are: 1) Distribute the deaminated residues (i.e., the G to a, or, C to t transitions) as evenly as possible along the length of the primer to maximize discrimination against amplification from nondeaminated DNA (i.e., avoid grouping transitions in one region of the primer, especially at the 5' end of the primer); 2) The T_m should be close to 60°C, as calculated according to Breslauer *(15)* (i.e., based on nearest-neighbor thermodynamic quantities). We use a program written in Basic language, which is available upon request, to calculate the T_m; 3) Avoid exceptionally long stretches of A or T (>7 bp); 4) Include a C (a1 and b1 primers) or G (a2 and b2 primers) at the 3' priming position; and 5) Avoid target CG sites, particularly at the 3' end of the primer. Con-

form to as many of these guidelines as possible. The last parameter can usually be disregarded because the total percentage of methylated sites at any given CG will be rather small (even if the site is in an accessible region) and, therefore, the majority cytosines at that position will be *fully* deaminated in the template molecules. Typical primers are approximately 30 bp in length and are usually chosen to amplify a region 600–800 bp in length. If a footprinting result is unsatisfactory, design another set of primers. The *GAL1* b1 and b2 primers shown in **Fig. 1** can be used as a positive control in the PCR amplication, yielding a 635 bp product, and in subsequent steps of the protocol.

14. We can usually directly electroporate 5 µL of digested miniprep DNA in 1X digestion buffer into yeast. However, if a high transformation efficiency is required (e.g., your strain has a revertable *lys2* marker) it may be necessary to desalt and concentrate the DNA by ethanol precipitation as follows: Add one-fourth volume 10.0 M NH$_4$OAc, vortex, and add 2.5 vol of absolute ethanol. Mix and incubate at least 5 min at –70°C and then centrifuge in a microcentrifuge at maximum speed (14,000 rpm = 16,000g) for a minimum of 15 min. Remove and discard the supernatant and wash the pellet by adding 0.3 mL 70% ethanol:30% 0.1X TE, vortex, centrifuge at 14,000 rpm for 2 min, and discard the supernatant. Air dry the pellet by centrifugation at 4,000 rpm for 2 min in a microcentrifuge, leaving the tube cap open. Resuspend the DNA in sterile 0.1X TE to a concentration of approximately 0.2 µg DNA/µL.

15. The smaller, approx 3 kb fragment contains a functional copy of *URA3* flanked by *Zygosaccharomyces rouxii* recombinase sites *(8)* replacing the *Bgl*II to *Hpa*I fragment of *LYS2*. Although we have typically gel purified this fragment for transforming yeast to Lys$^-$ by one-step gene replacement, the *Not*I/*Xho*I-digested plasmid can probably be electroporated directly (*see* **Note 14**).

16. It is necessary to remove the electrolytes from the cell suspension to achieve a good electroporation efficiency. We use a 100-mL capacity sterilization filtration unit with a 0.45 µm-pore membrane to remove wash solutions quickly. Resuspend the cells from the membrane surface with each wash, and with the third wash, transfer the cell suspension to a fresh 50-mL conical tube. Pellet the cells by centrifugation, decant the supernatant thoroughly, and resuspend the cells in the remaining trace supernatant by vortexing. The cell suspension should constitute a thick slurry and should be kept on ice.

17. It is unlikely, although possible, that Ura$^+$ Lys$^-$ colonies will have arisen from inappropriate integration of R-*URA3*-R elsewhere in the genome *and* a secondary mutation that occurred in a lysine biosynthetic gene other than *LYS2*. Correct integration should be verified by PCR from isolated total DNA, or alternately, by Southern blotting following integration of the MTase expression vector. If there is no need for a *ura3* marker, one can probably proceed to **Subheading 3.1.3.** However, we have always opted to reclaim the *ura3* marker (**Subheading 3.1.2., steps 14–17**).

18. Although there is no deleterious phenotype known to be caused by expression of M.*Sss*I in yeast, we maintain all M.*Sss*I$^+$ strains on glucose-containing medium

to repress the *GAL1* promoter. Cells are only transferred to galactose-containing medium when they are to be used in an experiment.

19. These conditions allow for fully induced expression of M.*Sss*I. However, the maximal level of accumulation of ^{5me}C, estimated to be one modification every 2–3 kb, is relatively low and produces no known phenotype.
20. Cells will accumulate a higher percentage of methylation as the cell density increases (*see* **Note 27**). Therefore, although the screen presented in this section is very sensitive and will yield satisfactory results with cells grown to mid-logarithmic phase, for best results, it is recommended to grow cells to $OD_{600} \geq 2$. Do not leave such high density cultures at 30°C for extended times, however, because loss of activation of the *GAL1* promoter will occur if the galactose is completely metabolized (the high concentration of galactose in the medium is included to help prevent this). Place the samples on ice temporarily if the DNA can not be isolated immediately.
21. An expedient way to do this is to cut a 1.5-mL microcentrifuge tube at the 0.15-mL mark and use the bottom as a 'measuring cup.' Wear gloves while adding the glass beads to each sample tube.
22. We use a thermocycler with an activated heated lid for the denaturation. To minimize evaporation, we close the heated lid following the addition of sodium metabisulfite to each tube. If 10–15 samples are worked with at a time, it does not appear necessary to stagger the denaturation start times of each sample (i.e., incubating some samples at 98°C longer than others will not compromise the experiment).
23. Under these reaction conditions, time course studies demonstrated that >95% of the cytosines were deaminated at 4 h. The deamination time can be increased to 16 h, but we and others *(16)* have noted significant degradation of DNA.
24. We usually expel a small bubble from the pipet tip after penetrating the oil and wipe the tip off with a fresh piece of a Kimwipe. The oil can also be removed following snap freezing *(7)*.
25. We recommend that cells from a relatively recent YPD plating (approx 1 mo old) be used for the inoculation. Inoculate the medium with an amount of cells that will lead to an OD_{600} approx 1 in the afternoon of the next day. In initial experiments, it is recommended to include an additional control of either SssI$^+$ or SssI$^-$ cells grown in YPD or YPG, respectively. These controls provide methylation-free DNA helpful in ascertaining the efficiency of cytosine deamination.
26. When the cells reach an OD_{600} approx 1, determine the volume of cells, which when resuspended in 5–10 mL, will give an OD_{600} of 1. Discard an appropriate volume of cells from each culture and pellet the cells by centrifugation at 1000*g* for 3 min. Decant the supernatant and resuspend the cell pellet in the determined volume of new medium. In general, the best results will be obtained with cells grown to high densities that have accumulated maximal levels of methylation. Reasonably good footprinting results are obtained with cultures seeded at densities as low as 0.5. There is a significant decrease in the accumulation of methylation in cultures seeded at $OD_{600} \leq 0.1$ *(5)*. In addition, the total amount of methylation at any particular CG target site is dependent on at least three main

factors *(5)*: 1) the preference of the enzyme for the site (i.e., nearest-neighbor sequences affect the rate of methylation in vitro), 2) the presence of histones in nucleosomes, and 3) the presence of non-histone proteins that may bind to, or in the vicinity of, the particular CG site. Removal by DNA excision-repair (*RAD1*, *RAD3*, and so forth) exerts a very nominal effect on the steady state levels of ^{5me}C produced by M.*Sss*I (MPK and RTS, unpublished results). Thus, footprinting data can be obtained in log phase cells if the enzyme exhibits a kinetic preference for a relevant site in protein-free DNA (*see* **Subheading 3.3.2.**).

27. Choose an enzyme that exhibits activity in NEB2 restriction enzyme buffer (e.g., *Hind*III), as this is the buffer used in the in vitro M.*Sss*I methylation reaction. The digestion reaction volume is limited to 10 µL so that if another buffer must be used, the differing salts will be diluted sufficiently in the *in vitro* methylation reaction.
28. It is essential to include Mg^{2+} in the reaction so that methylation is distributive rather than processive *(17)*.
29. The given, rapid DNA isolation procedure assures methylation levels that are indicative of those present in vivo, however, it yields a crude preparation that is not easily digested by a number of endonucleases. Therefore, if possible, use *Hind*III or *Eco*RI to digest the DNA.
30. Ideally, to control for the deamination and primer extension efficiencies, one should process in vitro-methylated DNA in parallel to in vivo samples in each deamination.
31. We find that 'hot start' PCR greatly improves the amplification reactions by eliminating non-specific bands and substantially increasing the yield of the desired product. A high quality PCR product can be purified directly by Wizard PCR Preps, thereby avoiding cumbersome techniques of gel purification.
32. If the products are not homogeneous, particularly if small products are visible, the samples must be gel purified to avoid production of unwanted run-off termination products during the thermal cycle sequencing that will co-migrate with the footprinted region on a gel. Occasionally, higher molecular weight products are observed, but these do not interfere with the assay.
33. Compare the fluorescent intensity of the PCR products relative to each other and to a band of similar size in a DNA size-marker that is run in parallel.
34. Since there is a low level of methylation by M.*Sss*I in vivo, the majority of cytosines in a given CG site will *not* be methylated in the population of cells, and consequently, not modified in the resulting PCR products. Therefore, cloning individual PCR products is of no use and the entire population of PCR products should be sequenced directly. For the sequencing reactions, one can use the same b1 primer used to amplify the PCR product from deaminated DNA or another one that anneals more 3' (e.g., primer walking). In general, we gel purify primers before kinasing, however, we have also obtained excellent results with crude oligonucleotides that are prepared as indicated in **Note 8**.
35. The bulk of the reaction volume will be taken up by the DNA and the radiolabeled primer. Therefore, the nucleotides are usually added to the sequencing mixture from high-concentration stocks to minimize added volume. Since, on

average, there is a low level of cytosine in the PCR products, the dGTP is deliberately omitted, and a high concentration of ddGTP is included, to achieve a high efficiency (>96%) of termination at cytosines.

Acknowledgments

We extend our gratitude to Bill Jack (New England Biolabs) for supplying the cloned M.*SssI* gene and *E. coli* GM2163. We also thank James C. Wang for generously supplying pJR-*URA3*, from which the R-*URA3*-R cassette was obtained, and pHM53. Supported by NIH Grant GM52908.

References

1. Singh, J. and Klar, A. J. S. (1992) Active genes in yeast display enhanced *in vivo* accessibility to foreign DNA methylases: a novel *in vivo* probe for chromatin structure of yeast. *Genes Dev.* **6,** 186–196.
2. Gottschling, D. E. (1992) Telomere-proximal DNA in *Saccharomyces cerevisiae* is refractory to methyltransferase activity *in vivo*. *Proc. Natl. Acad. Sci. USA* **89,** 4062–4065.
3. Kladde, M. P. and Simpson, R. T. (1994) Positioned nucleosomes inhibit Dam methylation *in vivo*. *Proc. Natl. Acad. Sci. USA* **91,** 1360–1365.
4. Renbaum, P., Abrahamove, D., Fainsod, A., Wilson, G., Rottem, S., and Razin, A. (1990) Cloning, characterization, and expression in *Escherichia coli* of the gene coding for the CpG DNA methylase from *Spiroplasma sp.* strain MQ-1 (M.*SssI*). *Nucleic Acids Res.* **18,** 1145–1152.
5. Kladde, M. P., M. Xu, and R. T. Simpson. (1996) Direct probing of DNA-protein interactions in repressed and active chromatin in living cells. *EMBO J.* **15,** 6190–6200.
6. Frommer, M., L. E. MacDonald, D. S. Millar, C. M. Collis, F. Watt, G. W. Grigg, P. L. Molloy, and C. L. Paul. (1992) A genomic sequencing protocol that yields a positive display of 5-methylcytosine residues in individual DNA strands. *Proc. Natl. Acad. Sci. USA* **89,** 1827–1831.
7. Clark, S. J., J. Harrison, C. L. Paul, and M. Frommer. (1994) High sensitivity mapping of methylated cytosines. *Nucleic Acids Res.* **22,** 2990–2997.
8. Roca, J., Gartenberg, M.R., Oshima, Y., and Wang, J.C. (1992) A hit-and-run system for targeted genetic manipulations in yeast. *Nucleic Acids Res.* **20,** 4671–4672.
9. Rose, M. D., F. Winston, and P. Hieter. (1990) *Methods in Yeast Genetics: A Laboratory Course Manual.* Cold Spring Harbor Laboratory Press, Cold Spring Harbor, NY.
10. Gilbert, D. M., Losson, R. and Chambon, P. (1992) Ligand dependence of estrogen receptor induced changes in chromatin structure. *Nucleic Acids Res.* **20,** 4525–4531.
11. Hill, J., Donal, I. G., and Griffiths, D. E. (1991) DMSO-enhanced whole cell yeast transformation. *Nucleic Acids Res.* **19,** 5791.
12. Herman, J. G., Graff, J. R., Myöhänen, S., Nelkin, B. D., and Baylin, S. B. (1996) Methylation-specific PCR: a novel PCR assay for methylation status of CpG islands. *Proc. Natl. Acad. Sci. USA* **93,** 9821–9826.

13. Budavari, S., ed. (1989) *The Merck Index: An Encyclopedia of Chemicals, Drugs, and Biologicals.* Merck, Rahway, NJ.
14. Sikorski, R. S. and Hieter, P. (1989) A system of shuttle vectors and yeast host strains designed for efficient manipulation of DNA in *Saccharomyces cerevisiae. Genetics* **122,** 19–27.
15. Breslauer, K. J., Frank, R., Blocker, H., and Markey, L. A. (1986) Predicting DNA duplex stablility from the base sequence. *Proc. Natl. Acad. Sci. USA* **83,** 3746–3750.
16. Raizis, A. M., Schmitt, F., and Jost, J.-P. A bisulfite method of 5-methylcytosine mapping that minimizes template degradation. *Anal. Biochem.* **226,** 161–166.
17. Matsuo, K., Silke, J., Gramatikoff, K., and Schaffner, W. (1994) The CpG-specific methylase *Sss*I has topoisomerase activity in the presence of Mg^{2+}. *Nucleic Acids Res.* **22,** 5354–5359.
18. Roth, S. Y., A. Dean, and R. T. Simpson. 1990. Yeast α2 repressor positions nucleosomes in TRP1/ARS1 chromatin. *Mol. Cell. Biol.* **10,** 2247–2260.

30

Restriction Nucleases as Probes for Chromatin Structure

Philip D. Gregory, Slobodan Barbaric, and Wolfram Hörz

1. Introduction

It has become increasingly clear over the last decade that chromatin structure and gene regulation are intricately intertwined. Different regulatory states of a given gene are frequently accompanied by changes in nuclease hypersensitive sites and nucleosome positioning *(1–4)*. Consequently, if one is interested in determining how gene activity is established on a given promoter or enhancer, it is worthwhile to determine the nucleosome structure of DNA sequences within and surrounding the regulatory element being studied. Nuclease digestion experiments with DNaseI and micrococcal nuclease originally uncovered details of the nucleosome substructure including hypersensitivity of a particular region. Hypersensitive regions appear to lack canonical nucleosomes and are the hallmarks of active genes. One of the inherent limitations of analyses with nonspecific nucleases has been the difficulty of quantitating accessibilities. A region is judged to be hypersensitive relative to the neighboring regions analyzed in a chromatin analysis, and variations in hypersensitivity are difficult to monitor by this method. We have therefore extensively used digestion with restriction nucleases as a complementary approach. The presence of a nucleosome strongly protects the underlying sequence against digestion with restriction nucleases while transcription factors which are often found associated with DNA within hypersensitive regions usually do not, under the rather extensive digestion conditions. One important advantage of a restriction analysis is therefore that the relative accessibility can be accurately quantitated.

From: *Methods in Molecular Biology, Vol. 119: Chromatin Protocols*
Edited by: P. B. Becker © Humana Press Inc., Totowa, NJ

We have used restriction nuclease analysis of chromatin mostly with yeast nuclei, but the method is equally applicable and has been used with nuclei from other organisms like, for example, *Drosophila* (5) and mammalian cells (6). In the following we describe our protocol for yeast, including a method for isolating nuclei suitable for restriction analysis.

2. Materials

1. Preincubation solution: 0.7 M ß-mercaptoethanol, 2.8 mM EDTA.
2. Sorbitol solution: 1 M sorbitol.
3. Lysis solution: 1 M sorbitol, 5 mM ß-mercaptoethanol.
4. Ficoll solution: 18% (w/v) Ficoll, 20 mM KH_2PO_4, pH 6.8, 1 mM $MgCl_2$, 0.25 mM EGTA, 0.25 mM EDTA.
5. Zymolyase solution: 20 mg/mL Zymolyase 100T (ICN) dissolved in water.
6. Digestion buffer: 10 mM Tris-HCl, pH 7.4, 50 mM NaCl, 10 mM $MgCl_2$, 0.5 mM spermidine, 0.15 mM spermine, 0.2 mM EDTA, 0.2 mM EGTA, 5 mM ß-mercaptoethanol.
7. Stop solution: 1 M Tris-HCl, pH 8.8, 0.24 M EDTA, pH 8.0.
8. Proteinase K solution: 10 mg/mL proteinase K, dissolved in 10 mM Tris-HCl, pH 8.0.
9. Chloroform solution: Chloroform/isoamylalcohol (24:1 v/v).
10. RNase solution: 5 mg/mL ribonuclease A (DNase-free) dissolved in 5 mM Tris-HCl, pH 7.5 and heated 10 min/100°C.

3. Methods

3.1. Isolation of Nuclei from S. cerevisiae (see Note 1)

1. A 1-L yeast culture is grown in the appropriate medium to early logarithmic phase (2–4×10^7 cells/mL) (*see* **Note 2**). Collect cells by centrifugation ($3000g$/10 min) (*see* **Note 3**). This is approx at an O.D.$_{600}$ of 2–4, although this measurement of cell density often varies between spectrophotometers (*see* **Note 4**).
2. Wash cells in ice cold water, and suspend in 50 mL water.
3. Transfer into preweighed 50 mL centrifuge tubes and centrifuge ($3000g$/5 min). Determine wet weight.
4. Add 2 mL preincubation solution per g of wet weight cells and shake for 30 min at 28°C (*see* **Note 5**).
5. Collect by centrifugation ($3000g$/5 min) and wash in 50 mL 1 M sorbitol.
6. Collect again ($3000g$/5 min) and resuspend in 5 mL lysis solution per 1 g of cells (wet weight).
7. Dilute 10-µL aliquots 200-fold in water and read optical density at 600 nm. This should be in the range between 0.5 and 1.
8. Add 1/50 vol of a freshly prepared zymolyase solution to the cells.
9. Incubate with slight agitation at 28°C.
10. Measure optical density at 600 nm after 15 and 30 min as in **step 7** with a 100-fold dilution in water. Zymolyase treatment is complete when values drop to 5–20% of the original measurement taken in **step 7** (*see* **Note 6**).

11. Centrifuge (3000g/5 min at 5°C) and wash in 50 mL ice-cold 1 M sorbitol.
12. Centrifuge (3000g/10 min at 5°C) and resuspend in 5 mL ice-cold Ficoll solution per 1 g cells (original wet weight). Cells lyse at this stage, but nuclei are stabilized by the Ficoll.
13. Distribute into as many aliquots as desired and centrifuge (30,000g/30 min at 5°C). Aliquots equivalent to 0.5 or 1 g wet weight cells are suitable for subsequent digestion experiments. 10 mL polypropylene centrifuge tubes are convenient.
14. Freeze the nuclear pellet in liquid nitrogen or dry ice/ethanol and store at −70°C.

3.2. Measuring Accessibility of Restriction Sites in Chromatin (see Note 7)

The strategy used is shown in **Fig. 1** with a representative result from the *PHO5* promoter. The results from an analysis using a large number of restriction enyzmes are shown in **Fig. 3** for the *PHO8* promoter and compared to an analysis with DNase I (*see* **Fig. 2**).

1. Suspend pelleted nuclei in digestion buffer by vortexing.
2. Centrifuge (2000g/5 min at 5°C) and resuspend in digestion buffer (*see* **Note 8**). Nuclei from approx 50 mg cells (wet weight) are used for one experiment and suspended in a total volume of 200 µL. Transfer to microfuge tubes.
3. Add restriction nuclease at two different concentrations (range between 150–1500 U/mL) (*see* **Note 9**). Incubate at 37°C for 60 min.
4. Terminate digestion by adding 10 µL stop solution, 5 µL 20% SDS, and 20 µL proteinase K solution. Incubate for 30 min at 37°C.
5. Add 1/5 vol 5 M NaClO$_4$, 1 vol phenol, vortex well, then 1 vol chloroform solution, vortex well.
6. Centrifuge for 5 min in a microfuge at maximum speed.
7. Take off supernatant, reextract with 1 vol chloroform solution (vortex well).
8. Take off supernatant and add 2.5 vol ethanol to precipitate nucleic acids.
9. Collect DNA by centrifugation (5 min) and resuspend in 125 µL TE.
10. Add 10 µL RNase solution, and incubate for 1 h at 37°C.
11. Add 5 µL 5 M NaCl and 0.6 vol isopropanol, and centrifuge immediately for 2 min at room temperature in a microfuge at maximum speed.
12. Wash the pellet with 70% ethanol, dry, and dissolve in 80 µL TE.
13. Use 10–20 µL for secondary digestion (*see* **Note 10**), indirect endlabeling, Southern transfer and hybridization (*see* **Note 11**).

3.3. Conclusion

We have in many instances used restriction nucleases to complement DNase I digestion experiments. Especially in cases where a given chromatin region undergoes structural transitions has digestion with restriction nucleases contributed valuable information as to the precise boundaries of such transitions and to the question if all or most cells in a population undergo the transition or only a small percentage. To elucidate the function and mechanism of a chro-

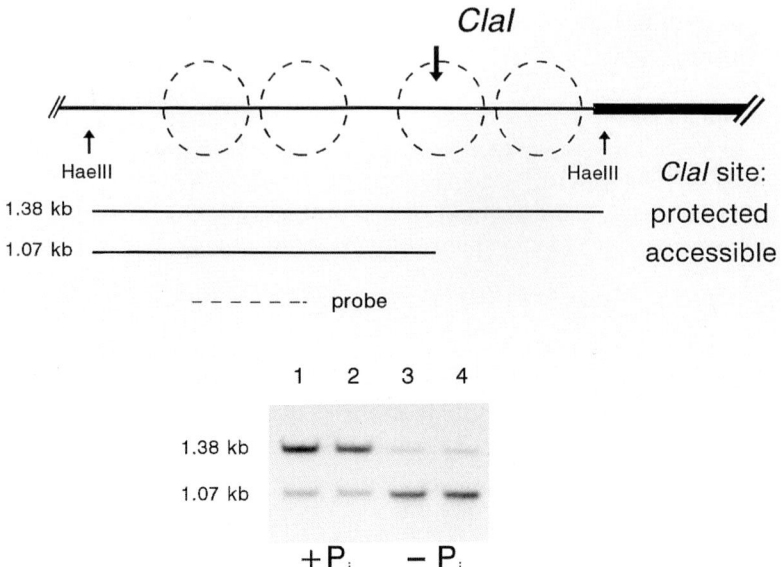

Fig. 1. Measuring the accessibility of a ClaI site at the *PHO5* promoter. In the repressed state, the *PHO5* promoter is organized in four positioned nucleosomes (broken circles) which are disrupted on activation of the promoter *(15)*. Nuclei from repressed (lanes 1, 2) and induced cells (lanes 3, 4) were digested with 100 U (lane 1, 3) or 300 U ClaI (lanes 2, 4). DNA was isolated, cleaved with HaeIII, analyzed in a 1% agarose gel, blotted and hybridized with probe D *(15)* as schematically shown at the top.

matin transition, it is ideal to study an inducible gene because active and inactive states can be studied in the same cell type. However, the same techniques can be applied to constitutively active genes in order to ascertain how the active state is established and maintained.

4. Notes

1. Our method for the isolation of nuclei is based on the original procedure of Wintersberger et al. *(7)*. The protocol is a balance between speed and purity. In this protocol, the nuclei are sufficiently purified so that the extent of nuclease digestion is reproducible between experiments. There are procedures to obtain highly pure yeast nuclei, but they involve substantially more time (**ref.** *8,9* and references therein). On the other hand, there are also other procedures that involve treating nuclei with nucleases in crude lysates *(10–13)*. The advantage of such procedures is that they are faster and loss of trans-acting factors is minimized. However, for the analysis of hypersensitivity, loss of factors is no draw back and if anything may even be advantageous.
2. Quantity of cells: We usually start with 1 L of a culture at 2×10^7 cells/mL. This gives us approx 2 g of cells (wet weight) and approx 0.2 mg of DNA.

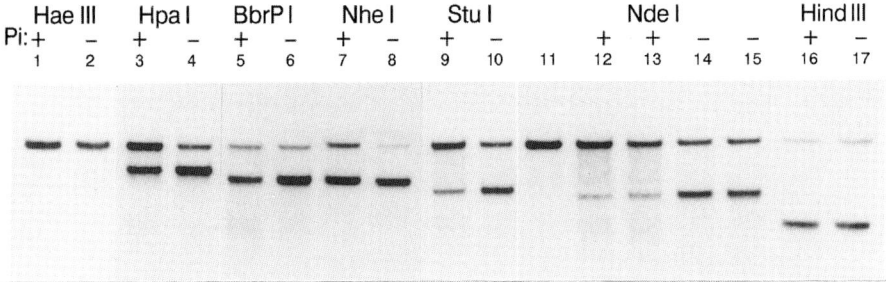
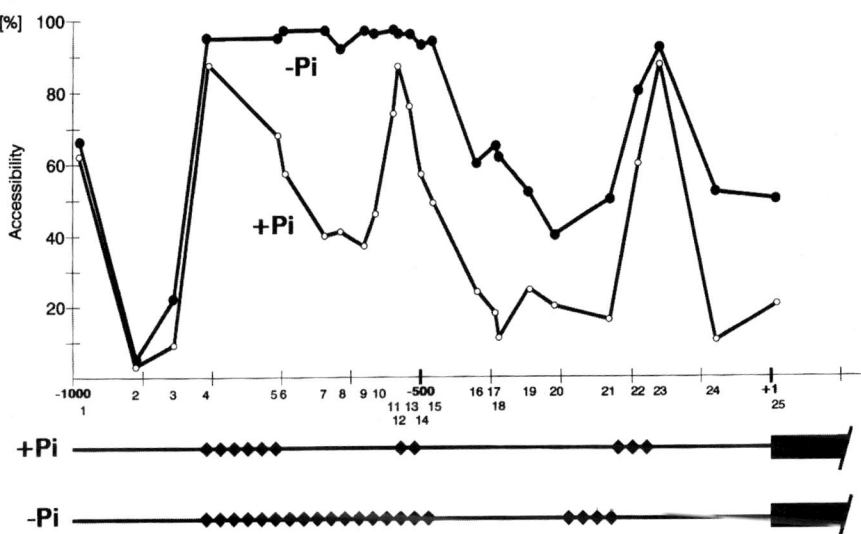

Fig. 2. Hypersensitive sites at the *PHO8* promoter. Nuclei isolated from cells grown in either high-phosphate (+Pi) or in no-phosphate medium (-Pi) were digested with DNase I. DNA was isolated and analyzed by indirect endlabeling. Hypersensitive sites at high phosphate and no phosphate conditions are shown schematically underneath. For details *see* **ref. 17**.

3. It is important to process a culture immediately after taking cells out of the incubator and not to store the culture first at low temperature, because subsequent lysis of the cells is greatly impeded otherwise.
4. Cell growth: For estimating cell density from absorbance measurements, it is best to determine the conversion factor for a particular spectrophotometer by counting cells.
5. Preincubation: This step facilitates digestion of the cell wall with Zymolyase. In the case of the acid phosphatase genes, it is fortunate that phosphate starvation is required for gene induction since all buffers used are phosphate starvation buff-

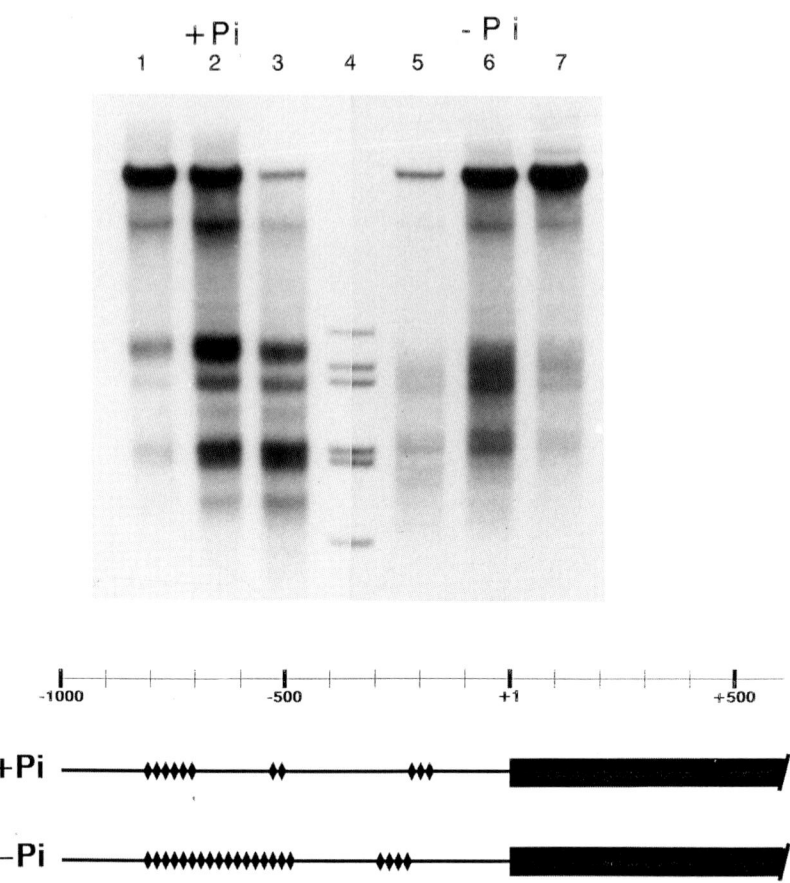

Fig. 3. Accessibility of the *PHO8* promoter to restriction nucleases. Nuclei were isolated from cells grown in either high-phosphate (+Pi) or in no-phosphate (-Pi) medium and digested with restriction nucleases. DNA was isolated and analyzed by indirect endlabeling as shown schematically in **Fig. 1**. Results for a number of restriction enzymes are shown at the top with those for *Nde*I, fully documented in lanes 11–15 (11, no enzyme added; 12 and 14, 60 U added; 13 and 15, 200 U added). For the other enzymes, results obtained with only the higher enzyme concentrations are shown. Accessibility for all restriction sites examined is shown underneath, as measured by determining the ratios of the band intensities in the autoradiograms. Restriction sites are designated by the numbers on the horizontal axis of the diagram as follows: *Eco*RI(1), *Sau*96I and *Hae*III(2), *Hin*fI(3), *Mbo*II(4), *Hin*fI(5), *Mbo*II(6), *Ban*II or *Hgi*AI(7), *Hpa*I and *Hin*dII(8), *Rsa*I(9), *Ban*II(10), *Sau*96I(11), *Bbr*PI(12), *Nhe*I(13), *Cfo*I(14), *Hin*dIII(15), *Stu*I and *Hae*III(16), *Hin*fI(17), *Nde*I(18), *Sau*96I(19), *Mbo*II(20), *Rsa*I(21), *Mbo*II(22), *Hin*dIII(23), *Hgi*AI(24), *Hin*fI(25). Sites underlined correspond to the examples shown at the top. The regions hypersensitive to DNase I at repressed and derepressed conditions are shown for comparison below (*see* **Fig. 2**).

ers. In other cases, it may be required to maintain induction or repression during preincubation and Zymolyase treatment by adding the appropriate inducer or repressor as discussed in **ref. 14**. It is also possible to treat cells with Zymolyase in medium supplemented with sorbitol and a reducing agent.

6. Lysis: This is strain dependent and also depends on growth conditions. Stationary cells are more difficult to lyse than dividing cells. The values as determined in **step 10** are only relative measures of lysis and do not reflect the actual percentage of unlysed cells. We have successfully used nuclei from cells that gave values for the OD_{600} at the end of Zymolyase treatment that only dropped to 60% of the starting value. In those cases we monitor the accessibility of a constitutively accessible restriction site in chromatin as a control *(see below)*. We have also used Lyticase (Boehringer Mannheim) instead of Zymolyase with similar results.

7. One of the main advantages of restriction nuclease analyses is that for any enzyme, different sites can be assayed in the same digest. They can serve as internal controls. Accessibility can be quantitated by determining the ratio of the two bands generated after secondary digestion. This is exemplified for the *PHO5* promoter in **ref. 15** and for the *PHO8* promoter in **Fig. 2**. In general, most sites will be largely inaccessible in the nucleus, because most sequences are incorporated into nucleosomes. This is only conclusive, however, if it can be shown that at least one site was cut by the enzyme in the particular digest, because some enzymes seem to work better than others in chromatin digestion. We have demonstrated constitutively accessible sites for a large number of enzymes in the *PHO5* **(16)**, *PHO3* **(16)**, *PHO8* **(17)**, and *TDH3* loci **(18)**. Probes suitable to assay these sites are available from the authors' laboratory and can serve as controls in such experiments.

8. The digestion buffer is a compromise between optimizing enzyme activity and preserving the chromatin structure. The addition of spermine and spermidine has been very useful since the polyamines stimulate activity of the vast majority of restriction nucleases. At the same time, endogenous, nonspecific nucleases are usually suppressed. The 50 mM NaCl salt concentration employed appears to work for almost all restriction enzymes in chromatin digestion.

9. We always use two nuclease concentrations that differ by a factor of 3–4. The results for both concentrations should be quite similar (from our experience they almost always are) which means that true plateau values were reached during restriction nuclease digestion of the chromatin and all accessible sites were cut.

10. There can be degradation of DNA during preparation of the nuclei. This is detected by the 0°C control incubation. By comparing the 37°C control with the 0°C control samples, one can also determine if endogenous nucleases are active during incubation with the restriction enzymes. It is useful to have the site used for secondary digestion not too far away from the restriction site actually assayed. This way the longer fragment, signaling protection, and the shorter one, reflecting accessibility, are of similar size. If limited degradation by endogenous nucleases does occur, comparison of the relative amounts of the restriction fragments is still meaningful because fragments of similar size should be similarly diminished by endogenous nucleases.

11. Typically intranucleosomal sites are approx 5 to 10% accessible. It is not clear if the residual accessibility reflects alternative nucleosome positions or cutting within the nucleosome. However, these values seem to be true plateau values since they do not change appreciably upon raising the nuclease concentrations by a factor of 2 to 4. Sites located within hypersensitive regions are usually 80–100% accessible, and again the significance of residual protection is not clear. At the *PHO5* promoter, sites located within short linker regions between positioned nucleosomes are about 50% accessible.

For certain nucleosomal regions, it appears that there is intermediate accessibility (40–60%) to restriction nucleases (e.g., **ref.** *17*). This may indicate that there exist two different populations of cells in which the nucleosome is present or absent. Such states may represent a true binding equilibrium in vivo, or intermediate states with rather different subunit compositions of the nucleosome. For example, loss of one or both H2A/H2B dimers could result in intermediate accessibility.

Acknowledgments

Work from the authors' laboratory was supported by the DFG (SFB 190), the European Commission Human Capital and Mobility Network (ERBCHRXCT940447) and Fonds der Chemischen Industrie.

References

1. Patterton, D. and Wolffe, A. P. (1996) Developmental roles for chromatin and chromosomal structure. *Dev. Biol.* **173,** 2–13.
2. Kingston, R. E., Bunker, C. A., and Imbalzano, A. N. (1996) Repression and activation by multiprotein complexes that alter chromatin structure. *Genes Dev.* **10,** 905–920.
3. Li, Q., Wrange, O. and Eriksson, P. (1997) The role of chromatin in transcriptional regulation. *Int. J. Biochem. Cell Biol.* **29,** 731–742.
4. Svaren, J. and Hoerz, W. (1996) Regulation of gene expression by nucleosomes. *Curr. Opin. Gen. Dev.* **6,** 164–170.
5. Lu, Q., Wallrath, L. L., and Elgin, S. C. R. (1995) The role of a positioned nucleosome at the *Drosophila melanogaster* hsp26 promoter. *EMBO J.* **14,** 4738–4746.
6. Archer, T. K., Lefebvre, P., Wolford, R. G., and Hager, G. L. (1992) Transcription factor loading on the MMTV promoter: a bimodal mechanism for promoter activation. *Science* **255,** 1573–1576.
7. Wintersberger, U., Smith, P., and Letnansky, K. (1973) Yeast chromatin: Preparation from isolated nuclei, histone composition and transcription capacity. *Eur. J. Biochem.* **33,** 123–130.
8. Aris, J. P. and Blobel, G. (1991) Isolation of yeast nuclei. *Methods Enzymol.* **194,** 735–749.
9. Lohr, D. (1988) Isolation of yeast nuclei and chromatin for studies of transcription-related processes, in *Yeast: A Practical Approach* (Campbell, I. and Duffus, J. H., eds.), IRL Press, Oxford, pp. 125–145.

10. Fedor, M. J., Lue, N. F., and Kornberg, R. D. (1988) Statistical positioning of nucleosomes by specific protein binding to an upstream activating sequence in yeast. *J. Mol. Biol.* **204,** 109–127.
11. Hull, M. W., Thomas, G., Huibregtse, J. M., and Engelke, D. R. (1991) Protein-DNA interactions in vivo—examining genes in *Saccharomyces cerevisiae* and *Drosophila melanogaster* by chromatin footprinting. *Methods Cell Biol.* **35,** 383–415.
12. Huibregtse, J. M. and Engelke, D. R. (1991) Direct sequence and footprint analysis of yeast DNA by primer extension. *Methods Enzymol.* **194,** 550–562.
13. Kent, N. A., Bird, L. E., and Mellor, J. (1993) Chromatin analysis in yeast using NP-40 permeabilised sphaeroplasts. *Nucleic Acids Res.* **21,** 4653,4654.
14. Schmid, A., Fascher, K. D., and Hörz, W. (1992) Nucleosome disruption at the yeast *PHO5* promoter upon *PHO5* induction occurs in the absence of DNA replication. *Cell* **71,** 853–864.
15. Almer, A., Rudolph, H., Hinnen, A., and Hörz, W. (1986) Removal of positioned nucleosomes from the yeast *PHO5* promoter upon *PHO5* induction releases additional upstream activating DNA elements. *EMBO J.* **5,** 2689–2696.
16. Almer, A. and Hörz, W. (1986) Nuclease hypersensitive regions with adjacent positioned nucleosomes mark the gene boundaries of the *PHO5/PHO3* locus in yeast. *EMBO J.* **5,** 2681–2687.
17. Barbaric, S., Fascher, K. D., and Hörz, W. (1992) Activation of the weakly regulated *PHO8* promoter in *S. cerevisiae*: chromatin transition and binding sites for the positive regulator protein Pho4. *Nucleic Acids Res.* **20,** 1031–1038.
18. Pavlovic, B. and Hörz, W. (1988) The chromatin structure at the promoter of a glyceraldehyde phosphate dehydrogenase gene from *Saccharomyces cerevisiae* reflects its functional state. *Mol. Cell. Biol.* **8,** 5513–5520.

31

Genomic Footprinting Using Nucleases

Lucia Cappabianca, Hélène Thomassin, Raymond Pictet, and Thierry Grange

1. Introduction

Gene expression is regulated by complex mechanisms involving dynamic interactions between cis-acting elements and trans-acting factors in a highly structured chromatin environment. Investigations of protein/DNA interactions in vitro may not have relevance to a living cell system. To analyze events occurring at the DNA level in a living cell, Church et al. introduced the genomic footprinting procedure *(1)*. The procedure was developed further *(2)* but ligation-mediated PCR (LM-PCR) was the breakthrough that rendered the technique accessible to many laboratories *(3)*.

Genomic footprinting comprises two main steps:

1. In vivo DNA modification using either chemicals (e.g., DMS, $KMnO_4$, Fe-EDTA, ortho-phenanthrolin copper) or enzymes (e.g., DNaseI, Micrococcal nuclease, restriction enzymes, exonuclease III).
2. Visualization of the modifications using purified cellular DNA.

The choice of the footprinting reagent and reaction conditions of use is the first crucial point of this technique. For the detection of footprints and their correct interpretation the best compromise must be found between the sensitivity provided by the selected reagent and the treatment necessary to render the DNA accessible to the reagent *(4,5)*. Several nucleases are well suited to study local chromatin structure as well as DNA-protein interactions, particularly DNaseI and Micrococcal Nuclease (MNase). DNaseI provides information on both binding of transcription factors and on the rotational positioning of nucleosomes whereas MNase is used to analyze translational positions of nucleosomes *(6–11)*. The drawback of nucleases as in vivo footprinting reagents is that they require either permeabilization of the cell membrane or cell lysis and isolation of nuclei. These two procedures obviously may interfere to a variable extent with nuclear processes. We describe two different cell permeabilization procedures that have given us satisfactory results depending on the nucleases used and a nuclei isolation procedure suitable for use with both nucleases (*see* **Note 1**).

The second step of the footprinting procedure is the visualization of the in vivo modifications using LM-PCR. The original LM-PCR procedure was devised to allow the mapping of any single-stranded cut in genomic DNA *(3)*. An adaptation to map exclusively the double-stranded cuts that have blunt ends has been proposed to visualize a subset of MNase cleavage sites *(8)*. We describe three alternative LM-PCR procedures that differ in the first step (**Fig. 1**). They allow the comparative mapping of the various types of cuts that can be generated in vivo by nucleases: single-stranded cuts (A in **Fig. 1**), blunt-ended double-stranded cuts with staggered ends (B and C in **Fig. 1**) and double-stranded cuts with blunt ends (D in **Fig. 1**). All procedures rely on the ligation of a double-stranded DNA linker onto double-stranded blunt ends corresponding to the cleavage sites of genomic DNA. The procedures differ in the way these blunt ends are generated. In the original LM-PCR procedure (upper central section of **Fig. 1**), all cuts are visualized because the DNA is denatured and double-stranded blunt ends are generated at the sites of cleavage by extension of a gene-specific primer with a DNA polymerase (Vent exo⁻ is the most suitable for this application; *4*). To map exclusively the double-stranded cuts with blunt ends in genomic DNA, the linker is ligated directly without prior denaturation and primer extension (upper right section of **Fig. 1**). In order to map staggered double-stranded cuts, ends are blunted without prior DNA denaturation using T4 DNA polymerase (left upper part of **Fig. 1**). The recessed 3'-ends (B in **Fig. 1**) are extended by the polymerase activity of this enzyme while the protruding 3'-ends (C in **Fig. 1**) are removed by its 3'-5' exonuclease activity. In both cases, the original position of the 5'-ends (those mapped by the LM-PCR procedure) is unaffected by the treatment. Once the blunt ends have ben generated, the linker is ligated and all subsequent steps follow the original LM-PCR procedure *(3,4)*.

It can prove useful to compare the various types of ends generated by MNase during a genomic footprinting reaction. MNase cleaves single-stranded DNA very efficiently and double-stranded DNA at a lower rate with a specificity that suggests it prefers transiently single-stranded regions *(12)*. In the initial phase of digestion, cleavages are predominantly endonucleolytic whereas later the proportion of exonucleolytic cleavages increases. When acting on chromatin, the enzyme cuts both linker and nucleosomal DNA but the rate of linker DNA cleavage is much faster. This later property has rendered MNase popular

Genomic Footprinting Using Nucleases

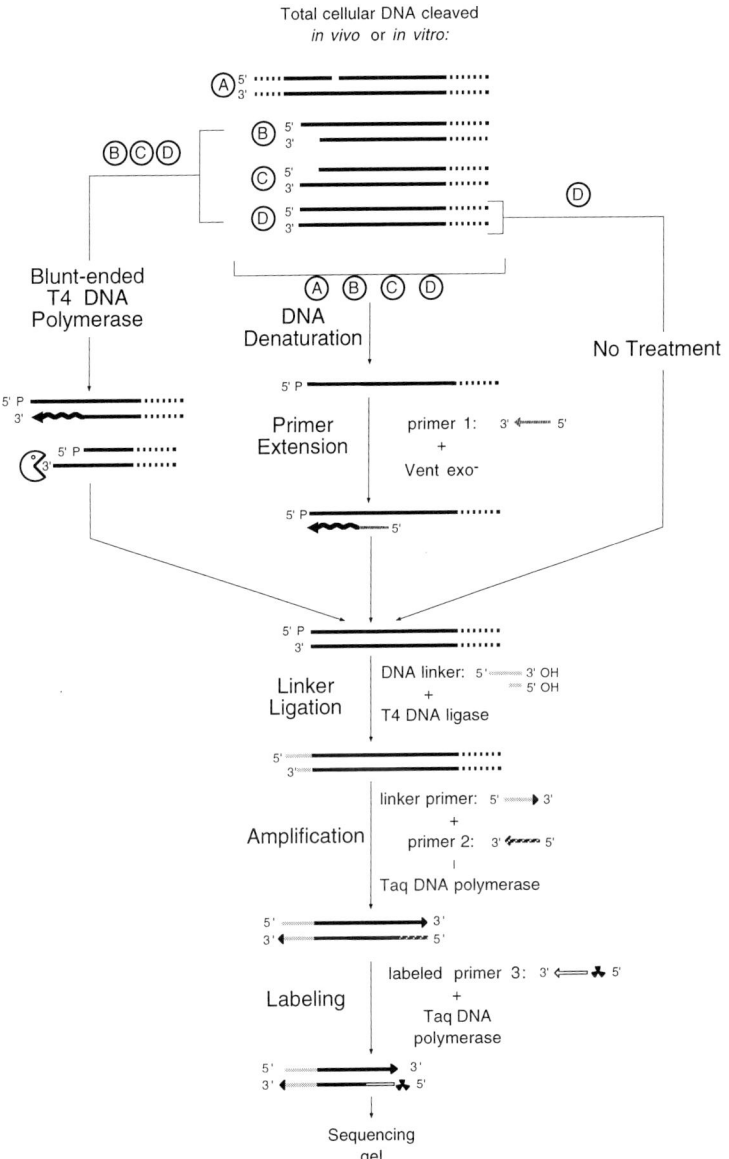

Fig. 1. Schematic description of the three LM-PCR procedures. The end of the DNA molecules that is going to be visualized is represented as a straight line, while the other end that does not participate in the reaction is shown as a dotted line. Circled letters indicate the different types of ends that are going to be mapped by the alternative procedures. Wavy lines indicate DNA polymerization whereas the Pacman™ indicates exonucleolytic cleavage. Since MNase cleavage leaves 5'-OH ends, it is necessary to phosphorylate this end using T4 PNK. This extra step has not been represented.

to map nucleosome boundaries (*3*). When cutting inside the nucleosome, MNase makes double-stranded cuts that have a 5'-end extended stagger of two nucleotides, (*14,15*). Once a staggered double-stranded cut is generated, one would predict it should be converted into a blunt end because of the preference of the enzyme for single-stranded DNA. It has been proposed that the high-resolution mapping of blunt ends should reveal preferentially the ends that cleaved the fastest and hence the linker position (*8*). **Figure 2** shows the comparative analysis of MNase cleavage patterns revealed by the three alternative procedures. The region analyzed corresponds to the -2500 glucocorticoid responsive unit (GRU) of the rat tyrosine aminotransferase (TAT) gene, where no strict nucleosomal positioning is clearly observed at high resolution. The results show that most of the MNase cuts of the naked DNA are readily converted into blunt ends, because the patterns are quite similar whichever LM-PCR adaptation was used (compare lanes 2, 4, and 6). However, the pattern derived form Mnase cleavage in nuclei varies depending on the visualisation strategy used. All cleavage positions are ultimately converted into blunt ends, but with variable efficiency. Cleavage at some positions occurs more readily in the form of staggered cuts and a subset of these are more readily converted into blunt ends (compare lanes 1, 3, and 5 together and with lanes 2, 4, and 6). The cuts that give rise to blunt ends are concentrated within a 90 bp region, whereas those that result in staggered cuts are found in an area of about 25 bp on both sides of this region. These results suggest that the 90 bp region is preferentially found within a linker, roughly in agreement with previous low resolution mapping of nucleosomal positioning (*16*). It is not clear whether the multiple bands observed within the 90 bp region correspond to multiple cleavages within an accurately positioned linker or whether they are due to the existence of multiple phases within the population and complete digestion of the entire linker region. We tend to favor the second interpretation. Indeed, such multiple phases have been described for another regulatory region using an alternative high resolution nucleosome mapping strategy (*7*). Furthermore, we have not observed a clear nucleosomal positioning using an alternative footprinting agent: ortho-phenanthrolin copper (OP_2Cu). In contrast, clearer nucleosomal

---

Fig. 2. *(continued on next page)* Comparative analysis of the three LM-PCR procedures for visualizing MNase cleavages in vitro and in vivo. The upper strand of the -2500 GRU of rat *Tat* gene is analyzed using the set of primers described in (*10*). In vivo indicates permeabilized rat fibroblasts cells FR3T3 (*22*) treated with 90 U/mL of MNase (lanes 1, 3, and 5); "naked DNA" indicates genomic DNA treated in vitro using 40 U/mL of MNase/100 µg DNA for 5 min at 0°C (lanes 2, 4, and 6); "Primer extension" indicates samples analyzed using the original LM-PCR procedure, i.e., by DNA denaturation followed by primer extension (lanes 1 and 2); "Direct linker Liga-

Genomic Footprinting Using Nucleases

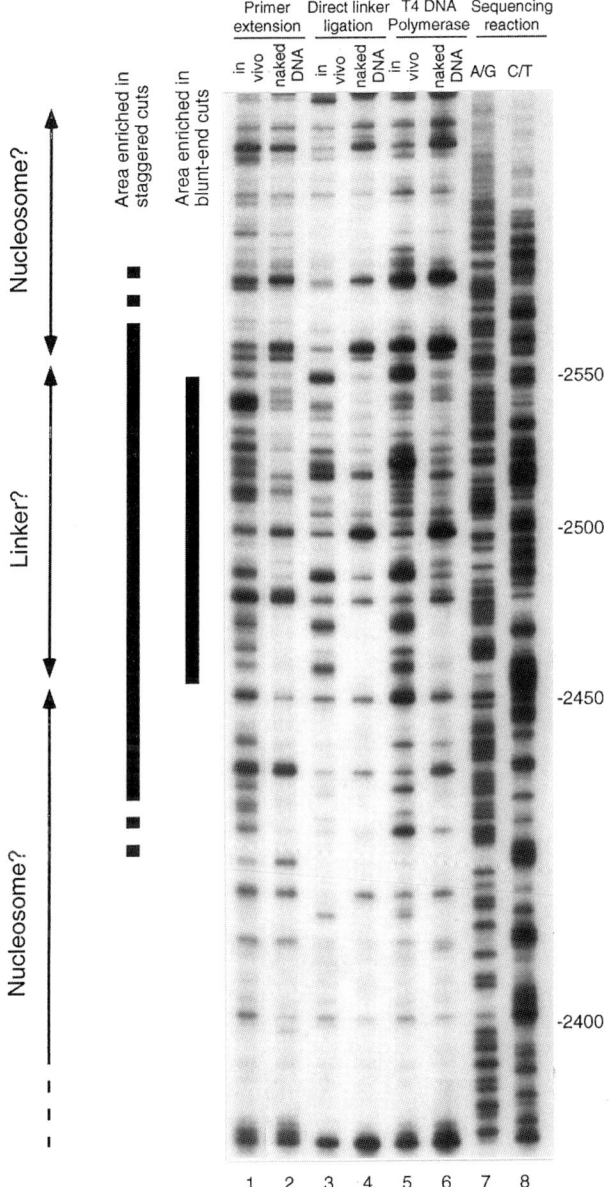

tion" indicates the procedure where these steps have been omitted (lanes 3 and 4); "T4 DNA polymerase" indicates the procedure that uses this enzyme to convert staggered ends into blunt ends (lanes 5 and 6); A/G and C/T indicate the purine and pyrimidine genomic sequences (lanes 7 and 8, respectively). To the left, areas enriched in double-stranded and staggered cuts after genomic footprinting with Mnase are indicated by thick lines, along with the deduced putative favored nucleosomal positions.

positioning has been observed when analyzing other regulatory regions with similar high resolution in vivo footprinting procedures using either MNase or OP_2Cu *(9,17)*.

We describe herein a detailed step-by-step protocol of the three mapping alternatives.

2. Materials

Solutions should be stored at the temperature indicated in brackets and, where indicated, should be sterilized by filtration through a 0.45-μm filter. The asterisk (*) indicates the products that should be added to the solutions just before use. To prepare all solutions we use sterile distilled water (dH_2O).

2.1. Nuclease Treatment In Vivo

2.1.1. Cell Permeabilization for DNaseI Treatment

1. DNaseI solution (Worthington, Lakewood, NJ). Dissolve at 1 mg/mL in distilled water. Store the stock frozen (–70°C) in small aliquots and use only once.
2. PBS 1X pH 7.5 (4°C).
3. 0.25 M PMSF in ethanol.
4. Trypsin solution (–20°C).
5. Buffer Ψ (filtered, -20°C): 11 mM phosphate buffer, pH 7.4, 108 mM KCl, 22 mM NaCl, 1 mM $MgCl_2$, 1 mM DTT, 1 mM ATP.
6. Lysis buffer: 50 mM Tris-HCl, pH 8.0, 20 mM EDTA, 1% SDS.
7. 10% NP-40.

2.1.2. Cell Permeabilization for MNase Treatment

1. MNase solution (Worthington). Dissolve at 15000 U/mL in distilled water. Store the stock frozen (–70°C) in small aliquots and use only once.
2. PBS 1X pH 7.5 (4°C).
3. Buffer A (filtered, –20°C): 150 mM sucrose, 15 mM Tris-HCl, pH 7.5, 15 mM NaCl, 60 mM KCl, 2 mM $CaCl_2$, 0.15 mM spermine*, 0.5 mM spermidine*.
4. Lysolecithin* in buffer A at 0.5 mg/mL (prepared just before use).
5. Stop solution: 40 mM Tris-HCl, pH 8.0, 40 mM EDTA, 150 mM NaCl, 2% SDS, 1.2 mg/mL Proteinase K*.

2.1.3. Nuclei Preparation

1. PBS, pH 7.5 (4°C).
2. 0.25 M PMSF in ethanol.
3. Trypsin solution (–20°C).
4. Buffer C (filtered, –20°C): 10 mM Tris-HCl, pH 7.5, 15 mM NaCl, 60 mM KCl, 1 mM EDTA pH 8.0, 10% sucrose, 0.15 mM spermine*, 0.5 mM spermidine*.
5. Buffer H (filtered, –20°C): 10 mM Tris-HCl, pH 7.5, 15 mM NaCl, 60 mM KCl, 1 mM EDTA pH 8.0, 0.2% Nonidet P-40, 5% sucrose, 0.15 mM spermine*, 0.5 mM spermidine*.

6. Buffer W (filtered, –20°C): 10 mM Tris-HCl, pH 7.5, 15 mM NaCl, 60 mM KCl, 0.15 mM spermine*, 0.5 mM spermidine*.
7. Lysis buffer: 50 mM Tris-HCl, pH 8.0, 20 mM EDTA, 1% SDS.
8. 0.1 M CaCl$_2$ or 0.1 M MgCl$_2$ (–20°C) (*see* **Subheading 3.1.3., step 16**).
9. MNase or DNaseI solutions (*see* **Subheadings 2.1.1.** and **2.1.2.**).
10. 1% SDS.
11. 30-mL Dounce homogenizer(s) with loose pestle.

2.2. DNA Preparation

1. Phenol saturated with Tris-HCl, pH 8.0 (4°C).
2. 100% and 70% ethanol (–20°C).
3. 4 M NH$_4$OAc (4°C).
4. TE (4°C): 10 mM Tris-HCl, pH 7.5, 1 mM EDTA.

2.3. LM-PCR

2.3.1. DNA Phosphorylation

1. 10 mM ATP (–20°C).
2. 10X Kinase buffer (–20°C): 500 mM Tris-HCl, pH 7.5, 100 mM MgCl$_2$, 50 mM DTT, 1 mM EDTA, pH 8.0, 1 mM spermidine.
3. T4 polynucleotide kinase 10 U/µL (–20°C).

2.3.2. Direct Linker Ligation and LM-PCR Steps Common to All Procedures

1. T4 DNA ligase (–20°C).
2. Taq DNA polymerase (Perkin Elmer, Norwalk, CT) (–20°C).
3. T4 polynucleotide kinase (–20°C).
4. Annealed linker solution (–20°C): 20 pmoles of each oligonucleotide that form the linker (*3*), 250 mM Tris-HCl, pH 7.5, 5 mM MgCl$_2$. After mixing, the solution is heated 5 min at 95°C, cooled at room temperature for 20 min, and aliquoted before freezing.
5. 5 pmole/µL primer #2 (–20°C) and 12 pmole/µL Primer #3 (–20°C) (*see* **Note 4**).
6. 1 M Tris-HCl, pH 7.5.
7. 1 M DTT (–20°C).
8. 10 mg/mL DNase free BSA (-20°C).
9. 100 mM ATP pH 7.5 (–20°C).
10. 3 M and 0.3 M NaOAc pH 5.5 (4°C).
11. 10X Taq buffer (filtered, –20°C): 650 mM Tris-HCl, pH 8.8, 400 mM NaCl, 100 mM β-mercaptoethanol.
12. 1 M MgCl$_2$ and 25 mM MgCl$_2$ (–20°C).
13. 10 mg/mL tRNA (–20°C).
14. 5 mM 4 dNTPs (–20°C).
15. 10X Kinase buffer (–20°C): 500 mM Tris-HCl, pH 7.5, 100 mM MgCl$_2$, 50 mM DTT, 1 mM EDTA pH 8.0, 1 mM spermidine.
16. γ-^{32}P-ATP 20 µCi/µL (–20°C).

17. Mineral oil.
18. Saturated phenol/chloroform (3/1) (4°C).
19. 100% and 70% ethanol (–20°C).
20. Formamide-loading buffer (–20°C): 95% formamide, 10 mM EDTA, 0.05% bromophenol blue, 0.05% xylene cyanol

2.3.3. DNA Denaturation and Primer Extension

1. Vent exo⁻ DNA polymerase (Biolabs, Beverly, MA) (–20°C).
2. 0.3 pmole/µL Primer #1 (–20°C) (*see* **Note 4**).
3. 5X First stand buffer (–20°C): 50 mM Tris-HCl, pH 8.9, 25 mM MgSO$_4$, 200 mM NaCl, 100 µg/mL BSA (DNase free).
4. 5 mM 4 dNTPs (–20°C).

2.3.4. Blunt-Ending with T4 DNA Polymerase

1. T4 DNA polymerase (Biolabs) (–20°C).
2. 10X T4 polymerase buffer (–20°C): 100 mM Tris-HCl, pH 7.9, 10 mM DTT, 500 mM NaCl, 100 mM MgCl$_2$, 500 µg/mL BSA (DNase free).
3. 5 mM 4 dNTPs (–20°C).

3. Methods
3.1. Nuclease Treatment In Vivo

The first two protocols are designed for $5\text{–}10 \times 10^6$ cells grown on 10-cm Petri dishes (approx to 50% confluence). It is recommended to test a range of nuclease concentrations when working with a new cell line (*see* **Note 2**).

3.1.1. Cell Permeabilization for DNaseI Treatment

1. Wash the cells with 10 mL of PBS (37°C).
2. Cover the cells with 1 mL of trypsin (37°C) and remove immediately and wait a few minutes until the cells start to detach.
3. Harvest the cells in 10 mL of cold PBS/0.25 mM PMSF (4°C) and collect by centrifugation, 3 min at 500g at 4°C.
4. Wash the cells in 1 mL of Ψ buffer (4°C).
5. Resuspend in 200 mL of Ψ buffer and put on ice.
6. To 100 µl of cells suspension, add 100 µL of DNaseI solution in Ψ buffer supplemented with 0.4% NP-40 (*see* **Note 2**).
7. Incubate 3 min at 4°C and stop the reaction by adding 5 mL of lysis buffer.
8. Let stand a few minutes at room temperature and continue with DNA purification (*see* **Subheading 3.2.**).

3.1.2. Cell Permeabilization for MNase Treatment

1. Wash the cells with 10 mL of PBS (37°C).
2. Cover the cells with 2 mL of buffer A (37°C) supplemented with 0.5 mg/mL lysolecithin, let stand for 1 min, and remove.

Genomic Footprinting Using Nucleases 435

3. Cover the cells with 2 mL of buffer A (37°C) containing MNase (*see* **Note 2**).
4. Incubate 2 min at 37°C and stop the reaction with 2 mL of stop solution.
5. Incubate for 1–3 h at 50°C and continue with DNA purification (*see* **Subheading 3.2.**).

3.1.3. Nuclei Preparation

The protocol is designed for approx 10–20 million cells grown on 15 cm Petri dishes (approx to 50% confluence). Before beginning, pre-cool the centrifuge at 4°C and put the Dounces, their pestles, the tubes and the indicated solutions on ice.

1. Wash the cells with 30 mL of PBS (37°C).
2. Cover the cells with 3 mL of trypsin (37°C) and remove immediately.
3. Wait a few minutes until the cells start to detach and harvest them in 20 mL of PBS/0.25 mM PMSF (4°C). From now on perform all steps at 4°C if not indicated otherwise.
4. Collect the cells by centrifugation, 3 min at 500g.
5. Wash the pellet with 10 mL of PBS (4°C) and repeat **step 4**.
6. During the centrifugations prepare a 4 mL cushion of buffer C in a 15 mL polypropylene tube and keep on ice.
7. Lyse the cells by gently resuspending the pellet in 7 mL of buffer H using a 10 mL plastic pipet (avoid bubbles!).
8. Transfer the suspension into a glass dounce homogenizer and homogenize it by three strokes with a loose pestle (check for lysis with a phase contrast microscope).
9. Carefully layer the nuclear suspension on top of the cushion.
10. Spin 5 min at 500g.
11. Aspirate off the supernatant with a Pasteur pipet connected to a vacuum starting from the top and rapidly invert the tube to eliminate the remaining supernatant, dry the tube wall with absorbent paper and keep the tube on ice.
12. Wash the pellet in 10 mL of cold buffer W, resuspending it by successive aspirations with a cooled plastic pipet (avoid bubbles!) and spin 5 min at 500g.
13. On ice, resuspend the nuclei in 1 mL of buffer W.
14. Measure the OD$_{260nm}$ as follows; dilute 50 µL of nuclear suspension in 450 µL of 1% SDS in a quartz cuvette. Cover with parafilm and invert several times, measure against a blank consisting of 50 µL of buffer W in 450 µL of 1% SDS.
15. Adjust the concentration to 15 OD/mL using buffer W.
16. To 1 mL of nuclear suspension, add 10 µL of 0.1 M MgCl$_2$ (or 10 µL of 0.1 M CaCl$_2$ for MNase) solution and a given amount of nuclease (*see* **Note 2**).
17. Incubate 10 min on ice (DNaseI) or 15 min at 25°C (MNase).
18. Stop the reaction by adding 5 mL of lysis buffer and continue with DNA purification (*see* **Subheading 3.2.**).

3.2. DNA Preparation

1. After nuclease digestion, add one volume of phenol, mix gently for 5–10 min and centrifuge 10 min at 5000g to separate the phases.

2. Transfer the aqueous phase into a fresh tube and repeat the phenol extraction.
3. Add 5 mL of NH$_4$OAc and 25 mL 100% ethanol to precipitate the DNA, incubate 15 min on ice and centrifuge 20 min at 5000g at 4°C.
4. Resuspend the pellet in 5 mL of TE and repeat the NH$_4$OAc/ethanol precipitation step.
5. Wash with 5 mL of 70% ethanol and dry the pellet.
6. Resuspend the DNA in 200 mL of TE, measure the OD$_{260}$. We assume 50 mg/OD$_{260}$ of nucleic acid and neglect the RNA contamination in the DNA amount indicated subsequently.
7. Check for a correct nuclease digestion pattern on a 1% agarose gel.
8. Store the DNA at –20°C.

3.3. LM-PCR

3.3.1. DNA Phosphorylation

When using MNase or any other nuclease that leaves 5'-OH ends, it is necessary to phosphorylate these ends with T4 Polynucleotide Kinase before proceeding through LM-PCR.

1. In a reaction tube, add 10 µg of DNA in 22 µL dH$_2$O.
2. Add 3 µL of 10 mM ATP, 3 µL of 10X kinase buffer and 2 µL of T4 polynucleotide kinase (10 U/µL).
3. Mix and incubate 1 h at 37°C.
4. Stop the reaction by incubation for 10 min at 75°C.
5. Add 20 µL of dH$_2$O to a final DNA concentration of 0.2 µg/µL and store it at –20°C.

3.3.2. Direct Linker Ligation and LM-PCR Steps Common to All Procedures

1. In a reaction tube add 5 µL of phosphorylated DNA (*see* **Subheading 3.3.1.**, or 1 µg of untreated DNA in 5 µL) to 25 µL of dH$_2$O.
2. Add 20 µL of Mix C containing per reaction: 2.2 µL of 1 M Tris-HCl, pH 7.5, 0.35 µL of 1 M MgCl$_2$, 1 µL of 1 M DTT, 0.25 µL of 10 mg/mL BSA, 16.2 µL of dH$_2$O, then add 25 µL of Mix D containing per reaction: 0.75 µL of 100 mM ATP, 0.25 µL of 1 M MgCl$_2$, 0.5 µL of 1 M DTT, 0.125 µL of 10 mg/mL BSA, 15.4 µL of dH$_2$O, 5 µL of annealed linker solution, 3 µL of T4 DNA ligase (1 U/µL), and incubate overnight at 17°C.
3. Next day put the tube on ice and precipitate the DNA by adding 8 µL of 3 M sodium acetate, pH 5.5 and 250 µL of 100% ethanol.
4. Incubate 30 min at 4°C, spin for 15 min, rinse with 70% ethanol, dry.
5. Resuspend pellet in 85 µL of Mix E containing per reaction: 10 µL of 10X Taq buffer, 10 µL of 25 mM MgCl$_2$, 2 µL of Primer #2 (5 pmole/µL), 1 µL of 10 mg/mL BSA, 62 µL of dH$_2$O.
6. Cover with mineral oil and incubate 2 min at 95°C.
7. Add 15 µL of Mix F containing per reaction: 1.5 µL of 10X Taq buffer, 1.5 µL of 25 mM MgCl$_2$, 8 µL of 5 mM dNTPs, 0.1 µL of 10 mg/mL BSA, 1 U of Taq DNA polymerase, made up to 15 µL with dH$_2$O.

Genomic Footprinting Using Nucleases

8. Amplify as follows: 4 min at 94°C then 40 s at 94°C, 2 min at $Td_{primer\ \#2}$ –5°C (*see* **Note 4**) and 3 min.at 76°C for 25 cycles, finish with a 7 min incubation at 76°C. Store at –70°C.
9. For primer labeling, mix: 1 µL of primer #3 (12 pmole/µL); 1 µL (20 µCi) of γ-^{32}P-ATP, 1 µL of 10X kinase buffer, 6.5 µL of dH$_2$O and 0.5 µL of T4 polynucleotide Kinase (10 U/µL). Incubate 30 min at 37°C and put on ice.
10. Cover 15 µL of the amplification reaction with mineral oil, incubate 2 min at 95°C and than add 5 µL of Mix G that contains per reaction: 0.25 µL of the primer labeling reaction, 0.5 µL of 10X Taq buffer, 0.4 µL of 25 mM MgCl$_2$, 1.5 µL of 5 mM dNTPs, 0.5–1 U of Taq DNA polymerase, made up to 5 µL with dH$_2$O.
11. Amplify as follows: 2 min at 94°C then 40 s at 94°C, 3 min at $Td_{primer\ \#3}$ –5°C and 5 min.at 76°C for 5 cycles.
12. Add 80 µL of 0.3 M sodium acetate, pH 5.5 containing tRNA at 0.1 µg/µL and transfer the aqueous phase (without mineral oil) in a fresh tube.
13. Extract with 100 µL of phenol/chloroform (3/1) and precipitate with 300 µL of 100% ethanol. Incubate on ice for 30 min. Spin down for 10 min rinse the pellet with 70% ethanol and dry.
14. Resuspend the pellet in 8 µL of formamide-loading buffer, incubate 3 min at 95°C and analyse 2 µL on a sequencing gel.

3.3.3. DNA Denaturation and Primer Extension

1. To 5 µL of phosphorylated DNA (*see* **Subheading 3.3.1.**, or 1 µg of untreated DNA in 5 µL), add 20 µl of Mix A containing per reaction: 5 µL of 5X first-strand buffer, 1 µL of primer #1 (0.3 pmole/µL) and 14 µL of dH$_2$O.
2. Incubate 5 min at 95°C then 30 min.at $Td_{primer\ \#1}$ –5°C.
3. Add 5 µL of Mix B containing per reaction: 1 µL of 5X first-strand buffer, 1.2 µL of 5 mM dNTPs, 2.5 µL of dH$_2$O and 0.25 µL Vent exo$^-$ DNA polymerase (2 U/µL).
4. Put the tube in a water bath set at $Td_{primer\ \#1}$ –5°C (*see* **Note 4**) for 5–10 min and raise the temperature to 76°C (it should take approx 10 min to reach 76°C) then incubate for 10 min at 76°C.
5. Put the tube on ice and continue with linker ligation (*see* **Subheading 3.3.2., step 2**).

3.3.4. Blunt-Ending with T4 DNA Polymerase

1. In a reaction tube, add 5 µL of phosphorylated DNA (*see* **Subheading 3.3.1.**, or 1 mg of untreated DNA in 5 mL) to 3 µL of 10X T4 polymerase buffer, 1 µL of 5 mM 4 dNTPs, 19 µL of dH$_2$O and 2 µL of T4 DNA polymerase (1 U/mL).
2. Mix and incubate 30 min at room temperature, then incubate 10 min at 75°C to stop the reaction.
3. Put on ice and continue with linker ligation (*see* **Subheading 3.3.2., step 2**).

4. Notes

1. Generally, cell permeabilization is a better choice than nuclei isolation because of the minimal delay separating cell membrane lysis and footprinting reaction. This can improve the detectability of some DNA-protein interactions when using

DNaseI *(5)*, although we have not observed differences in the MNase patterns obtained with permeabilized cells and nuclei. Depending on the cell type being analyzed, nuclei preparation might be preferred because of problems caused by some components of the cytoplasm, particularly endogenous nucleases or nucleases inhibitors. The cell permeabilization might result in the release of lysozomal nucleases leading to spurious degradation of the DNA that can complicate the analysis, particularly if the cells contain high levels of these nucleases (note that background cleavage due to endogenous nucleases is higher with NP-40 than with lysolecithin). Furthermore, nuclease inhibitors can cause problems in some cell types. Since actin inhibits DNaseI *(18)*, larger amounts of DNase is required when working with permeabilized cells than when working with nuclei. Muscle cells are likely to contain too much actin to permit the use of DNaseI on permeabilized cells. The homogeneity and the reproducibility of the permeabilization procedure can also be an important criterion of choice in some instances. This is why we prefer to use the NP-40 permeabilization procedure when we treat hepatoma cell lines with DNaseI *(6,10)*. Lysolecithin permeabilisation is variable with these cells and hence the determination of the optimal DNaseI amount is tricky (*see* **Note 2**). We believe this is due to the plate-to-plate variations in the amount of inhibitory actin released since lysolecithin treatment works well when the same cells are treated with MNase.

2. A careful titration of the nuclease amount must be performed to avoid overdigestion of the DNA, which may lead to the rearrangement of nucleoprotein complexes. Ideally, conditions should be adjusted to produce only a single hit in the area to be analysed. However, low levels of nuclease digestion may be difficult to distinguish from the background cleavage caused by endogenous nucleases (as discussed in **Note 1**, this level of background cleavage is influenced by the membrane disruption treatment). The presence of nuclease inhibitors affects the way the titration curve should be performed because it can result in a sharp threshold effect. If the amount of nuclease does not exceed the capacity of the inhibitors, almost no cleavage is observed but once this threshold is reached, overdigestion can rapidly occur. This is a problem when using DNaseI because of the aforementioned inhibition by actin whereas we have not observed such a behavior with MNase. This is why when working with permeabilized cells, we titrate the DNaseI in small increments. When a cell line is analysed initially, we suggest to test 0, 5, 10, 20, 40, and 80 µg of DNaseI for 5×10^6 cells in 200 µL of Ψ buffer + NP-40. When working with MNase, we cover a wider range of concentration: 0, 10, 30, 90, 270 U/mL of MNase in buffer A.. When working with nuclei, we recommend to test 0.5, 1, 2 , 4, 10, µg of DNAseI per mL of nuclei suspension at 15 OD_{260}/ml and 0, 10, 30, 90, 270, 800 U of MNase/mL of nuclei suspension at 15 OD_{260}/mL. The extent of the digestion of the genomic DNA is analyzed both on an agarose gel and by LM-PCR. We obtained satisfactory DNaseI footprinting results with DNA that begins to smear on the agarose gel. The MNase footprinting pattern varies slightly with the extent of digestion and we have not yet completely established the significance of these variations. Cur-

rent opinion is that all regions are not digested at the same rate and that the extent of opening of the chromatin structure affects this rate. This is potentially a source of erroneous interpretations. If there is heterogeneity in the cell population with a mixture of different degrees of opening, it is possible that only a subset of these can be detected at a given nuclease amount. Indeed a given amount might not cleave at all at a closed region whereas it digests optimally at an open one and conversely a higher amount that digests optimally at the closed region might degrade an open region entirely, effectively removing it from the population under analysis. If the PCR is performed at the plateau (*see* **Note 6**), differences in nuclease accessibility in a population might be missed, giving the erroneous impression that the chromatin structure is not modified.
3. Two kinds of problem can adversely affect the quality of the final results of the LM-PCR: lack of specificity (*see* **Note 4**) and lack of fidelity (*see* **Note 5**) of the reactions. As a control for specificity and we recommend to visualise two genomic sequence ladders as markers, one revealing all purines, the other all pyrimidines *(19)*. These can be performed as follows: Purine Ladder: To 10 µg of genomic DNA in 20 µL H_2O add 25 µL pure formic acid, mix with a pipetman and incubate 4 min at 20°C. Put on ice, immediately add 200 µL cooled Stop Solution (0.3 M NaOAc, pH 7.0, 1 mM EDTA), mix and precipitate with 750 µL ethanol. Resuspend the pellet in 250 µL 0.3 M NaOAc, pH 5.5 and precipitate again with 750 µL ethanol. Rinse the pellet with 1 mL 70% ethanol and air-dry. Proceed with piperidine cleavage *(see below)*. Pyrimidine Ladder: To 10 µg of genomic DNA in 20 µL H_2O add 30 µL hydrazine (highly toxic: use a fume hood and inactivate by incubating wastes in a 15% $FeCl_3$ solution), mix gently with a pipetman and incubate 4 min at 20°C. Put on ice, immediately add 200 µL cooled stop solution, mix and precipitate with 750 µL ethanol. Resuspend the pellet in 250 µL 0.3 M NaOAc, pH 5.5 and precipitate again with 750 µL ethanol. Rinse the pellet with 1 mL 70% ethanol and air-dry. Proceed with piperidine cleavage. Piperidine Cleavage: Dissolve the DNA in 100 µL of a fresh 1 M piperidine solution (toxic: use a fume hood) and incubate 30 min at 95°C. Use a heavy weight to keep the tube lids from opening as a result of increased vapor pressure. After incubation, keep the tubes on ice and spin down a few seconds. Add 1.2 mL of *n*-butanol and vortex vigorously for 15 s then centrifuge for 10 min at 12,000g. Remove the supernatant containing both piperidine and n-butanol and wash the pellet once with 1 mL 95% ethanol. Dry the pellet under vacuum and resuspend it in 100 µL of 1% SDS. Transfer the suspension into a fresh tube and repeat the *n*-butanol extraction, then wash the pellet twice with 1 mL 70% ethanol and dry under vacuum. Dissolve the pellet in 10 µL of TE. The sequences are visualized using the original LM-PCR procedure (using the DNA denaturation and primer extension steps) (*see* **Subheading 3.3.**).
4. The overall specificity of the LM-PCR procedure depends on the specificity of the PCR steps. In the original LM-PCR, the linker ligation step was designed so that it occurs specifically on the sequence of interest *(3)*, but we have no evidence that this step is really specific and that such specificity is necessary. In the two

alternative procedures described here, linker ligation is not specific and this does not adversely affect the final results. In the RL-PCR procedure we have developed for RNA molecules, linker ligation is also non-specific, and yet it works *(20)*. The crucial point is that the complement of the linker is synthesized specifically during the first and subsequent steps of the PCR amplification because this ensures that there are no sequences complementary to one of the amplification primers at undesired locations. Primer design is a key parameter to achieve specificity. We use a computer program (Oligo™) to design the base composition and length of our primers and estimate their Td using the nearest neighbor method. We have always followed the recommendations of the original LM-PCR procedure, i.e., nesting of primer 2 and 3 with a 15 base overlap and sequential use of primers with increasing Td *(3)*. The Td (°C) of the primers we have successfully used are 52 ± 6 for primer 1, 70 ± 3 for primer 2 and 77 ± 5 for primer 3 and the difference between the Td of primer 3 and 2 is kept at 7 ± 4 *(4)*. We usually perform the hybridization at a temperature corresponding to Td –5°C. In addition, as for every PCR, the 3' ends must not have a high propensity to engage in stable secondary structure with themselves (hairpin and primer dimer), with the linker primer, or with nonspecific sequences (i.e., the ΔG of the 3'-end should be as low as possible). Despite these precautions, we have encountered sets of primers that do not work with a frequency of approx 20%. Currently to overcome the difficulty, we simply redesign a new set of primers in the vicinity of the region of interest.

5. The final amplified material must reflect accurately the initial distribution of the various 5'-ends of the sequence of interest. This requires that the efficiency of the reaction at each step of the procedure is equivalent for all ends. Variation in this efficiency has been observed at the first primer extension step and at the following linker ligation step. Sequenase, the enzyme original recommended for primer #1 extension gave batch to batch variation in the efficiency of first strand synthesis that led to under-representation of some bands but results are much more consistent using Vent exo⁻ *(4)*. The ligation conditions described allow the linker to be ligated with an equal efficiency to most ends but underrepresentation of certain bands can be observed in G-rich regions presumably because of inefficient ligation *(5)*. For the amplification steps, we prefer Taq DNA polymerase to Vent exo⁻ because it is more efficient. Taq tends to stop before completing strand synthesis to every end but this is not a problem during the amplification and labeling steps of LM-PCR since the template sequence at the polymerization end is the same for all bands in the ladder (it is the linker sequence, *5*). One should keep in mind that deviations from the optimal conditions at any step can lead to underrepresentation of some bands. The use of good quality and properly stored enzymes and solutions minimizes this problem.

6. Differences in the amount of starting material might not be represented due to the so called "plateau effect" that is often encountered with PCR *(5,21)*. In the LM-PCR conditions that we currently use, the reactions are at or near the plateau. This allows to analyze on the same gel different extents of cleavage and to compensate for slight tube-to-tube variations in the efficiency of reactions. When a

quantitative comparison between samples is necessary, linearity of the PCR reaction can be achieved by withdrawing aliquots every 2 to 3 cycles after the 15th amplification cycle and by reducing the number of labeling cycles to 2 or 3. Because not all primers are incorporated with an identical efficiency, saturation is not reached at the same number of cycles for all primer combinations. As a consequence, quantitative in vivo footprinting represents a heavy investment that needs to be carefully evaluated.

Acknowledgments

We thank A. Hair for critical reading of the manuscript. Our work was supported by the CNRS and grants from the Association Française contre les Myopathies, the Ligue Nationale contre le Cancer and the Association de Recherche sur le Cancer. L. C. Was supported by fellowships from the E. C. (Human Capital and Mobility) and the Fondation MEDIC.

References

1. Church, G. M. and Gilbert, W. (1984) Genomic sequencing. *Proc. Natl. Acad. Sci. USA* **81,** 1991–1995.
2. Becker, P. B., Weih, F., and Schütz, G. (1993) Footprinting of DNA-binding proteins in intact cells. *Methods Enzymol.* **218,** 568–587.
3. Mueller, P. R. and Wold, B. (1989) In vivo footprinting of a muscle specific enhancer by ligation mediated PCR. *Science* **246,** 780–786.
4. Grange, T., Bertrand, E., Rigaud, G., Espinás, M. L., Fromont-Racine, M., Roux, J., and Pictet, R. (1997) *In vivo* footprinting of the interaction of proteins with DNA and RNA. *Methods: A Companion to Methods Enzymol.* **11,** 151–163.
5. Grange, T., Rigaud, G., Bertrand, E., Fromont-Racine, M., Espinás, M. L., Roux, J., and Pictet, R. (1997) *In vivo* footprinting of the interaction of proteins with DNA and RNA, in *In Vivo Footprinting*, vol. 21 (Cartwright, I. L., ed.), JAI Press, Greenwich, CT, pp. 73–109.
6. Espinás, M. L., Roux, J., Pictet, R., and Grange, T. (1995) Glucocorticoids and protein kinase A coordinately modulate transcription factor recruitment at a glucocorticoid-responsive unit. *Mol. Cell. Biol.* **15,** 5346–5354.
7. Fragoso, G., John, S., Roberts, M. S., and Hager, G. L. (1995) Nucleosome positioning on the MMTV LTR results from the frequency-biased occupancy of multiple frames. *Genes Dev.* **9,** 1933–1947.
8. McPherson, C. E., Shim, E.-Y., Friedman, D. S., and Zaret, K. S. (1993) An active tissue-specific enhancer and bound transcription factors existing in a precisely positioned nucleosomal array. *Cell* **75,** 387–398.
9. Quivy, J. P. and Becker, P. B. (1996) The architecture of the heat-inducible *Drosophila* hsp27 promoter in nuclei. *J. Mol. Biol.* **256,** 249–263.
10. Rigaud, G., Roux, J., Pictet, R., and Grange, T. (1991) In vivo footprinting of rat TAT gene: dynamic interplay between the glucocorticoid receptor and a liver-specific factor. *Cell* **67,** 977–986.

11. Truss, M., Bartsch, J., Schelbert, A., Haché, R. J., and Beato, M. (1995) Hormone induces binding of receptors and transcription factors to a rearranged nucleosome on the MMTV promoter in vivo. *EMBO J.* **14,** 1737–1751.
12. Anfinsen, C. B., Cuatrecasas, P., and Taniuchi, H. (1971) Staphylococcal nuclease, chemical properties and catalysis, in *The Enzymes*, 3rd ed. (Boyer, P. D., ed.), Academic Press, New York.
13. Thoma, F. (1992) Nucleosome positioning. *Biochim. Biophys. Acta* **1130,** 1–19.
14. McGhee, J. D. and Felsenfeld, G. (1983) Another potential artifact in the study of nucleosome phasing by chromatin digestion with micrococcal nuclease. *Cell* **32,** 1205–1215.
15. Sollner-Webb, B., Melchior, W. J., and Felsenfeld, G. (1978) DNAase I, DNAase II and staphylococcal nuclease cut at different, yet symmetrically located, sites in the nucleosome core. *Cell* **14,** 611–627.
16. Carr, K. D. and Richard-Foy, H. (1990) Glucocorticoids locally disrupt an array of positioned nucleosomes on the rat tyrosine aminotransferase promoter in hepatoma cells. *Proc. Natl. Acad. Sci. USA* **87,** 9300–9304.
17. Shimizu, M., Roth, S. Y., Szent, G. C., and Simpson, R. T. (1991) Nucleosomes are positioned with base pair precision adjacent to the alpha 2 operator in *Saccharomyces cerevisiae. EMBO J.* **10,** 3033–3041.
18. Lazarides, E. and Lindberg, U. (1974) Actin is the naturally occurring inhibitor of deoxyribonuclease I. *Proc. Natl. Acad. Sci. USA* **71,** 4742–4746.
19. Maxam, A. M. and Gilbert, W. (1980) Sequencing end-labeled DNA with base-specific chemical cleavage. *Methods Enzymol.* **65,** 499–560.
20. Bertrand, E., Fromont-Racine, M., Pictet, R., and Grange, T. (1993) Visualization of the interaction of a regulatory protein with RNA *in vivo. Proc. Natl. Acad. Sci. USA* **90,** 3496–3500.
21. Innis, M. A. and Gelfand, D. H. (1990) Optimization of PCRs, in *PCR Protocols: A Guide to Methods and Applications* (Innis, M. A., Gelfand, D. H., Sninsky, J. J., and White, T. W., eds.), Academic, San Diego, CA, pp. 3–12.
22. Espinás, M. L., Roux, J., Ghysdael, J., Pictet, R., and Grange, T. (1994) Participation of Ets transcription factor in the glucocorticoid response of rat tyrosine aminotransferase gene. *Mol. Cell. Biol.* **14,** 4116–4125.

32

In Situ Analysis of Chromatin Proteins During Development and Cell Differentiation Using Flow Cytometry

Didier Grunwald, Claude Gorka, Sandrine Curtet, and Saadi Khochbin

1. Introduction

General remodeling of chromatin is associated with events determining cell fate and the expression of specific genetic programs *(1,2)*. In almost every case there is a tight link between these chromatin remodeling events and a drastic modification of the cell cycle parameters. One of the most striking examples of this phenomenon is early embryonic development. Indeed, transition periods have been defined during development characterized by the modification of both chromatin constituents and the proliferative capacities of cells *(3–6)*. For instance, during *Xenopus laevis* early development, the midblastula transition (MBT) is characterized as a period after which the somatic replication-dependent type of H1 starts to accumulate and the embryonic H1 (B4) and HMG1 decrease dramatically *(7)*. Moreover, in vertebrates, later during development, other subtypes of H1 accumulate : H1° in almost all vertebrates, H1t in mammalian spermatogenic cells and H5 in avian erythrocytes, again associated with a drastic modification of proliferative capacities of cells *(7)*. To better understand the significance of these transitions, it is important to correlate chromatin remodeling events with cell cycle modification events *(8–10)*. For instance, we showed previously that during early *Xenopus* development, the type of H1 expressed appeared to be more related to the frequency of cell division than any other cellular event *(6,7)*.

Besides embryonic development, transitions in chromatin structure and function occur during different events in adult organisms, such as tissue regeneration or various ongoing differentiation programs, such as spermatogenesis, erythropoiesis, etc, that are again tightly linked to drastic modification of the proliferative capacities of cells.

Here we propose an approach based on a flow cytofluorimetric analysis of cells after *in situ* immunolabeling to analyze simultaneously cell cycle and chromatin proteins. A double staining procedure allows the simultaneous detection of a protein and DNA, a method that is very convenient to investigate cell cycle-dependent accumulation of a specific nuclear protein *(11)*. A computer program that we developed specially to analyze these data can eventually be used to better exploit the information obtained (*see* **Subheading 3.5.**). This methodology has been successfully applied to *Xenopus* embryos taken at different stages of development *(6)*, rat hepatocytes after partial hepatectomy *(12)* and to various cell lines in culture *(11,13)*. When a series of monoclonal antibodies against a chromatin protein are available, and their target epitopes mapped, interesting studies such as the cell cycle-dependent fine modulation of chromatin structure as a function of a particular cellular event can be investigated. The example of histone H1° is used to illustrate these possibilities. This technique is applicable to a variety of nuclear proteins provided good antibodies are available, and different questions regarding modification of the quantity/accessibility of a particular nuclear protein in relation to the modifications of cell cycle parameters during critical periods of cell life can be addressed *(13,14)*.

2. Materials
2.1. Flow Cytofluorimeter
1. We performed flow analysis on a FACStar$^+$ (Becton-Dickinson, San Jose, CA) using a dual laser configuration.

2.2. Buffers for Nuclei Extraction from Rat Liver or Cells in Culture
1. Buffer A (10X): 150 mM Tris-HCl, pH 7.6, 150 mM NaCl, 600 mM KCl.
2. Lysis buffer: Buffer A (1X), 0.34 M sucrose, 2 mM EDTA, 0.5 mM EGTA, 0.65 mM spermidine, 1 mM DTT, 0.1% Triton X100, 0.5 mM PMSF.
3. Buffer B1: Buffer A (1X), 0.34 M sucrose, 2 mM EDTA, 0.5 mM EGTA, 0.65 mM spermidine, 1 mM DTT, 0.5 mM PMSF.
4. Buffer B2: Buffer A (1X), 1.5 M sucrose, 0.65 mM spermidine, 1 mM DTT, 0.5 mM PMSF.
5. Buffer D: Buffer A (1X), 0.34 M sucrose, 0.65 mM spermidine, 1 mM DTT, 0.5 mM PMSF.

2.3. Buffers for Nuclei Extraction from Xenopus Embryos
1. Buffer L (10X): 150 mM Tris-HCl, pH 7.4, 150 mM NaCl, 600 mM KCl, 20 mM EDTA, 5 mM EGTA.

2. Buffer L1 (to be prepared extemporally): Buffer L (1X), 0.65 mM spermidine, 15 mM Thioglycerol, 1 mM PMSF.
3. Ringer solution: add 6 g NaCl, 0.075 g KCl, 0.1 g CaCl$_2$, and 0.1 g NaHCO$_3$ to 1 L of H$_2$O.

2.4. Buffers and Reagents for Immunodetection

1. Bovine serum albumin (BSA; Boehringer).
2. Paraformaldehyde (Fluka).
3. Hoechst 33258 (Sigma).
4. Antibodies:
   a. Monoclonal antibodies (MAbs) specific for H1°: clone 34 and clone 27 (hybridoma cell culture supernatant).
   b. Goat FITC-conjugated F(ab')2 fragment against Fc fragment mouse IgG (Jackson Immunoresearch Lab, Westgrove, PA).
5. Phosphate-buffered saline (PBS): 140 mM NaCl, 10 mM KCl, 8 mM Na phosphate, 10 mM KCl, pH 7.4.
6. Permeabilization buffer: PBS pH 7.4, 0.25% Triton X-100.
7. Antibody buffer: PBS pH 7.4, 3% BSA.
8. Rinse buffer: PBS pH 7.4, 0.05% BSA.
9. Fixer: PBS pH 7.4, 3 % paraformaldehyde. Stock solution, 7X at –20°C.
10. DNA staining: PBS pH 7.4, Hoechst 33258 (2 µg/mL). Stock solution, 10X at –20°C.

3. Methods

3.1. Optimal Configuration of FACS for Dual Fluorescence Analysis

Two types of FACS machines are available using either one or two lasers. A dual fluorescence analysis (cell cycle vs a specific protein), can be performed with both machines. Indeed, with a single argon laser tuned at 488 nm, it is possible to excite simultaneously a DNA-specific fluorochrome, such as propidium iodide (PI), and a protein-specific antibody labeled with FITC (Fluorescein IsoThioCyanate). However, the emission spectra of FITC and PI are close and overlapping. In addition, PI (as well as ethidium bromide) fluorescence is not absolutely specific for DNA, and a RNase treatment is necessary before the use of these fluorochromes for the measurement of DNA content.

We therefore prefer a flow cytometer configuration with two lasers at different wavelengths, allowing the use of dyes with well separated emission spectra. The first argon laser is tuned on the u.v. lines with 200 mW power, for excitation of the DNA-specific fluorochrome Hoechst 33258 (HO), which has an emission peak at 450 nm; the second laser is tuned at 488 nm (100 mW), for excitation of the FITC-labeled antibodies; the emission peak is 520 nm. The HO fluorescence was collected through a 450 ± 10 nm pass-band filter, and the FITC fluorescence through a 520 ± 10 nm pass-band filter. The debris and cell-doublets were eliminated by gating on a DNA fluorescence pulse-height versus pulse-area dual parameter histogram.

The nozzle used in the flow cytometer has a 70 µm diameter convenient for most of the experiments, using both cells or nuclei. But in the case of nuclei prepared from *Xenopus* embryos, the samples may contain debris (especially yolk) which form clumps capable of clogging the nozzle. In this case, a 100 µm diameter nozzle reduces the problem (*see* **Note 1**).

Every analysis is performed at a speed of 200–1500 cells/s. For each file, the signals of size, DNA fluorescence (height and area), and antibody fluorescence from 10,000 or 20,000 cells are recorded. Results are displayed as dual parameter histograms. The relative importance of the different phases of the cell cycle, is estimated from univariate histograms of the DNA content.

3.2. Preparation of Nuclei from Cells in Culture or Tissues from Adult Organisms for Immunostaining (see Note 2)

3.2.1. From Adult Organism Tissues (Rat Liver)

1. Cut liver lobes into small pieces and perform all the following steps at +4°C (*see* **Note 3**).
2. Homogenize with a grinder of Potter Elvehjem in buffer B1 (approx 7 mL/g of liver) at 1000 runs/min.
3. Filter the homogenate through three layers of gauze.
4. Centrifuge at $3000g$ for 15 min.
5. Resuspended the pellet in half volume of buffer B2 (3.5 mL/g) and homogenize again as above.
6. Centrifuge at $12,000g$ for 15 min.
7. Homogenize the pellet in half volume of buffer D (3.5 mL/g) and centrifuge at $3000g$ for 5 min (*see* **Note 4**).
8. Resuspend the pellet in a small volume of buffer D and add the suspension dropwise into a 3% paraformaldehyde solution in PBS at approx 10^6 nuclei/mL.
9. Incubate at room temperature with occasional mixing for 30 min.
10. Centrifuge at $2500g$ for 5 min.
11. Resuspend the pellet in PBS and store at 4°C.

3.2.2. From Xenopus Embryos

1. Collect *X. laevis* embryos and wash briefly in 1/3 Ringer solution (*see* **Note 5**).
2. Transfer embryos to buffer L1 + 0.34 M sucrose, and homogenize using a small 2 mL Potter.
3. Centrifuge homogenates for 15 min at $5000g$.
4. Resuspend the pellet in buffer L1 + 1.5 M sucrose.
5. Centrifuge for 15 min at $20,000g$.
6. Wash the pellet twice in buffer L1 + 0.34 M sucrose and centrifuge at $2500g$ for 15 min.
7. Resuspend the pellet in 500 µL buffer L1 and add the suspension dropwise, under mild vortex, into 10 mL of 70% ethanol.

8. Incubate at room temperature for 30 min.
9. Centrifuge at 2500g for 15 min.
10. Resuspend the pellet in 0.6X PBS and store at 4°C.

3.2.3. From Tissue Culture Cells (Murine Erythroleukemia Cells)

1. Collect cells by centrifugation at room temperature (200g for 5 min).
2. Decant and washed with PBS (5 mL/10^7 cells).
3. Centrifuge as above, decant the supernatant.
4. Resuspend the pellet in lysis buffer (2 mL/10^7 cells).
5. Homogenize the suspension by passing the pellet through a pipet and centrifuge at 1500g for 5 min at +4°C.
6. Wash the nuclei with buffer D (2 mL), centrifuge and decant as above (*see* **Note 4**).
7. Resuspend the nuclei in buffer D and add the suspension dropwise into a 3% paraformaldehyde solution in PBS at approx 10^6 nuclei/mL (*see* **Note 6**).
8. Incubate at room temperature with occasional mixing for 30 min.
9. Centrifuge at 2500g for 5 min.
10. Resuspend the pellet in PBS and store at 4°C.

3.3. Preparation of Whole Cells for Immunostaining

1. Collect cells (cells growing in suspension were collected by centrifugation, attached cells were harvested after trypsination).
2. Wash the suspension of cells twice with PBS by centrifugation (200g, 5 min).
3. Resuspend the cell pellet in a small volume of PBS and add the suspension dropwise into a 3% solution of paraformaldehyde in PBS at approx 10^6 cells/mL.
4. Incubate at room temperature with occasional mixing for 30 min.
5. Centrifuge at 2500g for 5 min.
6. Resuspend the pellet in PBS and store at 4°C.

3.4. Immunostaining Protocols

1. Take 2 × 10^6 cells or nuclei per assay (*see* **Note 7**).
2. Wash the cells by centrifugation at 200g for 5 min and resuspend in PBS. Repeat twice.
3. Vortex the pellet and add 2 mL of permeabilization buffer and incubate at room temperature for 5 min.
4. Wash twice with 5 mL of rinse buffer (centrifuge at 300g for 5 min).
5. Resuspend cells in an appropriate amount of the primary antibody diluted in the antibody buffer (150 µL).
6. Incubate at 37°C for 1 h with occasional shaking.
7. Wash the cells three times with the rinse buffer (centrifuge at 300g for 5 min).
8. Add the appropriate amount of the secondary fluorescent antibody diluted in PBS-BSA (150 µL).
9. Incubate at room temperature for 30 min.
10. Wash the cells three times with the rinse buffer (centrifuge at 300g for 5 min).
11. Resuspend cells with PBS to 10^6/mL and add an Hoechst 33258 solution up to 2 µg/mL.

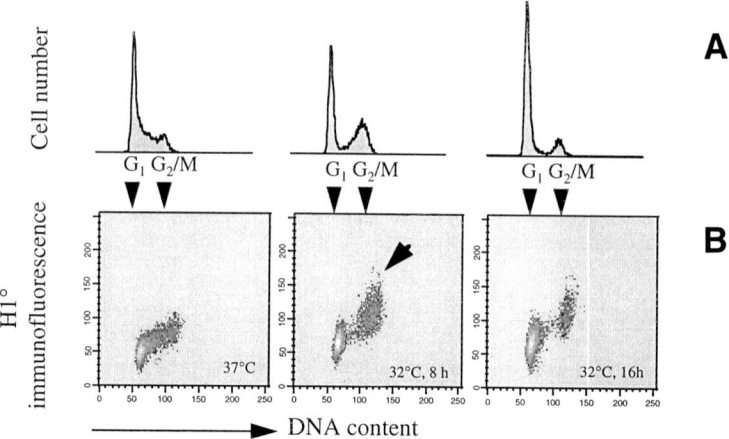

Fig. 1. Cell cycle-dependent modulation of chromatin structure during the induced arrest of cell proliferation. Clone 6 cells expressing a thermosensitive P53 were shifted to 32°C for the indicated times (8 h, 16 h), and the control cells were maintained at 37°C during this period. (**A**) For each temperature, the modification of the cell-cycle parameters as a function of time, is visualized in histograms. These histograms represent the number of cells present at different positions in the cell cycle (DNA content). (**B**) Cells are doubly stained for H1°, by indirect immunofluorescence, and for DNA, with the DNA-specific dye Hoechst 33258. This method (dual analysis of cells by cytofluorimetry and dot plot representations), allows one to monitor H1° immunoreactivity as a function of the position of cells in the cell cycle. The *arrow* shows the enhanced immunodetection of H1° in the G2/M cell population.

3.5. Output of the Analysis and Interpretation of the Data

A dot-plot representation is usually the output of the doubly stained cell analysis. In this representation, a dot represents a cell or a nucleus having a given intensity of DNA-specific dye fluorescence and immunofluorescence (X axis and Y axis respectively). The intensity of DNA specific dye fluorescence is proportional to the amount of DNA in each cells. Therefore, if the fluorescence for cells in G1 is recorded in channel N, the DNA fluorescence of cells in G2/M population will fall in channel 2N. Channels in between N and 2N represent cells in different stages of the S phase of the cell cycle (*see* **Fig. 1B**). In this two parameter representation, it is possible to estimate the intensity of immunofluorescence corresponding to each DNA channel or, in other words, to evaluate the variation of immunofluorescence as a function of the position of cells in the cell cycle. Another parameter representing the number of cells or nuclei having a given value of both DNA fluorescence and immunofluorescence is also recorded. Based on the consideration of this third parameter we developed a computer program that is able to calculate, for each DNA channel, the number of cells having a given immunofluorescence intensity, obtain a histogram and compute the mean immunofluorescence value *(11)*. In this way the dot plot can be transformed into a representation showing the mean fluorescence value as a function of DNA content (position in the cell cycle). The interest of this computation is the availability of mean fluorescence values which allows for quantitation *(13,14)*. But in most cases the dot-plot output is sufficient to show essential features of the distribution of immunofluorescence during the cell cycle and in examples chosen below we will show this representation (ex. #1 and 3).

Additionally, if the mean immnuofluorescence of the total cell population, without any special interest in the cell cycle, is the parameter to be considered, a histogram can show the distribution of cells as a function of the intensity of the immunofluorescence (this type of representation is shown in ex. #2).

Several examples of the flow cytofluorimetric analysis of histone H1° accumulation or accessibility toward monoclonal antibodies are used here to illustrate potential applications of this methodology.

1. The proliferation arrest signal is associated with an alteration of chromatin structure that occurs first in the G2/M phase cell population before the arrest of cells. We developed monoclonal antibodies raised against histone H1° and precisely mapped the recognized epitope for several clones *(15)*. Histone H1° present in chromatin can be recognized very efficiently using one of these antibodies interacting specifically with the amino acids 20–30 region of the protein. This antibody was used to monitor H1° recognition in chromatin during the P53 mediated arrest of cell prolifcration. The model system was constituted of clone six cells, a cell line transformed by a thermosensitive P53 that shows a high rate of cell proliferation at 37°C. At 32°C, P53 is able to control cell proliferation and participate in the process of the arrest of cell division *(16)*. The flow cytofluorimetric analysis of H1° recognition by our antibody showed an enhanced recognition of the protein in the G2/M phase of the cell-cycle, 8 h after the shift of temperature (**Fig. 1B**, 32°C 8 h). After this mitosis the majority of cells stay in the G1 phase and the enhanced immunodetection of the H1° is maintained (32°C, 16 h). This experiment showed that the arrest of cell proliferation is accompanied by a chromatin remodeling process, causing enhanced accessibility of the AA 20–30 region of H1° to a monoclonal antibody. This remodeling process is visible first in the G2/M cell population after the propagation of the arrest signal and before the arrest itself. After mitosis, the modified chromatin structure is inherited by daughter cells.
2. Enhanced recognition of H1° by a monoclonal antibody during cell differentiation and chromatin hyperacetylation.
  One interesting feature of FACS analysis is the relative quantitative character of the fluorescence recorded. We took advantage of this property to analyze the *in*

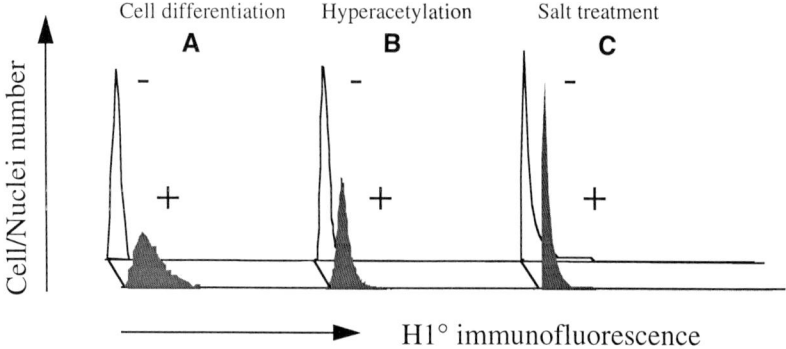

Fig. 2. H1° recognition by a monoclonal antibody after cell differentiation, histone hyperacetylation, and salt treatment. An anti-H1° monoclonal antibody recognizing the AA 24–30 region of the protein is used to monitor *in situ* protein recognition by flow cytofluorimety. **(A)** Murine erythroleukemia (MEL) cells were induced to differentiate after the treatment of cells by hexamethylene bis acetamide. Undifferentiated cells and cells induced for 48 h were used for an *in situ* immunodetection of H1°. (–) undifferentiated cells, (+) induced cells. Histograms show the distribution of cells as a function of immunofluorescence. **(B)** FM3A cells, a mouse mammary gland tumor cell line, were treated with trichostatin A (an inhibitor of histone deacetylases, 100 ng/mL, for 6 h), and the immunostaining and FACS analysis were performed as above (–) untreated cells, (+) treated cells. **(C)** Isolated nuclei from undifferentiated MEL cells were incubated in a buffer containing 100 mM NaCl (–) and 200 mM NaCl (+) and fixed. Immunostaining and FACS analysis was performed as above.

situ recognition ability of one monoclonal antibody recognizing the amino-acids 24–30 region of histone H1°, located at the entry of the globular domain of the protein (clone 27, *15*). The analysis of clone 27-related immonufluorescence was performed after the immunostaining of H1° in undifferentiated and differentiated murine erythroleukemia (MEL) cells. In undifferentiated cells, H1° is not recognized *in situ* by this antibody. Indeed, the intensity of the immunofluorescence recorded from these cells is close to the background value (fluorescence recorded when the anti-H1° antibody is omitted, not shown). After the differentiation, an enhanced recognition of H1° by the antibody is observed (**Fig. 2A**, notice the wider distribution of cells along the immunofluorescence axis).

Treatment of cells by an inhibitor of histone deacetylase, trichostatin A (TSA), also enhances the recognition of H1° by the antibody (**Fig. 2B**, compare [–] untreated, [+] treated histograms). However, compared to cell differentiation, chromatin hyperacetylation has a relatively modest effect on H1° recognition. Finally, the treatment of isolated nuclei with 200 mM NaCl, while enhancing the recognition of H1° by other monoclonal antibodies (not shown), does not allow the recognition of H1° in undifferentiated MEL cell nuclei by this specific antibody (**Fig. 2C**, compare [–] untreated, [+] treated histograms).

These data suggest that a modification of the chromatin structure and/or H1° conformation occurs specifically during cell differentiation.

3. H1° accumulation during *Xenopus* embryonic development and in rat liver after partial hepatectomy.

   We have chosen two examples to illustrate the fact that *in situ* immunodetection and FACS analysis can be applied to organisms: whole embryos as well as adult tissues.

   a. During *Xenopus laevis* embryogenesis, *in situ* immunodetection of histone H1° on cryosections as well as Western blot showed that the major accumulation of H1° occurs during the tailbud-tadpole transition period *(6)*. Therefore, it appeared interesting to understand if this phenomenon is associated with modification of proliferative capacities of embryonic cells. Nuclei were prepared from embryos taken at stage 36, 41, and 45, a period characterized by the major accumulation of H1° in the nucleus of the embryonic cells *(6)*. Histograms representing the distribution of cells in the cell cycle showed that cell proliferation ceased during this period: at stage 45, more than 90% of cells are in the G0/G1 phase of the cell cycle (**Fig. 3**, I, panel **A**). We then performed double staining for DNA and H1°, to evaluate the relationship between cell cycle and H1° accumulation. A gradual increase in the intensity of H1°-related immunofluorescence is observed as the development proceeds. Surprisingly, the accumulation of the protein does not occur only in the G0/G1 cells but is observed in all phases of the cell cycle (**Fig. 3**, I, panel **B**). We have already observed the presence of H1° in different phases of the cell cycle in tissue culture cells, as shown in **Fig. 1**, but the in vivo situation was not known. This in vivo study confirmed the observations described above and showed that in organisms, H1° expression and cell proliferation are not incompatible. We therefore suggested that in vivo, H1° accumulation is associated with the lengthening of the cell cycle rather than with the arrest of cell proliferation.

   b. Using another in vivo system consisting of rat hepatocytes after partial hepatectomy, we were able to gain more information concerning this issue *(12)*. After partial hepatectomy, waves of cell proliferation occur to regenerate the tissue *(17)*. Adult hepatocytes are arrested cells with 2N and 4N DNA content; 48 h after partial hepatectomy, approx 27% of the cells actively replicate their DNA (estimated by measuring the percentage of *Brd*U incorporating cells). The double-staining analysis for DNA and H1° showed that there is a general decrease in H1° content that occurs in cells in every phase of the cell cycle and not only in DNA replicating cells (**Fig. 3**, II), showing again that H1° accumulation is not restricted to arrested cells but is controlled in such a manner that the protein accumulates in slowly dividing cells and decreases in rapidly growing cells.

4. Notes

1. For flow analysis, we recommend the use of distilled water as sheath fluid. Indeed, we have observed a better hydrodynamic focalization, which improves the analysis, and reduces the risk of clogging the nozzle. In any case, it is funda-

I - Xenopus embryonic development (stage)

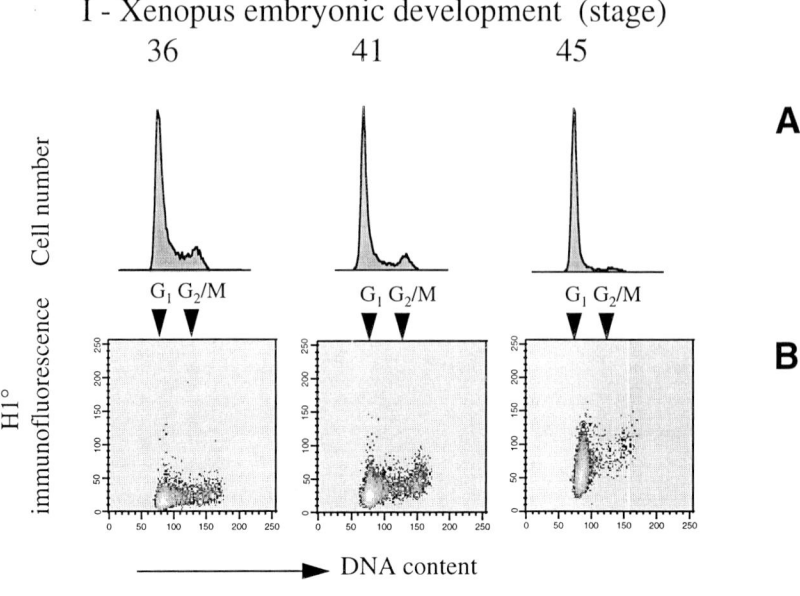

II - Rat liver regeneration after partial hepatectomy

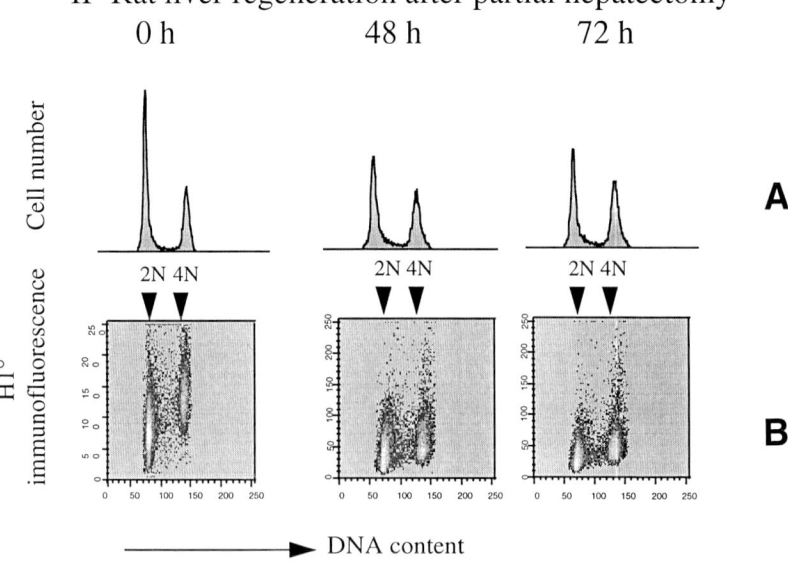

Fig. 3. *(continued on next page)* Cell-cycle-dependent accumulation of H1°. I—Nuclei isolated from *Xenopus* embryos at the indicated stages of development are doubly stained for H1° by indirect immunofluorescence, and for DNA with the DNA-specific dye Hoechst 33258. (**A**) For each stage, the modification of the cell-cycle parameters

mental to have a sheath fluid with a density equal to or lower than the density of the sample liquid.
2. The immunostaining should be performed on isolated nuclei or cells, depending on the original material: organisms (tissues, embryos, etc.) or cell lines.
   a. In order to perform the immunolabeling using organisms, the nuclei should first be prepared and fixed.
   b. In the case of cell lines, it is not necessary to prepare nuclei, unless treatment is envisaged to modify chromatin, i.e., salt treatment (*see* **Note 6** and ex. 2).
3. Forty-day-old Wistar male rats were partially hepatectomized as described by Gorka et al., *(12)* and were inoculated i.p. with BrdU (50 mg/kg), 1 h before sacrifice.
4. The wet pellet obtained at this stage could be kept at −80°C.
5. *X. laevis* eggs are fertilized in vitro, and embryos are prepared as described *(6)*. The number of embryos collected depends on how early they are taken: 40 for the early stages (stage 9 until 18/20) and 10 for the latest. Except when indicated, all steps should be realized at 4°C.
6. When treated with salt, purified nuclei were resuspended in 20 mM Tris-HCl pH 7.4, 6 mM MgCl$_2$ containing either 100 mM or 200 mM NaCl, then fixed with 3% paraformaldehyde in the same buffer for 1 h at +4°C.
7. The same immunostaining procedure is used for fixed cells or isolated fixed nuclei.

References

1. Patterton, D., and Wolffe, A. P. (1996) Developmental role of chromatin and chromosomal structure. *Dev. Biol.* **173**, 2–13.
2. Wolffe, A. P. (1996) Chromatin and gene regulation at the onset of embryonic development. *Reprod. Nutr. Dev.* **36**, 581–606.
3. Andrews, M. T., Loo, S., and Wilson, L. R. (1991) Coordinate inactivation of class III genes during the gastrula-neurula transition in *Xenopus*. *Dev. Biol.* **146**, 250–254.
4. Newport, J. and Kirschner, M. (1982) A major developmental transition in early *Xenopus* embryo: I. Characterization and timing of cellular changes at midblastula transition stage. *Cell* **30**, 675–686.
5. Newport, J. and Kirschner, M. (1982) A major developmental transition in early *Xenopus* embryo: II. Control of the onset of transcription. *Cell* **30**, 687–696.
6. Grunwald, D., Lawrence, J. J., and Khochbin, S. (1995) Accumulation of histone H1° during early *Xenopus laevis* development. *Exp. Cell Res.* **218**, 586–595.
7. Khochbin, S. and Wolffe, A. P. (1994) Developmentally regulated expression of linker-histone variants in vertebrates. *Eur. J. Biochem.* **225**, 501–510.

---

as a function of time, is visualized in histograms as in **Fig. 1**. **(B)** H1° immunodetection and analysis was performed as described in **Fig. 1**. II—Nuclei were prepared from adult rat liver (0 h) or from liver at indicated times after partial hepatectomy. **(A)** Histograms show the modification of the cell-cycle parameters as a function of time. Arrows indicate the position of 2N and 4N cell populations. **(B)** The relative amount of H1° as a function of the position of cells in the cell cycle is shown.

8. Telford, N. A., Watson, A. J., and Schultz, G. A. (1990) Transition from maternal to embryonic control in early mammalian development: a comparison of several species. *Mol. Reprod. Dev.* **26,** 90–100.
9. Kane, D. A. and Kimmel, C. B. (1993) The zebrafish midblastula transition. *Development* **119,** 447–456.
10. Lehner, C. F., Lane, M. E. (1997) Cell cycle regulators in *Drosophila:* downstream and part of developmental decisions. *J. Cell Science* **110,** 523–528.
11. Khochbin, S., Chabanas, A., Albert, P., and Lawrence, J. J. (1989) Flow cytofluorimetric determination of protein distribution throughout the cell cycle. *Cytometry* **10,** 484–489.
12. Gorka, C., Lawrence, J. J., and Khochbin, S. (1995) Variation of H1° content throughout the cell cycle in regenerating rat liver. *Exp. Cell Res.* **217,** 528–533.
13. Khochbin, S., Principaud, E., Chabanas, A., and Lawrence, J. J. (1988) Early event in murine erythroleukemia cells induced to differentiate: accumulation and gene expression of the transformation-associated cellular protein p53. *J. Mol. Biol.* **200,** 55–64.
14. Khochbin, S., Chabanas, A., and Lawrence J. J. (1988) Early event in murine erythroleukemia cells induced to differentiate: variation of the cell cycle parametres in relation to p53 accumulation. *Exp. Cell Res.* **179,** 565–574.
15. Gorka, C., Brocard, M. P., Curtet, S., and Khochbin, S. (1998) Differential recognition of histone H1° by monoclonal antibodies during cell differentiation and arrest of cell proliferation. *J. Biol. Chem.* **273,** 1208–1215.
16. Michalovitz, D., Halevy, O., and Oren, O. (1990) Conditional inhibition of transformation and of cell proliferation by a temperature-sensitive mutant of p53. *Cell* **62,** 671–680.
17. Alison, M. R. (1986) Regulation of hepatic growth. *Physiologic. Rev.* **66,** 499–541.

33

Mapping DNA Target Sites of Chromatin Proteins In Vivo by Formaldehyde Crosslinking

Helen Strutt and Renato Paro

1. Introduction

The method described here is based on a technique developed to analyze the chromatin structure of the SV40 origin of replication, and also alterations in nucleosomal structures in the *hsp70* promoter of *Drosophila* after heat shock *(1,2)*. It relies on the ability of formaldehyde to crosslink proteins and nucleic acids in living cells, from which chromatin can be isolated and protein-DNA interactions probed by immunoprecipitation using specific antibodies.

Formaldehyde is a very reactive dipolar compound that reacts with the amino groups of proteins and amino acids *(3,4)*. It shows no reactivity however towards free double stranded DNA, and thus does not cause the extensive DNA damage seen after prolonged exposure to other crosslinking reagents such as UV. Each formaldehyde molecule has the capacity for interaction with two amino groups. Therefore, DNA-protein, protein-protein and RNA-protein crosslinks are rapidly induced after formaldehyde treatment, creating a stable structure which prevents redistribution of cellular components. Furthermore, a simple heat treatment is sufficient to reverse the reaction equilibrium, and to allow isolation of pure DNA for further analysis *(1)*.

A schematic drawing of the crosslinking and immunoprecipitation technique is shown in **Fig. 1**. Tissue culture cells are crosslinked in vivo with formaldehyde, and then sonicated to produce sheared, soluble chromatin of an average size of 1 kb. Chromatin is purified on a caesium chloride gradient, and specific antibodies are used to immunoprecipitate DNA which is covalently crosslinked to the protein of interest. Formaldehyde crosslinks are reversed, and the DNA purified. This procedure allows the isolation of a small quantity of DNA, which is enriched for the specific protein-associated elements. If potential target

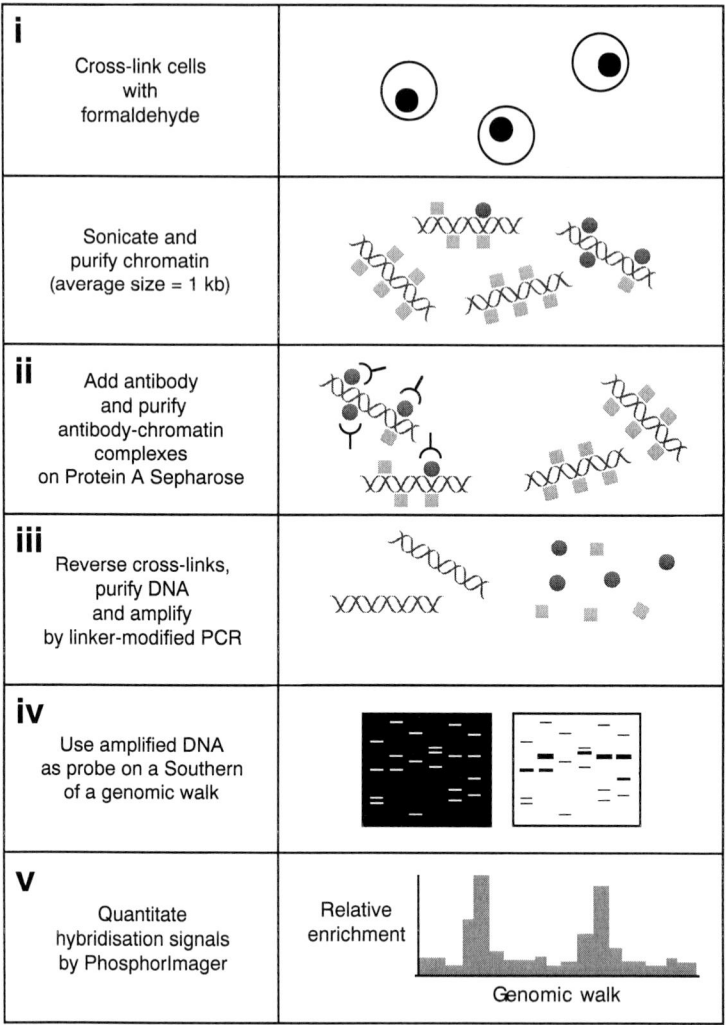

Fig. 1. The formaldehyde crosslinking and immunoprecipitation technique. (i) *Drosophila* Schneider cells are crosslinked with formaldehyde before sonicating to produce soluble chromatin of an average size of 1 kb, in which proteins (squares/circles) are covalently crosslinked to DNA. Chromatin is purified on a caesium chloride gradient. (ii) Purified chromatin is incubated with antibodies that recognize a particular DNA-binding or chromatin-associated protein (black circles). Immuno-complexes are then purified on Protein A Sepharose. (iii) The crosslinks are reversed and the coimmunoprecipitated DNA is purified. Approximately 1 ng DNA is isolated, and this DNA is subjected to linker-modified PCR amplification. (iv) Amplified DNA is used as a probe on a Southern blot. (v) Hybridization of the immunoprecipitated probe to each restriction fragment of a genomic walk is quantitatively analyzed and plotted according to position on the genomic walk.

sequences are available, they can be used to probe a slot blot containing immunoprecipitated DNA, and to determine enrichments over control immunoprecipitations without antibody.

A relatively short exposure to formaldehyde (8 min) was sufficient to detect DNA-histone interactions. However, the analysis of non DNA-binding, chromatin-associated factors could be acheived by utilizing longer crosslinking times *(5,6)*. A modification of the technique was introduced by Orlando and Paro *(5,7)*, in which immunoprecipitated DNA is amplified by linker-modified PCR, and used as a probe to map elements bound by Polycomb protein in an extended genomic walk.

We have now further modified the PCR amplification step of this technique to allow higher resolution mapping of protein binding sites *(8)*. In the original study DNA was digested with restriction endonucleases and a cohesive linker attached to create sites for annealing PCR primers. However, small restriction fragments are amplified much more efficiently than larger fragments, which thus become underrepresented in the final probe *(9)*. Here we describe the direct ligation of blunt-end linkers to the immunoprecipitated DNA fragments; such fragments have a random size distribution with respect to the genomic walk and amplify approximately linearly.

The method is divided into five sections:

1. The in vivo formaldehyde crosslinking of whole cells and the subsequent purification of crosslinked chromatin.
2. Immunoprecipitation with specific antibodies from chromatin.
3. Reversal of crosslinks and PCR amplification of purified DNA.
4. Use of the immunoprecipitated DNA as a probe on a Southern of a genomic walk.
5. Interpretation of the results of a Southern analysis.

2. Materials
2.1. Crosslinking and Chromatin Purification

1. 1 mCi/mL (methyl ^{3}H)-thymidine, cell-culture grade (*see* **Note 1**).
2. Fixation solution: 11% formaldehyde, 100 mM NaCl, 50 mM Tris-HCl, pH 8.0, 1 mM EDTA, 0.5 mM EGTA. Add formaldehyde immediately before use, from a 37% stock solution stabilized with 10% methanol.
3. Glycine.
4. PBS (on ice).
5. Wash solution A: 10 mM Tris-HCl, pH 8.0, 10 mM EDTA, 0.5 mM EGTA, 0.25% Triton X-100.
6. Wash solution B: 200 mM NaCl, 10 mM Tris-HCl, pH 8.0, 1 mM EDTA, 0.5 mM EGTA.
7. Sonication buffer: 10 mM Tris-HCl, pH 8.0, 1 mM EDTA, 0.5 mM EGTA.

8. 10% sarkosyl.
9. Caesium chloride, optical grade.
10. Dialysis buffer: 5% glycerol, 10 mM Tris-HCl, pH 8.0, 1 mM EDTA, 0.5 mM EGTA.

2.2. Chromatin Immunoprecipitation

1. Protein A Sepharose CL4B (PAS, Sigma, *see* **Note 2**). Equilibrate in RIPA buffer *(below)* by mixing at 4°C for 30–60 min. 100 mg PAS equilibrated in 1 mL RIPA buffer results in a 50% v/v suspension. After equilibration, PAS is stable for up to 1 wk at 4 °C.
2. RIPA buffer: 140 mM NaCl, 10 mM Tris-HCl, pH 8.0, 1 mM EDTA, 1% Triton X-100, 0.1% SDS, 0.1% sodium deoxycholate, 1 mM PMSF (on ice). Add PMSF immediately before use from a 100 mM stock in isopropanol.
3. Stocks for adjusting chromatin samples to RIPA conditions: 10% Triton X-100, 1% SDS, 1% sodium deoxycholate, 1.4 M NaCl, 0.1 M Tris-HCl, pH8.0, 10 mM EDTA.
4. LiCl buffer: 250 mM LiCl, 10 mM Tris-HCl, pH 8.0, 1 mM EDTA, 0.5% NP-40, 0.5% sodium deoxycholate (on ice).
5. TE buffer: 10 mM Tris-HCl, pH 8.0, 1 mM EDTA.

2.3. Purification and Amplification of Immunoprecipitated DNA

All reagents should be clean to minimize the risk of contamination during the PCR (*see* **Note 3**).

1. 5 mg/mL RNase A (DNase-free).
2. 10% SDS.
3. 10 mg/mL proteinase K (in TE buffer, stored at –20°C).
4. 1 µM linker DNA: Two oligonucleotides annealed:
   a. a 20-mer of sequence 5'-AGA AGC TTG AAT TCG AGC AG (phosphorylated at 5'-end);
   b. a 20-mer of sequence 5'-CTG CTC GAA TTC AAG CTT CT. Store in small aliquots at –20 °C.
5. 10X ligation buffer: 0.5 M Tris-HCl, pH 7.6, 125 mM MgCl$_2$, 250 mM DTT, 12.5 mM ATP. Store at –20°C in small aliquots.
6. T4 DNA ligase, 4 U/µL.
7. Taq polymerase and buffer (Boehringer).
8. 250 µM dGTP, dATP, dCTP, dTTP (diluted from commercially available 10 mM stocks in 10 mM Tris-HCl, pH 8.0, and stored in small aliquots at –20°C).
9. PCR primer: linker oligonucleotide (ii), at 100 µM, stored in small aliquots at –20°C.
10. Phenol-chloroform, chloroform, 3 M sodium acetate pH 5.2, 100% ethanol (on ice), 20 mg/mL glycogen (Boehringer), 70% ethanol, TE buffer, distilled water.
11. *Hin*dIII restriction endonuclease and corresponding buffer.
12. PCR purification columns (e.g., Qiagen).

2.4. Southern Analysis

1. Standard materials for agarose gel electrophoresis and Southern blotting onto positively charged nylon membrane (e.g., Genescreen Plus, NEN).
2. Phosphorimager apparatus.
3. Standard materials for random prime DNA labeling with $\alpha(^{32}P)$-dATP.
4. Hybridization buffer: 0.5 M NaHPO$_4$ pH 7.2, 7% SDS, 1 mM EDTA, 1% BSA. A 1 M NaHPO$_4$, pH 7.2, stock is 0.5 M Na$_2$HPO$_4$ containing 4 mL orthophosphoric acid per liter (*see* **Note 4**).
5. Wash buffer 1: 40 mM NaHPO$_4$ pH 7.2, 5% SDS, 1 mM EDTA, 0.5% BSA.
6. Wash buffer 2: 40 mM NaHPO$_4$ pH 7.2, 1% SDS, 1 mM EDTA.

3. Methods

3.1. Formaldehyde Crosslinking and Chromatin Purification

1. To approx 10^9 tissue culture cells in suspension add (methyl-^{3}H)-thymidine (*see* **Note 5**) to a final concentration of 1 µCi/mL and continue growing cells for 36–48 h.
2. Add one-tenth volume fixation solution. The optimal fixation time varies between cell types (*see* **Note 6**), but for *Drosophila* Schneider cells, fixation is carried out for approx 60 min at 4 °C.
3. Stop fixation by addition of solid glycine to 125 mM, and mix until it is completely dissolved. Centrifuge at 4°C to pellet the cells, remove the supernatant and wash cells briefly by resuspending in 100–200 mL ice-cold PBS.
4. Re-centrifuge and resuspend the cell pellet in 20 mL wash solution A. Mix gently at room temperature for 10 min and recentrifuge, before resuspending in 20 mL wash solution B. Again mix gently for 10 min at room temperature.
5. Centrifuge and resuspend the cell pellet in 12–15 mL sonication buffer.
6. Sonicate in 5–6-mL aliquots. Aliquots are sonicated in a 15-mL tube on ice-water to prevent heating of the sample. We find that four 30-s bursts of a Branson Model 250 sonicator microtip at maximum setting produces chromatin of an average size of 1 kb or less. The sonication efficiency is critical for creating chromatin fragments of a small enough size (*see* **Note 6**); for best results immerse the tip 10–15 mm under the liquid surface (*see* **Note 7**).
7. Adjust samples to 0.5% with sarkosyl and mix for 10 min at room temperature. Centrifuge at 15,000g for 15 min at room temperature to remove cellular debris.
8. Add caesium chloride: the final volume should be 20 mL, with caesium chloride at a final density of 1.42 g/cm^3. Adjust the volume of the sample to 16–17 mL before adding the caesium chloride (11.36 g), as high concentrations of caesium chloride cause precipitation of the sample. Finally adjust the sample to exactly 20 mL with TE-sarkosyl buffer.
9. Centrifuge at 40000 rpm (200,000g) for 72 h at 20°C in a Beckmann SW55Ti rotor. A 20-mL sample will thus be split into four 5-mL gradients (*see* **Note 8**).
10. Collect 0.4-mL fractions from the bottom of each gradient. Identify peak DNA fractions by scintillation counting 10 µL of each fraction. Most of the ^{3}H label

should be present in 3–4 fractions in the middle of the gradient (*see* **Note 9**). This should correspond to a density of 1.39 g/cm^3, which can be checked if a refractometer is available.
11. Pool the chromatin fractions, and dialyse overnight at 4°C into 2 L of dialysis buffer. Freeze samples in 500-µL aliquots in liquid nitrogen and store at -70°C. Each aliquot should contain 30–60 µg DNA; this can be tested by reversing the crosslinks and purifying the DNA from one aliquot (*see* **Subheading 3.3.**). This step is also useful for estimation of the average length of chromatin fragments (*see* **Note 10**).

3.2. Immunoprecipitation from Crosslinked Chromatin

1. Thaw a 500-µL aliquot of chromatin, and adjust to 1 mL RIPA buffer, by sequential addition of 100 µL 10% Triton X-100, 100 µL 10% sodium deoxycholate, 100 µL 1% SDS and 100 µL 1.4 *M* NaCl. Allow 2 min gentle mixing between additions for equilibration of the chromatin into the new conditions. Finally add 50 µL 0.1 *M* Tris-HCl, pH 8.0, 50 µL 10 m*M* EDTA, and 10 µL 100 m*M* PMSF.
2. Add 30–40 µL of the 50% v/v PAS suspension (*see* **Note 2**) to the chromatin sample. Incubate for 1 h at 4°C, before removing the PAS by centrifugation in a microfuge at top speed for 30 s. This acts as a preclearing step to reduce nonspecific binding to the protein A Sepharose.
3. Remove the chromatin sample to a new tube, and add 2–5 µg antibody. Incubate overnight at 4°C, with gentle mixing. The optimal amount of antibody may need to be determined empirically, and a control immunoprecipitation without antibody should be carried out in parallel. Mock immunoprecipitations isolate DNA nonspecifically, but specific DNA fragments should be several-fold enriched in antibody immunoprecitations.
4. Purify immunocomplexes by adding 30–40 µL 50% (v/v) PAS, and incubating for 3 h at 4°C, again with gentle mixing.
5. Wash PAS-antibody-chromatin complexes five times for 10 min each in RIPA buffer, one time in LiCl buffer, and two times in TE buffer. All wash steps should be carried out at 4°C using 1 mL wash buffer, and between washes centrifuge at full speed for 20 s to pellet the PAS before removing the supernatant.
6. Finally resuspend PAS complexes in 100 µL TE buffer.

3.3. Purification and Amplification of Immunoprecipitated DNA

1. Add RNase A to 50 µg/mL and incubate for 30 min at 37°C.
2. Adjust samples to 0.5% SDS, 0.5 mg/mL proteinase K and incubate overnight at 37°C, followed by 6 h at 65°C.
3. Phenol-chloroform extract the sample, and back-extract the lower phenol phase by adding an equal volume of 50 m*M* Tris-HCl, pH 8.0, mixing and centrifuging. Combine the aqueous phases from the phenol extraction and the back-extraction, and chloroform extract.
4. Precipitate by adding 1 µL 20 mg/mL glycogen (as carrier), 1/10 vol 3 *M* sodium acetate, pH 5.2, and 2 volumes ice-cold 100% ethanol. Store on ice for 30 min,

before centrifuging at 4°C for a further 30 min. Wash the DNA pellet in 70% ethanol, air dry, and resuspend in 20 µL distilled water. Store at –20°C.
5. Ligate linker directly to 7 µL of the sample by adding 1 µL 10X ligation buffer, 1 µL 1 µM linker, and 1 µL (4 U) T4 DNA ligase. Incubate overnight at 4 °C. The use of too little immunoprecipitated DNA in the ligation can cause problems with the PCR (see **Note 11**).
6. The ligation mixture is amplified directly, without intermediate purification. Make the volume of the sample up to 68.5 µL with distilled water, and add 10 µL 10X Taq polymerase buffer, 2.5 µL each 250 µM dNTP, 1 µL 100 µM primer and 0.5 µL Taq polymerase.
7. PCR amplify using the following cycles:
1 cycle of 94°C for 2 min.
35 cycles of 94°C/1 min; 55°C/1 min; 72°C/3 min.
1 cycle of 94°C/1 min; 55°C/1 min; 72°C/10 min.
8. Five microliters of amplified sample may be checked on a 1.2% agarose gel; the product should be a smear ranging between 200–500 bp. Obvious bands in the smear may be a result of contamination (see **Note 3**).
9. Purify the PCR products by phenol-chloroform and chloroform extraction, followed by ethanol precipitation. Most of the linker sequences may then be digested away with *Hin*dIII restriction enzyme, and the linker fragments removed with commercially available PCR purification kits, as instructed by the manufacturer. The expected yield from the PCR reaction is approximately 5 µg, and amplification of DNA from control immunoprecipitations should be as efficient as of antibody immunoprecipitated DNA (see **Note 12**).

3.4. Analysis of Immunoprecipitated DNA

This basic protocol describes the analysis of DNA-protein interactions over an extended genomic walk. If association of the protein of interest with a small defined target element is being examined slot-blot analysis may be more suitable (see **Note 13**).

1. Prepare a Southern filter of a genomic walk which is a potential target of the immunoprecipitated protein, using standard methods. Prehybridize for 2–4 h at 65°C in hybridization buffer (see **Note 4**).
2. Take 50–100 ng of amplified DNA and random prime label with α-(^{32}P)-dATP.
3. Denature the probe by boiling for 5 min, and cool in ice, before adding to 5–10 mL hybridization buffer and incubating overnight at 65°C.
4. Wash the filter once for 10 min at 65°C in wash solution A, and 3–4 times for 5 min at 65°C in wash solution B. Expose the filter overnight to a Phosphorimager screen.

3.5. Interpretation of Southern Results

1. Determine the intensity of hybridisation to each restriction fragment of the genomic walk using the Phosphorimager software. Intensity is proportional to molecular weight, and the resulting values should be normalized with regard to molecular weight if the relative enrichments of different DNA fragments are

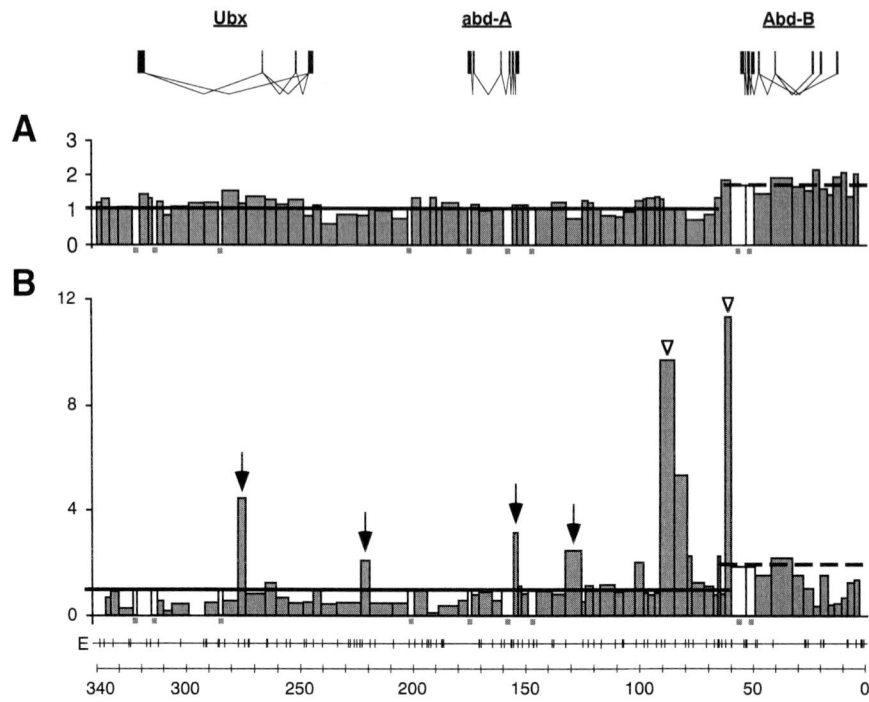

Fig. 2. Hybridization of control and GAGA factor immunoprecipitated DNA to the bithorax complex of *Drosophila*. Mock immunoprecipitated DNA (**A**) or DNA immunoprecipitated with GAGA factor (**B**) was used as a probe on a Southern containing bacteriophage DNA covering the *Drosophila* bithorax complex. The hybridization of probe to each *Eco*RI fragment was quantitated. As hybridization intensity is proportional to molecular weight, the resulting values were normalized to hybridization signal per kilobase, before plotting on a map of the genomic region. The exon structure of the three homeotic genes *Ubx*, *abd-A*, and *Abd-B* is shown at the top of the figure. Below is shown the position of the *Eco*RI (E) sites, and the coordinates of the genomic walk *(13)*. Hybridization intensity is shown as gray boxes, and the scale bar to the left is in arbitrary units. Repetitive elements cannot be quantitated, and are shown as white boxes with a gray spot below. (**A**) Control immunoprecipitated DNA. The mean hybridization signal is depicted by the thick black line across the profile. Note that the probe hybridizes more strongly to DNA in the region of the *Abd-B* gene, which is expressed *(5)*. (**B**) GAGA factor immunoprecipitated DNA *(8)*. Note the strong enrichment of two distinct restriction fragments (arrowheads) and the weaker enrichment of four fragments (arrows). The level of nonspecific enrichment was determined by comparing plus and minus antibody immunoprecipitations of various restriction fragments by slot blot, and is shown by the thick black line across the profile.

being compared. For example, the amount of signal per kb of DNA in each fragment may be calculated and plotted onto a map of the genomic region (shown in **Fig. 2B** for GAGA factor; *see* **ref. 8**).

2. As a control for the experiment, the mock immunoprecipitated DNA probe should also be hybridized to a Southern of the genomic region: this DNA should hybridize approximately uniformly to all fragments (dependent on molecular weight), and can be visualised after a long (48 h) phosphorimager exposure. As the method is just semi-quantitative some sequence-specific differences in amplification may occur, but amplification of different fragments generally varies by no more than 30% from the mean (**Fig. 2A**). This degree of error must therefore be assumed for all experiments. If only a few random restriction fragments of a genomic walk hybridize to the control immunoprecipitated DNA probe, it is likely that too little input DNA was added to the ligation reaction (*see* **Subheading 3.3.**), or that ligation occurred at low efficiency (*see* **Note 11**).
3. Repetitive elements are always strongly enriched in immunoprecipitations and therefore hybridize strongly to all immunoprecipitated DNA probes. These elements can be identified by hybridizing the filter to genomic DNA. Whereas most DNA fragments give low-level hybridization (visible with a long phophorimager exposure), repetitive elements are those fragments that hybridize much more strongly.
4. The DNA that is purified nonspecifically by control immunoprecipitations contributes to the final hybridization signal from antibody immunoprecipitation probes. If the protein of interest is a transcription factor or a protein with a few strong binding sites (e.g., GAGA factor; *see* **Fig. 2B** and **ref. *8***), those fragments which are specifically enriched are easy to identify. If the protein is associated with many DNA fragments in a genomic region (e.g., a component of heterochromatin) it may be more difficult to distinguish specific enrichments (*see* **Note 14**).
5. Chromatin isolated from active or inactive regions of the genome may behave differently in the immunoprecipitation and amplification procedures (*see* **Note 15** and **Fig. 2**). This effect is too small to influence the overall outcome of experiments, but care must be taken in comparing relative enrichments.
6. The resolution of the technique in practice appears to be lower than the size of the sonicated chromatin fragments (i.e., less than 1 kb). This is probably due to the fact that in a mixed population of differently sized DNA fragments, smaller fragments are preferentially amplified, as previously reported *(9)*. As the final PCR products are 200–500 bp in length, this step presumably sets the resolution of the whole technique.

4. Notes

1. Appropriate care must be taken in the handling and disposal of all plasticware and liquids in contact with ^{3}H, until the chromatin has been purified by dialysis.
2. Protein A Sepharose is the reagent of choice for isolating rabbit polyclonal antibodies, but it does not efficiently bind mouse monoclonal antibodies *(10)*; magnetic Dynabeads (Dynal) or GammaBind Plus (Pharmacia) could be used instead.
3. All reagents for the steps between immunoprecipitation and PCR amplification should be kept "clean" and used only for this purpose. We would recommend the use of aerosol-free pipet tips, and the storage of nucleotides, linkers, primers, etc.

in small aliquots to prevent contamination of valuable reagents (in addition to preventing frequent freeze-thawing which may cause destabilisation/inactivation of buffers). PCR contamination has been a frequent problem with this technique, as the blunt-end ligation allows amplification of many contaminating sequences; in particular compare carefully Southern hybridizations with control and antibody immunoprecipition DNA probes for spurious signals.

4. The hybridization procedure used is based on a previously described stringent hybridisation protocol (11).
5. ^{3}H is incorporated into replicating DNA, and is used as a marker for the position of crosslinked chromatin in the caesium chloride gradient.
6. Crosslinking time effects the efficiency of sonication. Material that has been crosslinked too much cannot be sheared to a sufficiently small size, even with extensive sonication. The efficiency of sonication can be estimated by examining 10 µL under phase contrast microscopy. If large particles or intact nuclei are still visible, try reducing the crosslinking period.
7. The position of the sonicator tip is critical for efficient sonication. If it is too close to the surface of the liquid excessive foaming can occur which results in loss of energy. Conversely, the tip should not be too deep in the solution, as the energy radiates from the horn and thus most material will not receive treatment but will circulate around the sample tube.
8. Loading too much material on each gradient should be avoided. After centrifugation a broad sarkosyl/lipid/protein aggregate is present at the top of the gradient, and overloading can result in poor resolution of the gradient, and thus poor quality chromatin.
9. Most of the ^{3}H signal should be present near the middle of the caesium chloride gradient, corresponding to a density of approx 1.39 g/cm^3, at which DNA-protein crosslinked material is found. Noncovalently crosslinked DNA equilibrates to the bottom of the gradient (density approx 1.6 g/cm^3). If the majority of ^{3}H signal is in the lower fractions, this indicates that the sample was under crosslinked, and a longer crosslinking period should be chosen.
10. For optimal performance in immunoprecipitation and PCR amplification the average size of chromatin produced should be no more than 1 kb: by gel electrophoresis a smear will be seen ranging from 300 bp to 5 kb, which is most intense at 1–2 kb (*see* **Fig. 3A** for an example). A balance must be sought between degree of crosslinking and chromatin size (*see* **Notes 6** and **9**).
11. The efficiency of ligating linker directly to the sonicated DNA fragments purified in the immunoprecipitation is presumably low, as the DNA contains a mixture of ends which are blunt, 5'-overhangs, 3'-overhangs, phosphorylated and non-phosphorylated. We had problems in maintaining genome complexity if there was too little input DNA in the ligation (thus reducing the amount of ligatable material), and only a small number of random restriction fragments on a Southern hybridized to the amplified probe. This could be a greater problem if the procedure is carried out in cells from organisms with a higher genome complexity than *Drosophila*. Because of the dangers in reduced complexity, it is essential

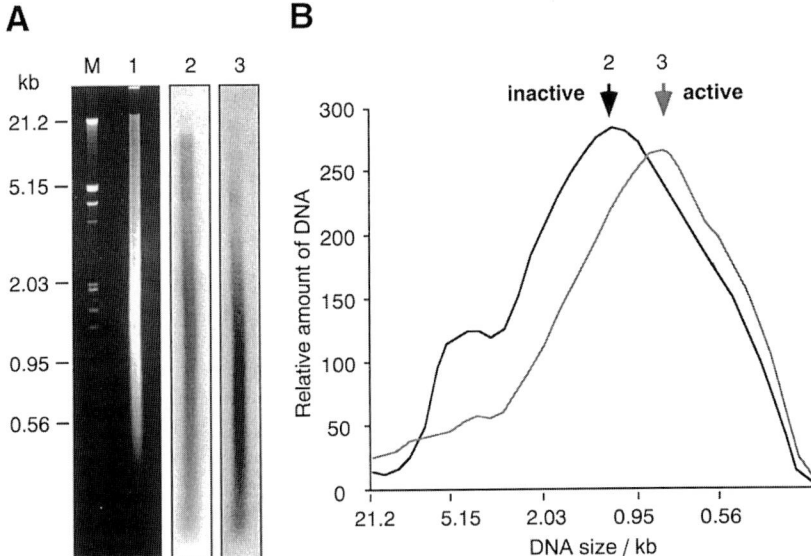

Fig. 3. Sonification efficiency of chromatin from active and inactive genes. **(A)** 5 μg DNA from an aliquot of sheared, crosslinked chromatin from *Drosophila* Schneider cells was purified by reversing the crosslinks, and separated on a 0.5% agarose gel (lane 1). Identical Southern filters of this gel were hybridized either to a probe from an inactive genomic region of the bithorax complex (lane 2), or to the expressed gene *Abd-B* (lane 3). Lane M is a molecular weight marker of λDNA digested with *Eco*RI/*Hin*dIII: the molecular weights of some DNA fragments are shown to the left. **(B)** The signal intensity of the smears in lanes 2 and 3 was quantitated by phosphorImager and plotted with respect to molecular weight. The signal from the inactive gene probe *(black line)* is most intense at 1.2 kb *(arrow)*, whereas the signal from the expressed gene probe *(gray line)* is most intense at 0.8 kb *(arrow)*.

that all experiments are carried out several times independently, to ensure that all apparently enriched fragments are real and not such a PCR artifact.

12. PCR amplification is carried out until nucleotides become limiting. Therefore, amplifications should result in a similar amount of product, regardless of the input. The amount of DNA produced from immunoprecipitations is in the range of 1 ng, and therefore difficult to measure accurately. Assuming that control immunoprecipitations isolate only 50% of the DNA of antibody immunoprecipitations, the control may go through an extra amplification cycle.

13. If binding of the protein of interest to an already defined target element is being investigated, it may not be appropriate to carry out Southern analysis, but slot blot analysis may be more efficient and accurate for determining relative enrichments over control immunoprecipitations. 100 ng of control and immunoprecipitated DNA is immobilized on a nylon filter by slot blot, and hybridized with

DNA probes derived from potential target sequences. In some cases the enrichment may be great enough that signal can be seen before PCR amplification; in this case we would recommend putting at least 50% of the DNA recovered from the immunoprecipitation on the slot blot.
14. Antibody immunoprecipitated DNA probes will hybridize to all fragments of a genomic region to some extent (although enriched fragments hybridize much more strongly). True enrichments due to the antibody may be distinguished from background immunoprecipitation by slot blot analysis. Typically 100–200 ng of DNA from control and antibody immunoprecipitations is immobilized on nylon membrane by slot blot and hybridized to a number of probes derived from the target DNA of interest. The resulting signals are quantitated and the actual enrichment accurately determined. Comparison between a number of fragments allows the setting of a "background" level, and only hybridisation signals above this level are considered to be enriched.
15. We found that hybridization of mock immunoprecipitated DNA to the *Abdominal-B* gene, which is expressed in our Schneider cell line, was 40–50% more than to inactive genes in the same region (**Fig. 2**). Control experiments revealed that sonication of active regions of the genome was more efficient than silent domains (**Fig. 3**), perhaps because they consist of a more open chromatin structure, which is more accessible to shearing *(12)*.

References

1. Solomon, M. J. and Varshavsky, A. (1985) Formaldehyde-mediated DNA-protein crosslinking: a probe for *in vivo* chromatin structures. *Proc. Natl. Acad. Sci. USA* **82,** 6470–6474.
2. Solomon, M. J., Larsen, P. L., and Varshavsky, A. (1988) Mapping protein-DNA interactions in vivo with formaldehyde: evidence that histone H4 is retained on a highly transcribed gene. *Cell* **53,** 937–947.
3. McGhee, J. D. and von Hippel, P. H. (1975) Formaldehyde as a probe of DNA structure. I. Reaction with exocyclic amino groups of DNA bases. *Biochemistry* **14,** 1281–1296.
4. McGhee, J. D. and von Hippel, P. H. (1975) Formaldehyde as a probe of DNA structure. II. Reaction with endocyclic imino groups of DNA bases. *Biochemistry* **14,** 1297–1303.
5. Orlando, V. and Paro, R. (1993) Mapping Polycomb-repressed domains in the bithorax complex using *in vivo* formaldehyde crosslinked chromatin. *Cell* **75,** 1187–1198.
6. Zhao, K., Hart, C. M., and Laemmli, U. K. (1995) Visualisation of chromosomal domains with boundary element-associated factor BEAF-32. *Cell* **81,** 879–889.
7. Orlando, V., Strutt, H., and Paro, R. (1997) Analysis of chromatin structure by *in vivo* formaldehyde crosslinking. *Methods: A Companion to Methods Enzymol.* **11,** 205–214.
8. Strutt, H. L., Cavalli, G., and Paro, R. (1997) Co-localisation of Polycomb protein and GAGA factor on regulatory elements responsible for the maintenance of homeotic gene expression. *EMBO J.* **16,** 3621–3632.

9. Lüdecke, H.-J., Senger, G., Claussen, U., and Horsthemke, B. (1989) Cloning defined regions of the human genome by microdissection of banded chromosomes and enzymatic amplification. *Nature* **338,** 348–350.
10. Harlow, E. and Lane, D. (1988) *Antibodies: A Laboratory Manual.* Cold Spring Harbor Laboratory Press, Cold Spring Harbor, New York.
11. Church, G. M. and Gilbert, W. (1984) Genomic sequencing. *Proc. Natl. Acad. Sci. USA* **81,** 1991–1995.
12. Strutt, H. (1997) *In vivo mapping of Polycomb and trithorax group proteins in chromatin of Drosophila melanogaster.* PhD thesis, The Open University, UK.
13. Martin, C. H., Mayeda, C. A., Davis, C. A., Ericsson, C. L., Knafels, J. D., Mathog, D. R., Celniker, S. E., Lewis, E. B., and Palazzolo, M. J. (1995) Complete sequence of the bithorax complex of *Drosophila. Proc. Natl. Acad. Sci. USA* **92,** 8398-8402.

34

Mapping DNA Interaction Sites of Chromosomal Proteins

Crosslinking Studies in Yeast

Andreas Hecht, Sabine Strahl-Bolsinger, and Michael Grunstein

1. Introduction

Eucaryotes use a common theme for packaging genomic DNA: 146 bp of DNA are wrapped around an octameric protein core consisting of two molecules each of the histones H2A, H2B, H3, and H4. Although this nucleosomal arrangement is ubiquitous throughout the genome, different chromosomal portions are nonetheless functionally and structurally distinct. This is owing to regional interactions between regulatory chromosomal proteins (i.e., transcription factors, structural components, histone-modifying enzymes) and specific portions of the chromosome. Therefore it is important to determine the natural sites of action of these chromatin (re-)modeling factors and to gain insight into the parameters governing their association with chromatin in vivo. To address this problem, Orlando and Paro *(1)* have used formaldehyde crosslinking to prevent redistribution of chromatin components during chromatin preparation, immunoprecipitation to isolate the chromosomal factor(s) to be investigated and hybridization to enable the detection of genomic sequences contacted by the factor(s) (*see* Chapter 33). A similar approach has been used by Braunstein et al. *(2)*, employing antibodies against sites of histone acetylation, to determine the acetylation state of histones in chromatin sites. We have streamlined and further developed this protocol and included a polymerase chain reaction (PCR) step to more sensitively and rapidly map the chromosomal distribution of regulatory proteins in *Saccharomyces cerevisiae* (**Fig. 1**). The use of mutations in components of the chromosomal complex in question further allows the identification of protein sequences which mediate formation of the complex.

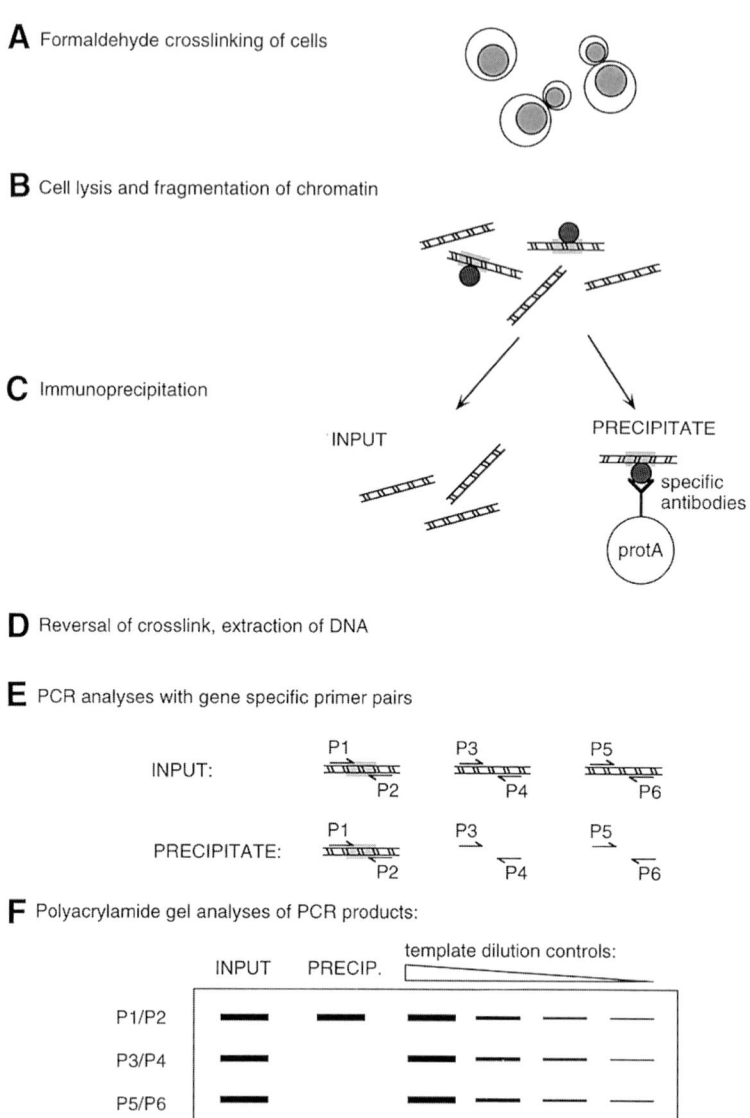

Fig. 1. *(continued on next page)* Strategy for the mapping of chromosomal interaction sites of nonhistone chromatin associated factors. Yeast cells are crosslinked with formaldehyde in vivo (**A**) and subsequently disrupted mechanically. Chromatin fragments released from the cells are sheared by sonication (**B**) to an average size suitable for the desired accuracy and resolution of the mapping experiment. Antibodies recognizing a chromosomal non-histone protein are added and protein-DNA complexes are recovered by adsorption to proteinA sepharose beads (**C**). Samples from the INPUT and PRECIPITATE are heat-treated to reverse the crosslinks and DNA is isolated (**D**). The purified DNA is then used as template in multiplex PCR reactions with primer pairs derived from different

The initial step of the mapping procedure is the crosslinking of living cells using formaldehyde (FA), which is a reagent particular useful for these studies *(3,4)* It targets primary amino groups bridging distances of about 2Å. Both protein-protein and protein-DNA crosslinks are induced. An important feature is that both types of adducts can be reversed selectively. FA readily penetrates biological membranes and does not require special buffer or temperature conditions allowing the crosslinking of live cells (except that one should avoid buffer substances containing primary amines such as Tris-HCl). Therefore, it is not necessary to lyse cells prior to crosslinking which circumvents the potential problem of redistribution or reassociation of chromosomal proteins during the preparation. After FA crosslinking, the cells are broken and chromosomal DNA is fragmented by sonication. Specific chromosomal proteins together with DNA crosslinked to them are subsequently immunoprecipitated from the cell lysate. After reversal of the FA-induced crosslinks the DNA is purified and used as template in PCR reactions to detect a genomic region of interest. The representation of specific sequences in the precipitated material thus is taken as indication for the association of the immunoprecipitated protein with that particular chromosomal area in vivo.

Figure 2 shows an example for the application of the protocol to the study of SIR3 association with heterochromatin-like regions in the yeast genome. SIR3 is a factor required for the transcriptional repression of subtelomeric regions and the silent mating type loci *HML* and *HMR* *(5,6)*. After FA-crosslinking and immunoprecipitation of epitope-tagged SIR3 we performed PCR reactions with primer pairs from *HML* , *ADH4, ACT1, PHO5*, and *GAL1*. DNA from *HML* and *ADH4* can be readily detected in the immunoprecipitate whereas DNA from the *ACT1, PHO5*, or *GAL1* genes, which are not known to be subject to SIR regulation, are much less abundant. The control experiments shown include immunoprecipitation from unfixed cells and from cells expressing WT SIR3, which is not recognized by the antibody against the epitope tag. In both controls *HML* and *ADH4* are absent in the immunoprecipitate demonstrating the stringency and specificity of the method. In additional experiments we used this method to map interactions of SIR2, SIR3 and SIR4 with subtelomeric heterochromatin *(7)*, to analyze the involvement of histone H4 in

---

chromosomal regions. Control PCR reactions with decreasing amounts of template are analyzed in parallel to assure that the PCR was still in the exponential phase to allow for a quantitative interpretation of the results (**E**). (**F**) Polyacrylamide gel electrophores's is used to separate and visualize different sized PCR products. Bound as well as unbound sequences are equally represented in the genome and will be amplified accordingly from the INPUT sample. Only sequences specifically bound by the regulatory protein in vivo coimmunoprecipitate with this factor (the shaded area in the P1/P2 region symbolizes an interaction site) and will give rise to PCR products from the PRECIPITATE (**F**).

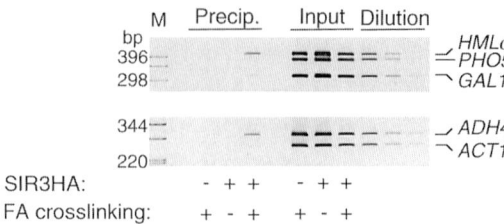

Fig. 2. Example for an experiment using the method described to analyze the association of SIR3 with silent chromosomal regions. Yeast cells expressing WT or epitope-tagged SIR3 were treated with or without formaldehyde. Cell lysates were prepared and subjected to immunoprecipitation with an antibody directed against the HA-epitope in the tagged SIR3 protein. DNA from INPUT and PRECIPITATE was isolated and analyzed in multiplex PCR with primer pairs from the *HML*, *PHO5*, *GAL1*, *ADH4*, and *ACT1* loci. Only *HML* and the telomeric *ADH4* sequences which are known targets for SIR3 are enriched in the PRECIPITATE. When the crosslinking step is omitted or when the epitope for the antibody used is absent from SIR3 *HML* and *ADH4* are no longer detectable in the PRECIPITATE. This demonstrates the stringency and specificity of the procedure.

the association of SIR3 with chromatin *(8)*, and to determine the acetylation state of histone H4 at specific yeast genes *(9)*. Recently, even the occupation of a single chromosomal site (the ARS1 origin of replication) by ORC, CDC6 and MCM7 proteins in a cell cycle dependent fashion could be shown *(10)*, providing another example for the power of the method.

Compared to the original version of Orlando and Paro *(1)* (*see* also Chapter 36) our protocol is simpler, more versatile and less time consuming. This is primarily because we take advantage of certain peculiarities of the yeast system. The relative small size of the yeast genome provides a better ratio of specific (bound) versus nonspecific (unbound) DNA that allows us to omit the CsCl gradient step for the separation of the protein-DNA adducts and noncrosslinked material. Moreover, the entire yeast genome has been sequenced. This information can be exploited to design primers for virtually any genomic region and thus all of the yeast genome is accessible for investigation. By subcloning and sequencing of the immunoprecipitated DNA it should also be possible to modify the procedure for the identification of unknown interaction sites. Finally, the ease with which yeast genes can be genetically altered provides unequaled opportunities to study the influence of targeted mutations on the chromosomal distribution and assembly of chromatin associated protein complexes.

2. Materials

2.1. Growth, In Vivo Crosslinking, Harvest, Lysis of Yeast Cells

1. Equipment, supplies and media for the growth of yeast *(11)*.
2. 37% formaldehyde (Merck).

3. 2.5 M glycine (Sigma) in double distilled H_2O.
4. Phosphate buffered saline (PBS): 140 mM NaCl, 2.5 mM KCl, 8.1 mM Na_2HPO_4, 1.5 mM KH_2PO_4, pH 7.5.
5. Lysis buffer: 50 mM HEPES/KOH pH 7.5, 140 mM NaCl, 1 mM EDTA, 1% Triton X-100, 0.1 % sodium-deoxycholate.
6. 100X stock solutions of protease inhibitors (Sigma; 100 mM PMSF freshly prepared in isopropanol, 100 mM benzamidine in H_2O stored at $-20°C$, 5 mg/mL TPCK in ethanol stored at $-20°C$, 25 mM TLCK in 50 mM sodium-acetate pH 5.0 stored at $-20°C$.
7. 1000X stock solutions of protease inhibitors (Sigma): 1 mg/mL aprotinin in 10 mM HEPES/KOH pH 8.0, 1 mg/mL Leupeptin in H_2O, 1 mg/mL PepstatinA in ethanol, 2 mg/mL Antipain in H_2O. All solutions are stored at $-20°C$.
8. Glass beads (0.45–0.52 mm diameter) (Thomas Scientific). Before use, the glass beads are acid washed by soaking in concentrated nitric acid for 1 h, extensive rinsing with distilled water, a final wash with 70% ethanol and baking at 80°C until dry *(12)*.

2.2. Immunoprecipitation and DNA Isolation

1. Affinity-purified, specific antibodies *(13)*.
2. 50% (v/v) suspension of proteinA-sepharose beads (Pharmacia), swollen according to the manufacturer and equilibrated in lysis buffer.
3. Wash buffer 1: equals lysis buffer but with 500 mM NaCl
4. Wash buffer 2: 10 mM Tris-HCl pH 8.0, 0.25 M LiCl, 0.5% NP-40, 0.5% sodium-deoxycholate, 1 mM EDTA.
5. TE: 10 mM Tris-HCl pH 8.0, 1 mM EDTA.
6. TE/1%SDS.
7. 20 mg/mL glycogen (Boehringer).
8. 20 mg/mL proteinase K (Merck).
9. PCI solution (phenol:chloroform:isoamylalcohol 25:24:1).
10. Reagents for DNA precipitation: 5 M LiCl, 50 mM Tris-HCl pH 8.0; absolute ethanol; 70% ethanol.
11. 10 mg/mL RNaseA (DNase-free; Sigma).

2.3. PCR Analyses and Gel Electrophoresis

1. Oligonucleotides: 24-mers with approx 50% GC content and melting temperatures as similar as possible.
2. 10X PCR buffer: 0.2 M Tris-HCl pH 8.3, 0.5 M KCl, 0.015 M $MgCl_2$, 0.5% Tween 20, 1 mg/mL gelatin.
3. 10X dNTP mixture containing 2mM dATP, dGTP, dCTP, dTTP each (Pharmacia).
4. Recombinant Taq polymerase (Gibco/BRL) 5 U/µL.
5. Supplies and reagents for polyacrylamide gel electrophoresis, 40% (19:1) acrylamide/bisacrylamide solution (Accugel, National Diagnostics), 10X TBE buffer, TEMED (Bio-Rad), 10% w/v ammonium persulfate in H_2O, 10 mg/mL ethidium bromide in H_2O.

3. Methods
3.1. Growth, In Vivo Crosslinking, Harvest, Lysis of Yeast Cells

1. Grow overnight cultures of the strains to be analyzed in 5 mL liquid media (*see* **Note 1**).
2. Measure the OD_{600} of the cultures and dilute into 50 mL of fresh media. Adjust the OD_{600} so that the culture will have reached a density of approx 1.5 the next morning.
3. After the cultures have reached the desired density, add formaldehyde to a final concentration of 1%. Mix rapidly and incubate for 15 min at room temperature with constant mixing. Avoid skin contact or breathing formaldehyde fumes (*see* **Note 2**).
4. Quench the crosslinking reaction by adding 2.5 mL 2.5 M glycine (final concentration 125 mM). Continue the incubation for 5 min.
5. Harvest the cells by centrifugation at 1500g using a centrifuge cooled to 4°C. Pour off the supernatant and place the cell pellets on ice.
6. Resuspend the cells in ice-cold PBS and pellet again.
7. Repeat **step 6**.
8. Pour off the supernatant, keep the tube inverted and wipe the walls with a tissue to remove residual liquid.
9. Add 400 µL ice-cold lysis buffer with protease inhibitors and resuspend the cell pellet by pipeting up and down several times (*see* **Note 3**).
10. Transfer the cell suspension to a 1.5-mL Eppendorf tube.
11. Add a volume of glass beads equal to the volume of the cell suspension (approx 500 µL).
12. Lyse cells by vortexing on an Eppendorf shaker model 5432 for 45 min at 4°C.
13. After cell lysis briefly spin in a table top centrifuge to remove any liquid from the top of the tubes.
14. With the lid of the Eppendorf tube open puncture the tubes at the bottom using a red hot 26G (0.45-mm) needle.
15. Place the punctured tube in a second (collection) Eppendorf tube and close the lid of the upper tube. Some liquid will leak out.
16. Centrifuge the two tubes for 5 s at full speed. Make sure that the tube assembly doos not obstruct rotation. If necessary leave open the lid of the centrifuge, use a different centrifuge or rotor. Cell debris and the crude cell lysate will be captured in the bottom tube, while the glass beads remain behind. Place tubes with the cell lysate on ice.
17. Shear chromatin by sonication with two pulses of 10 s each with a 20 s rest interval while cooling the samples in a ice-water bath (*see* **Note 4**).
18. Centrifuge the lysate from **step 17** at 10,000g for 5 min at 4°C.
19. Transfer as much as possible of the supernatant to a fresh tube. Spin again at 10,000g for 15 min at °C.
20. Transfer the supernatant to a fresh tube once more. This is the crude cellular extract (WCE).
21. Set aside a 20-mL aliquot as INPUT material. Store on ice until further processing.

3.2. Pilot Experiment to Titrate the Amount of Antibodies Needed for Quantitative Immunoprecipitation of Your Protein of Interest

1. Prepare crude cell extract from six 50-mL cultures as outlined in **Subheading 3.1.**
2. Place 400-mL aliquots of the cell lysates into 1.5-mL Eppendorf tubes.
3. Add increasing amounts of affinity purified antibodies to the cell lysates. Use 0.1 mg/mL as lowest amount and increase the antibody concentration in fourfold increments up to approx 25 mg/mL. Include one control sample which does not receive antibody (see **Note 5**).
4. Adjust the volumes of all samples by adding lysis buffer as required.
5. Incubate the samples at 4°C on a nutator for 3 h.
6. Add 60 µL of the proteinA sepharose slurry and continue the incubation for 1 h (see **Note 6**).
7. Pellet the protein A sepharose beads with the immunoprecipitate for 5 s at 10,000g.
8. Transfer the supernatant to a fresh tube. Avoid any carry over of the immunoprecipitate.
9. Perform Western blot analyses on aliquots of the supernatant to monitor the presence or absence of the protein of interest. The amount of antibodies that completely removes the protein from the lysate will be used in the experiments below. If necessary, repeat the titration with lower or higher antibody concentrations.

3.3. Immunoprecipitation and DNA Isolation

1. Prepare crude cellular extracts as outlined in **Subheading 3.1.**
2. To 400 µL aliquots of the lysates, add the amount of antibodies determined in **Subheading 3.2.**
3. Incubate the samples on a nutator for 3 h at 4°C.
4. Add 60 µL of the proteinA sepharose slurry and continue the incubation for 1 h.
5. Pellet the proteinA sepharose beads with the immunoprecipitate for 5 s at 10,000g.
6. Transfer the SUPERNATANT to a fresh tube and keep on ice for further processing if required (see **Note 7**).
7. Add 1 mL of lysis buffer to the proteinA sepharose beads and incubate for 5 min on a nutator (see **Note 3**).
8. Pellet the proteinA sepharose beads as in **step 5**.
9. Remove the supernatant and repeat **steps 7** and **8**.
10. Add 1 mL of wash buffer 1 to the protein A sepharose beads from **step 9** and incubate for 5 min at 4°C with agitation.
11. Pellet the protein A sepharose beads, add 1 mL of wash buffer 2 and incubate for 5 min at 4°C with agitation.
12. Pellet the protein A sepharose beads, add 1 mL TE. and incubate for 5 min at 4°C with agitation.
13. Pellet the protein A sepharose beads. Remove as much of the supernatant as possible.
14. Spin again and remove as much of the residual liquid as possible without losing beads (a Hamilton syringe or a glass capillary with a narrow tip are convenient to use for this step).

15. Add 60 μL TE/1%SDS, resuspend the beads and incubate the samples at 65°C for 10 min.
16. Centrifuge the beads for 2 min at 10,000g. Transfer the supernatant to a fresh Eppendorf tube. This sample is the PRECIPITATE.
17. Add 100 μL TE/1%SDS to 20 μL of the PRECIPITATE and the INPUT. Incubate at 65°C at least 6 h up to overnight to reverse the DNA-protein crosslinking. Keep the remainder of the PRECIPITATE at 4°C for additional analyses if desired (*see* **Note 7**).
18. Allow the samples to cool and collect the condensate at the bottom of the tubes by brief centrifugation.
19. Add 120 μL TE, 20 μg glycogen, and 100 μg proteinase K to the PRECIPITATE, and 120 μL TE, 2 μg glycogen, and 100 μg proteinase K to the INPUT. Keep at 37°C for 2 h.
20. Add 250 μL PCI solution and 25 μL 5 M LiCl, 50 mM Tris-HCl pH 8.0 to the samples. Vortex vigorously for 10 s.
21. Separate aqueous and organic phases by centrifuging for 5 min at 10,000g.
22. Transfer the upper (aqueous) layer to a fresh tube and add 750 μL abs. ethanol. Mix carefully.
23. Pellet the nucleic acids by centrifugation for 20 min at 10,000g.
24. Add 100 μL TE, 10 μg RNase A, and incubate for 1 h at 37°C.
25. Add 2 μg glycogenen, 10 μL 5 M LiCl, 50 mM Tris-HCl, pH 8.0. Vortex vigorously for 10 s.
26. Pellet nucleic acids by centrifugation for 20 min at 10,000g or more.
27. Wash pellet once with 70% ethanol, spin again for 5 min at 10,000g.
28. Discard the ethanol and dry the pellet.
29. Resuspend DNA from the INPUT in 20 μL TE, from the PRECIPITATE in 200 μL.

3.4. PCR Analyses and Gel Electrophoresis

1. Set up PCR reactions in 50 μL volume with 50 pMol of each primer, 0.2 mM dNTPs and 1.25 U Taq polymerase. Prepare sufficient PCR premix for all samples by combining appropriate amounts of 10X PCR buffer, primers, dNTPs, Taq polymerase and H$_2$O. The minimal number of samples includes a control without DNA, one reaction with the PRECIPITATE-DNA and five reactions with dilutions of the INPUT material as template. Place 50-μL aliquots of the PCR premix in thermal cycler tubes (*see* **Note 8**).
2. Prepare a 1:120 dilution of the INPUT-DNA by adding 2 μl of the DNA to 238 μL TE. From this sample prepare 4 serial 2.5-fold dilutions up to 1:39.000 (*see* **Note 9**).
3. To each tube add 4 μL of the various dilutions of the INPUT-DNAs and 4 μL of the undiluted PRECIPITATE-DNA. These amounts correspond to 1/13,500 or less of the INPUT and 1/50 of the PRECIPITATE. Include an additional reaction to which no DNA is added. This serves as a control for DNA contaminations by carryover.
4. If using a thermal cycler without heated cover, overlay the samples with mineral oil, place the reactions in the cycler and perform PCR using the following program: 2 min initial denaturation at 95°C followed by 25 cycles with 30 s at 95°C

(denaturation), 30 s at 55°C (annealing), 60 s at 72°C (elongation), and a final extension step of 5 min at 72°C.
5. Load 20 µL from each reaction along with a size standard on 6% polyacrylamide/ 0.5X TBE gels (20 cm long). Run gels at 150 V constant voltage until the bromophenol blue dye has reached the end of the gel. Disassemble and stain the gel in a 0.1 µg/mL ethidium bromide solution. Photograph the gel using a Polaroid camera or use another gel documentation system.

4. Notes

1. In particular when handling several strains in parallel it is desirable to work with synchronous growing cultures and to harvest all cultures in the same growth phase, preferably log phase. Wildtype strains and their mutant derivatives often differ in their growth properties. In addition, the type of growth media, complex media like YEPD or synthetic selective media, influence the growth rate of your strains and thus the time needed to obtain a sufficient number of cells. We therefore recommend that the doubling time of your strains is determined in a pilot experiment. This allows one to calculate the size of the inoculum from the preculture required to achieve the desired cell density after a given period of time. As the entire protocol from crosslinking through cell lysis, immunoprecipitation and clean up of the precipitated DNA adds up to a quite lengthy procedure we usually start our cultures in the late afternoon and adjust them in such a way that the cells are ready for harvesting the next morning.
2. Especially when comparing mapping experiments from different strains it is important to harvest the same numbers of cells from cultures with different OD's after the crosslinking step. Alternatively, one can grow all strains to the same OD_{600} before starting with the crosslinking step. If not all strains are ready at the same time keep those cultures on ice which are, while the other strains still grow to the desired density.
3. The buffer composition during cell lysis, immunoprecipitation and washing determines the stringency of the analyses. Increasing salt concentrations or including 0.1% SDS in the lysis and wash buffers would increase the stringency and could reduce nonspecific precipitation, but might also be harmful for the antibody. Nonetheless, varying the nature and concentration of the detergent and salt in the lysis and wash buffers may be necessary steps to optimize the procedure for a particular application. Additional tests to see whether a particular DNA sequence co-precipitates specifically, can be omission of the antibody or the crosslinking step. Another possibility is to compare precipitation from WT and mutant strains with a known phenotype.
4. The extent of chromatin fragmentation will influence the resolution of the mapping experiment. The smaller the DNA fragments the closer can neighboring chromosomal regions be located to determine whether a protein is associated with them or not. To establish the conditions for sonication with a particular sonicator prepare some WCE and subject it to an increasing number of sonication cycles. After each pulse remove a small aliquot of WCE. Process all collected

aliquots for DNA clean up as outlined in **Subheading 3.3, steps 17–26**. Analyze the DNA by agarose gel electrophoresis and determine how many cycles of sonication are needed to shear the chromatin to a certain size range. The conditions given in **Subheading 3.1., step 17** yield DNA fragments the majority of which are between 0.5–1.0 kb in size and refer to the model W-375, Heat Systems Ultrasonic Inc., equipped with the microtip and operated at setting 3. More extensive sonication can result in DNA fragments even smaller (300–500 bp; **ref. 6**). As average fragment size approaches the distance between the PCR primers used, amplification efficiency may decrease. In that case DNA dot blots and hybridization methods may be used as detection system *(1)*.

5. The quality of the antibodies used is of prime importance for the outcome of the experiment. Ideally, the antibodies should possess a high affinity for the antigen, withstand the most stringent conditions during the immunoprecipitation and they should not cross-react with other proteins. However, polyclonal and even monoclonal antibodies often react with several proteins, not just the one in which you are interested. Moreover, polyclonal antisera from various sources quite often react randomly against various yeast proteins. Therefore, when starting an immunization one should test the preimmune sera from a number of animals and select the ones which show the lowest nonspecific background. To further avoid problems due to nonspecific reactivity of polyclonal antisera one should affinity purify specific antibodies by absorption to the antigen used for immunization *(11)*.

6. Not all antibodies bind efficiently to proteinA. In cases where the antibodies cannot be recovered with proteinA sepharose beads the use of proteinG beads may help *(11)*.

7. Although the prime interest of the experiment is to identify the immunoprecipitated DNA it is a good idea to monitor efficiency of the precipitation by analyzing the protein content of INPUT, SUPERNATANT and PRECIPITATE in Western blots. For this purpose, aliquots from the INPUT, SUPERNATANT and PRECIPITATE are heated to 95°C in the presence of 0.5 M ß-mercaptoethanol for 30 min to reverse protein-protein crosslinks prior to SDS-polyacrylamide gel electrophoresis *(2,3)*.

8. We have chosen multiplex PCR as the method to detect representation of specific genomic regions in the immunoprecipitate. This requires simultaneous amplification of several PCR products from different primer pairs in a single test tube. To this end oligonucleotides were typically designed as 24-mers with 50% GC content to allow for similar annealing temperatures to be used for all primers. The length of the PCR products was in the range between 200 bp and 400 bp with a size difference of approx 20 bp between individual products. Thus, amplification efficiency should be minimally biased by length differences. Individual PCR product can easily be distinguished from each other on polyacrylamide gels or high resolution agarose gels. However, to allow unequivocal identification of PCR products, we recommend that all primer pairs are tested in separate reactions and that the products are analyzed side-by-side before setting up multiplex PCR. In addition, Southern blot hybridizations can be performed to verify the identity of the PCR products.

9. Whether a chromosomal protein is associated with a particular genomic region is deduced by comparing the relative abundance of PCR products from the region in question relative to a reference. Therefore it is important that the PCR products are analyzed, while the PCR is still in the exponential phase. Hence the amount of PCR product should be proportional to the amount of template DNA added to the reaction. The dilution series of the INPUT DNA serves to control this relationship. Specific applications may require increasing or decreasing the number of PCR cycles.

References

1. Orlando, V. and Paro, R. (1993) Mapping Polycomb-repressed domains in the bithorax complex using *in vivo* formaldehyde cross-linked chromatin. *Cell* **75**, 1187–1198.
2. Braunstein, M., Rose, A. B., Holmes, S. G., Allis, C. D., and Broach, J. R. (1993) Transcriptional silencing in yeast is associated with reduced nucleosome acetylation. *Genes Dev.* **7**, 592–604.
3. Solomon, M. J. and Varshavsky, A. (1985) Formaldehyde-mediated DNA-protein crosslinking: a probe for *in vivo* chromatin structures. *Proc. Natl. Acad. Sci. USA* **82**, 6470–6474.
4. Jackson, V. (1978) Studies on histone organization in the nucleosome using formaldehyde as a reversible cross-linking agent. *Cell* **15**, 945–954.
5. Laurenson, P. and Rine, J. (1992) Silencers, silencing, and heritable transcriptional states. *Microbiologic. Rev.* **56**, 543–560.
6. Thompson, J. S., Hecht, A., and Grunstein, M. (1993) Histones and the regulation of heterochromatin in yeast. *Cold Spring Harbor Symp. Quant. Biol.* **58**, 247–256.
7. Strahl-Bolsinger, S., Hecht, A., and Grunstein, M. (1997) SIR2 and SIR4 interactions differ in core and extended telomeric heterochromatin in yeast. *Genes Dev.* **11**, 83–93.
8. Hecht, A., Strahl-Bolsinger, S., and Grunstein, M. (1996) Spreading of transcriptional repressor SIR3 from telomeric heterochromatin. *Nature* **383**, 92–96.
9. Rundlett, S. E., Carmen, A. A., Suka, N., Turner, B. M., and Grunstein, M. (1998) Transcriptional repression by UME6 involves deacetylation of lysine 5 of histone H4 by RPD3. *Nature* **392**, 831–835.
10. Tanaka, T., Knapp, D., and Nasmyth, K. (1997) Loading of an MCM protein onto DNA replication origins is regulated by CDC6 and CDKs. *Cell* **90**, 649–660.
11. Rose, M. D., Winston, F., and Hieter, P. (1990) *Methods in Yeast Genetics: A Laboraotry Manual.* Cold Spring Harbor Laboratory Press, Cold Spring Harbor, NY.
12. Hoffman, C. S. and Winston, F. (1987) A ten-minute DNA preparation from yeast efficiently releases autonomous plasmids for transformation of *Escherichia coli*. *Gene* **57**, 267–272.
13. Harlow, E. and Lane, D. (1988) *Antibodies: A Laboratory Manual.* Cold Spring Harbor Laboratory Press, Cold Spring Harbor, NY.

35

UV Laser Footprinting and Protein–DNA Crosslinking

Application to Chromatin

Dimitri Angelov, Saadi Khochbin, and Stefan Dimitrov

1. Introduction

Protein–nucleic acid complexes play a crucial role in the events involved in gene expression and regulation. Direct and powerful approaches in studying this regulation are footprinting and protein–DNA crosslinking. Protein–DNA crosslinking detects the presence of a protein on a given DNA sequence, and footprinting allows one to judge the mechanism and details of protein–DNA interactions.

1.1. UV Laser-Induced Protein–DNA Crosslinking

Irradiation with conventional UV light sources has been widely used in studies on protein–DNA crosslinking. However, to achieve an acceptable degree of crosslinking, irradiation times ranging from minutes to several hours were required *(1–3)*. Such prolonged irradiation allows for redistribution of proteins and it also precludes kinetic studies. These drawbacks are overcome by the help of UV lasers.

UV laser protein–DNA crosslinking is a biphotonic process and the induced crosslinks have for precursors radical cations of nucleic bases. The chemical reaction of laser protein–DNA crosslinking is poorly understood. Nevertheless, this method has been widely used in studying protein–DNA interactions both in vitro and in vivo *(4–14)*. The laser crosslinking technique has several advantages over techniques using conventional UV light sources: very high rapidity (crosslinking is induced for less than 1 µs), high quantum yield (exceeding 50–100 times that obtained with conventional sources) and induction of crosslinks, which are unable to be produced by irradiation with conventional UV light *(4–7)*. According to literature data, by using UV laser, 5–15% crosslinking of irradiated protein–DNA complexes is achieved. *(4,6,7)*. Laser irradiation of multiprotein–DNA complexes induces protein–DNA crosslinks only: no protein–protein crosslinks are formed *(4,6)*.

1.2. Laser Protein–DNA Photofootprinting

UV irradiation of protein–DNA complexes results in different types of lesions in DNA, the lesion spectrum depending on the presence and type of proteins in these complexes *(8,15–19)*. This dependence is determined by local conformational changes in DNA induced by protein–DNA interactions. In fact, UV light "feels" local DNA structure. Thus, it can be used as a probing agent for analysis of both protein–DNA interactions and DNA conformation. The method developed for this analysis is called "photofootprinting." Here, again, the use of UV lasers has many advantages compared to conventional light sources. With a single UV laser pulse a footprint of the protein is achieved. Additionally, high intensity laser irradiation, contrary to conventional light sources, induces specific biphotonic lesions in DNA. These lesions are extremely sensitive to local DNA structure and can be easily detected by treatment with chemical reagents or enzymatic digestion.

Irradiation of DNA with conventional UV light sources produces mainly two types of monophotonic lesions: cyclobutane pyrimidine dimers (CPDs) and pyrimidine *(6–4)* pyrimidone monoadducts *[6–4]* PPs). The efficiency of induction of these photolesions might be affected by the presence of specifically bound proteins. This has been exploited to map *in vivo* the interactions of different transcription factors with their cognate DNA sequence *(15–18)*. UV irradiated DNA was digested with T4 endonuclease V (for detection of CPD's) or treated with hot piperidine (for *[6–4]* PP detection). The sites of the cleavage were found by ligation mediated PCR, followed by separation on a sequencing gel. The above two types of monophotonic lesions can also be induced by using UV laser irradiation, and this was used by Geiselmann and coworkers *(19)* to study the interactions of integration host factor of *E. coli* with DNA.

Laser irradiation induces both monophotonic and laser-specific oxidative lesions. However, by changing irradiation conditions one can create one type of lesions only. For example, during high-intensity laser irradiation essentially oxidative lesions are formed, the quantum yield of monophotonic lesions being negligible.

Two well studied laser specific guanine lesions are: 8-oxodG and oxazolone *(20,21)*. These lesions are very sensitive to local helical DNA conformation and they can be analyzed and quantified at the nucleotide level by treatment with specific reagents. 8-oxodG is quantitatively cleaved by Fpg protein, whereas oxazolone is removed during hot piperidine treatment *(20)*. Separation of the cleaved irradiated DNA allows one to find the positions of the lesions *(21)*. The chemical mechanism of formation of these two types of lesions is well documented. These lesions originate from the same type of initial radical cations and are induced by a competitive transformation process: hydration at position 8 of guanine radical cation leads to the formation of 8-oxodG, while oxazalone is formed upon radical cation deprotonation. The sum of quantum yields of both lesions is practically equal to the guanine photoionization quantum yield, i.e., the yield of other guanine oxidative lesions is less than 10% *(22)*.

1.3. Application of UV Laser Photofootprinting in Studies on Four-Way Junction DNA Structure and Four Way Junction DNA-Histone H1° Complexes

Four-way junction (4WJ) DNA is the core structure of cruciform DNA and of Holiday junction. 4WJ DNA showed a specific secondary and 3D structure *(23)*. Its 3D structure seemed to be similar to DNA crossovers *(23)*. Linker Histones H1, H5, and H1° interact preferentially with 4WJ DNA *(24,25)*. Here we will illustrate the power of UV laser photofootprinting in studying both 4WJ DNA structure and 4WJ DNA-H1° complexes.

In **Fig. 1** are presented the results of the strand 2 (4WJ2) laser footprinting of 4WJ DNA compared to those for the respective duplex DNA. As seen, UV laser footprinting is sensitive to DNA secondary structure. The intensity of the bands of guanines located close to the central part of the 4WJ DNA and containing nonpaired bases, clearly differed from those of the corresponding double stranded form (for secondary 4WJ structure *see* **Fig. 2**). For example, in the case of both piperidine and Fpg treatment an increased band intensity of guanines G9 and G10 for 4WJ2 DNA is observed. On the contrary, upon Fpg treatment G5, G6 and G8 band intensities of duplex DNA are decreased compared to those of 4WJ2 DNA. These data reflect radical cation yield differences at specific guanine bases in 4WJ structure relative to duplex DNA *(21,22)*. This effect might be determined by different structural dependent redistribution of nucleic base excitation, induced by charge and/or energy migration phenomena in the two samples *(21,22)*. The data presented demonstrated that UV laser footprinting can discriminate secondary DNA structures.

Binding of proteins induces local structural changes in DNA, which can be detected by UV laser footprinting. A typical example of such detection in the case of 4WJ DNA-H1° complex is shown in **Figs. 2** and **3**. Binding of H1° to 4WJ DNA results in decrease of guanine photoreactivity, the highest decrease being observed for guanines located in very close vicinity to the central part of 4WJ DNA. The above finding is compatible with specifically induced deformations in 4WJ DNA in the central part of this structure, suggesting that this 4WJ DNA part is specifically recognized by histone H1°.

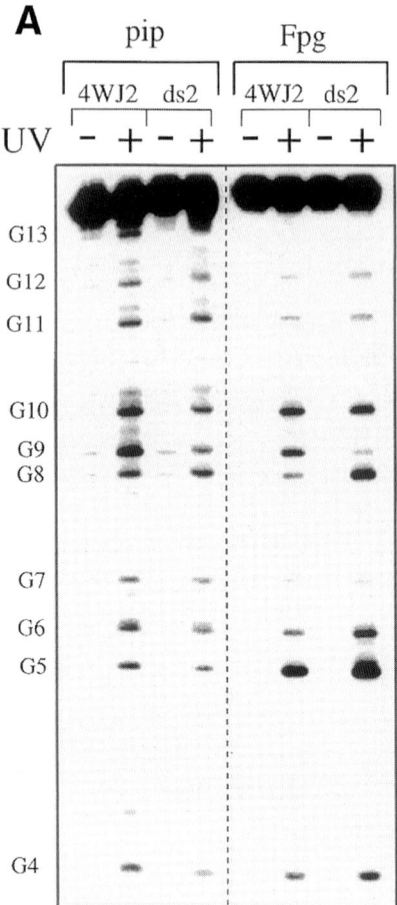

Fig. 1. UV laser footprinting of 4WJ DNA and duplex DNA. (**A**) Strand 2 of 4WJ DNA (4WJ2) was 5' end-labeled and the four oligonucleotides were then annealed to form 4WJ structure. Alternatively, the labeled strand 2 was annealed with its complementary strand to build duplex DNA. Both samples were irradiated with a high-intensity single UV laser pulse and treated with hot piperidine or digested with Fpg. Nonirradiated control samples, submitted to identical treatment, are also shown. The positions of guanine bases are shown on the right part of the pannel.

It should be noted that such photofootprinting for either naked or protein-complexed DNA cannot be obtained by using irradiation with conventional UV light sources, since under the latter conditions no oxidative DNA lesions are formed. However, UV laser irradiation induced the same monophotonic lesions as those induced by conventional light sources. Thus, a combination of both analyses for oxidative UV laser specific lesions (8-oxodG and oxazolone) and for monophotonic lesions

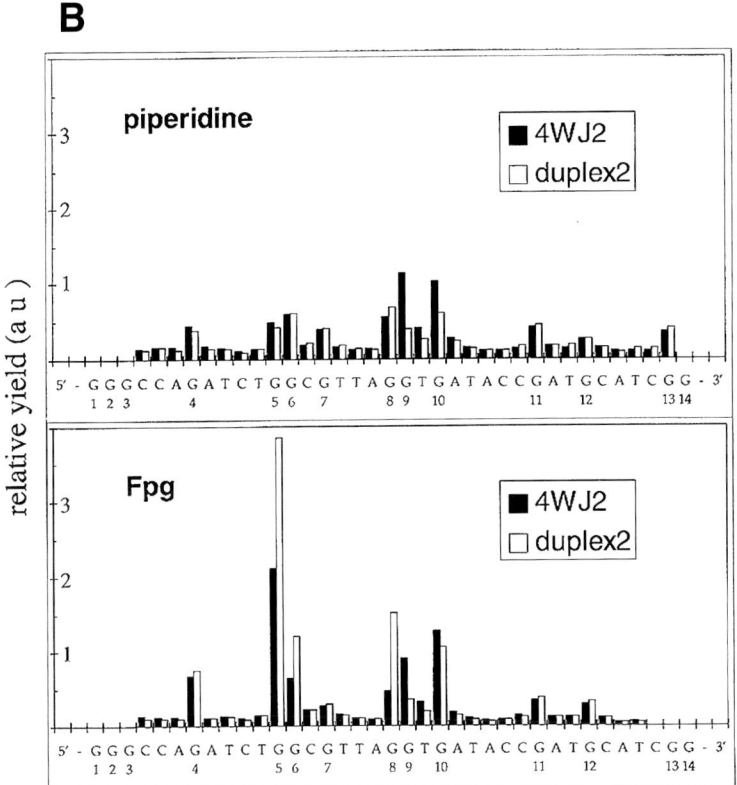

Fig. 1. *(continued)* **(B)** Relative yield (in arbitrary units) for lesions sensitive to hot piperidine or to Fpg for both duplex DNA and 4WJ2 DNA strand.

CPD's and *(6–4)* PP's will give a very detailed picture of DNA structural transitions and protein–DNA interactions. Because these phenomena can be detected by a single laser pulse, this will also allow one to study their kinetics.

2. Materials
2.1. UV Laser (see Note 1)

For UV irradiation we usually used a single pulse from fourth harmonics (266 nm) of a Surelite II (Continuum, USA) Nd:YAG laser (maximum energy 60 mJ, pulse duration 5 ns). The pulse energy was measured with calibrated pyroelectrical detector (Ophir Optronics Ltd.) using 8% deviation beam splitter. Typical irradiation doses (energy divided by beam cross-section surface) used were about 1kJ/m^2.

In the case of photofootprinting studies, the size of the laser beam was adjusted by means of a set of circular diaphragms to perfectly fit the surface

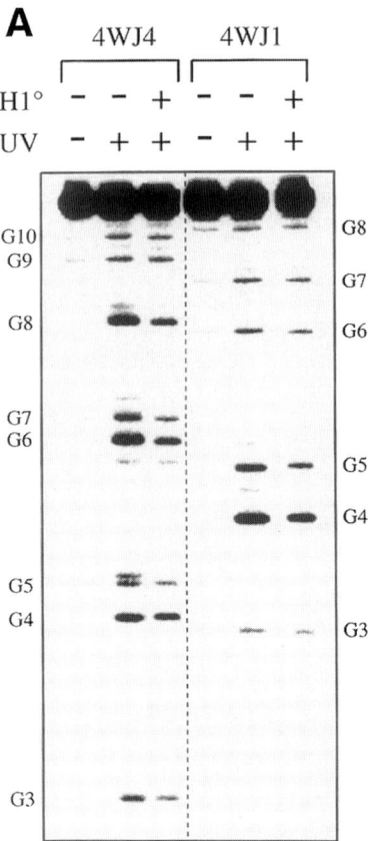

Fig. 2. UV laser footprinting for 4WJ1-H1° and 4WJ4- H1° complexes. (**A**) 4WJ DNA containing either 5'-labeled strand 1 (4WJ1) or strand 4 (4WJ4) were complexed with histone H1° and irradiated with a single laser pulse. The irradiated samples were treated with hot piperidine and the cleaved products separated on sequencing gel. The pattern of cleavage for non irradiated complexes and irradiated naked 4WJ DNA are also shown.

area of the sample. Typical irradiation doses (energy divided by beam cross-section surface) used were about 1kJ/m^2.

2.2. Laser Footprinting

1. TBE buffer: 89 mM Tris base, 89 mM boric acid, 2 mM EDTA, pH 8.0.
2. Acrylamide/bis-acrylamide (w/w), 50% urea in TBE buffer.
3. Polynucleotide kinase buffer.
4. Polynucleotide kinase.
5. [γ-^{32}P]ATP.
6. 10 M piperidine.
7. Formamidopyrimidine DNA glycosylase (Fpg protein).

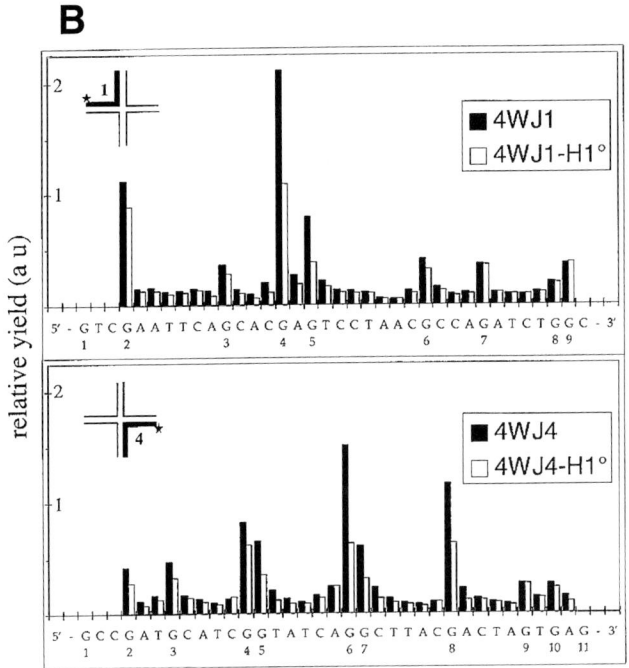

Fig. 2. *(continued)* **(B)** Relative yield (in arbitrary units) for hot piperidine sensitive lesions of naked 4WJ DNA and 4WJ DNA-H1° complexes.

8. Phenol.
9. TE buffer: 10 mM Tris-HCl, 1 mM EDTA, pH 7.6, 10 mM NaCl.
10. 0.65 µL siliconized Eppendorf tubes.

2.3. Piperidine and Fpg Treatment of Irradiated Samples

1. TE, 35 mM NaCl.
2. 10 M piperidine.
3. Formamide loading buffer: deionized formamide, containing 1 mM EDTA, 0.1% xylene cyanol and 0.1% bromphenol blue.
4. Proteinase K (10 mg/mL stock solution).
5. Bovine serum albumin (BSA; 10 mg/mL stock solution).
6. Fpg protein (5 mg/mL stock solution).
7. Stock acrylamide/urea solution (*see* **Subheading 2.2.2.**).

2.4. UV Laser Protein–DNA Crosslinking

2.4.1. Purification of the Covalent Protein–DNA Complexes on CsCl Gradients

1. 8 M urea, 1% sarkosyl.
2. Solutions of CsCl with ρ = 1.76, 1.57, 1.54, and 1.32 g/cm^3.

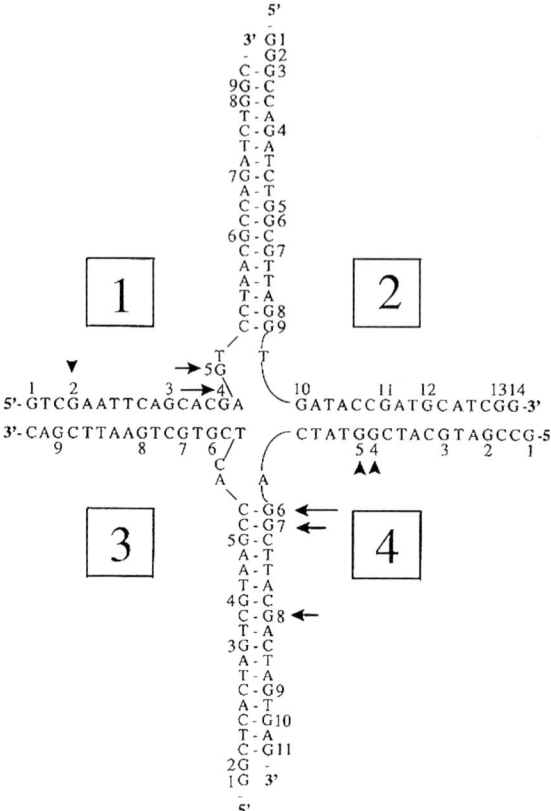

Fig. 3. Schematic 4WJ DNA representation. Arrow heads show the guanine bases with decreased photoreactivity resulting from histone H1° binding. The length of the arrow reflects the extent of decreased guanine base photoreactivity.

3. Immunoprecipitaion buffer: 50 mM HEPES, pH 7.5, 2 M NaCl, 0.1% SDS, 1% Triton X-100, 1% deoxycholate, 5 mM EDTA.

2.4.2. Immunoprecipitation of the Crosslinked Protein–DNA Complexes

1. IgGsorb (The Enzyme Center, Malden, MA).
2. 1% bovine serum albumin (BSA) in PBS.
3. Immunoprecipitation buffer (*see* **Subheading 2.4.1.**).
4. 50 mM HEPES, pH 7.5, 0.15 M NaCl, 5 mM EDTA.
5. 1% SDS.
6. Phenol.
7. Ethanol.
8. TE (*see* **Subheading 2.2.**).

UV Laser Footprinting and Crosslinking

2.4.3. Hybridization Analysis of Immunoprecipitated DNA

1. 0.5 N NaOH.
2. Hybond N (Amersham) blotting membranes.
3. [α-^{32}P] dCTP.
4. Random priming kit (Boehringer).
5. Hybridization buffer: 0.75 M NaCl, 0.075 M Na citrate, 5X Denhard's solution, 0.5% SDS, 50 mM Phosphate buffer, pH 7.0, 200 µg/mL denatured salmon sperm.
6. Washing buffer 1: 0.3 M NaCl, 0. 03 M Na citrate, 1% SDS .
7. Washing buffer 2: 0.15 M NaCl, 0.015 M Na citrate, 0.1% SDS.

3. Methods

3.1. UV Laser Photofootprinting

Below we describe a detailed procedure for UV laser guanine photofootprinting. This procedure is quite reproducible. It consists of the following steps:

1. Preparation of DNA samples.
2. Sample irradiation.
3. Piperidine and Fpg cleavage of irradiated samples.
4. Separation of the cleaved products on a sequencing gel.
5. Determination of the yield of lesions at individual sites.

3.1.1. DNA Sample Preparation (see **Note 2**)

1. Purify oligonucleotides either with HPLC or on 15–20% polyacrylamide sequencing gel in TBE and 5' label them with ^{32}P according to standard procedures (*see* **Note 3**) *(16)*.
2. Repurify labeled oligonucleotides on a small size (10 × 10 cm) 15% gel from preexisting oxidative lesions after either piperidine treatment or Fpg digestion of the hybridized duplex form (*see below* for details of digestion).
3. Remove piperidine by phenol treatment followed by ethanol precipitation. Dissolve the material in TE, 10 mM NaCl (*see* **Note 4**).
4. Anneal labeled oligonucleotide to nonlabeled complementary strand (at a molar ratio 1:1.3) . Heat at 90°C in a heater block for 3 min, switch off the heater and leave the material to cool down slowly to room temperature. After checking the efficiency of annealing by 12–15% native polyacrylamide gel, aliquot the material and store it at –20°C.

3.1.2. Sample Irradiation

1. Deposit 10–20 µL sample on the bottom of a small (0.65-µL) siliconized Eppendorf tube (*see* **Note 6**). Each sample to be studied has to be prepared in duplicate: one as a control and second to be irradiated. All samples have to contain the same amount of radioactively labeled DNA.
2. Irradiate with a single laser pulse with a dose not exceeding 1kJ/m^2 (*see* **Notes 5, 7**, and **8**).

3.1.3. Piperidine and Fpg Treatment of Irradiated Samples

For piperidine and Fpg treatment samples have to be in TE buffer solution, pH 7.6, containing 35 mM NaCl.

3.1.3.1. PIPERIDINE TREATMENT
1. Add 1.1 µL of 10 M piperidine to 10 µL of irradiated samples and to a nonirradiated control.
2. Incubate at 90°C for 30 min.
3. Remove piperidine from samples by five successive evaporations in speed-vac under heating, every time redissolving the dried material in 30 µL H$_2$O.
4. Dissolve the material in 3 µL formamide loading buffer. At this stage, if necessary, you can freeze the samples until loading on the gel.

3.1.3.2. FPG DIGESTION

If the sample is a protein–DNA complex, dilute it to 100 µL and remove proteins by proteinase K (100 µg/mL final concentration) for 15 min at 37°C, followed by phenol extraction and ethanol precipitation. Carry out the Fpg digestion as follows:

1. Add to samples BSA to 200 µg/mL.
2. Mix gently 1 µL of Fpg protein (5 µg/mL) to 10 µL of sample and incubate at 37°C for 30–40 min.
3. Dry the material in speed-vac and resuspend it in 3 µL of formamide loading buffer. The material can be kept at –20°C until use.

3.1.4. Separation of the Cleaved Products on a Sequencing Gel
1. Pour 15% sequencing gel (19:1 ratio of acrylamide:bis-acrylamide).
2. Prerun the gel for 45 min at 60 W.
3. Heat the samples for 3 min at 90–95° C and keep them on ice until loading.
4. Load the whole sample (10–20000 cpms) on the gel. Make sure (by using a bench counter) that all labeled material is loaded.
5. Run the gel, dry it, and expose it on Phosphorimager screen overnight.

3.1.5. Determination of the Yield of Individual Site Lesions

The relative yield of individual site lesions is defined as: $r_i = R_i/R_0$, where R_i is the freed from background radioactivity signal of a band i and R_0 is the total radioactivity signal in the lane.

Determine the relative yield as follows:

1. Scan the Phosphorimager screen by using Image Quant Software (Molecular Dynamics).
2. Measure total loaded radioactivity signal R_0 for both control and irradiated samples by integrating the respective whole lines
3. Design a series of integration rectangles, each one covering the band to be quantified
    a. Integrate respective rectangles for each band corresponding to both irradiated and control sample.

b. Determine the total loaded radioactivity R_0 by integrating the whole line for both control and irradiated samples.
   c. Normalize the radioactivity signal of each band by dividing it by the total lane signal.
   d. Find r_i by substracting the normalized band signals of controls from those of irradiated ones.
   e. Plot the distribution of relative yields, using as an abscissa the oligonucleotide sequence (*see* **Figs. 1B** and **2B**).

3.2. UV Laser Protein–DNA Crosslinking

Below we give a detailed protocol for UV laser protein–DNA crosslinking in nuclei and whole cells. Immunoprecipitation of individual protein–DNA complexes with specific antibodies allows one to select the whole set of DNA sequences to which the protein of interest was crosslinked, i.e., the sequences with which this protein interacts in vivo. Hybridization with selected DNA sequences permits one to find whether these sequences exist as in vivo complexes with the studied protein.

The protocol consists of the following steps:

1. UV laser sample irradiation.
2. Purification of the crosslinked protein–DNA complexes.
3. Immunoprecipitation of the crosslinked protein–DNA complexes.
4. Identification of the DNA sequence crosslinked to the studied protein.

3.2.1. Sample Irradiation

1. Irradiate the material at 2–5 Hz (1500–1800 J per optical unit of sample) in rectangular fused silica cuvets under continuous stirring. The optical density of the solution has to be in the range $3 < A_{260} < 5$ (optically thick sample).

3.2.2. Purification of the Covalent Protein–DNA Complexes on CsCl Gradients

1. To reduce DNA size, resuspend the irradiated nuclei (after centrifugation on a top bench centrifuge) in 1% sarkosyl, 8 M urea in an Eppendorf tube and sonicate them with several bursts in an ice bath, until DNA molecular weight is decreased down to 200–300 bp.
2. Load the material on preformed CsCl gradients (four layers with $\rho = 1.76$, 1.57, 1.54, and 1.32 g/cm³, each layer being 2.5 mL) and run in SW41 Beckman rotor at 35,000 rpm (150,000g) for 42 h at 15°C.
3. Collect eight-drop fractions and measure their optical densities at 260 nm.
4. Draw the sedimentation profile of the gradient and pool the fractions of the peak. Dialyze extensively against TE.
5. Precipitate the material with 3 vol of ethanol and wash it with 70% cold ethanol. Air dry the sample and dissolve it in Immunoprecipitaion buffer at a concentration 4–5 A_{260} units

3.2.3. Immunoprecipitation of the Crosslinked Protein–DNA Complexes (see **Note 9**)

1. Suspend 50 µL IgGsorb in 1% BSA in PBS. Shake gently for 3–4 h at room temperature in order to block the sites of nonspecific absorption (*see* **Note 9**).
2. Centrifuge for 30 s on a top bench centrifuge. Suspend the pellet in Immunoprecipitation buffer, centrifuge again for 30 s and discard the supernatant.
3. Resuspend the pelleted IgGsorb in 0.5 mL of CsCl purified material (approx 2 optical units, i.e., 100 µg of sample measured as DNA) in immunoprecipitation buffer. Add to this solution the specific antibody (also in immunoprecipitation buffer) to a sample:antibody ratio of 2.5:1 (*see* **Note 12**).
4. Shake gently for 4 h at room temperature.
5. Centrifuge for 30 s, take off the supernatant (the unbound material) and keep it at 4°C.
6. Resuspend the pellet in 1 mL of immunoprecipitation buffer and vortex it for 1 min at maximum speed.
7. Centrifuge for 30 s and discard the supernatant. Repeat the washing (**steps 6 and 7**) four more times.
8. Repeat the washing three more times, but with a solution of 50 mM HEPES, pH 7.5, 0.15 M NaCl, 5 mM EDTA.
9. Elute the bound material with 300 µL 1% SDS on gentle shaking for 10 min and incubate it with pronase (1 mg/mL) for 4 h at 37°C.
10. After phenolyzation, precipitate the material with 3 vol of ethanol and dissolve it in TE. Measure fluorimetrically, if possible, the amount of the precipitated DNA.

3.2.4. Dot Hybridization Analysis of Immunoprecipitated DNA from Crosslinked Protein–DNA Complexes

1. Alkali denature in 0.5 M NaOH at 37°C for 10 min: a) DNA isolated from the immunoprecipitated complexes (antibody bound DNA), and b) total (nonfractionated) DNA.
2. Prepare three identical filters (Hybond N, Amersham) for hybridization, each one containing three dots of immobilized total DNA (1.0, 0.5, and 0.25 µg DNA) and three dots of plasmid containing as an insert the DNA sequence of interest (corresponding to 1.0, 0.5, and 0.25 µg insert).
3. Label total DNA, bound DNA and unbound DNA with ^{32}P label by random priming using a Boehringer random priming kit as suggested by the manufacturer.
4. Prepare three hybridization mixtures, each one containing the respective probe in hybridization buffer. These three hybridization mixtures have to contain equal quantities of DNA (estimated fluorimetrically) as well as equal amounts of ^{32}P-radioactivity in equal final volumes.
5. Prehybridize the filters for 2 h at 62°C in hybridization buffer.
6. Carry out the hybridization for 16 h at 62°C.
7. Following hybridization wash filters twice in washing buffer 1 for 15 min at 65°C and twice in washing buffer 2 for 15 min at 65 °C.
8. Dry the filters and autoradiograph them.

4. Notes

1. Usually we used nano- or picosecond Nd:YAG lasers. Another type of laser that might be used is a nanosecond excimer KrF, operating at 248 nm. However, this laser requires gas devices which makes its handling more difficult.
2. **Subheading 3.1.1.** describes the procedures to follow when working with oligonucleotides. Elimination of preexisting oxidative lesions cannot be carried out with double stranded DNA fragments and thus, special care has to be taken in this case to avoid the induction of such lesions. Labeled stock DNA has to be kept at −20°C and multiple thawings has to be avoided, since this could enhance the background due to autoradiolysis.
3. When working with multiple samples that have to be compared, use samples with the same cpm's and carry out all procedures in the same tube.
4. In the case of piperidine treatment of protein–DNA complexes phenolization is not necessary. For some protein–DNA complexes even Fpg treatment might not require phenolization, but this should be checked for every individual complex.
5. Make sure that the laser beam crosssection area to coincide with that of the irradiated sample. Otherwise you will not measure correctly the dose of irradiation.
6. Avoid the presence of air bubbles in the irradiated material since bubbles impede sample irradiation. Very much attention should be paid when working with protein–DNA complexes, since formation of bubbles is relatively easy in these samples
7. In single pulse experiments optically thin samples ($E_{260}<0.1$) should be used.
8. Avoid using laser pulse dose exceeding 1 kJ/m^2 for irradiation of oligonucleotides with 30–40 bp length: higher doses might lead to the production of single stranded oligonucleotides containing more than one hit (single hit conditions).
9. The delicate point in this approach is the immunoprecipitation procedure and thus, it should be carried out carefully and strictly following the above protocol. Two important notes are listed below.
10. We usually blocked IgGsorb sites of non-specific absorption by BSA treatment. However, in some cases such blocking was not sufficient: a high background was observed. This background can be decreased by incubating the IgGsorb in laser irradiated *E. coli* DNA (1 mL of DNA at 0.5 mg/mL).
11. For immunoprecipitation we always used immunopurified polyclonal antibodies. If such antibodies are not available we recommend a cocktail of monoclonal antibodies to be used.

Acknowledgments

We thank Dr. Jean-Jacques Lawrence for support throughout the course of this work. This work was supported by grants from CNRS, INSERM (contract #4E006B), the Foundation de la Recherche Medicale, and the Region Rhône-Alpes.

References

1. Markovitz, S. (1972) Ultraviolet light-induced stable complexes of DNA and DNA polymerase. *Biochem. Biophys. Acta* **281**, 522–534.
2. Welsh, J. and Cantor, C. R. (1984) Protein–DNA crosslinking. *TIBS* **9**, 505–507.

3. Gilmor, D. S. and Lis., J. T. (1984) Detecting protein–DNA interactions *in vivo*: distribution of RNA polymerase on specific bacterial genes. *Proc. Natl. Acad. Sci. USA* **71**, 947–951.
4. Hockensmith, J. W., Kubasek, W. L., Vorachek, W. R., and Von Hippel, P. H. (1986) Laser crosslinking of nucleic acids to proteins. Methodology and first applications to phage T4 DNA replication system. *J. Biol. Chem.* **261**, 3512–3518.
5. Harrison, C. A., Turner, D. H., and Hinkle, D. C. (1982) Laser crosslinking of E. Coli RNA polymerase and T7 DNA. *Nucleic Acids Res.* **10**, 2399–2414.
6. Pashev, I.G., Dimitrov, S.I. and Angelov, D. (1991)) Laser-induced protein–DNA crosslinking. *TIBS.* **16**, 323–326.
7. Angelov, D., Stefanovsky, V. Yu., Dimitrov, S. I., Russanova, V. R., Keskinova, E., and Pashev, I. G. (1988) A picosecond UV laser induced protein–DNA crosslinking in reconstituted nucleohistones, nuclei and whole cells. *Nucleic Acids. Res.* **16**, 4525–4538.
8. Angelov, D., Berger, M., Cadet, J., Marion, C., and Spassky, A. (1994) High-intensity ultraviolet laser probing of nucleic acids. *Trends Photochem. Photobiol.* **3**, 643–663.
9. Budowsky, E. I., Axentyeva, M. S., Abdurashidova, G. G., Simukova, N. A., and Rubin, L. B. (1986) Induction of polynucleotide-protein crosslinkages by ultra-violet irradiation. Peculiarities of the high-intensity laser pulse irradiation. *Eur. J. Biochem.* **159**, 95–101.
10. Dobrov, E. N., Arbieva, Z. K., Timofeeva, E. K., Esenaliev, R. O., Oraevsky, A. A., and Nikogosyan, D. N. (1989) UV laser induced RNA-protein crosslinks and RNA chain breaks in tobacco mosaic virus RNA *in situ. Photochem. Photobiol.* **49**, 595–598.
11. Buckle, M., Geiselmann, J., Kolb, A., and Buc, H. (1991) Protein–DNA crosslinking at the lac promoter. *Nucleic Acids Res.* **19**, 833–840.
12. Stefanovsky, V. Y., Dimitrov, S. I., Russanova, V. R., Angelov, D., and Pashev, I. G. (1989) Laser-induced crosslinking of histones to DNA in chromatin and core particles: implications in studying histone-DNA interactions. *Nucleic Acids Res.* **17**, 10,069–10,081.
13. Stefanovsky, V. Yu., Dimitrov, S. I., Angelov, D., and Pashev, I. G. (1989) Interactions of acetylated histones with DNA as revealed by UV laser- induced histone-DNA crosslinking. *Biochem. Biophys. Res. Commun.* **164**, 304–310.
14. Dimitrov, S. I., Stefanovsky, V. Y., Karagyozov, L., Angelov, D., and Pashev, I. G. (1990). The enhancers and promoters of the *Xenopus* laevis ribosomal spacer are associated with histones upon active transcription of the ribosomal genes. *Nucleic Acids. Res.* **18**, 6393–6397.
15. Becker, M. M. and Wang, J. C. (1984) Use of light for footprinting DNA *in vivo*. *Nature* **309**, 682–687.
16. Pfeifer, G. P., Drouin, R., Riggs, A. D., and Holmquist, G. P. (1992) Binding of transcription factors creates hot spots for UV photoproducts *in vivo*. *Mol. Cell. Biol.* **12**, 1798–1804.
17. Pfeifer, G. P., Drouin, R., and Holmquist, G. P. (1993) Detection of DNA adducts at the DNA sequence level by ligation-mediated PCR. *Mutat. Res.* **288**, 39–46.

18. Becker, M. M., Lesser, D., Kurpiewski, M, Baranger, A., and Jen-Jacobson, L. (1988) Ultraviolet footprinting accurately maps sequence-specific contacts and DNA linking in the EcoRI endonuclease-DNA complex. *Proc. Natl. Acad. Sci. USA* **85,** 6247–6251.
19. Engelhorn, J., Boccard, F., Murtin, C., Prenki, P., and Geiselmann, J. (1995) *In vivo* interaction of the Escherichia coli integration host factor with its specific binding site. *Nucleic Acids Res.* **23,** 2959–2965.
20. Cadet, J., Berger, M., Douki, T., and Ravanat, J. L. (1997) Oxidative damage to DNA: formation, measurement and biological significance. *Rev. Physiol. Biochem. Pharmacol.* **131,** 1–87.
21. Spassky, A. and Angelov, D. (1997) Influence of the local helical conformation on the guanine modifications generated from one-electron DNA oxidation. *Biochemistry* **36,** 6571–6576.
22. Angelov, A., Spassky, A., Berger, M., and Cadet, J. (1997) High-intensity UV laser photolysis of DNA and purine 2'-deoxyribonucleosides: formation of 8-oxopurine damage and oligonucleotide strand cleavage as revealed by HPLC and gel electrophoresis studies. *J. Am. Chem. Soc.,* in press.
23. Lilley, D. M. J. (1992) *Nature (London)* **357,** 282–283.
24. Varga-Weisz, P., van Holde, K., and Zlatanova, J. (1993) Preferential binding of histone H1 to four-way helical junction DNA. *J. Biol. Chem.* **268,** 20,699,20,700.
25. Varga-Weisz, P., Zlatanova, J., Leuba, S. H., Schroth, G. P., and van Holde, K. (1994) Binding of histones H1 and H5 and their globular domains to four-way junction DNA. *Proc. Natl. Acad. Sci. USA* **91,** 3525–3529.

36

An In Vivo UV Crosslinking Assay That Detects DNA Binding by Sequence-Specific Transcription Factors

Alan Carr and Mark D. Biggin

1. Introduction

In vivo UV crosslinking permits direct analysis of protein–DNA interactions in intact cells. This technique has been used to study DNA binding by a wide variety of proteins including RNA Polymerase II, Topoisomerase I, and sequence specific transcription factors such as Even-Skipped, Zeste, and GAGA *(1–4)*. For many of these proteins, the pattern of DNA binding discovered by UV crosslinking differs dramatically from that predicted by earlier indirect approaches. This has led to fundamental reassessments of how these proteins act and illustrates the importance of examining protein–DNA interactions in vivo.

The two principal approaches used to study sequence-specific protein–DNA interactions in vivo are footprinting and crosslinking. In vivo footprinting generates a high resolution map of DNA sequences bound by proteins *(5)*. However, it cannot identify the bound proteins. This is a serious limitation when studying DNA sites bound by families of proteins with similar DNA binding specificities. In vivo crosslinking, on the other hand, makes use of immunoprecipitation to isolate DNA fragments bound to a specific protein. Additionally, crosslinking, unlike footprinting, can detect binding to a DNA site that is not occupied in a large percentage of cells.

The two principal crosslinking agents used to date are UV light and formaldehyde *(1,6–8; see* also Chapters 33 and 35). Formaldehyde can give 100-fold higher levels of protein–DNA crosslinking in vivo than UV light (Toth and Biggin, unpublished data). However, formaldehyde also efficiently crosslinks proteins to other proteins. Because of this, it is conceivable that formaldehyde crosslinking assays may detect indirect as well as direct protein–DNA interactions. UV light is therefore likely to be a better crosslinking agent for studying direct protein–DNA interactions.

The UV crosslinking method described here is more sensitive than the original technique developed by Gilmour et al. *(9)* and can detect crosslinking of sequence-specific DNA binding proteins to DNA. This protocol was designed specifically for the quantitative analysis of DNA binding by proteins in *Drosophila* embryos. It can, however, be modified to study protein–DNA interactions in other organisms or in tissue culture cells (*8,10*; *see* **Note 4**).

The protocol involves the following steps: Live *Drosophila* embryos are irradiated with UV light for 30 min. Chromatin is extracted and then purified by buoyant density ultracentrifugation. After dialysis, the chromatin is restriction digested. An affinity purified antibody directed against the protein of interest is used to immunoprecipitate DNA restriction fragments crosslinked to the protein. These DNA fragments are then separated on an agarose gel, transferred to a nylon membrane, and detected by hybridization to a specific DNA probe. Note that, because of the high sensitivity of our Southern blot protocol, an intermediate PCR amplification step is not necessary. Such amplifications, which are frequently employed in formaldehyde crosslinking assays, can misrepresent the apparent relative levels of protein-DNA crosslinking *(11)*.

Several steps must be taken to ensure that an immunoprecipitation signal is due to the protein–DNA interaction being studied. We suggest including as many of the following controls as possible. First, perform a mock immunoprecipitation with a nonspecific antibody. We use a generic rabbit anti-mouse antibody from Promega (Madison, WI). Second, measure protein binding to different DNA fragments. Be sure to examine binding at sequences not expected to be associated with the protein of interest. Third, perform separate immunoprecipitations with a distinct antibody that recognizes different region of the protein of interest. This control is especially important when the protein under investigation contains a domain found in several other proteins. Fourth, confirm that chromatin derived from cells lacking the protein of interest cannot be immunoprecipitated. For example, if a protein is expressed at a certain time of development, measure binding in chromatin prepared from embryos of different ages. If working with a cell culture, use chromatin from a cell line that does not express the protein. Fifth, perform an immunoprecipitation of chromatin extracted from cells that have not been irradiated. Lastly, although we have not observed significant variation, test different batches of chromatin.

A few shortcomings of using UV light as a crosslinking agent warrant consideration. First, not all proteins crosslink to DNA with equal efficiency. A specific protein-DNA configuration as well as a DNA binding site rich in thymidine are probably required for optimal crosslinking *(12,13)*. For this reason, it may be sensible to determine how well a protein of interest crosslinks to DNA in vitro prior to commencing an in vivo analysis *(8)*. For the proteins we have studied, the relative levels of UV crosslinking in vitro accurately reflect the relative levels of binding to different DNA fragments *(14)*. Second, due to the relative inefficiency of UV crosslinking in vivo, a substantial amount of chromatin is required. The protocol described here calls for 450 µg of chromatin per immunoprecipitation. Probably at least 10% of the cell population being irradiated must be expressing the protein of interest to obtain a clearly detectable signal.

2. Materials

2.1. Embryo Collection and Irradiation

1. Fine and coarse polyamide nylon mesh (Tetko, Switzerland, Nitex; #3-125 [fine] and #3-500 [coarse]).
2. 50% Clorox bleach (2.6% hypochlorite solution).
3. 0.1% (v/v) Tween-20 (Sigma).
4. DNA transfer lamp (Fotodyne, Hartland, WI, model 2-1500; 4 × 15 W 254 nm UV bulbs).
5. Plastic storage box (10 × 19 × 13 cm) filled with ice; Irradiation tray (inverted lid from storage box) lined with aluminum foil placed on bed of ice so that embryos are 3 cm below bulbs.

2.2. Chromatin Preparation

1. Nuclear incubation buffer: 15 mM Tris-HCl, pH 7.5, 0.3 M sucrose, 15 mM NaCl, 5 mM MgCl$_2$, 60 mM KCl, 0.1 mM EDTA, pH 8.0, 0.1 mM EGTA. Filter-sterilize (0.22 µm) and store at 4°C. Add fresh immediately before use: 0.5 mM dithiothreitol (DTT), 1 mM PMSF (Sigma, St. Louis, MO, stock solution: 200 mM PMSF in 100% ethanol).
2. Motorized dounce homogenizer (Thomas *Teflon Pestle* tissue homogenizer, Thomas, Swedesboro, NJ, #3431-E25).
3. Miracloth (Calbiochem, San Diego, CA, #475855).
4. 20% (v/v) Triton X100 (Sigma).
5. Nuclear lysis buffer: 10 mM Tris-HCl, pH 8.0, 100 mM NaCl, 1 mM EDTA, pH 8.0, 0.1% (v/v) NP-40. Filter-sterilize (0.22 µm) and store at 4°C. Add fresh immediately before use: 1 mM PMSF.
6. Glass B dounce (Bellco #1984-40040).
7. 20% N-lauroylsarcosine (Sigma: Sarkosyl).
8. 18-gauge 1 1/2" needles and 25-gauge 1 1/2" needles.
9. SW28 Ultracentrifuge tubes, 25 × 89 mm polyallomer (Beckman, Somerset, NJ).
10. CsCl buffer solution: 0.5% (w/v) Sarkosyl, 1 mM EDTA, pH 8.0. Filter-sterilize (0.22 µm) and store at room temperature. Add fresh immediately before use: 1 mM PMSF.

11. CsCl gradient solutions: 1.75 g/mL: 100.0 g CsCl, 75.0 mL CsCl buffer solution. 1.50 g/mL: 66.7 g CsCl, 83.3 mL CsCl buffer solution. 1.30 g/mL: 40.0 g CsCl, 90.0 mL CsCl buffer solution.
12. Dialysis tubing (Spectrum, Houston, TX, Spectra/por™ Spec 2).
13. 10X Dialysis buffer (pH 8.0): 0.5 M Tris, 20 mM EDTA. Combine dry ingredients and adjust pH with HCl. Autoclave and store at 4°C. For 1X dialysis buffer: Dilute in autoclaved water. Add fresh immediately before use: 1 mM PMSF.

2.3. Restriction Digest and Immunoprecipitations

2.3.1. Restriction Digest

1. 10X Restriction enzyme buffer (*see* **Note 1**).
2. 5 mg/mL BSA (Sigma: Fraction V RIA Grade #A-7888).
3. 1 M DTT.
4. 20% (v/v) Triton X-100.
5. 200 mM PMSF in 100% ethanol.
6. 500 µg/mL RNase (Boehringer Mannheim, Indianapolis, IN, DNase Free, #1119-915).
7. Restriction enzymes (New England Biolabs, Beverly, MA).
8. 20% (w/v) Sarkosyl.
9. 0.5 M EDTA, pH 8.0.

2.3.2. Immunoprecipitation

1. 0.22-µm syringe tip filter units (Millipore, Bedford, MA, Millex-GP).
2. Wash solution 1: 50 mM Tris-HCl, pH 8.0, 2 mM EDTA, pH 8.0, 0.2% Sarkosyl. Add fresh immediately before use (do not add when preparing 20% staph A cell stock): 1 mM PMSF, 0.5 µL/mL aprotinin (Sigma), 2.5 µg/mL leupeptin (Boehringer Mannheim; Stock solution: 5 mg/mL in H$_2$O), 0.25 µg/mL pepstatin (Boehringer Mannheim; Stock solution: 0.5 mg/mL in ethanol)
3. Fixed, Killed *Staphylococcus aureus* Protein A-positive cells (staph A cells); lyophilized suspension (Boehringer Mannheim; *see* **Note 2**). Prepare 20% stock of staph A cells in advance: Suspend lyophilized staph A cells in 10 mL wash solution 1 (without protease inhibitors). Centrifuge and wash once more with wash solution 1. Resuspend in two volumes of 1X PBS, 3% SDS and 10% 2-mercaptoethanol. Heat for 30 min in a boiling water bath. Wash twice in wash solution 1 and then resuspend in wash solution 1 to give a 20% suspension. Freeze in liquid nitrogen in 100 µL aliquots and store at –70°C. These are stable for at least 18 mo. Do not refreeze.
4. Nutator (Clay Adams).
5. Wash solution 2: 100 mM Tris-HCl, pH 9.0, 500 mM LiCl, 1% (v/v) NP-40, 1% (w/v) Deoxycholic Acid (Sigma). Filter sterilize. Store at 4°C. Add fresh immediately before use: 1 mM PMSF.
6. Elution buffer: 50 mM NaHCO$_3$-NaOH, pH 10.0, 1% SDS. 1.5 µg/mL Sonicated Calf Thymus carrier DNA (average size under 1 kb). Filter sterilize (0.22 µm) and store at room temperature.

7. Multitube Vortexer (Troemner, Philadelphia, PA).
8. Protease solution: 50 mM Tris-HCl, pH 7.5, 10 mM EDTA, pH 8.0, 0.3% SDS. Add immediately before use: 1 mg/mL proteinase K (Boehringer Mannheim).
9. 3 M Sodium acetate, pH 5.3.
10. 20 mg/mL yeast RNA (Sigma: Type VI from Torula Yeast).
11. 5X Loading dye: 20% (w/v) sucrose, 1X TBE, 0.1% (w/v) Bromophenol blue, 5 mM EDTA, pH 8.0. Filter-sterilize (0.22 μm) and store at room temperature. Add 1 μL/mL RNase (Boehringer Mannheim: DNase free, #1119-915) when diluting to 1X loading dye

2.4. Southern Transfer and Hybridization

1. 20X SSPE, pH 8.0: 0.2 M NaH$_2$PO$_4$, 3.6 M NaCl, 20 mM EDTA. Add dry ingredients and adjust pH with NaOH.
2. Denaturation solution: 1.5 M NaCl, 0.5 M NaOH.
3. Neutralization solution (pH 7.2): 0.5 M Tris, 1.5 M NaCl, 1 mM EDTA. Add dry ingredients and adjust pH with HCl.
4. Uncharged nylon membrane (Amersham, UK, Hybond-N).
5. Prehybridization solution: 50% Formamide (Fluka, Ronkonkoma, NY, #47670), 6X SSPE, 90 μg/mL Sonicated/Half denatured Calf thymus DNA (Sonicate DNA to an average size of 1 kb; Heat half of the solution in a boiling water bath for 15 min and then recombine.), 10 μg/mL sheared *E. coli* genomic DNA, 5X Denhardts, 10% (w/v) dextran sulphate (Pharmacia, Sweden), 5% (w/v) SDS, 1% (w/v) Sarkosyl. Prepare fresh (requires heating to dissolve) and filter with a 5.0-μm syringe tip filter unit (Millipore: Millex SV).
6. 50X Denhardts: 1% Ficoll 400 (Sigma), 1% polyvinylpyrrolidone (Sigma), 10 mg/mL BSA (Sigma). Filter sterilize (0.22 μm) and store at –20°C in 5–10 mL fractions; Do not refreeze.
7. Amersham Random Prime Label Kit (RPN 1607).

3. Methods

3.1. Embryo Collection and Irradiation

1. Collect embryos of desired age from standard size population cages using standard techniques (*see* **Note 3**). Use the coarse mesh to remove adult flies and the fine mesh to retain embryos. Blot the embryos dry and weigh. Dechorionate embryos by stirring them for 2 min in 50% Clorox bleach. Rinse with water over fine Nitex mesh to remove bleach. Save a few embryos for visual inspection if necessary. From this point on, work in a cold room.
2. Resuspend embryos in cold 0.1% (v/v) Tween-20. Pour embryos onto irradiation tray and spread evenly (1.5–3.5 g embryos and 20–30 mL solution per tray).
3. Irradiate embryos for a total of 30 min. Redistribute the embryos every 5 min (*see* **Note 4**).
4. Drain embryos over fine Nitex mesh and freeze immediately in liquid nitrogen. Store at –70°C.

Table 1
Chromatin Purification Parameters

Age of embryos (h)	Amount of embryos to be processed per SW28 tube (g)	Expected chromatin yields per gram of embryos (µg)
2–3	25–35	50
4–5	15–25	150–200
5–7	10–15	400–500
8–10	5–8	600–800

3.2. Chromatin Preparation

1. Carry out all chromatin preparation steps in a cold room. Glassware must be clean and prechilled to 4°C (*see* **Note 5**). Work quickly and carefully. Perform the chromatin purification in the afternoon so that the centrifugation will end in the morning two days later. Older embryos contain significantly more chromatin than younger embryos: Consult **Table 1** for information on how to compensate for this difference during the ultracentrifugation step. For **steps 2–4**, work in 5–7 g aliquots (stage 10–11 embryos) or 10–13 g aliquots (stage 4–5 embryos).
2. Partially thaw each tube of embryos in 37°C water. Empty embryos into the tissue homogenizer dounce. If necessary, break up large frozen clumps of embryos with a spatula. Rinse out remaining embryos from thawed storage tubes with 35 mL nuclear incubation buffer and add to dounce.
3. Homogenize embryos with motorized dounce at 8k rpm and then twice at 7k rpm.
4. Pour homogenate over prewetted Miracloth filter and collect filtrate in an appropriate-sized beaker. This removes intact embryos and adult debris.
5. After homogenization and filtration of all embryos, gently lyse cells by adding 20% Triton X-100 to filtrate to a 0.3% final concentration. Stir for one minute with a magnetic stir bar and pour into SS34 tubes. Centrifuge in a Sorvall at 4°C for 15 min at 4k rpm (2000g).
6. Aspirate supernatant. For each CsCl gradient (*see* **Table 1**), add 8.1 mL of nuclear lysis buffer to the pelleted nuclei. Gentle swirling or vortexing on low setting should free the pellet. Pour into a glass B dounce and homogenize. Transfer to a 50-mL Falcon tube.
7. Lyse the nuclei by adding 0.9 mL of 20% sarkosyl (1/10 vol). Vortex vigorously immediately to ensure complete lysis.
8. Shear DNA by pulling chromatin into syringe through an 18-gauge needle. Empty the syringe through the same needle and then repeat. Shear twice with a 25-gauge needle by pouring chromatin into syringe and pushing through the needle.
9. Prepare CsCl gradient by carefully layering 18.5 mL of 1.75 g/mL CsCl, 6.0 mL of 1.50 g/mL CsCl, and 3.5 mL of 1.3 g/mL CsCl in an SW28 ultracentrifuge tube. Apply 9 mL of the chromatin to the top of the gradient and centrifuge at 25k rpm for 40 h at room temperature.

10. At end of centrifuge run, remove chromatin fraction from gradient. Cover tube with a rubber stopper to slow flow from the tube while draining. Insert 18-gauge needle 1.5 cm beneath the fine white line of material in the gradient located toward the bottom of the centrifuge tube. Slowly collect twelve 1-mL fractions. If the solution becomes too viscous, use hand pressure or the rubber stopper to force the CsCl/chromatin solution through the needle.
11. Determine which fractions contain the most chromatin by running 2 µL from each on a 1% agarose gel. Pool the peak fractions—usually 4–6 mL per gradient (*see* **Note 6**).
12. Dialyze the purified chromatin against 2 L of dialysis buffer three times for 2 h each at 4°C. Centrifuge briefly to remove insoluble material and freeze the supernatants in liquid nitrogen in 5–10 mL fractions. Store at –70°C.

3.3. Restriction Digest and Immunoprecipitations

3.3.1. Restriction Digest and Antibody-Chromatin Incubation

1. Each restriction digest contains 450 µg of chromatin and is carried out in a 1.7 mL total volume (Thaw chromatin in 4°C water or on ice. Refreeze excess chromatin in liquid nitrogen and store at –70°C). Reactions contain the following: 1X Restriction enzyme buffer, 100 µg/mL BSA, 0.01% (v/v) Triton X-100, 1 m*M* PMSF and 1 µL RNase. Add 450 U of restriction enzyme, mix well by inversion, and incubate overnight at 37°C. A second addition of restriction enzyme (225 U) should be made four hours into the reaction or no later than 1 h prior to the end. Verify digestion by examining 2µL of the reaction on a 1% agarose gel.
2. Stop the reaction by adding EDTA to a 20 m*M* final concentration, Triton X-100 to a 0.3% final concentration, and Sarkosyl to a 0.05% final concentration.
3. Centrifuge in a microcentrifuge at full speed for 15 min at 4°C and carefully transfer supernatant to a new 2-mL tube.
4. Add 1–2 µg of affinity purified antibody and nutate at 4°C for 3 h (*see* **Note 7**).

*3.3.2. Immunoprecipitation (see **Note 10**)*

1. Prepare staph A cells: Thaw 20% staph A cell aliquots on ice and add two volumes of sonicated UV-crosslinked chromatin (Blocking Chromatin). This prevents nonspecific chromatin-staph binding during the immunoprecipitation steps. We generally use 200 µL of 500 µg/mL of UV-crosslinked chromatin from older embryos (8–12 h) per 100 µL of 20% staph A cells. Blocking chromatin should be sonicated to less than 1 kb in length. Nutate at 4°C for 2–3 h. Centrifuge 1 min at full speed in a microcentrifuge and aspirate supernatant. Wash once in wash buffer 1 (with protease inhibitors) and resuspend in the original volume of the 20% staph A cell solution with wash buffer 1.
2. Centrifuge chromatin-antibody solution at full speed for 15 min at 4°C in a microcentrifuge. All subsequent work can be carried out at room temperature. Filter supernatant through a 3-mL syringe with a 0.22-µm filter tip.
3. Add 25 µL of 20% staph A cell solution to chromatin-antibody solution and nutate at room temperature for 15 min (*see* **Notes 8** and **14**).

4. Centrifuge 1 min and save supernatant on ice for use as Total DNA.
5. Resuspend pellet in 200 µL wash solution 1 (with protease inhibitors) by pipetting up and down. Then transfer to a new 2-mL Eppendorf tube. Rinse pipet tip with another 200 µL of wash solution 1 and empty into the new Eppendorf tube. Add an additional 1 mL of wash solution 1 and mix. Centrifuge at full speed for 1 min in a microcentrifuge at room temperature and aspirate supernatant.
6. Wash staph A cells once more with 1.7 mL wash solution 1 using the same tip rinsing technique. Do not transfer to a new Eppendorf tube. Centrifuge at full speed for 1 min.
7. Wash staph A cells 4 times with 1.4 mL wash solution 2. Transfer the immunoprecipitation solution to a new 1.4-mL Eppendorf tube during the last wash.
8. Resuspend the Staph A cells in 100 µL elution buffer and vortex for ten min. in a multitube vortexer. Centrifuge and save supernatant for proteolytic digestion. Be careful not to remove any staph A cells. Repeat the elution step twice more.
9. Add 200 µL Protease solution to the 300 µL of eluted chromatin-antibody solution and incubate overnight at 65°C.
10. Concurrently, 1/100 of the total DNA (generally around 15 µL) should be protease digested in 300 µL of elution buffer and 200 µL of Protease solution. This solution containing 1% of total chromatin will be further diluted and used as a standard on the Southern blot to determine the efficiency of the immunoprecipitation. Incubate overnight at 65°C (*see* **Note 15**).
11. Precipitate the chromatin from the proteolytic digestion reactions by adding 40 µg yeast RNA, 50 µL 3 M sodium acetate, pH 5.3, and 1.25 mL ethanol. Freeze until solid at –70°C (generally 45 min). Centrifuge 15 min at 4°C. Wash pellets with –20°C 75% ethanol and then dry in a Speed Vac. Resuspend immunoprecipitation in 20 µL and the 1% of total DNA in 1 mL of Loading dye.

3.4. Southern Transfer and Hybridization
3.4.1. Agarose Gel Electrophoresis and Southern Transfer

1. Prepare a 0.7% agarose 1X TBE gel without ethidium bromide. Use 1 × 6 mm combs to accommodate 20 µL samples.
2. Prepare Total DNA dilution series. All of the proteins we have studied precipitate between the background limit (approx 0.00005%) and 0.005% of total DNA. Therefore, as a starting point, we suggest generating a dilution series containing 0.005%, 0.001%, 0.0005%, and 0.0001% of total DNA. Load the total DNA standards (20 µL vol each) adjacent to the immunoprecipitated chromatin sample. Run gel at 5 V/cm until the dye has migrated at least 6 cm.
3. Stain gel by soaking for 5 min in at least three gel volumes of 1X TBE containing 1.5 µg/mL ethidium bromide. Destain in three volumes of 1X TBE for 10 min. Photograph gel and cut away excess agarose. For a detailed discussion of Southern transfers, see **ref. 15**.
4. Gently rock gel for 30 min in excess denaturation solution. Rock gel for 15 min in neutralization solution twice.
5. Wet the nylon filter in distilled/deionized water for 10 min and then in 20X SSPE for another 10 min.

6. Assemble Southern transfer apparatus and transfer DNA to nylon filter using 20X SSPE. Allow DNA to transfer for at least 12 h.
7. Dismantle apparatus. Transfer of DNA can be verified by restaining the agarose gel with ethidium bromide. Bake filter for 30 min at 80°C in a vacuum oven. UV irradiate the dry filter for 5 s with the Fotodyne DNA Transfer Lamp. Optimal irradiation times may vary between filter lots.

3.4.2. Hybridization

1. Prehybridize blot overnight (no less than 8 h) at 42°C in at least 0.2 mL/cm² of prehybridization solution (*see* **Note 16**).
2. Label gel purified DNA probes with the random prime label kit and add to prehybridization solution. Approximately $1-2 \times 10^7$ cpm/ml of probe with a specific activity of about 5×10^6 cpm/ng should be added (*see* **Note 12**). Unincorporated label must be separated from the probe because it causes high filter background. Hybridize overnight at 42°C (*see* **Notes 9** and **13**).
3. Wash blot twice for 15 min in 2X SSPE 0.1% SDS at room temperature. Wash for 15 min in 1X SSPE 0.1% SDS at 65°C and then for at least 1 h at 65°C in 0.1X SSPE 1% SDS. If background is above 1–5 cps, continue washing for up to another 15 h (*see* **Note 13**).
4. Expose for 7–10 d on film or for 24 h on an imaging plate (Fuji) (*see* **Note 11**).

3.4.3. Stripping and Reprobing Southern Blots

1. Southern blots may be stripped and reprobed to examine binding by the same protein at different DNA sites. Boil a 0.1% SDS solution and remove from heat. Add blot immediately and shake until solution comes down to room temperature. Rinse blot well with room temperature water to remove remaining SDS. Sensitivity will gradually decline because chromosomal DNA is partially removed during the stripping process. On average, a blot can be probed 3–6 times.

4. Notes

1. We do not use commercial restriction enzyme buffers. The buffer contents, such as NaCl concentration, are adjusted according to the needs of the enzymes in the reaction.
2. Protein A agarose is not compatible with this technique: It fails to precipitate crosslinked chromatin. Protein A sepharose is unable to bind large restriction fragments because much of the protein A is embedded within the sepharose.
3. We use six population cages ($36 \times 40 \times 30$ cm), each containing approx 100 mL of Canton S flies. Food trays should be changed twice over the 2-h period prior to collecting embryos to reduce contamination by retained embryos. At least 0.5 g of embryos per cage per hour can be collected.
4. This is the optimal exposure time for irradiation of *Drosophila* embryos. DNA-protein interactions can also be studied in tissue culture cells. In this case, $2-3 \times 10^8$ cells should be irradiated for 4–6 min. Each immunoprecipitation requires chromatin from $4-5 \times 10^7$ cells. See **refs.** *8*, *10*, and *14* for more information on UV crosslinking in tissue culture cells.

5. Contaminating proteases can reduce immunoprecipitation signals significantly. To prevent this, all equipment including glassware, pipetmen, and tips should be kept clean. We have found that it is extremely important to use the purest reagents. Solutions should be autoclaved or filter sterilized whenever possible to reduce the risk of bacterial or fungal growth. A simple and effective way to test for the presence of proteases in stock solutions is to incubate a known protein in the suspicious solution overnight at 37°C. (We use 10 ng/µL of Even-skipped protein; BSA is not as sensitive to the proteases we have encountered and is therefore not helpful.) The status of the protein can then be examined by SDS-PAGE the next morning. In our experience, however, some proteases are activated only in the presence of detergents. For a more rigorous protease assay, combine chromatin and all solutions included in the restriction digest and immunoprecipitation reaction and incubate overnight at 37°C with a test protein.
6. Gradient fractions can be tested for the distribution of chromatin and free protein. UV-crosslinked chromatin should migrate to approximately 1.66 g/mL CsCl. Free protein should be found around 1.3 g/mL.
7. We suggest using affinity purified polyclonal antibodies. The use of unpurified serum requires that more staph A cells be used during the immunoprecipitation step. This leads to higher background signals in the negative control lanes. The likelihood of crossreaction with other proteins is also reduced when affinity purified antibodies are used.

   It should be noted that staph A cells do not bind antibodies from all species with the same affinity. Staph A cells bind rabbit polyclonal antibodies. Consult the Boehringer Mannheim catalogue for details on antibody-staph cell interactions. A secondary antibody can be added to the binding reaction if the primary antibody does not efficiently bind staph A cells (*see* **Note 10b**). In this case, add the second antibody 3 h after the first one is added. Allow the reaction to proceed for 1 h.
8. The amount of staph A cells and length of incubation can be varied if necessary. However, negative control background signals should be closely monitored when such changes are made.
9. In our experience, non-radioactive detection systems are not as sensitive as the radioactive detection technique we use on our Southern blots. As little as 10 fg of a 4 kb plasmid or a restriction fragment from 225 pg of genomic DNA (0.00005% of 450 µg of chromatin) can be detected with this radioactive system.
10. A low or nonexistent immunoprecipitation signal may be due to any of the following:
    a. Contaminating proteases could be degrading protein crosslinked to the chromatin. Assay all solutions as described in **Note 5** and take extra care to use clean glassware. It may be helpful to prepare important solutions (i.e., restriction enzyme buffer, detergents, wash buffers) in larger batches and to store them frozen in smaller aliquots.
    b. The primary antibody-antigen interaction or the antibody-staph A cell interaction may be weak. A small scale in vitro immunoprecipitation can be carried out to test for these possibilities. Nutate protein and antibody for 3 h at

4°C in the same solutions used in the full scale immunoprecipitations (i.e., 1x restriction enzyme buffer, 100 µg/mL BSA, 0.3% Triton X-100, etc.). Add staph A cells at the proper scale and incubate an additional 15 min at room temperature. Wash the staph A cells with both wash buffers twice. Elute antibody and protein from the staph A cells using a scaled amount of elution buffer and analyze the sample by SDS-PAGE. If neither antibody nor protein is present, the antibody-staph A cell interaction is weak. Consider using a secondary antibody (*see* **Note 7**). If antibody but no protein is present, the antibody-protein interaction is weak. In this case, a different antibody, possibly raised against a different epitope, should be used.

    c. The chromatin was not sufficiently crosslinked. This may be due to overcrowding during the irradiation step. Approximately 2.5 g of embryos should be distributed over a 190 cm² area.

    d. The protein may not crosslink to DNA. An in vitro UV crosslinking test described in **ref.** *8* can be performed to determine how well the protein crosslinks to DNA.

11. The Southern assay is extremely sensitive. Small amounts of contaminating plasmid DNA may be detected on the Southern blot as one or more extraneous bands.

12. If the immunoprecipitation and Total DNA signals are weak, the specific activity of the probe may be low. Consult the Amersham RPN 1601 manual for improving radioactive label incorporation. Alternatively, there may be too little chromatin on the Southern blot. An insufficient amount of chromatin may have been used in the immunoprecipitation. Confirm the concentration of the DNA or use more chromatin in the next experiment. Chromatin could have been degraded by contaminating nucleases or it may not have efficiently transferred to the nylon filter. Be sure to bake and UV crosslink filters immediately after the Southern transfer.

13. Dark spots on the blot exposure can arise from hybridizing the blot in too little hybridization solution. A diffuse high background on the Southern blot can be reduced by washing it for longer periods of time in the 0.1X SSPE 1% SDS solution at 65°C. The blot can be washed overnight. Consult the Amersham Hybond Blotting Guide if filter background troubles persist.

14. High background signals in the negative control lane may be attributed to troubles with the staph A cells. Be sure to centrifuge and filter the antibody-chromatin solution before adding staph A cells. Aggregates formed during the antibody-chromatin incubation step may nonspecifically bind the staph A cells. Different batches of staph A cells may give different negative control signal backgrounds. Finally, check both Wash solutions for correct contents and pH.

15. Immunoprecipitated DNA may migrate as a smear at an anomolously high molecular weight (10–30 kb). Proteinase K may be associating with the DNA. Phenol/chloroform extract the chromatin after the proteinase K digestion step.

16. If the prehybridization solution does not contain properly prepared carrier DNA (10 µg/mL sheared *E. coli* DNA and 90 µg/mL sheared/half denatured calf thymus DNA), smears may appear in the immunoprecipitation lanes. Store the half denatured Calf thymus DNA in small aliquots at –20°C.

References

1. Gilmour, D. S. and Lis, J. T. (1986) RNA polymerase II interacts with the promoter region of the noninduced hsp70 gene in *Drosophila melanogaster* cells. *Mol. Cell. Biol.* **6,** 3984–3989.
2. Gilmour, D. S., Pflugfelder, G., Wang, J. C., and Lis, J. T. (1986) Topoisomerase I interacts with transcribed regions in *Drosophila* cells. *Cell* **44,** 401–407.
3. Walter, J., Dever, C. D., and Biggin, M. D. (1994) Two homeodomain proteins bind with similar specificity to a wide range of DNA sites in *Drosophila* embryos. *Genes Dev.* **8,** 1678–1692.
4. O'Brien, T., Wilkins, R. C., Giardina, C., and Lis, J. T. (1995) Distribution of GAGA protein on *Drosophila* genes *in vivo*. *Genes Dev.* **9,** 1098–1110.
5. Bossard, P., McPherson, C. E., and Zaret, K, S. (1997) *In vivo* footprinting with limiting amounts of embryo tissues: a role for C/EBP beta in early hepatic development. *Methods* **11,** 180–188.
6. Solomon, M. and Varshavsky, A. (1988) Mapping protein-DNA interactions *in vivo* with formaldehyde: evidence that histone H4 is retained on a highly transcribed gene. *Cell* **53,** 937–947.
7. Hecht, A., Strahl-Bolsinger, S., and Grunstein, M. (1996) Spreading of transcriptional repressor SIR3 from telomeric heterochromatin. *Nature* **383,** 92–96.
8. Walter, J. and Biggin, M. D. (1997) Measurement of *in vivo* DNA binding by sequence-specific transcription factors using UV crosslinking. *Methods* **11,** 215–224.
9. Gilmour, D. S., Rougvie, A. E., and Lis, J. T. (1991) Protein-DNA cross-linking as a means to determine the distribution of proteins on DNA *in vivo*. *Methods Cell Biol.* **35,** 369–381.
10. Boyd K. E., and Farnham, P. J. (1997) Myc versus USF: Discrimination of the *cad* gene is determined by core promoter elements. *Mol. Cell. Biol.* **17,** 2529–2537.
11. Strutt, H., Cavalli, G., and Paro, R. (1997) Co-localization of Polycomb protein and GAGA factor on regulatory elements responsible for the maintenance of homeotic gene expression. *EMBO J.* **16,** 3621–3632.
12. Hockensmith, J. W., Kubasek, W. L., Vorachek. W. R., Evertz, E. M. and von Hippel, P. H. (1991) Laser cross-linking of protein-nucleic acid complexes. *Methods Enzymol.* **208,** 211–235.
13. Blatter, E. E., Ebright, Y. W., and Ebright, R. H. (1992) Identification of an amino acid base contact in the GCN4-DNA complex by bromouracil-mediated photocrosslinking. *Nature* **359,** 650–652.
14. Walter, J. and Biggin, M.D. (1996) DNA binding specificity of two homeodomain proteins in vitro and in *Drosophila* embryos. *Proc. Natl. Acad. Sci. USA* **93,** 2680–2685.
15. Sambrook, J., Fritsch, E. F., and Maniatis, T. (1989). *Molecular Cloning: A Laboratory Manual.* Cold Spring Harbor Laboratory Press, Cold Spring Harbor, NY.

Index

A

Acetylation,
    core histones, 207, 311, 343
    SDS-PAGE-based enzyme activity assay,
        materials, 346
        methods, 348
Acetyltransferase activity gel assay, 344
Agarose gel electrophoresis
    chromatin analysis, 113–124
    DNA topology,
        materials, 382
        methods, 383–385
    rigid spheres equation, 115
    UV crosslinking assay,
        materials, 501
        methods, 504, 505
    quantitative agarose gel electrophoresis, 116
Amino-terminal tails,
    core histones,
        acetylation, 311
Analytical ultracentrifugation,
    Beckman Instruments, 130
    chromatin, 127–139
    materials, 131, 132
    methods, 131, 132
APB,
    cysteine-substituted protein,
        modification materials, 29
        modification method, 34, 35
    photocrosslinking,
        histone–DNA contact mapping materials, 31

        histone–DNA interaction methods, 38–40
Atomic force microscope, *see* Scanning force microscopy
ATP-dependent chromatin remodeling complexes,
    nucleosome disruption analysis, 319–330
    reconstituted chromatin templates,
        materials, 320–322
        methods, 322–327
    inactivation by Sarkoyl treatment, 338
ATP-regenerating system, 176
Drosophila embryos, 333, 334
4-azidophenacylbromide, *see* APB

B

Base-pair resolution mapping,
    nucleosome, 45–59
    site-directed hydroxyl radical method,
        materials, 50–52
        methods, 52–56
Beckman Instruments,
    analytical ultracentrifugation, 130
Biotin,
    removal method, 199
Bisulfite deamination,
    5-methylcytosine,
        detection, 396f
Blunt ending,
    T4 DNA polymerase,
        materials, 434

methods, 437
BPM, *see* APB

C

Cell cycle,
 chromatin remodeling events, 443
Cell differentiation,
 flow cytofluorimetric analysis, 443–453
Cell-free system,
 chromatin assembly, 175–185, 187–194, 219–228
 chromatin hyperacetylation, 207
 DNA repair, 231–242
 replication, 297, 298
 transcription, 261–286, 333–340
Centrifuge cell,
 components, 134f, 135f
Chelex resin,
 trace metal ion removal materials, 52
Chemical probes,
 SV40 MCs, 276
Chimeric DNA-binding activator GAL4-HSF, 333
Chromatin,
 analytical ultracentrifugation, 127–139
 conformational and configurational changes, 113
 nucleosome spacing, 175, 176
 purification, 319, 320,
  spin column method, 337, 338
 schematic illustration, 128f
Chromatin accessibility
 measurements, 419
Chromatin analysis,
 micrococcal nuclease digestion, 214, 215, 210f, 248
 quantitative agarose gel electrophoresis, 113–124
  materials, 115–117
  methods, 117–121

multigel, 115
scanning force microscopy, 143–159
 website, 158
Chromatin assembly,
 Drosophila embryo extracts, 208
 Xenopus oocyte extracts, 175
 nucleotide excision repair, 231–242
 posttranslational acetylation,
  core histone amino-terminal tails, 343
 solid-phase approach, 195–206
  materials, 198
  methods, 202, 203
Chromatin assembly extracts,
 materials, 178–180
 methods, 180–183
 preblastoderm *Drosophila* embryos, 187–194
  materials, 188–190
  methods, 190–192
 preparation,
  Xenopus oocytes, 175–185
 preparation prerequisite, 188
 testing,
  materials, 189, 190
  methods, 191
Chromatin assembly reaction,
 histone depletion, 210, 211
 time-dependent process, 176
Chromatin fibers,
 image analysis method, 152
 imaging protocols, 143–159
Chromatin gel,
 M_T and M_O reconstitution products analysis, 90
Chromatin immunoprecipitation,
 formaldehyde crosslinking,
  materials, 458
  methods, 460
Chromatin proteins,
 analysis, 443–453
 analysis materials, 198

Index

analysis methods, 202, 203
Chromatin purification, 195
  linker histone binding analysis,
    materials, 104
    methods, 106
  UV crosslinking assay,
    materials, 499, 500
    methods, 502, 503
Chromatin reconstitution, *see*
  Chromatic assembly
Chromatin remodeling, 333
  cell cycle modification events, 443
  chromatin structure, 292
  NURF, 339, 340
Chromatin structure,
  cell cycle dependent modulation,
    448f
  chromatin remodeling factors, 292
  CPDs repair,
    photolyase, 248, 250f, 251f
  nuclease digestion and indirect
    endlabeling, 367–369
  photolyase,
    yeast, 245–258
  topological measurements,
    379, 380
Chromatin templates,
  in vitro replication, 291–300
    assay method, 296–298
    materials, 293, 294
    methods, 294–299
Chromatin transcription,
  nucleosome remodeling factor
    NURF, 333–340
  oocyte extract, 176
Chromosomal fibers,
  schematic illustration, 128f
Chromosomal proteins,
  DNA interaction site mapping,
    469–479
Chromosome,
  histone extraction, 219
  mitotic assembly,

Xenopus egg extract, 219–228
Chromosome assembly assay,
  materials, 222, 223
  methods, 226
Circularization,
  DNA minicircles,
    production and purification
      method, 86–88
    schematic, 88f
Cleavage site mapping,
  cleavage reagents, 364–366
  indirect endlabeling, 366
Core histones, *see* Histones
Counterions, 175, 176
CPDs, 245, 246
  DNA fragments, 248
  indirect endlabeling
    mapping methods, 256, 257
    materials, 253, 254
CPDs repair,
  chromatin structure,
    photolyase, 248, 250f, 251f
Crosslinking,
  and immunoprecipitation technique,
    schematic drawing, 456f
CsCl gradients,
  UV laser-induced protein–DNA
    crosslinking,
    materials, 487–489
    methods, 491
Cyclobutane pyrimidine dimers, *see*
  CPDs
Cysteine substituted protein,
  construction materials, 28, 29
  expression and purification methods,
    31–33
  reconstituted nucleosomes,
    binding, 37
  reduction and modification
    materials,
    EPD, 29
Cytoplasmic type B HAT, *see* HAT
Cytosine-5-MTases, 397

D

Dam MTase,
  mapping, 161, 162, 395
Demembranated sperm nuclei,
  preparation,
    materials, 222
    methods, 225
Denaturing PAGE,
  base-pair resolution mapping, 55, 56
Derivatization method,
  histone octamer,
    base-pair resolution mapping,
      52–54
Dilution transfer,
  nucleosome reconstitution, 324
  reconstitution materials,
    HeLa oligonucleosomes, 321
    histones octamers, 321
Dinucleosome,
  linker histone binding analysis,
    reconstituted characterization,
      108, 109
    reconstitution method, 107, 108
Direct linker ligation,
  nuclease genomic footprinting,
    materials, 433, 434
    methods, 436, 437
DNA,
  biotinylation method, 199
  coupling methods,
    dynabeads, 199, 200
  damaged,
    UVC, 232
  denaturing PAGE analysis method,
    base-pair resolution mapping,
      55, 56
  histones interactions,
    site-directed chemical probing,
      27–42
  psoralen crosslinking technique, 163f
  plasmid,
    isolation method, 255, 256
    purification materials, 252, 253
  purification and quantitation,
    materials, 357
    methods, 359, 360
DNA content,
  immunofractionated nucleosomes,
    analysis materials, 306
    analysis methods, 307, 308
DNA crosslinking reagent APB,
  modification method,
    cysteine-substituted protein,
      34, 35
DNA deamination,
  bisulfite treatment,
    materials, 400, 401
    methods, 406, 407
  hot start PCR amplification,
    *Sss*I yeast strain identification
      method, 407, 408
  PCR amplification,
    footprinting materials, 402, 403
    footprinting methods, 409
    materials, 401, 402
    methods, 407, 408
DNA denaturation,
  nuclease genomic footprinting,
    materials, 434
    methods, 437
DNA extractions, *see* DNA isolation
DNA fragments,
  base-pair resolution mapping,
    materials, 51
  circularization, 79, 80
  CPDs, 248
  linker histone binding analysis,
    materials, 104
    radiolabeling method, 105
  positioned nucleosome,
    preparation and analysis, 17–25
DNA immobilization, 196f
  solid-phase approach,
    chromatin reconstitution analysis
      materials, 197, 198

Index

methods, 198–200
DNA interaction site mapping,
  chromosomal proteins, 469–479
DNA isolation,
  formaldehyde crosslinking/PCR,
    materials, 473
    methods, 475, 476
  nuclease genomic footprinting,
    materials, 433
    methods, 435, 436
  nucleosome position mapping,
    materials, 369, 370
    methods, 372
  yeast,
    materials, 381
    methods, 383–387
DNA lesions, 245, 246
DNA linking number change,
  rule of thumb, 83f
DNA methylation,
  M.*SssI*,
    materials, 402
    methods, 408
DNA methyltransferases,
  probes,
    chromatin structure, 395–415
DNA minicircles, 79–98
  production and purification,
    materials, 81, 82
    methods, 82–93
  reconstitution particles, 80
DNA phosphorylation,
  nuclease genomic footprinting,
    materials, 433
    methods, 436
DNA photolyase, 245, 246
DNA preparation, *see* DNA isolation
DNA purification,
  materials, 381, 382
DNA repair-linked chromatin assembly,
  materials, 234–237
  methods, 237–239
DNaseI,

cell treatment,
  materials, 356, 357
  methods, 359
factor binding,
  stimulation materials, 322
  stimulation measurement, 327
nucleosome disruption,
  analysis materials, 321
  analysis methods, 324–327
  detection, 326f
nucleosome position mapping,
  materials, 370
  plasmid chromatin, 373
  plasmid DNA, 373
DNaseI footprinting assay, 163
DNaseI hypersensitive sites,
  permeabilized cell cleavage,
    355–361
  mapping, 357f, 358f
NP-40,
  materials, 356–359
  methods, 359–361
DNaseI sensitivity,
  analysis method,
    chromatin hyperacetylation, 215
DNaseI treatment,
  permeabilized cell,
    materials, 432
    methods, 434
DNA site exposure,
  nucleosome, 68–74
DNA supercoiling,
  chromatin assembly extracts,
    processing materials, 180
    processing methods, 183
  chromatin hyperacetylation,
    analysis method, 215
  DNA minicircles, 80–82
  nucleosome formation, 176
DNA synthesis,
  radioactive labeled dATP, 298, 299
DNA target site mapping,
  chromatin proteins,

formaldehyde crosslinking,
455–466
DNA templates,
  DNA repair-linked chromatin
    assembly,
      preparation and UV treatment
        materials, 234, 235
      preparation and UV treatment
        methods, 237
DNA topoisomers,
  gel electrophoresis, 381
DNA topology,
  analysis,
    yeast chromatin, 379–391
      analysis materials, 381–383
      analysis methods, 383–389
DNA twisting, 79, 80
Drosophila, 232, 418, 455
  chromatin assembly extracts,
    harvesting materials, 188
    harvesting methods, 190
    preparation, 187–194
  chromatin hyperacetylation,
    207, 208
  chromatin reconstitution analysis,
    195, 337
    materials, 197, 198
    methods, 200–202
  preblastoderm embryos,
    characterization, 187
    collection methods, 190, 191
    preparation materials, 189
    washing and settling
      materials, 189
Dyad axis,
  histone octamer, 45, 46
Dynabeads,
  coupling methods,
    DNA, 199, 200
Dynamic nucleosome stability,
  equilibrium, 61–74
Dynamic site exposure,
  nucleosome, 68–74

E

E. coli polymerase I,
  Klenow fragment,
    labeling DNA method, 20
EDTA cyst reagents,
  base-pair resolution mapping,
    materials, 50, 51
  hydroxyl radicals, 48
Electron microscopy,
  chromatin analysis, 162
  M_T and M_O reconstitution
    products analysis, 90
  psoralen crosslinking technique and
    band retardation assay, 162
Electrophoresis,
  chromatin analysis, 113–124
  M_T and M_O reconstitution products
    analysis, 90
Electrophoretic patterns,
  M_T and M_O particles, 92f
Elimination method,
  histone–DNA contact mapping
    materials, 31
  histone–DNA interactions, 40
Elution,
  salt gradients,
    histones, 8t
    nucleosome, 11t
Enzymatic probes,
  SV40 MCs, 276
Enzyme restriction,
  NCP mapping method, 21, 22
  permeabilized cell cleavage
    materials, 358
    methods, 360
Enzymes,
  DNA minicircles,
    production and purification
      materials, 81
  NCP mapping materials, 19
  SDS-PAGE-based enzyme activity
    assay,

Index

preparation methods, 346–348
EPD,
  cysteine-substituted protein,
    reduction and modification
      materials, 29
  histone–DNA contact mapping
    materials, 30
  modification method,
    cysteine-substituted protein, 34
Equilibrium,
  definition, 63
  dynamic nucleosome stability, 61–74
Ethidium bromide, 445
Expression,
  cysteine-substituted protein,
    construction materials, 29
    methods, 31–33
  histones,
    materials, 2
    methods, 4, 5
Extraction method,
  genomic DNA, 374
  plasmid DNA, 372
Extract preparation,
  chromatin assembly extracts,
    materials, 189
    methods, 191

F

FACS,
  dual flow cytofluorimetric analysis,
    methods, 445, 446
Factor binding,
  DNaseI digestion, 322
  gel shift assay, 322
Fenton cycle, 47, 48
Fenton reaction, 108
FITC (Fluorescein IsoThioCyanate), 445
Flow cytofluorimetric analysis,
  cell cycle and chromatin proteins,
    analysis output, 448–451
    data interpretation, 448–451
    materials, 444, 445
    methods, 445–451
  cell differentiation, 443–453
Fluorography,
  postreaction processing method, 349
Footprinting,
  DNA methyltransferases, 395–415
  M.*Sss*I,
    materials, 402, 403
    methods, 408–410
  nucleases, 322, 327
  photofootprinting, 483–485
Formaldehyde crosslinking,
  chromosomal proteins,
    materials, 472, 473
    methods, 474–477
  DNA target site mapping,
    chromatin proteins, 455–466
  and immunoprecipitation technique,
    schematic drawing, 456f
  mapping chromatin proteins,
    materials, 457, 458
    methods, 459–463
  and PCR, 469
Four-way junction DNA
  schematic representation, 488f
  UV laser photofootprinting, 483–485
Fpg treatment,
  UV laser-induced protein–DNA crosslinking,
    materials, 487
    methods, 490

G

GAL4-HSF, 333
Gel analysis,
  H1°aC-EPD cleavage, 40
  histone–DNA contact mapping
    materials, 31
Gel electrophoresis,

chromatin analysis, 113–124
DNA topoisomers, 381
formaldehyde crosslinking/PCR,
  materials, 473
  methods, 476, 477
linking numbers,
  plasmid chromatin, 388f
nucleosome purification,
  materials, 3, 4
  method, 10, 11
permeabilized cell cleavage,
  materials, 358
  methods, 361
psoralen crosslinking technique and
    electron microscopy,
  chromatin analysis, 162, 163
replication products, 292
  analysis method, 298
SDS-PAGE-based enzyme activity
    assay,
  materials, 346
  methods, 348
Gel filtration,
  histone purification method, 6, 7
Gel image digitizing,
  quantitative agarose gel
    electrophoresis,
  chromatin analysis methods, 120
Gel shift assay,
  factor binding,
    stimulation materials, 322
    stimulation measurement, 327
  methods,
    linker histone binding analysis,
      109, 110
    nucleosome, 11–13
Gene activation,
  histone acetylation, 311
Gene regulation, 319
Genomic chromatin,
  nucleosome position mapping,
    preparation materials, 370
    preparation methods, 373, 374

Genomic DNA,
  extraction method, 374
  MNase digestion, 374
  nucleosome position mapping,
    preparation materials, 370
    preparation methods, 374
  packaging, 469
Genomic footprinting,
  nuclease,
    materials, 432–434
    methods, 434–437
  nucleases, 427–441
Glutaraldehyde fixation,
  chromatin analysis,
    scanning force microscopy
      method, 144

H

H1°,
  cell cycle dependent accumulation,
    452f, 453f
H1°aC-EPD cleavage,
  gel analysis method, 40
HAT, 343
  activity gel assay procedure,
    schematic diagram, 345f
  binding,
    Nickel agarose resin, 311
  classes, 311
  identification and analysis,
    311–316
  identification and detection,
    SDS-PAGE-based enzyme
      activity assay, 343–351
  polypeptides,
    acetyltransferase activity gel
      assay, 344
  separation, 312f
HAT activities,
  materials, 313
  methods, 314
  separation methods, 314

Index

HAT assays,
  materials, 313
  methods, 315
HeLa cells,
  cytosolic S100 replication extract,
    preparation methods, 294, 295
  whole-cell extract,
    transcription method,
      SV40 MCs, 276, 277
HeLa core histones,
  reconstituted chromatin templates,
    purification materials, 320, 321
    purification methods, 323, 324
HeLa oligonucleosomes,
  chromatin preparation,
    methods, 322, 323
  dilution transfer, 320, 321
  reconstituted radiolabeled
    DNAs, 324
Histone acetylation,
  gene activation, 311
  Trition-Acid-Urea (TAU) gel
    system,
      analysis materials, 211, 212
      analysis methods, 214
Histone acetyltransferases, *see* HAT
Histone analysis,
  Trition-Acid-Urea (TAU) gel
    system, 208, 209f
Histone deacetylase (HD), 343
Histone depletion,
  assembly extract, 210, 211
  chromatin assembly reaction,
    210, 211
Histone–DNA interactions,
  mapping contacts,
    materials, 28–31
    methods, 31–40
Histone extraction,
  chromosomes, 219
Histone fractionation,
  linker histone binding analysis,
    materials, 105

    methods, 107
Histone octamers,
  derivatization method,
    base-pair resolution mapping,
      52–54
  dilution transfer,
    reconstitution materials, 321
  dyad axis, 45, 46
  free-energy preference
    measurements, 65, 66
  reconstitution methods, 324
Histone purification,
  chromatin hyperacetylation,
    materials, 209, 210
  methods,
    linker histone binding
      analysis, 106
  Trichostatin A, 208
Histones,
  amino-terminal tails,
    acetylation, 311
  chromatin assembly extracts,
    purification materials, 179
    purification methods, 182
  concentration ranges, 62
  dissociation/association, 65
  elution,
    salt gradients, 8t
  expression,
    materials, 2
    methods, 4, 5
  and purification,
    cysteine-substituted protein
      construction materials, 29
  molecular weights and extinction
    coefficients, 7t
  nucleosomal assay reconstitution,
    325f
  posttranslational modifications, 311
  purification, 2, 3
    materials, 2, 3
    methods, 5–7
  recombinant, 1–15

refolding, 7, 8
regulating transcription, 175
Histone subunits,
  free exchange, 64, 65
HMG-14/-17,
  analysis, 303–309,
    materials, 305, 306
    methods, 306–308
  clustering,
    quantitative analysis
      methods, 308
Hooke's Law, 381
Hot-start PCR amplification,
  DNA deamination,
    SssI yeast strain identification
      method, 407, 408
H1°, 450f
Hybridization analysis,
  DNA topology,
    materials, 382, 383
    methods, 385
  immunoprecipitated DNA,
    materials, 489
    methods, 492
  permeabilized cell cleavage,
    materials, 358
    methods, 361
  SV40 MCs,
    methods, 277–279
  UV crosslinking assay,
    materials, 501
    methods, 504, 505
Hydroxyl radical cleavage,
  histone–DNA interaction, 37, 38
  methods,
    DNA, 37, 38
    linker histone binding analysis,
      108, 109
Hydroxyl radicals,
  EDTAcyst reagents, 48
Hyperacetylated chromatin,
  reconstitution and analysis,
    207–216

I

Image analysis,
  chromatin fibers, 152
  in air,
    scanning force microscopy
      method,
        blotting process, 147f
        software controls, 150–152
  in liquid,
    scanning force microscopy
      method, 152
Immunoaffinity chromatography,
  purification method,
    SV40 T-Ag, 295, 296
Immunodepletion,
  mitotic chromosome assembly,
    materials, 223
    methods, 226, 227
Immunodetection,
  cell cycle and chromatin proteins,
    buffer and reagent materials, 445
    buffer and reagent methods, 447
Immunofluorescence,
  mitotic chromosome assembly,
    materials, 223
    methods, 226
Immunofractionated nucleosomes,
  protein and DNA content,
    analysis materials, 306
    analysis methods, 307, 308
Immunoprecipitated DNA,
  hybridization analysis,
    materials, 489
    methods, 492
  linker-modified PCR, 457
  purification and amplification,
    analysis methods, 461
    materials, 458
    methods, 460, 461
Immunoprecipitation,
  crosslinked protein–DNA
    complexes,

Index

materials, 488
methods, 492
and crosslinking,
   schematic drawing, 456f
UV crosslinking assay,
   materials, 500, 501
   methods, 503, 504
Inclusion body preparation,
   histone purification method, 6
Incorporated nucleotides,
   TCA precipitation,
      measurement method, 298, 299
Indirect endlabeling,
   chromatin structure interpretation, 367–369
   cleavage site mapping, 366
   CPDs,
      mapping methods, 256, 257
      materials, 253, 254
      nucleosome position mapping, 364f, 365f
      materials, 371
      methods, 375
Ion exchange chromatography,
   histone purification method, 7
   nucleosome purification materials, 3
   nucleosome purification method, 10

K

Kinetic analysis, 380, 381
Klenow fragment,
   E. coli polymerase I
      labeling DNA method, 20

L

Labeling DNA,
   methods, 20
Laser protein–DNA photofootprinting, 482, 483
Ligation,
   PCR insert,

cysteine substituted protein
   construction materials, 29
Linker histones,
   nucleosome binding,
      analysis,
         materials, 104, 105
         methods, 105–110
      mono- and dinucleosomes, 103–112
   SV40 MCs,
      preparation method, 275
Linker-modified PCR,
   immunoprecipitated DNA, 457
Linking number difference,
   DNA minicircles,
      production and purification method, 82–84
Linking number paradox, 79, 80
LM-PCR,
   nuclease genomic footprinting,
      materials, 433, 434
      methods, 436, 437
   procedures,
      comparative analysis, 430f, 431f
      schematic description, 429f

M

Macromolecular radius,
   quantitative agarose gel
      electrophoresis,
         chromatin analysis methods, 121
Maxam-Gilbert G specific reaction,
   histone–DNA contact mapping
      materials, 30, 36, 37
5-methylcytosine,
   detection,
      bisulfite deamination, 396f
Mica preparation,
   scanning force microscopy
      method, 145
Micrococcal nuclease, *see* MNase
Microdialysis apparatus,

preparation materials,
    base-pair resolution mapping, 52
Midblastula transition,
    Xenopus laevis, 443
Mitotic chromosome assembly,
    Xenopus egg extract, 219–228
    materials, 221–223
    methods, 223–227
Mitotic chromosomes,
    structure, 219
Mitotis promoting factor (MPF), 220
MNase,
    chromatin analysis, 248
    chromatin digestion, 77f
        materials, 305
        methods, 306
    NCP mapping method, 23
MNase assay and repair synthesis,
    232–234
    DNA repair-linked chromatin
        assembly,
        materials, 236, 237
        methods, 238, 239
MNase cleavage assay, 188f
MNase digestion, 177f
    chromatin reconstitution analysis,
        methods, 214, 215
    genomic DNA, 374
    methods,
        histone–DNA interaction
            boundaries, 109
        nucleosome position mapping,
            materials, 370
        solid-phase approach,
            materials, 198
MNase footprinting,
    CPDs repair, 247f
    nucleosome mapping, 246, 247
    and photolyase,
        comparison, 247f
MNase treatment,
    permeabilized cell,
        materials, 432

methods, 434, 435
solid-phase approach,
    methods, 203
Mononucleosome,
    linker histone binding analysis,
        reconstituted characterization,
            108, 109
        reconstitution method, 107, 108
M.*Sss*I,
    chromatin analysis,
        materials, 397–403
        methods, 403–410
    chromatin structure probe, 395, 396
    gene encoding, 395
MTase expression plasmid,
    integration,
        materials, 399, 400
        methods, 405
Multigel,
    chromatin analysis, 113
    electrophoresis apparatus,
        chromatin analysis materials,
            115, 116
    quantitative agarose gel
        electrophoresis, 120, 121
        chromatin characterization, 115
        schematic illustration, 114f
        gel pouring, 117–119,
    gel running, 119–120
Mu (μ_o) determination,
    quantitative agarose gel
        electrophoresis, 20–22

N

Nascent RNA strands, 162
Native PAGE analysis method,
    nucleosome,
        base-pair resolution mapping, 54
Nickel agarose resin,
    HAT binding, 311
Nonhistone chromosomal proteins
    HMG-14/-17, *see* HMG-14/-17

Index

Nonidet P-40 (NP-40),
  DNaseI hypersensitive sites,
    materials, 356–359
    methods, 359–361
Nuclease,
  digestion, 355
    chromatin structure interpretation, 367–369
    genomic footprinting, 427–441
      materials, 432–434
      methods, 434–437
Nuclei extraction,
  cell cycle and chromatin proteins,
    buffer materials, 444
    buffer methods, 446, 447
  Xenopus laevis,
    buffer materials, 444, 445
    buffer methods, 446, 447
Nuclei isolation, *see* Nuclei preparation
Nuclei preparation,
  nuclease genomic footprinting,
    materials, 432, 433
    methods, 435
  Saccharomyces cerevisiae,
    methods, 418, 419
Nucleoprotein gel preparative,
  isolation method,
    NCP, 21
Nucleosomal ladder,
  purification and gel separation,
    materials, 190
    methods, 192
Nucleosome,
  base-pair resolution mapping, 45–59
  assembly method, 54
  binding,
    cysteine-substituted protein, 37
  dissociative behavior, 62
  DNA site exposure, 68–74
  dynamic association/disassociation equilibrium, 63, 64
  electrophoretic patterns, 92f
  elution,
    salt gradients, 11t
    equilibrium stability paradox, 64, 65
  linker histone binding, 103–112
    materials, 105
    methods,
      gel shift assay, 11–13
      purification, 9–11
      reconstitution, 8, 9
  occupancy probability,
    specific position, 68
  oocyte extract, 176
  PAGE analysis method,
    base-pair resolution mapping, 54
  physiological solution conditions,
    behaviors, 61–74
  structure and dynamics, 79–98
  structure and function, 63–64
Nucleosome assembly equilibria, 61–66
Nucleosome assembly extract,
  histone depletion, 210, 211
Nucleosome complex fraction,
  chromatin assembly extracts,
    isolation materials, 179
    isolation methods, 181, 182
Nucleosome core particle preparation,
  linker histone binding analysis,
    materials, 104
    methods, 106, 107
Nucleosome cores, 363
Nucleosome disruption,
  DNaseI digestion,
    analysis,
      ATP-dependent chromatin remodeling complexes, 319–330
      materials, 321
      methods, 324–327
    detection, 326f
Nucleosome dissociation, 61, 62
Nucleosome formation,
  DNA supercoiling assay, 176
Nucleosome mapping,
  MNase footprinting, 246, 247

psoralen crosslinking technique,
    low resolution, 164
psoralen–DNA crosslinking and
    primer extension, 161–172
schematic representation,
    psoralen crosslinking technique
    and primer extension, 165f
Nucleosome mobility, 67
  methods,
    linker histone binding analysis,
      109, 110
Nucleosome positioning, 66, 67, 245,
    363, 364
  base-pair resolution mapping, 54
  chromosomal gene, 368f
  detection, 397f
  DNA sequence, 364
  free energy DNA binding, 67, 68
  indirect endlabeling, 364f, 365f
  mapping, 1, 2
    materials, 18, 19
    methods, 20–23
  psoralen crosslinking technique,
    examples and interpretation,
      164–167
  statistical, 67, 68
  SV40 MCs, 276
  translational positioning, 45
    determination method, 46, 47
  yeast, 363–376
    materials, 369–371
    methods, 371–375
Nucleosome reconstitution,
  DNA minicircles,
    analysis method, 90
    production and purification
      method, 88–90
  histone–DNA contact mapping
    materials, 30
  recombinant histones,
    expression and purification,
      1–15
  salt gradient dialysis,

solid-phase approach
    materials, 198
salt step dialysis method, 35, 36
slow assembly kinetic reactions,
    62, 63
supercoiling assay, 176
Nucleosome remodeling factor NURF
    activity,
  MNase digestion assay,
    method, 338, 339
  chromatin remodeling,
    in vitro transcription method,
      339, 340
  chromatin transcription, 333–340
    materials, 334, 335
    methods, 336–340
Nucleotide excision repair
  chromatin assembly, 231–242
  and photoreactivation, 246
NURF, *see* Nucleosome remodeling
    factor

O

Oligo directed labeling, 168, 169
Oligonucleosome fractionation,
  HMG-14/-17
    materials, 305
    methods, 306, 307
Oocyte extract,
  chromatin assembly extracts,
    testing materials, 179, 180
    testing methods, 182
  chromatin transcription studies, 176
  nucleosome, 176

P

P55,
  in-gel histone acetyltransferase
    assay, 344
PAGE,
  base-pair resolution mapping,

DNA, 55, 56
  nucleosome, 54
PCR amplification,
  DNA deamination,
    footprinting materials, 402, 403
    footprinting methods, 409
PCR analysis,
  formaldehyde crosslinking/PCR,
    materials, 473
    methods, 476, 477
Permeabilized cell,
  DNaseI hypersensitive sites,
    analysis, 355–361
  DNaseI treatment,
    materials, 432
    methods, 434
  MNase treatment,
    materials, 432
    methods, 434, 435
PHO8 promoter,
  accessibility measurements, 422f
  hypersensitive sites, 421f
Photocrosslinking,
  APB,
    histone–DNA contact mapping
      materials, 31
    histone–DNA interaction
      methods, 38–40
Photolyase,
  chromatin structure,
    yeast, 245–258
  CPDs repair, 248, 250f, 251f
  DNA repair,
    materials, 249–254
    methods, 254–257
  and MNase footprinting,
    comparison, 247f
Photoreactivation,
  and nucleotide excision repair, 246
  yeast cells,
    materials, 249, 252
    methods, 254
Piperidine,

UV laser-induced protein–DNA
  crosslinking,
    materials, 487
    methods, 490
Plasmid,
  isolation,
    method, 255, 256, 372
    materials, 252, 253
  linking numbers,
    gel electrophoresis, 388f
  nucleosome position mapping,
    DNaseI digestion method, 373
    partial purification method,
      371, 372
    preparation materials, 369
Positioned nucleosome,
  preparation and analysis, 17–25
Preblastoderm *Drosophila* embryos,
  chromatin assembly extracts,
    187–194
Primer extension, 163
  extension product analysis,
    materials, 168
    methods, 170
  nuclease genomic footprinting,
    materials, 434
    methods, 437
  nucleosome mapping, 161–172
Probe synthesis,
  methods, 170, 171
Propidium iodide (PI), 445
Proteinase K treatment,
  cell treatment,
    materials, 356, 357
    methods, 359
Protein–DNA crosslinking,
  UV laser footprinting, 481–493
Psoralen crosslinking technique,
  DNA,
    materials, 167, 168
    methods, 169
    physical map and electron
      micrographs, 163f

with electron microscopy and band
retardation assay,
chromatin analysis, 162
gel electrophoresis and electron
microscopy,
chromatin analysis, 162, 163
nucleosome mapping, 161–172
materials, 167–169
methods, 169–171
nucleosome positioning,
examples and interpretation,
164–167
yeast cells,
materials, 167
methods, 169
Purification,
cysteine-substituted protein,
31–33
DNA minicircles, 81–93
histones, 2, 3, 5–7
immunoaffinity chromatography,
SV40 T-Ag, 295, 296
nucleosome, 3, 9–11
QIAGEN protocol,
purification materials, 252
purification methods, 254, 255

Q

QIAGEN protocol,
DNA,
purification materials, 252
purification methods, 254, 255
Quantitative agarose gel
electrophoresis,
chromatin analysis, 113–124
materials, 115–117
methods, 117–121
principles, 114, 115

R

Radiolabeling method,
DNA,

histone–DNA contact mapping
materials, 30
method, 35
linker histone binding
analysis, 105
DNA synthesis, 298, 299
Recombinant GAL4 activator proteins,
E. coli,
expression, 336
induction, 336, 337
purification, 336, 337
Recombinant histones,
nucleosome reconstitution,
base-pair resolution mapping, 50
expression and purification, 1–15
Recombinant yeast Gcn5, 311
Reconstituted chromatin,
in vitro replication, 292
Reconstituted chromatin templates,
ATP dependent chromatin
remodeling complexes,
materials, 320–322
methods, 322–327
Reconstituted radiolabeled DNA's,
dilution transfer,
HeLa oligonucleosomes, 324
Reconstituted SV40 MCs,
diagrams, 268f
linker histones,
preparation method, 275
preparation methods, 267–272
Reconstitution,
DNA minicircles, 80
mono- and dinucleosomal templates,
linker histone binding analysis
method, 107, 108
NCP,
salt/urea dialysis method, 20, 21
nucleosomes, 3, 8, 9
salt step dialysis method, 35, 36
histone-DNA contact mapping
materials, 30
Refolding,

Index 525

histones, 7, 8
Relaxation,
  M_T and M_O particles, 90, 91
Repair synthesis,
  DNA repair-linked chromatin
    assembly,
      materials, 236, 237
      methods, 238, 239
    and MNase assay, 232–234
    and supercoiling assay, 232–234
      DNA repair-linked chromatin
        assembly,
          materials, 235, 236
          methods, 237, 238
Replicated chromatin,
  analysis, 298, 299
Replication products,
  gel electrophoresis, 292
    analysis method, 298
Restriction digest,
  UV crosslinking assay,
    materials, 500
    methods, 503
Restriction nucleases,
  chromatin accessibility, 417–424
    materials, 418
    methods, 418, 419
RNA analysis,
  method,
    SV40 MCs, 277–279
  primer extension assay,
    analyzation, 339, 340
R-value,
  psoralen crosslinking technique, 162

S

Saccharomyces cerevisiae, 245, 311,
  469
  MTase strain expression, 415
  nuclei isolation,
    methods, 418, 419
Salt-Jump method,
  DNA minicircles,

production and purification,
  89, 90
Salt dialysis method,
  nucleosome reconstitution, 35, 36
Salt/urea dialysis method,
  nucleosome reconstitution, 20, 21
Sarkoyl treatment,
  ATP-dependent nucleosome
    remodeling activities,
    inactivation, 338
Scaffold, 219
Scanning force microscopy,
  chicken erythrocyte chromatin fibers
    image, 155f
  chromatin analysis, 143–159
    materials, 144
    methods, 144–152
  nucleosome image, 157f
  website, 158
SDS-PAGE-based enzyme activity
  assay,
  histone acetyltransferase, 343–351
    materials, 345, 346
    methods, 346–350
Sedimentation analysis,
  SV40 MCs, 275, 276
Sedimentation equilibrium,
  128–130, 129f
Sedimentation velocity, 128–130, 129f
Sequence-specific transcription
  factors,
  DNA binding,
    UV crosslinking assay, 497–507
Simian virus 40, *see* SV40
SIR3, 472f
Site-directed chemical probing,
  histone-DNA interactions, 27–42
Site-directed hydroxyl radical method,
  base-pair resolution mapping,
    materials, 50–52
    methods, 52–56
  chemistry, 48f
  cutting sites,

schematic diagrams, 49f
vs free solution hydroxyl radical method, 47
Site exposure,
nucleosome, 68–70
binding semiquantitatively, 71, 72
facile even, 71
mobility and transcription, 74
multiple protein binding, 72–74
Software,
scanning force microscopy, 150–152, 158
Solid-phase approach,
chromatin reconstitution analysis, 195–206
materials, 197, 198
methods, 198–203
Southern analysis,
mapping chromatin proteins, interpretation methods, 461–463
materials, 459
Southern blotting,
DNA topology,
materials, 382
methods, 383–385
permeabilized cell cleavage,
materials, 358
methods, 361
UV crosslinking assay,
materials, 501
methods, 504, 505
Spin column method,
chromatin assembly, purification, 337, 338
SssI yeast strain identification, 407, 408
Sucrose gradients,
oligonucleosome fractionation,
materials, 305
methods, 306, 307
Supercoiling, *see also* DNA supercoiling,

Supercoiling assay and repair synthesis, 232–234
DNA repair-linked chromatin assembly,
materials, 235, 236
methods, 237, 238
SV40 MCs, 455
abundance, isolation, and purification, 262
acetylated core histones preparation method, 275
biochemical properties, 263, 264
chromatin induction transcription,
materials, 264–267
methods, 267–279
compaction state, 275, 276
core,
isolation method, 275
preparation method, 275
hybridization and RNA analysis, methods, 277–279
isolation materials, 265
isolation method, 273, 274
model system, 261, 262
native or salt-treated, 296
preparation method, 273–275
propagation method, 271, 272
structural analysis materials, 266
structural analysis method, 275, 276
sucrose gradient profile, 274f
transcriptional analysis materials, 266, 267
transcriptional analysis method, 276–279
transcriptional and structural analysis, 261–286
transcriptional properties, 263
in vitro replication, 292
in vitro replication system, 291
SV40 T-antigen (T-Ag),
preparation method, 291, 295, 296

Index

T

TCA precipitation, 292
  incorporated nucleotides, 298, 299
T4 DNA polymerase blunt ending,
  materials, 434
  methods, 437
Tetrahymena thermophila, 311, 344
Tetramer (M_T) particle,
  electrophoretic patterns, 92f
  reconstitution,
    DNA minicircles,
      analysis method, 90
      production and purification method, 88–90
  relaxation,
    DNA minicircles,
      analysis method, 91–93, 97f
      production and purification method, 90, 91
Topoisomer,
  distribution,
    center location, 386f
  DNA minicircles,
    production and purification method, 86–88
  linking number,
    definition, 82
    equation for, 82
Topological measurements,
  chromatin structure, 379, 380
  linking numbers, 387–389
Transcription reactions,
  methods,
    linker histone binding analysis, 110
Transitional positioning, *see* Nucleosome positioning
Trichostatin A,
  histone purification, 208
Trimethylpsoralen (TMP), 162
Trition-Acid-Urea (TAU) gel system,
  histone analysis, 208, 209f

U

UV crosslinking,
  DNA,
    nucleosome core particles, 39
    nucleosome reconstituted, 38, 39
    sequence-specific transcription factors, 497–507
  DNA damaged, 232
  histone–DNA contact mapping
    materials, 31
  laser-induced protein–DNA crosslinking, 481, 482
  materials, 485–489
  methods, 489–492
  protein to DNA in vivo,
    materials, 499–501
    methods, 501–505
UV irradiation, 485
  yeast cells,
    materials, 249, 252
    methods, 254
UV laser footprinting,
  four-way junction DNA and histone complexes, 483–485
  protein–DNA crosslinking, 481–493

W

Website,
  scanning force microscopy, 158
Whole cell extracts,
  yeast, 311, 312

X

Xenopus egg extract, 232
  collection,
    materials, 221
    methods, 223
  dejellying,
    materials, 222
    methods, 223, 224

isolation protocol, 220
mitotic chromosome assembly,
    219–228
preparation,
    materials, 222
    methods, 224
Xenopus laevis, 176, 220
    erythrocyte nuclei,
        structural changes, 221f
    midblastula transition, 443
Xenopus mitotic extracts,
    mitotic chromosome assembly, 220
Xenopus oocytes,
    chromatin assembly extracts,
        preparation materials, 178, 179
        preparation methods, 180, 181
    chromatin reconstituted,
        in vitro replication method, 297
    chromatin reconstitution analysis,
        195
Xenopus sperm,
    chromatin structural changes, 221f
    preparation, 220
Xenopus sperm nuclei,
    preparation,
        materials, 222
        methods, 225

Y

Yeast cells,
    chromatin structure,
        photolyase, 245–258
    culturing,
        total DNA isolation,
            materials, 402
            methods, 408
        formaldehyde crosslinking/PCR,
            growth, harvest, and lysis,
                materials, 472, 473
                methods, 474
        nucleosome position mapping,
            363–376
        UV irradiation and
            photoreactivation,
                materials, 249, 252
                methods, 254
        whole cell extracts, 311, 312
Yeast chromatin,
    DNA topology,
        analysis, 379–391
Yeast construction,
    DNA methyltransferases expression,
        materials, 398–400
        methods, 403–405
Yeast minichromosomes,
    analysis, 166f
    mapping, 163
Yeast strains expression,
    cytosine-5-DNA MTases,
        screening,
            materials, 400–402
            methods, 405–408